RNA Technologies

For further volumes:
http://www.springer.com/series/8619

Volker A. Erdmann · Jan Barciszewski
Editors

Non Coding RNAs in Plants

Springer

Editors
Prof. Dr. Volker A. Erdmann
Free University Berlin
Institute of Chemistry/Biochemistry
Thielallee 63
14195 Berlin
Germany
erdmann@chemie.fu-berlin.de

Prof. Dr. Jan Barciszewski
Polish Academy of Sciences
Institute of Bioorganic Chemistry
Z. Noskowskiego 12/14
61-704 Poznan
Poland
jan.barciszewski@ibch.poznan.pl

ISBN 978-3-642-19453-5 e-ISBN 978-3-642-19454-2
DOI 10.1007/978-3-642-19454-2
Springer Heidelberg Dordrecht London New York

Library of Congress Control Number: 2011933481

Cover design: deblik, Berlin

Printed on acid-free paper

Springer is part of Springer Science+Business Media (www.springer.com)

Preface

Non coding endogenous RNAs were first discovered in the last decade of the previous century. These new discoveries changed our views of the transcriptome landscape of plant genomes and paradigms of the regulation of gene expression. With the beginning of this century, we have witnessed an explosion of studies on small regulatory RNAs that has yielded a basic understanding of the many types of small RNAs in diverse eukaryotic species and how they are functioning as RNA–protein complexes along the RNA silencing pathways.

While reading this book, the reader will realize that much more remains to be learned about the non coding RNAs and their complex regulatory mechanisms, and we are sure that many more discoveries in this field will be made concerning, so far not even imagined, most interesting and complex regulation principles, based on the structure and function of ncRNAs. Non coding RNAs might introduce another level of mate selection through the epigenetic regulation of genes mediating self-incompatibility. They mediate regulation of dominant–recessive patterns of Mendelian inheritance. ncRNAs act in trans in heterozygous genomes to regulate transcriptional gene silencing through DNA methylation and provide new insights into monoallelic transcriptional control.

Ever since the discoveries by Andrew Z. Fire and Craig C. Mello of the RNA interference mechanisms in 1998, the field of gene regulation by RNA interference has been developing with unforeseen speed. RNA silencing is a widespread mechanism of gene regulation in all eukaryotes. At the core of all RNA silencing pathways lie small RNAs (20–30 nt in length) associated with the Argonaute family of proteins. Non coding RNAs provide the specificity of regulation by base-pairing to the target nucleic acids, while the Argonaute proteins execute the silencing effects.

In the most recent years, each of the RNA silencing pathways of plants has appeared to generate ncRNAs with dedicated functions, specialized biological activities, and specific functional scopes.

RNA silencing plays a crucial role in coordinating the expression, stability, protection, and inheritance of eukaryotic genomes. It comprises several mechanisms that invariably depend on core small non coding RNAs and that achieve dedicated sequence-specific functions. RNA silencing has been recognized to carry out critical developmental, stress response and bodyguard functions, thus coordinating the expression, protection, stability, and inheritance of virtually all eukaryotic genomes. Thus, the ncRNAs encompass a wide set of mechanisms that achieve specialized functions. It still seems very surprising that all of these regulatory functions are carried out by double-stranded RNA molecules, which are only 20–30 nucleotides in length.

As we know that not even 30% of the plant genome is coding for the 40,000 different plant proteins, we have begun to realize how diversified and complex the regulatory mechanisms of the non coding RNAs must be, if they are primarily encoded in the remaining 70% of the plant genome.

For this book, we have tried to get some of the world's best known scientists to write chapters on their respective research in the area of non coding RNAs in plants.

We are certain that many books concerning the structure and function of non coding RNAs in plants will follow before we understand the complexity of regulation in this class of eukaryotic organisms.

Berlin, Germany — Volker A. Erdmann
Poznan, Poland — Jan Barciszewski

Contents

Contributors

Muhammad Asif Arif Institute for Genetics, Heinrich-Heine-University Düsseldorf, Universitätsstraße 1, 40225 Düsseldorf, Germany

Kyohei Arita Laboratory of Biomolecular Function Chemistry, Department of Molecular Engineering, Kyoto University, Kyoto 615-8510, Japan

Sebastian Asurmendi Instituto de Biotecnología, CICVyA, INTA Castelar, Dr. N. Repetto y Los Reseros s/n, CP 1686 Hurlingham, Buenos Aires, Argentina

Consejo Nacional de Investigaciones Científicas y Técnicas (CONICET), Buenos Aires, Argentina

Department of Genetics, Yale University School of Medicine, New Haven, CT 06510, USA

Carol Auer Department of Plant Science, University of Connecticut, Storrs, CT 06269, USA
e-mail: carol.auer@uconn.edu

Ariel A. Bazzini Instituto de Biotecnología, CICVyA, INTA Castelar, Dr. N. Repetto y Los Reseros s/n, CP 1686 Hurlingham, Buenos Aires, Argentina

Consejo Nacional de Investigaciones Científicas y Técnicas (CONICET), Buenos Aires, Argentina

Department of Genetics, Yale University School of Medicine, New Haven, CT 06510, USA
e-mail: ariel.bazzini@yale.edu

Zhu Benzhong Laboratory of Fruit Biotechnology, College of Food Science and Nutritional Engineering, China Agricultural University, Beijing 100083, China
e-mail: cauzbz@yahoo.com.cn

Nicolás G. Bologna Instituto de Biología Molecular y Celular de Rosario, Suipacha 531, 2000 Rosario, Argentina

Lewis H. Bowman Department of Biological Sciences, University of South Carolina, Columbia, SC 29208, USA
e-mail: lbowman@mailbox.sc.edu

Andrea Carra Research Unit for Wood Production Outside Forests, CRA, Strada Frassineto 35, 15033 Casale Monferrato, Italy
e-mail: andrea.m.carra@gmail.com

Ming Chen College of Life Sciences, Zhejiang University, Hangzhou 310058, China
e-mail: mchen@zju.edu.cn

Z. Jeffrey Chen Section of Molecular Cell and Developmental Biology, Center for Computational Biology and Bioinformatics, and Institute for Cellular and Molecular Biology, The University of Texas at Austin, One University Station, A-4800, Austin, TX 78712, USA
e-mail: zjchen@mail.utexas.edu

Sinda Chiibi Centro de Protección Vegetal y Biotecnología, Instituto Valenciano de Investigaciones Agrarias (IVIA), Carretera Moncada-Náquera, Km 4.5, 46113 Moncada, Valencia, Spain

Martin Crespi Institut des Sciences du Végétal, Centre National de la Recherche Scientifique (C.N.R.S.), 91198 Gif-sur-Yvette, Cedex, France
e-mail: Martin.Crespi@isv.cnrs-gif.fr

Shaun J. Curtin Department of Agronomy and Plant Genetics, University of Minnesota, 1991 Upper Buford Circle, St. Paul, MN 55108, USA

Hugh Dickinson Department of Plant Sciences, University of Oxford, South Parks Road, OX1 3RB, Oxford, UK
e-mail: hugh.dickinson@plants.ox.ac.uk

Biao Ding Department of Molecular Genetics, The Ohio State University, 207 Rightmire Hall, 1060 Carmack Road, Columbus, OH 43210, USA
e-mail: ding.35@osu.edu

Andrew L. Eamens School of Molecular Bioscience, University of Sydney, Sydney, NSW 2006, Australia

H. Alexander Ebhardt Institute of Molecular Systems Biology, Eidgenössische Technische Hochschule (ETH) Zürich, Wolfgang-Pauli-Str. 16, CH-8093 Zürich, Switzerland
e-mail: ebhardt@imsb.biol.ethz.ch

Carmen Fagoaga Centro de Protección Vegetal y Biotecnología, Instituto Valenciano de Investigaciones Agrarias (IVIA), Carretera Moncada-Náquera, Km 4.5, 46113 Moncada, Valencia, Spain

Isam Fattash Department of Plant Biotechnology, University of Freiburg, Schänzlestraße 1, 79104 Freiburg, Germany

Ricardo Flores Instituto de Biología Molecular y Celular de Plantas (CSIC-UPV), Universidad Politécnica, Avenida de los Naranjos, 46022 Valencia, Spain

Wolfgang Frank Department of Plant Biotechnology, University of Freiburg, Schänzlestraße 1, 79104 Freiburg, Germany
e-mail: wolfgang.frank@biologie.uni-freiburg.de

Dominique Gagliardi Institut de Biologie Moléculaire des Plantes du CNRS, Université de Strasbourg, 12 rue du Général Zimmer, 67084 Strasbourg, France

Giorgio Gambino Plant Virology Institute, Grugliasco Unit, CNR, Via L. da Vinci 44, 10095 Grugliasco, Italy

Shang Gao Department of Plant Pathology and Microbiology, Center for Plant Cell Biology, Institute for Integrative Genome Biology, University of California, 3234B Genomics Building, Riverside, CA 92521, USA

Robert Grant-Downton Department of Plant Sciences, University of Oxford, South Parks Road, OX1 3RB, Oxford, UK

Xingyi Guo Department of Neurology, Albert Einstein College of Medicine, 1300 Morris Park Avenue, Bronx, NY 10461, USA

Yoshiki Habu Division of Plant Sciences, National Institute of Agrobiological Sciences, Tsukuba 305-8602, Japan
e-mail: habu@affrc.go.jp

Yousef Haj-Ahmad Norgen Biotek Corporation, 3430 Schmon Parkway, L2V 4Y6, Thorold, ON, Canada

Caroline Hartmann Institut des Sciences du Végétal, Centre National de la Recherche Scientifique (C.N.R.S.), 91198 Gif-sur-Yvette, Cedex, France

Université Paris Diderot-Paris 7, 5 rue Thomas Mann, 75205 Paris, Cedex 13, France

Elodie Hudik Institut des Sciences du Végétal, Centre National de la Recherche Scientifique (C.N.R.S.), 91198 Gif-sur-Yvette, Cedex, France

Hailing Jin Department of Plant Pathology and Microbiology, Center for Plant Cell Biology, Institute for Integrative Genome Biology, University of California, 3234B Genomics Building, Riverside, CA 92521, USA
e-mail: hailingj@ucr.edu

Cameron Johnson Section of Plant Biology, College of Biological Sciences, University of California Davis, Davis, CA 95616, USA
e-mail: camjohnson@ucdavis.edu

Richard A. Jorgensen Department of Plant Sciences, University of Arizona, Tucson, AZ 85721-0036, USA

BIO5 Institute, University of Arizona, Tucson, AZ 85721-0240, USA

Tatsuo Kanno School of Natural Sciences, National University of Ireland, Galway, Ireland

Ghazanfar Abbas Khan Institut des Sciences du Végétal, Centre National de la Recherche Scientifique (C.N.R.S.), 91198 Gif-sur-Yvette, Cedex, France

Basel Khraiwesh Department of Plant Systems Biology, Flanders Institute for Biotechnology (VIB), and Department of Plant Biotechnology and Genetics, Ghent University, Technologiepark 927, 9052 Gent, Belgium

Yukio Kurihara Plant Biology Laboratory, The Salk Institute for Biological Studies, 10010 North Torrey Pines Road, La Jolla, CA 92037, USA
e-mail: ykurihara@salk.edu

Bernard Chi-Hang Lam Norgen Biotek Corporation, 3430 Schmon Parkway, L2V 4Y6, Thorold, ON, Canada

Heike Lange Institut de Biologie Moléculaire des Plantes du CNRS, Université de Strasbourg, 12 rue du Général Zimmer, 67084 Strasbourg, France
e-mail: heike.lange@ibmp-cnrs.unistra.fr

Christine Lelandais-Brière Institut des Sciences du Végétal, Centre National de la Recherche Scientifique (C.N.R.S.), 91198 Gif-sur-Yvette, Cedex, France

Université Paris Diderot-Paris 7, 5 rue Thomas Mann, 75205 Paris, Cedex 13, France

Carmelo López Instituto de Biología Molecular y Celular de Plantas (CSIC-UPV), Universidad Politécnica, Avenida de los Naranjos, 46022 Valencia, Spain

Jie Lu Section of Molecular Cell and Developmental Biology, Center for Computational Biology and Bioinformatics, and Institute for Cellular and Molecular Biology, The University of Texas at Austin, One University Station, A-4800, Austin, TX 78712, USA

Qing-Jun Luo Department of Biological Sciences, Texas Tech University, Lubbock, TX 79409-3131, USA

Department of Pediatrics, Stanford University, Stanford, CA 94305-5208, USA

Chonglie Ma Department of Plant Sciences, University of Arizona, Tucson, AZ 85721-0036, USA

Ruth C. Martin USDA-ARS, National Forage Seed Production Research Center, Corvallis, OR 97331, USA

Cristina Martínez-Andújar Department of Horticulture, Oregon State University, Corvallis, OR 97331, USA

Andrea Masotti Gene Expression – Microarrays Laboratory, IRCCS-Bambino Gesù Children's Hospital, P.za S. Onofrio 4, 00139 Rome, Italy
e-mail: andrea.masotti@opbg.net; andrea_maso@yahoo.it

Julieta L. Mateos Instituto de Biología Molecular y Celular de Rosario, Suipacha 531, 2000 Rosario, Argentina

Max-Planck-Institute for Plant Breeding Research, Carl-von-Linné-Weg 10, 50829 Cologne, Germany

Yijun Meng College of Life Sciences, Zhejiang University, Hangzhou 310058, China

Pedro Moreno Centro de Protección Vegetal y Biotecnología, Instituto Valenciano de Investigaciones Agrarias (IVIA), Carretera Moncada-Náquera, Km 4.5, 46113 Moncada, Valencia, Spain

Rebecca A. Mosher The School of Plant Sciences, The University of Arizona, P.O. Box 210036, Tucson, AZ 85721-0036, USA
e-mail: rmosher@cals.arizona.edu

Magdalena Musialak-Lange Max-Planck Institute for Molecular Plant Physiology, Am Mühlenberg 1, 14476 Potsdam, Germany

Luis Navarro Centro de Protección Vegetal y Biotecnología, Instituto Valenciano de Investigaciones Agrarias (IVIA), Carretera Moncada-Náquera, Km 4.5, 46113 Moncada, Valencia, Spain

Giuseppe Nervo Research Unit for Wood Production Outside Forests, CRA, Strada Frassineto 35, 15033 Casale Monferrato, Italy

Hiroyuki Nonogaki Department of Horticulture, Oregon State University, Corvallis, OR 97331, USA
e-mail: hiro.nonogaki@oregonstate.edu

Przemyslaw Nuc Max-Planck Institute for Molecular Plant Physiology, Am Mühlenberg 1, 14476 Potsdam, Germany

Institute of Molecular Biology and Biotechnology, Adam Mickiewicz University, Poznan, Poland

Fangqian Ou-Yang Department of Biological Sciences, Texas Tech University, Lubbock, TX 79409-3131, USA

Javier F. Palatnik Instituto de Biología Molecular y Celular de Rosario, Suipacha 531, 2000 Rosario, Argentina
e-mail: palatnik@ibr.gov.ar

Alessia Palma Genomic Core Facility, IRCCS-Bambino Gesù Children's Hospital, P.za S. Onofrio 4, 00139 Rome, Italy

Bikram Datt Pant Max-Planck Institute for Molecular Plant Physiology, Am Mühlenberg 1, 14476 Potsdam, Germany

Leandro Peña Centro de Protección Vegetal y Biotecnología, Instituto Valenciano de Investigaciones Agrarias (IVIA), Carretera Moncada-Náquera, Km 4.5, 46113 Moncada, Valencia, Spain
e-mail: lpenya@ivia.es

Katherine A. Petsch Department of Plant Biology, Cornell University, Ithaca, NY 14853, USA

Wioletta E. Pluskota Department of Plant Physiology and Biotechnology, University of Warmia and Mazury, Oczapowskiego 1A, 10-718 Olsztyn, Poland

Gail Pruss Department of Biological Sciences, University of South Carolina, Columbia, SC 29208, USA
e-mail: pruss@biol.sc.edu

Nezar Rghei Norgen Biotek Corporation, 3430 Schmon Parkway, L2V 4Y6, Thorold, ON, Canada

Christopher D. Rock Department of Biological Sciences, Texas Tech University, Lubbock, TX 79409-3131, USA
e-mail: chris.rock@ttu.edu

Letizia Da Sacco Gene Expression – Microarrays Laboratory, IRCCS-Bambino Gesù Children's Hospital, P.za S. Onofrio 4, 00139 Rome, Italy

Michael J. Scanlon Department of Plant Biology, Cornell University, Ithaca, NY 14853, USA
e-mail: mjs298@cornell.edu

Wolf-Rüdiger Scheible Max-Planck Institute for Molecular Plant Physiology, Am Mühlenberg 1, 14476 Potsdam, Germany
e-mail: scheible@mpimp-golm.mpg.de

Motoaki Seki Plant Genomic Network Research Team, RIKEN Plant Science Center, 1-7-22 Suehiro-cho, Tsurumi-ku, Yokohama-shi 230-0045, Japan

Kihara Institute for Biological Research, Yokohama City University, 641-12 Maioka-cho, Totsuka-ku, Yokohama 244-0813, Japan
e-mail: mseki@psc.riken.jp

Nuria Soler Centro de Protección Vegetal y Biotecnología, Instituto Valenciano de Investigaciones Agrarias (IVIA), Carretera Moncada-Náquera, Km 4.5, 46113 Moncada, Valencia, Spain

Céline Sorin Institut des Sciences du Végétal, Centre National de la Recherche Scientifique (C.N.R.S.), 91198 Gif-sur-Yvette, Cedex, France

Université Paris Diderot-Paris 7, 5 rue Thomas Mann, 75205 Paris, Cedex 13, France

Nathaniel Robert Street Department of Ecology and Environmental Science, Umeå Plant Science Centre, Umeå University, 901 87 Umeå, Sweden
e-mail: nathaniel.street@plantphys.umu.se

Guiling Sun Department of Biology, East Carolina University, Greenville, NC 27858, USA

Yuan Tian Department of Molecular Genetics, The Ohio State University, 207 Rightmire Hall, 1060 Carmack Road, Columbus, OH 43210, USA

Simona Urso CRA-GPG Genomic Research Centre, Via S. Protaso 302, 29017 Fiorenzuola d'Arda, Italy

Vicki Vance Department of Biological Sciences, University of South Carolina, Columbia, SC 29208, USA
e-mail: vvance@mailbox.sc.edu

Qinglian Wang Henan Institute of Science and Technology, Xinxiang, Henan 453003, China

Peter M. Waterhouse School of Molecular Bioscience, University of Sydney, Sydney, NSW 2006, Australia
e-mail: peter.waterhouse@sydney.edu.au

Ping Wu College of Life Sciences, Zhejiang University, Hangzhou 310058, China

Zhixin Xie Department of Biological Sciences, Texas Tech University, P.O. Box 43131, Lubbock, TX 79409, USA
e-mail: zhixin.xie@ttu.edu

Xiaorui Yang Department of Molecular Genetics, The Ohio State University, 207 Rightmire Hall, 1060 Carmack Road, Columbus, OH 43210, USA

Manabu Yoshikawa Division of Plant Sciences, National Institute of Agrobiological Sciences, Tsukuba 305-8602, Japan

Baohong Zhang Department of Biology, East Carolina University, Greenville, NC 27858, USA

Henan Institute of Science and Technology, Xinxiang, Henan 453003, China
e-mail: zhangb@ecu.edu

Deyou Zheng Department of Neurology, and Departments of Genetics and Neuroscience, Albert Einstein College of Medicine, 1300 Morris Park Avenue, Bronx, NY 10461, USA
e-mail: deyou.zheng@einstein.yu.edu

Quantification of Small Non Coding RNAs May Allow Accurate Comparisons of MiRNA Expression Profiles from Plant Specimens

Letizia Da Sacco, Alessia Palma, Bernard Chi-Hang Lam, Yousef Haj-Ahmad, Nezar Rghei, and Andrea Masotti

Contents

Abstract MicroRNAs (miRNAs) are highly conserved ~22-mer RNA molecules, encoded by plants and animals that regulate the expression of genes binding to the 3′-UTR of specific target mRNAs or to mRNA itself. The amount of miRNAs in a total RNA sample depends on the recovery efficiency that may be significantly

A. Masotti (✉) and L. Da Sacco
Gene Expression – Microarrays Laboratory, IRCCS-Bambino Gesù Children's Hospital, P.za S. Onofrio 4, 00139 Rome, Italy
e-mail: andrea.masotti@opbg.net; andrea_maso@yahoo.it

A. Palma
Genomic Core Facility, IRCCS-Bambino Gesù Children's Hospital, P.za S. Onofrio 4, 00139 Rome, Italy

B.C.-H. Lam, Y. Haj-Ahmad, and N. Rghei
Norgen Biotek Corporation, 3430 Schmon Parkway, L2V 4Y6, Thorold, ON, Canada

V.A. Erdmann and J. Barciszewski (eds.), *Non Coding RNAs in Plants*, RNA Technologies, DOI 10.1007/978-3-642-19454-2_1,

affected by the different purification methods employed. Traditional approaches may be inefficient at recovering small RNAs, and common spectrophotometric determination is not adequate to quantify selectively these low-molecular-weight (LMW) species from total RNA samples. Here, we describe the use of qualitative and quantitative *lab-on-a-chip* tools for the analysis of these LMW RNA species in plant RNA samples. The same concepts apply to human samples, and our previously published data emphasized the close correlation of LMW RNAs with the expression levels of some human miRNAs. We also applied our result to perform a comparison of some miRNA expression profiles in different tissues. The methods we propose allowed the analysis of the efficiency of extraction protocols, to study the small (but significant) differences among various preparations and to allow a proper comparison of some miRNA expression profiles in various specimens. Therefore, by applying the same concepts and methodologies used for human samples, plant molecular biologists will be able to perform suitable comparisons and methodologically correct miRNA expression profiling studies.

Keywords Small RNAs • miRNAs • Quantification • Endogenous control • 2100 Bioanalyzer • RealTime qPCR • Tissue expression

1 Introduction

In the last few years, a new class of highly conserved ~22-mer non coding RNAs, microRNAs (miRNAs), has emerged as an important player in posttranscriptional gene expression control in eukaryotes (Bartel 2004). A large proportion of miRNAs are highly conserved among distantly related species, from worms to mammals in the animal kingdom (Bartel 2004), and from mosses to high flowering eudicots among plants (Axtell et al. 2007; Zhang et al. 2006). In plants, miRNAs are fundamental for genome stability, development and differentiation, cellular communication, signaling, adaptive responses to biotic and abiotic stress and have been shown to regulate floral patterning and floral timing (Bartel and Bartel 2003; Mallory and Vaucheret 2006; Zhang et al. 2007; Yu and Wang 2010; Aukerman and Sakai 2003; Palatnik et al. 2003).

To date, the miRBase Database counts up to 15,172 annotated miRNA entries (Release 16.0, September 2010) from vertebrates, flies, worms, plants, and viruses (Griffiths-Jones et al. 2006). For plants (*Viridaeplantae*), miRBase reports 3,070 known miRNAs. In animals, miRNAs appear predominantly to inhibit translation by targeting partially complementary sequences located within the 3′-untranslated regions (3′-UTR) of target mRNAs, promoting either mRNA degradation or translation arrest (Nilsen 2007; Pillai et al. 2007; Eulalio et al. 2008; Jackson and Standart 2007). The mechanisms of translational inhibition by plant miRNAs are largely unknown, but existing evidence has indicated that plants and animals share some mechanistic similarity of translational inhibition. One of the differences between plant and animal miRNAs is that the regulatory targets of plant miRNAs can be convincingly predicted simply by identifying mRNAs with near-perfect complementarity (Rhoades et al. 2002). Therefore, the transcript cleavage by miRNAs in plants occurs thanks to the high degree of complementarity between miRNAs and their targets (Aukerman and Sakai 2003; Palatnik et al. 2003). In animals, each miRNA may control the activity of many genes, and almost 30% of the genome could be regulated in such a way, which renders these small molecules as important as the transcription factors (Lewis et al. 2003; John et al. 2004).

Hence, the imperfect miRNA/mRNA pairing generally leads to a regulation occurring mostly through translational inhibition. Owing to these "relaxed" base-pairing requirements, individual metazoan miRNAs may have dozens of target transcripts. By contrast, since plant miRNAs regulate transcripts by single, highly complementary target sites in coding regions, a large number of plant miRNAs function through "slicing," constituting an efficient means of "mRNA clearance." Consequently, plant miRNAs are predicted to have only a limited number of mRNA targets (Voinnet 2009). However, in *Arabidopsis thaliana* the 199 known miRNAs (miRBase 15.0) have been predicted to regulate the expression of more than 600 genes (Alves et al. 2009), and at least 225 genes are validated targets (Backman et al. 2008). Most of the miRNA targets are important transcription factors that play important roles in patterning the plant form (Chen et al. 2005) or genes involved in the response to environmental stresses (Sunkar and Zhu 2004).

Owing to their extreme importance as regulators, the complexity in miRNA modes of action, isolation, precise quantification, and reliable detection techniques of these tiny molecules in specific tissues are, therefore, fundamental for better understanding of miRNA-mediated gene regulation. Although miRNA represent a relatively abundant class of transcripts, their expression levels vary greatly among different cells and tissues. Unlike mammals, which have relatively simple small RNA populations comprising mainly miRNAs and no siRNAs (Mineno et al. 2006), plants have a hugely complex small RNA fraction. It is comprised of both miRNAs and endogenous siRNAs derived from repetitive sequences, intergenic regions, and genes (Llave et al. 2002; Lu et al. 2005). This complexity renders miRNAs highly underrepresented in the small RNA fraction and further affects detection methods such as cloning and microarray hybridization.

Purity and integrity are other two essential requirements not only for total RNA but also for these small species. Therefore, RNA extraction protocols have to maximize their recovery. Nowadays, it is well established that the traditional glass-fiber total RNA extraction protocol may be inefficient at recovering small RNAs. In addition, the common spectrophotometric determination of total RNA is not adequate to quantify low-molecular-weight (LMW) species selectively.

We and other authors have found previously that the recovery of LMW RNA species is significantly affected by the specific purification process (Masotti et al. 2009; Ach et al. 2008; Ibberson et al. 2009). Our results showed that different extraction strategies lead to significantly different recovery of LMW species including miRNAs. We also demonstrated that using the same amount of total

RNA (from different tissues), different amounts of miRNAs may be obtained. Moreover, the amount of LMW RNA species does not perfectly parallel that of miRNAs: even with the same extraction protocol, the concentration of miRNAs may differ significantly among various tissues. These observations are particularly relevant for plant researchers involved in complex studies (i.e., cell-type specific gene expression studies) (Galbraith and Birnbaum 2006), since in plants small RNA fraction is complex and also contains other small RNAs (siRNA) other than miRNAs.

2 Materials and Methods

2.1 Total RNA Extraction and Small RNAs Enrichment Protocols

Total RNA was extracted using four different methods: an acid phenol/guanidine isothiocyanate solution (TRIzol Reagent – Invitrogen), two glass-fiber kits such as MirVana™ miRNA Isolation Kit (Ambion) and the RNEasy Mini Kit (Qiagen) and the Norgen's Total RNA extraction kit that consists in a solid-phase filtration column supporting a silicon carbide-based resin (Norgen Biotech Corporation – Thorold, CANADA). All extractions were performed according to manufacturer's instructions.

2.2 Plant Tissues and Cell Lines

Arabidopsis thaliana (*L.*) leaves were collected, washed with deionized water, and immediately subjected to grinding under liquid nitrogen. About 100 mg of grinded plant tissue was used for each extraction. Three different cell lines (HeLa, COS-1, and a lymphoblastoid cell line [LCL]) were cultured using standard procedures, trypsinized (if necessary), and pelleted by centrifugation. Approximately 10^7 cells for each extraction were resuspended in the appropriate lysis solution contained in the RNA extraction kit and treated according to manufacturer's instructions.

2.3 RNA Integrity Analysis by Gel Electrophoresis and Lab-on-a-Chip Technology

The integrity of RNA samples was checked by gel electrophoresis (agarose 1%) stained with ethidium bromide. Gel images were acquired and analyzed with the Quantity One (software Ver.2.0 – Biorad). Total RNA samples were also analyzed with the Total RNA 6000 Nano Kit (Vers. II), specifically optimized for total RNA analysis with the Agilent 2100 Bioanalyzer. For miRNA quantification, we used the dedicated Small RNA kit. The instrument uses fluorescence detection, monitoring the emission between 670 and 700 nm. The run was performed according to manufacturer's instructions. Electropherograms were analyzed using the Agilent 2100 Expert B.02.06 software that includes data collection, presentation, and interpretation functions.

2.4 Real-Time PCR

For the small RNAs recovery, efficiency evaluation of different extraction procedures, hsa-miR-21, and small nucleolar Z30 (snoZ30) were assayed by Real-Time PCR (TaqMan – Applied Biosystems), according to manufacturer's instructions (Chen et al. 2005). For the analysis of miRNA expression in different tissues, three miRNAs (hsa-miR-26a, hsa-miR-26b, and hsa-miR-134) and two controls (U6 and snoZ30) were arbitrarily chosen as model miRNAs. A panel of five tissues (brain, skeletal muscle, heart, liver, and uterus) was chosen for evaluation, and the corresponding total RNA (Clontech – BD Biosciences) was analyzed with Agilent 2100 Bioanalyzer. Briefly, from five to ten nanograms of each RNA samples were reverse-transcribed to cDNA using the High-Capacity cDNA Archive Kit (Applied Biosystems) with specific primers. The principle of TaqMan MicroRNA Assays is a specific stem-loop reverse transcription (RT) primer. The short length of mature miRNAs (~22 nt) prohibits conventional design of a random-primed RT step followed by a specific real-time assay. In the former case, the resulting RT amplicon is a suitable template for standard real-time PCR with TaqMan assays. Reactions were performed by incubating samples for 30 min at 16°C, 30 min at 42°C, 5 min at 85°C and finally cooled on ice. PCR products were assayed with specific probes using the TaqMan Universal PCR Master Mix, No AmpErase UNG (Applied Biosystems) following the manufacturer's protocol. PCR reactions were performed incubating samples for 10 min at 95°C, then for 15 s at 95°C and 60 s at 60°C for 45 cycles by means of ABI PRISM® 7900HT Sequence Detection System. Data were analyzed using the SDS software (Version 2.1).

3 Results

3.1 The Recovery of Low Molecular Weight (LMW) RNAs is Affected by Different Extraction Protocols

RNA samples extracted from plant tissues were run on agarose gel to visualize the differences between various extraction methods. Plant RNA samples, extracted with TRIzol reagent, MirVana™, and Norgen's kit clearly showed the

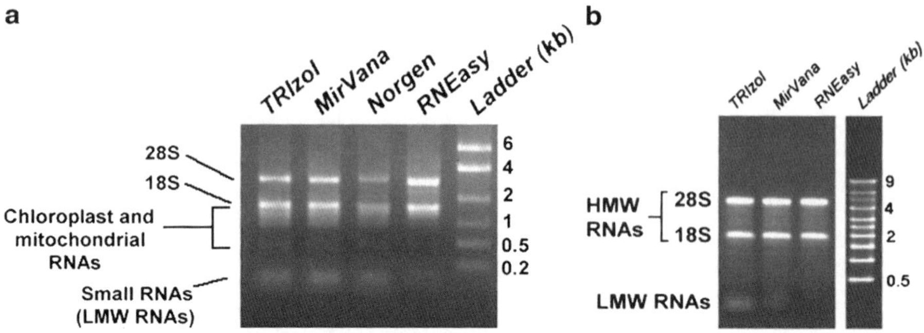

Fig. 1 (**a**) Gel electrophoresis (agarose 1% stained with ethidium bromide) of RNA samples (from *Arabidopsis thaliana L.*) extracted with TRIzol reagent, MirVana™ kit, Norgen's kit, and RNEasy kit. (**b**) Gel electrophoresis (agarose 1% stained with ethidium bromide) of RNA samples (from COS-1) extracted with TRIzol reagent, MirVana™ kit, and RNEasy kit

high molecular weight (HMW) 28S and 18S rRNA bands together with chloroplast and mitochondrial RNAs, while small RNAs (LMW RNAs) are visualized as faint, smeary bands (Fig. 1a, lanes 1–4). Plant RNA extracted with RNeasy kit displayed only the HMW RNA bands (28S and 18S) and very little amount of LMW RNAs compared to the other three kits (Fig. 1a, lane 4). Total plant and human RNA have a similar pattern on agarose gel (Fig. 1b), apart from the chloroplast and mitochondrial RNA bands.

The same samples were also checked with Agilent 2100 Bioanalyzer, which is one of the most versatile microfluidics-based platforms for the analysis of DNA, RNA, proteins, and cells. In all electropherograms, the 28S and 18S RNAs are represented on the right side together with chloroplast and mitochondrial RNAs (HMW RNAs) (Fig. 2a), and the smaller species (LMW RNAs) are present at a very low concentration and are distinguishable on the left side of the profile (see the magnification of the LMW RNAs region in Fig. 2b). Electropherograms emphasize that all samples present a similar HMW profile region, while the major differences are localized in the LMW region (Fig. 2b). Total RNA from human origin has a similar aspect (Fig. 3). Although total plant RNA recovery is quite modest and similar for the four protocols (from 22 to 45 ng/μL), the LMW fractions are substantially different (from 2 to 11 ng/μL), as shown in Table 1. In particular, Norgen's kit allowed the highest LMW RNA recovery (27.3% of total RNA), while RNEasy Mini Kit the lowest (5.4%). TRIzol and MirVana™ miRNA Isolation Kit gave good yields for LMW RNA species (24.4 and 23.5%, respectively). While LMW RNA species extracted with TRIzol, MirVana™, and Norgen have comparable profiles (Fig. 2b), RNEasy kit retains only one RNA peak in comparable concentrations to the others. This peak has been previously recognized to be the 5.8S peak (Masotti et al. 2009). Again, similar results have been obtained for

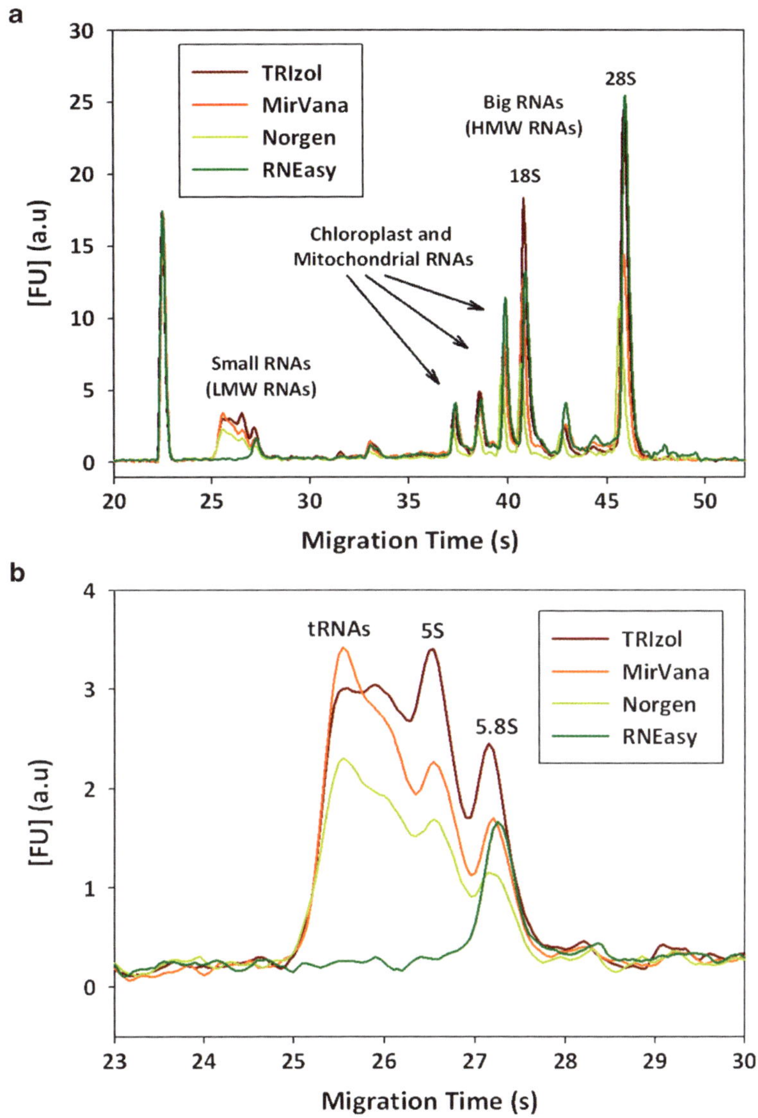

Fig. 2 (**a**) Agilent 2100 Bioanalyzer electropherogram profiles of total RNA samples (from *A. thaliana* (*L.*)) extracted with TRIzol reagent (*brown*), MirVana™ kit (*orange*), Norgen's kit (*yellow*), and RNEasy kit (*green*). (**b**) Magnification of small RNA profiles for the four samples (between 23 and 30 s)

human specimens (Table 2 and Fig. 3). Therefore, the *lab-on-a-chip* analysis is a useful tool to quantify precisely the amount of LMW RNAs of samples extracted with different protocols.

Since we have previously obtained similar results using human cell lines (Masotti et al. 2009), we are inclined to think that our reasoning is still valid for plant tissues

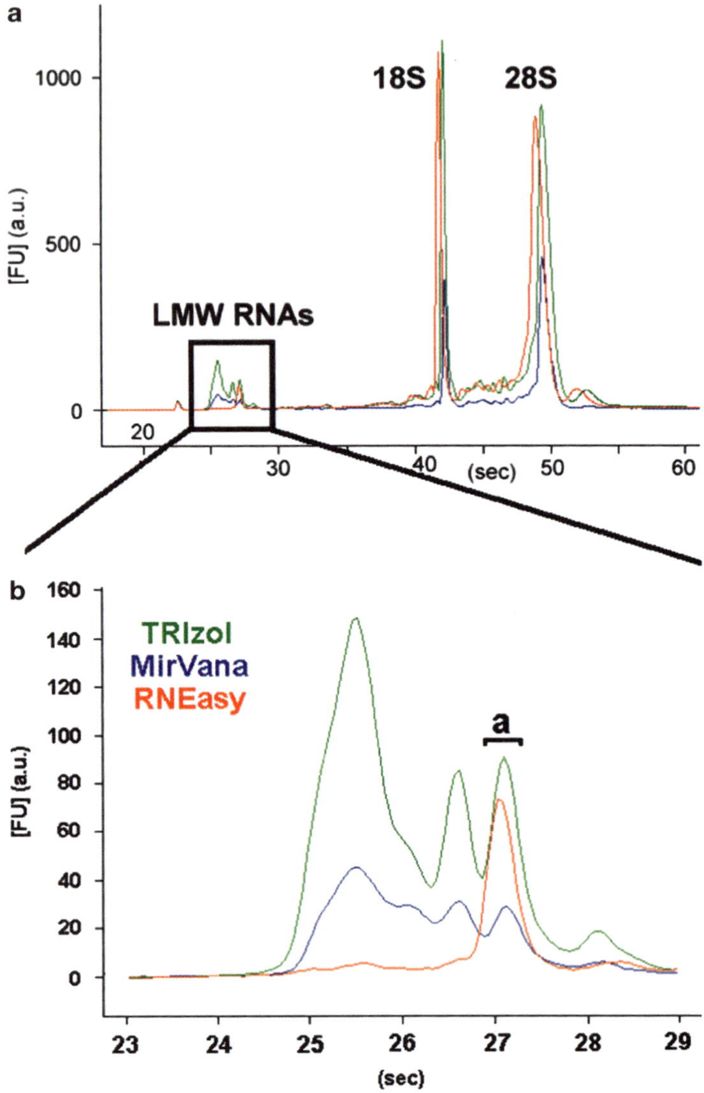

Fig. 3 Agilent 2100 Bioanalyzer electropherogram profiles of (**a**) Total RNA samples (HeLa cells) extracted with TRIzol reagent (*green*), MirVana™ kit (*blue*), and RNEasy kit (*red*). (**b**) Magnification of small RNA profiles for the three samples (between 23 and 29 s)

and that our methodology could be easily applied also in the field of plant molecular biology. Therefore, from this point onward we extensively illustrate results obtained with human cell lines emphasizing differences and similarities with *A. thaliana* (*L.*) samples.

Table 1 Low molecular weight (LMW) RNA mean concentration (% with respect to total RNA) for *A. thaliana* (*L.*) RNA samples extracted with different RNA extraction protocols evaluated with Agilent 2100 Bioanalyzer

Extraction method (total RNA)	HMW RNA concentration (ng/μL)	LMW RNA concentration (ng/μL)	LMW/HMW ratio (%)
TRIzol reagent	45	11	24.4
MirVans kit	34	8	23.5
Norgen's kit	22	6	27.3
RNEasy kit	37	2	5.4

Table 2 Low-molecular-weight (LMW) RNA mean concentration (% with respect to total RNA) for HeLa, COS-1 and LCL extracted with different RNA extraction protocols evaluated with Agilent 2100 Bioanalyzer

	LMW RNA mean concentration (%)		
Extraction method (total RNA)	HeLa	COS-1	LCL
TRIzol reagent	24 (±3)	34 (±2)	22 (±3)
MirVana kit	16 (±1.5)	19 (±1)	19 (±1)
RNEasy kit	2.5 (±0.5)	3 (±0.5)	3 (±0.5)

Standard deviations of at least three independent extractions are reported in parentheses

3.2 *MicroRNAs Amount Correlates with the LMW RNA Fraction*

In order to quantify miRNAs extracted with different protocols and to study the correlation between LMW RNAs and miRNAs, we carried out a TaqMan quantitative assay for two representative human miRNAs: a control non coding RNA (hsa-snoZ30) and a miRNA (hsa-miR-21). A target-specific stem-loop adapter technology was employed to obtain the corresponding cDNA (Fig. 4) (Chen et al. 2005). We started the RT reaction with 10 ng of total RNA from each human RNA sample and quantified the absolute expression level of each miRNAs through the analysis of cycle threshold (Ct) values. Ct is the PCR cycle at which the sample reaches the level of detection above the background. LCL RNA samples extracted with different protocols showed different Ct values for snoZ30 (Fig. 5a) and for miR-21. A similar behavior was also obtained by using HeLa and COS-1 cell lines with both probes (data not shown).

Then, we repeated the RT reaction using 10 ng of LMW RNA calculated from the Bioanalyzer electropherogram. As expected, using the same LMW RNA amount, similar Ct values both for miR-21 and snoZ30 between the various samples were obtained (Fig. 5b). This demonstrates that the amount of miR-21 and snoZ30 correlates only to the LMW RNA fraction and not to total RNA amount. In fact, Fig. 5a clearly shows that using the same amount of total RNA different Ct values can be obtained.

Ct differences (ΔCt) between total RNA and LMW RNA Ct values for each extraction protocol and for each cell line are very similar both for hsa-miR-21 and

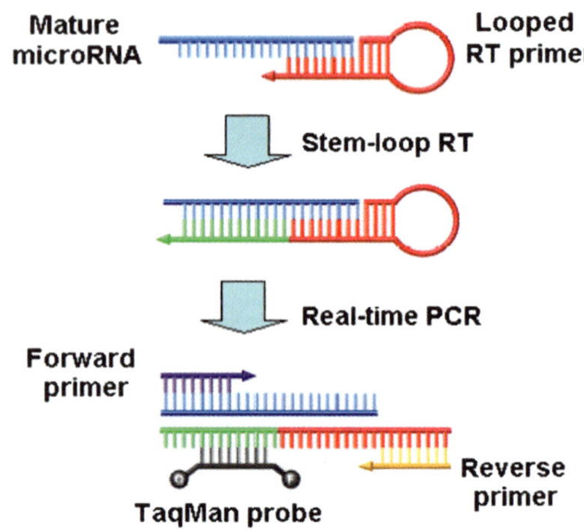

Fig. 4 Principle of microRNA RT-PCR assay method using stem-loop primers

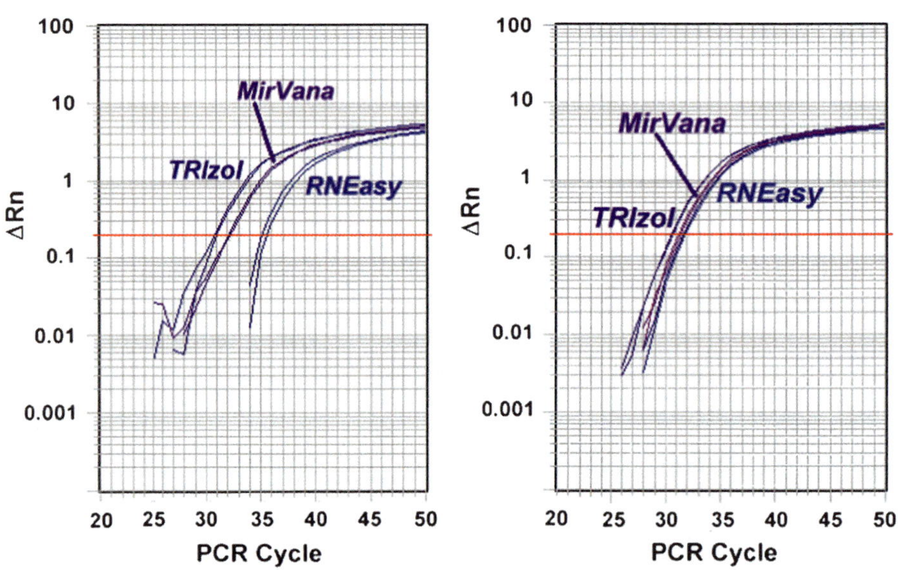

Fig. 5 Small nucleolar Z30 real-time PCR assay in duplicate of RNA samples (LCL) obtained starting from 10 ng of total RNA samples (**a**) and from 10 ng of LMW RNA (**b**) extracted with different protocols. In the latter case, Ct values (in log scale) of samples from different extractions are more reproducible, indicating a strict correlation between the amount of miRNAs and the LMW RNA fraction

snoZ30 (Table 3). This demonstrates that ΔCt differences parallel the amount of input RNA that is ultimately linked to the amount extracted by various protocols. Table 3 reports also the calculated LMW RNA concentration (expressed in %) calculated from ΔCt values applying the formula $2^{-\Delta Ct}$. MiRNA concentrations calculated from real-time assays are, as expected, in keeping with LMW RNAs concentrations evaluated with the *lab-on-a-chip* technology.

Table 3 Real-time PCR data obtained for HeLa and LCL were compared calculating the ΔCt differences between Ct (10 ng of total RNA) and Ct (10 ng of LMW RNA) values for miR-21 and snoZ30 (indicated as ΔCt (total RNA-LMW RNA))

		ΔCt (total RNA-LMW RNA)		Calculated LMW RNA
Cell type	Extraction method	miR-21	snoZ30	(%) ($2^{-\Delta Ct(snoZ30)} \cdot 100$)
HeLa	TRIzol reagent	1.8 (2)	2.1 (1)	23 (±2)
	MirVana kit	2.5 (1)	2.9 (1)	13 (±1)
	RNEasy kit	3.5 (4)	5.1 (3)	3 (±0.5)
LCL	TRIzol reagent	2.3 (1)	2.4 (1)	19 (±1)
	MirVana kit	2.2 (1)	2.1 (4)	23 (±5)
	RNEasy kit	4.2 (2)	4.6 (4)	4 (±1)

Standard deviations of at least three independent assays are reported in parentheses. LMW RNA concentrations (%) calculated using the formula $2^{-\Delta Ct(snoZ30)} \times 100$ are also reported where ΔCt are referred to snoZ30 values. Standard deviations of at least three independent assays are reported in parentheses

3.3 Different Tissues Express a Different Amount of Small- and Micro-RNAs

We showed that LMW RNAs and miRNAs amount are closely correlated. Therefore, we asked if different tissues might also express different amounts of LMW and miRNAs and if it might be possible to quantify them individually with *lab-on-a-chip* technology. For this reason, we selected five commercial RNA samples from different tissues extracted with the same protocol (according to manufacturer information). Total RNA from brain, skeletal muscle, heart, liver, and uterus were run on RNA 6000 Nano kit to quantify the total and the LMW RNA fractions. Total RNA concentrations were quite homogeneous brain (1,159 ng/mL) > heart (1,106 ng/mL) > uterus (1,080 ng/mL) > liver (963 ng/mL) > skeletal muscle (890 ng/mL). These values are in good agreement with the nominal manufacturer's concentration of 1,000 ng/mL. For all samples, the concentration of LMW RNA fraction was quite homogeneous and varied in the following order: liver (43 ng/mL) > brain (29 ng/mL) > heart (26 ng/mL) > uterus (25 ng/mL) > skeletal muscle (23 ng/mL). Electropherograms of LMW RNAs of different tissues are displayed in Fig. 6a. Calculating the percentage of LMW RNAs with respect to the whole total RNA amount, we found the following: liver (4.4%) > skeletal muscle (2.6%) > brain (2.5%) > heart (2.3%) > uterus (2.3%). Only some minor differences may be observed among various tissues. Given that the extraction protocol employed by the manufacturer was the same for all samples, we could hypothesize that the only observed difference for liver might depend on the different global expression of small RNAs in this tissue.

We ran the total RNA samples on Agilent Small RNA kit, specifically designed for the evaluation of miRNAs, to investigate if the difference in LMW RNA amount may also pertain to the miRNA fraction. Figure 6b shows a magnification of the electropherogram profile obtained by running total RNA samples from different tissues. The displayed region (from 35 to 45 s) is specific to the miRNA region (as indicated by the manufacturer). MiRNA concentrations were as follows: liver

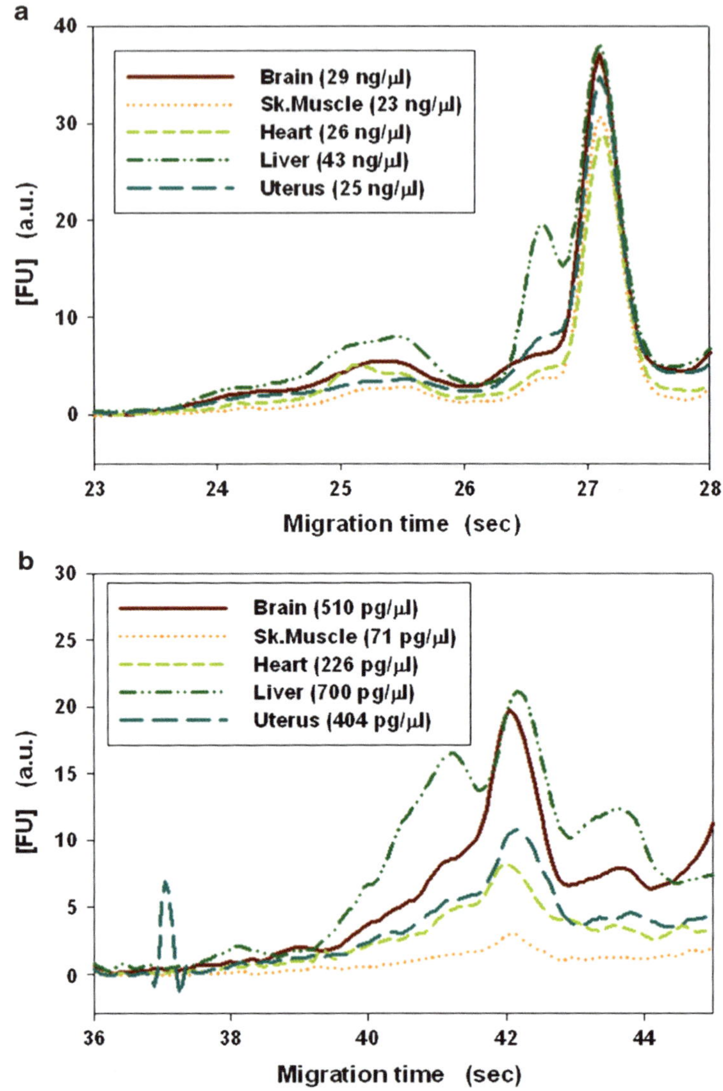

Fig. 6 (**a**) Electropherogram profile of LMW RNA species of five different tissues (brain, skeletal muscle, heart, liver, and uterus). (**b**) Electropherogram profile of miRNA species of the same tissues

(700 pg/mL) > brain (510 pg/mL) > uterus (404 pg/mL) > heart (226 pg/mL) > skeletal muscle (71 pg/mL). The miRNA concentration, expressed in percentage, with respect to LMW RNAs resulted in the following: brain (1.7%) > liver (1.6%) = uterus (1.6%) > heart (0.9%) > skeletal muscle (0.3%). From this *lab-on-a-chip* quantification it was possible to conclude that the miRNAs amount does not parallel that of LMW RNAs in the same tissue. Moreover, the variation is greater than that displayed by LMW RNAs among different tissues. Again, assuming that the extraction

efficiency is the same for all samples, the observed differences are only owed to the different global expression of miRNAs in these tissues.

3.4 Evaluation of Small and MiRNAs in Plants with Lab-on-a-Chip Technology

In the case of plant RNA samples, the amount of small RNAs and miRNAs were dependent on the protocol employed for total RNA extraction. Figure 7 shows the electropherogram profiles of the various plant RNA samples obtained with the Small RNA *lab-on-a-chip* kit. MiRNA concentrations were as follows: MirVana™ (3.4 ng/μL) > TRIzol (2.6 ng/μL) > Norgen (2.1 ng/μL) > RNEasy (0.4 ng/μL). However, the miRNA concentration, expressed in percentage respect to LMW RNAs resulted: Norgen (21.4%) > MirVana™ (20.2%) > TRIzol (16.4%) > RNEasy (10.8%) (Table 4).

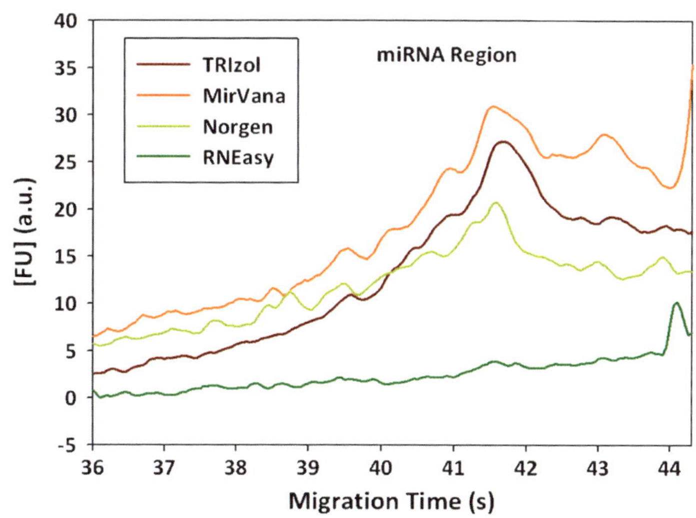

Fig. 7 Electropherogram profiles of miRNA species (Agilent Small RNA kit) of total plant RNA extracted with TRIzol reagent (*brown*), MirVana™ kit (*orange*), Norgen's kit (*yellow*), and RNEasy kit (*green*)

Table 4 Small RNA and miRNA mean concentrations (ng/μL) for *A. thaliana* (*L.*) RNA samples extracted with different RNA extraction protocols evaluated with Agilent 2100 Bioanalyzer

RNA extraction method	Small RNA (ng/μL)	miRNA (ng/μL)	miRNA/small RNA ratio (%)
TRIzol reagent	15.9	2.6	16.4
MirVans kit	16.8	3.4	20.2
Norgen's kit	9.8	2.1	21.4
RNEasy kit	3.7	0.4	10.8

The ratio miRNA/small RNA is reported as percentage

3.5 Evaluation of Endogenous Controls Reliability and Their Use for Expression Profile Comparison

The Small RNA kit, specifically designed for the identification and quantification of miRNA species, allowed us to know the exact amount of miRNAs present in the samples used for RT reactions. This information is useful, since the Ct values obtained from real-time assays are directly proportional to the amount of miRNAs. In fact, the Ct differences observed for the same miRNA in different tissues may owe essentially (1) to the intrinsic nature of the samples and/or (2) to dilution that generates a shift in Ct absolute values. To eliminate the dilution problem that occurs when absolute quantification is performed, a relative quantification with respect to an endogenous control is commonly followed. In fact, the endogenous control must have a constant expression in all samples and hence it may be used to normalize the expression of the other miRNAs. For a methodologically correct comparison, however, one must be confident that the control (i.e., nuclear and/or nucleolar small RNAs) really does have a constant expression in considered samples. The *lab-on-a-chip* technology that we used is able to give an estimate of this variation and let the researcher choose the right endogenous control (the one that does not significantly vary) from an adequate selection. Therefore, this validation ensures that the miRNA expression comparison among considered samples is methodologically correct.

The expression of three miRNAs (hsa-miR-26a, hsa-miR-26b, and hsa-miR-134) and two small RNAs (U6 and snoZ30) from the same five tissues analyzed before was assayed with real-time PCR. According to the manufacturer's suggestion, we started the RT reactions with 10 ng of total RNA. Cycle threshold values for all tissues are reported in Table 5. As expected, different Ct values for endogenous and other miRNAs were obtained. These values reflect the absolute concentrations of these miRNAs in various samples. Since most studies aim to discover differences in expression levels of miRNAs and not absolute levels of expression, the use of an endogenous control is needed. In order to assess if the differences in Ct values of the controls we used (U6 and snoZ30) are because of a different starting concentration or a real differential expression, we corrected the obtained values by taking into account the concentration of miRNAs previously obtained with the *lab-on-a-chip* technology. Hence, we considered that for double the concentration a correction of one Ct value should be applied. This preliminary correction eliminated the intrinsic variability owing to different sample concentrations and allowed us to estimate the reliability of the selected endogenous controls. We observed slight variation for Ct values of U6 (Average $Ct = 33.5 \pm 0.8$), while Z30 displayed a higher variability (Average $Ct = 34.1 \pm 1.1$) even after the applied correction. This means that U6 is constitutively expressed, at least in these tissues. Certainly, the lower the Ct difference, the more reliable are the results.

Therefore, we concluded that U6 is a more reliable endogenous control than snoZ30 for miRNA expression profile comparison in the analysed tissues. The miRNA expression profile comparison is reported in Fig. 8a.

Table 5 List of miRNAs expressed in different tissues (brain, muscle, heart, liver and uterus) with their average Ct values

micro-RNA	Average Ct	Corrected Ct
Brain		
U6	34.16	33.70
Z30	35.98	35.52
miR-26a	29.77	29.31
miR-26b	33.45	32.99
miR-134	33.67	33.21
Muscle		
U6	36.55	33.24
Z30	37.35	34.04
miR-26a	31.74	28.43
miR-26b	34.40	31.09
miR-134	38.23	34.92
Heart		
U6	36.15	34.52
Z30	36.28	34.65
miR-26a	32.60	30.97
miR-26b	36.21	34.58
miR-134	37.93	36.30
Liver		
U6	32.37	32.37
Z30	32.85	32.85
miR-26a	28.90	28.90
miR-26b	30.70	30.70
miR-134	37.30	37.30
Uterus		
U6	34.61	33.81
Z30	34.12	33.32
miR-26a	30.67	29.87
miR-26b	33.28	32.48
miR-134	38.77	37.97

Corrected Ct values represent the correction made after the precise quantification of miRNA species with the *lab-on-a-chip* technology

We then compared our data with those reported in the literature (Hsu et al. 2006, 2008). Figure 8b shows the expression values of hsa-miR-26a, hsa-miR-26b, and hsa-miR-134 compared with the expression values of some tissue-specific miRNAs (hsa-miR-1 for heart and muscle, hsa-miR-122a for liver, and hsa-miR-124a for brain). Expression values of miRNAs are expressed as copies per ng of RNA.

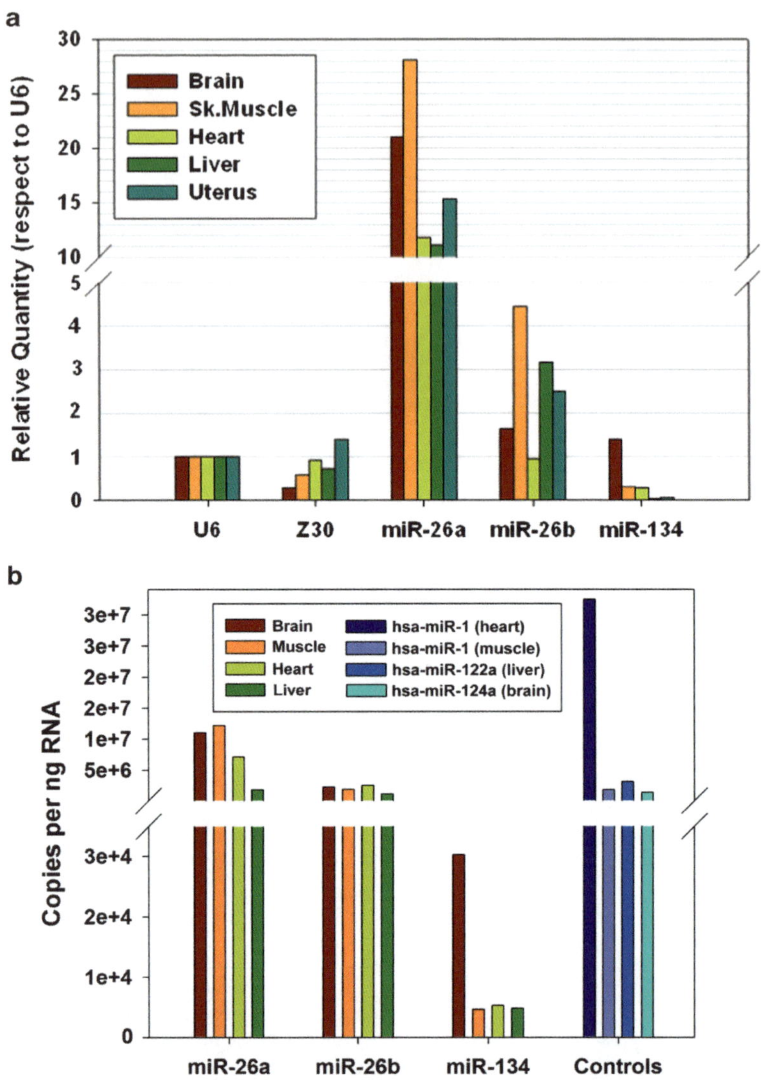

Fig. 8 (**a**) Corrected comparison of the relative expression of a miRNA selection in five different tissues (brain, skeletal muscle, heart, liver, and uterus) with respect to the U6 endogenous control. (**b**) Expression values of hsa-miR-26a, hsa-miR-26b, and hsa-miR-134 reported in the literature compared with the expression values of tissue-specific miRNAs

4 Discussion

One of the most frequent problems when dealing with miRNAs concerns the efficiency of quantitative and qualitative recovery after total RNA extraction from cells or tissues. In some cases, traditional total RNA extraction protocols are not efficient methods for extracting both HMW and LMW RNAs. Moreover, conventional quantification methods (i.e., spectrophotometric measurements) giving an overall quantification of total RNA concentration are therefore inadequate for these small RNA species. Polyacrylamide gel electrophoresis (PAGE) and northern blotting are two alternative viable techniques employed to visualize and evaluate small RNA molecules, particularly miRNAs. However, their low abundance sometimes prevents the use of such techniques. On the contrary, the *lab-on-a-chip* technology we employed is an alternative and valuable tool for the precise quantification of small RNAs and miRNAs present in total RNA samples.

We initially compared four common RNA extraction protocols to examine and quantify the recovery of HMW and LMW RNA species: the classic acid phenol/guanidine isothiocyanate solution, two glass-fiber filtration protocols, and a novel extraction kit based on a silicon-carbide purification resin. Actually, the Norgen's kit is the best performer compared to the other kits in terms of percentage of LMW RNA recovery (27.3%) (Table 1) with respect to total RNA and of miRNA with respect to small RNA ratio (21.4%) (Table 4). On the contrary, for human specimens, the acid phenol/guanidine isothiocyanate solution (Table 2), maximized not only the recovery of HMW RNA fractions but also the LMW RNAs (from 22 to 34%), as previously observed (Masotti et al. 2009). Then, we assessed by real-time PCR how significantly affected the recovery of miRNA species was as a function of the total RNA extraction protocol used on human RNA samples. The expression levels of a miRNA (hsa-miR-21) and a small nucleolar RNA (snoZ30) were evaluated by real-time PCR specific assays (Chen et al. 2005). This technique represents an effective alternative to Northern blotting for miRNA detection and quantification. Real-time quantification results (Table 3) demonstrated that the amount of miRNAs correlates better with the amount of LMW RNAs than with total RNA.

Another aspect that we critically analyzed concerned the correlation between LMW RNA species and miRNAs. The presence of LMW RNAs is not always directly correlated with that of miRNAs. Both plant and human specimens that we analyzed are a clear evidence for this. Moreover, we proved that even with the same extraction protocol, the concentration of miRNAs may differ considerably between various preparations (i.e., extraction from different tissues or plants) at least for human RNA samples. Different plant specimens extracted with the same protocol could behave similarly to what observed for human tissues.

As a potential application of our results, we applied our findings to the comparison of some miRNA expression in different human tissues, even if similar evaluations could be done also when dealing with plants. We assessed the expression of human miR-26a, miR-26b, and miR-134 as model miRNAs and two small RNAs (U6 and snoZ30) as controls. The expression profile comparison is correct only if we assume that the endogenous control (U6 or others) has a constant and equal expression in all the considered tissues. The *lab-on-a-chip* technology we used allowed precise quantification of input miRNAs, ultimately leading to a fine correction of real-time PCR Ct data for eventual variations. Therefore, we obtained a reliable and correct estimate of the relative quantities of miRNAs present in various tissues. Surprisingly, the expression of miR-26a was the highest of all the miRNAs in all the tissues considered (fold change >10 respect to U6). Although our data are in good agreement with those reported in the literature (Chen et al. 2005) (Fig. 8a), the high expression values found for miR-26a will deserve future investigations. However, the expression values of miR-26a reported in the literature (Hsu et al. 2006, 2008) are also very high (and higher than those of miR-26b) and above the values of those tissue-specific miRNAs considered as highly expressed (i.e., miR-1 for heart and muscle, miR-124a for brain and miR-122a for liver) (Fig. 8b).

5 Conclusions

In conclusion, all the methods illustrated in this chapter could help plant molecular biologists to study accurately the efficiency of extraction protocols, to analyze the small (but significant) differences between various preparations, and to suggest a methodologically correct method for the comparison of miRNA expression profiles in various plant tissues. Though we assessed our methodology only on a single plant species (*A. thaliana*), any vegetal species may be assessed with this *lab-on-a-chip* technology specifically designed to evaluate the quality and quantity of RNA, DNA, and proteins. We firmly think that the *lab-on-a-chip* technology that we used may be of help also to plant molecular biologists involved in miRNA profiling studies or in the analysis of non coding RNAs in the identification and quantification of small RNA species of interest.

Acknowledgments This work was supported by grants of the Italian Ministry of Health.

References

Ach RA, Wang H, Curry B (2008) Measuring microRNAs: comparisons of microarray and quantitative PCR measurements, and of different total RNA prep methods. BMC Biotechnol 8:69

Alves L Jr et al (2009) Comprehensive prediction of novel microRNA targets in *Arabidopsis thaliana*. Nucleic Acids Res 37:4010–4021

Aukerman MJ, Sakai H (2003) Regulation of flowering time and floral organ identity by a MicroRNA and its APETALA2-like target genes. Plant Cell 15:2730–2741

Axtell MJ, Snyder JA, Bartel DP (2007) Common functions for diverse small RNAs of land plants. Plant Cell 19:1750–1769

Backman TW et al (2008) Update of ASRP: the Arabidopsis Small RNA Project database. Nucleic Acids Res 36:D982–D985

Bartel DP (2004) MicroRNAs: genomics, biogenesis, mechanism, and function. Cell 116:281–297

Bartel B, Bartel DP (2003) MicroRNAs: at the root of plant development? Plant Physiol 132:709–717

Chen C et al (2005) Real-time quantification of microRNAs by stem-loop RT-PCR. Nucleic Acids Res 33:e179

Eulalio A, Huntzinger E, Izaurralde E (2008) Getting to the root of miRNA-mediated gene silencing. Cell 132:9–14

Galbraith DW, Birnbaum K (2006) Global studies of cell type-specific gene expression in plants. Annu Rev Plant Biol 57:451–475

Griffiths-Jones S et al (2006) miRBase: microRNA sequences, targets and gene nomenclature. Nucleic Acids Res 34:D140–D144

Hsu PW et al (2006) miRNAMap: genomic maps of microRNA genes and their target genes in mammalian genomes. Nucleic Acids Res 34:D135–D139

Hsu SD et al (2008) miRNAMap 2.0: genomic maps of microRNAs in metazoan genomes. Nucleic Acids Res 36:D165–D169

Ibberson D et al (2009) RNA degradation compromises the reliability of microRNA expression profiling. BMC Biotechnol 9:102

Jackson RJ, Standart N (2007) How do microRNAs regulate gene expression? Sci STKE 367:re1

John B et al (2004) Human MicroRNA targets. PLoS Biol 2:e363

Lewis BP et al (2003) Prediction of mammalian microRNA targets. Cell 115:787–798

Llave C et al (2002) Endogenous and silencing-associated small RNAs in plants. Plant Cell 14:1605–1619

Lu C et al (2005) Elucidation of the small RNA component of the transcriptome. Science 309:1567–1569

Mallory AC, Vaucheret H (2006) Functions of microRNAs and related small RNAs in plants. Nat Genet 38:S31–S36

Masotti A et al (2009) Quantification of small non coding RNAs allows an accurate comparison of miRNA expression profiles. J Biomed Biotechnol 2009:659028

Mineno J et al (2006) The expression profile of microRNAs in mouse embryos. Nucleic Acids Res 34:1765–1771

Nilsen TW (2007) Mechanisms of microRNA-mediated gene regulation in animal cells. Trends Genet 23:243–249

Palatnik JF et al (2003) Control of leaf morphogenesis by microRNAs. Nature 425:257–263

Pillai RS, Bhattacharyya SN, Filipowicz W (2007) Repression of protein synthesis by miRNAs: how many mechanisms? Trends Cell Biol 17:118–126

Rhoades MW et al (2002) Prediction of plant microRNA targets. Cell 110:513–520

Sunkar R, Zhu JK (2004) Novel and stress-regulated microRNAs and other small RNAs from Arabidopsis. Plant Cell 16:2001–2019

Voinnet O (2009) Origin, biogenesis, and activity of plant microRNAs. Cell 136:669–687

Yu B, Wang H (2010) Translational inhibition by microRNAs in plants. Prog Mol Subcell Biol 50:41–57

Zhang B et al (2006) Conservation and divergence of plant microRNA genes. Plant J 46:243–259

Zhang B, Wang Q, Pan X (2007) MicroRNAs and their regulatory roles in animals and plants. J Cell Physiol 210:279–289

Host Small RNAs and Plant Innate Immunity

Shang Gao and Hailing Jin

Contents

Abstract Small non-coding RNAs, including microRNAs (miRNAs) and small interfering RNAs (siRNAs), are crucial gene expression modulators in most eukaryotes. Increasing evidence indicates that host small RNAs also play a critical role during the plant immune responses. In this chapter, we discuss the functions of these pathogen-responsive endogenous small RNAs and the silencing pathway components during pathogen–host interactions.

Keywords Argonaute proteins • Dicer and Dicer-like proteins • miRNAs • RNA-dependent RNA polymerases • siRNAs

H. Jin (✉) and S. Gao
Department of Plant Pathology and Microbiology, Center for Plant Cell Biology, Institute for Integrative Genome Biology, University of California, 3234B Genomics Building, Riverside, CA 92521, USA
e-mail: hailingj@ucr.edu

V.A. Erdmann and J. Barciszewski (eds.), *Non Coding RNAs in Plants*, RNA Technologies, DOI 10.1007/978-3-642-19454-2_2,

1 Introduction

Plants have evolved innate immune systems to protect themselves from invading microorganisms (Chisholm et al. 2006; Jones and Dangl 2006; Bent and Mackey 2007). Plant extracellular surface receptors, so-called pattern recognition receptors (PRRs), can recognize conserved microbe-associated molecular patterns (MAMPs) or pathogen-associated molecular patterns (PAMPs) and trigger general defense responses, which are referred to as PAMP-triggered immunity (PTI), mainly through activating MAP kinase cascade and WRKY transcription factors. Many microbes have developed countermeasures by delivering effector proteins into the plant cell and suppressing host PTI pathways. Many plant hosts have subsequently evolved resistance (*R*) genes to overcome the suppression of effectors by triggering so-called effector-triggered immunity (ETI). ETI responses are more robust and rapid than PTI. In ETI, the pathogens are often repelled by localized cell death at the site of infection, and the whole plants become totally immune to the pathogen.

Host endogenous small RNAs have been recognized as important regulators in gene expression reprogramming during both PTI and ETI responses (Jin 2008; Voinnet 2008; Padmanabhan et al. 2009). In this chapter, we discuss the roles of host endogenous small RNAs and small RNA pathway components in plant immunity.

2 Small RNAs

2.1 miRNAs

microRNAs (miRNAs) are generated from single-stranded RNA precursors with hairpin structures by Dicer or Dicer-like (DCL) proteins, a class of RNase III type endoribonucleases. Emerging evidence indicates that some miRNAs function in plant defense responses against pathogen attacks. These miRNAs are upregulated or downregulated during pathogen infection to suppress negative regulators or to release positive regulators of immune responses, respectively (Fig. 1).

Arabidopsis miR393 was the first example of miRNA that regulates PTI in antibacterial defense (Navarro et al. 2006). miR393 is induced by a bacterial elicitor flagellin-derived MAMP flg22 and targets Auxin receptors, the F-box genes *TIR1*, *AFB2*, and *AFB3*. In miR393 overexpressing lines, the growth of bacterium *Pseudomonas syringae* pv. *tomato* (*Pst*) DC3000 was reduced, indicating that miR393-mediated suppression of auxin signaling contributes to PTI. AFB1, a third paralog of TIR1, is partially resistant to miR393-directed cleavage. The virulent *Pst* DC3000 can accumulate 20-fold higher in AFB1-Myc-overexpressing *Arabidopsis* in a *tir1-1* mutant background compared to the *tir1-1* plants. This enhanced disease symptom is mainly caused by the constitutive overexpression of AFB1, which confirmed a negative role of auxin signaling in plant immune responses. However, no difference was observed when the same plants were inoculated with avirulent *Pst* DC3000 carrying a type-III effector gene, *AvrRpt2*, which can trigger race-specific resistance ETI. Thus, auxin signaling has antagonistic effect mainly in PTI, but not

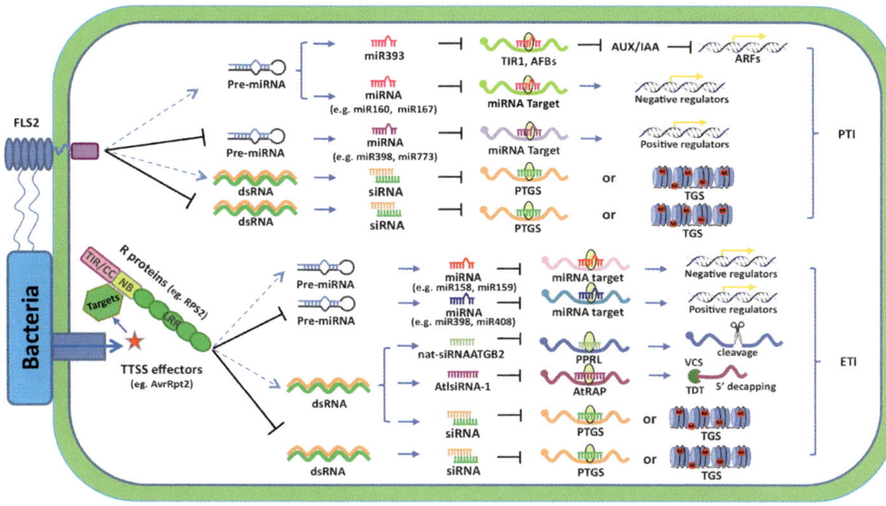

Fig. 1 miRNAs and siRNAs involved in the fine-tuning of plant immunity entail a combination of silencing and translational repression through PTI and ETI. *Upper panel*: after the elicitation of the extracellular domain of FLS2 receptor by bacterial flagellin, an unidentified signaling pathway will either upregulate the level of some miRNAs (eg. miR393, miR160 and miR167) and some siRNAs or downregulate the level of some other small RNAs (eg. miR398 and miR773). miR393 induction during PTI can suppress auxin receptor genes TIR1 and AFBs. The accumulated Aux/IAA inactivates auxin response factors (ARFs) and then suppresses the auxin responsive gene expression to enhance the basal defense. *Lower panel*: after the delivery of the bacteria TTSS effectors (eg. AvrRpt2) into the plant cell, the cognate R proteins (eg. RPS2) activates a suite of genes, which lead to up-regulation of some miRNAs (eg. miR158 and 159) and some endogenous siRNAs (eg. nat-siRNAATGB2 and AtlsiRNA-1) or down-regulation of some small RNAs (eg. miR398 and miR408). Induction of nat-siRNAATGB2 and AtlsiRNA-1 during RPS2-mediated ETI specifically down-regulate PPRL or AtRAP genes, respectively, through either target cleavage or mRNA degradation

in ETI. Fahlgren et al. (2007) further confirmed that miR393 can be strongly induced by more than ten-fold at 3 h post inoculation (hpi) by nonpathogenic *Pst* DC3000 *hrcC*, a strain that was mutated in the type III secretion system (TTSS). In addition, miR160 and miR167 can also be upregulated by *Pst* DC3000 *hrcC* at 3 hpi by five- and six-fold respectively, they both target Auxin-responsive factors (ARF) in auxin signaling pathway (Rhoades et al. 2002). Thus, at least three bacteria-induced miRNA families repress the auxin signaling and contribute to the PTI in plants. Auxin is a plant growth-promoting hormone that is antagonistic to salicylic acid (SA)-mediated defense pathways. Upon perceiving the pathogen PAMPs, these miRNAs are induced to rapidly repress the auxin signaling and shift the energy from plant growth to defense responses. By profiling RNA silencing effector AGO1-bound small RNAs, Li et al. identified that miR158, miR160, miR161.2, miR169a, miR391, miR396a, miR399f, miR822, miR824, and miR1888 are also induced by flg22 (Li et al. 2010). Further study with the transgenic *Arabidopsis* that overexpressed these miRNAs confirmed that miR160, but not miR158a positively regulate PTI responses.

During the pathogen attacks, a group of miRNAs is also downregulated to accumulate certain target mRNAs, which may contribute positively to defense responses. miR398 is downregulated in response to avirulent strains of *Pst* (*avrRpm1*) or *Pst* (*avrRpt2*) at 12 hpi and continued until 24 hpi (Jagadeeswaran et al. 2009). Only a small reduction was observed for miR398 after infection of virulent *Pst* DC3000 at 24 hpi. The targets of miR398 are Cu/Zn superoxide dismutases 1 and 2 (CSD1 and CSD2). Both targets were regulated by miR398 when *Arabidopsis* seedlings were exposed to high Cu^{2+} or high Fe^{3+} (Sunkar et al. 2006). However, during the *Pst* infection, only CSD1 level was upregulated and negatively correlated with miR398 levels, but not CSD2. These data suggest that miRNAs are likely to regulate different subgroup of target genes under different conditions. It has been found that miR398 negatively regulated PAMP-induced callose deposition (Li et al. 2010). In addition, expression of miR156, miR168, and miR773 was also reduced upon flg22 treatment (Li et al. 2010). Stable transgenic plants overexpressing miR398b and miR773 showed enhanced susceptibility to *Pst hrcC*, which indicated their negative roles of these two miRNAs in PTI defense.

The profile of miRNA expression in response to virulent *Agrobacterium* (*A.*) *tumefaciens*-infection was different from that in the tumors induced by *A. tumefaciens* (Dunoyer et al. 2006; Pruss et al. 2008). The levels of several conserved miRNAs, such as miR171, and *Arabidopsis*-specific miRNAs, such as miR163, were reduced by about two-fold in the tumors. Moreover, the levels of miR393 and miR167 were repressed to the limit of detectable level in the tumors, which led to the derepression of the auxin signaling pathway and promote tumor growth. Interestingly, similar phenomenon was also observed in human malignancies. miR-103/107 could attenuate miRNA biosynthesis by targeting Dicer to benefit the metastasis in human breast cancer (Martello et al. 2010). Inhibition of miR-103/107 opposes migration and metastasis of malignant cells. These studies suggest that global downregulation of miRNA may be a common feature of tumor growth, with only a few induced miRNAs.

2.2 *siRNAs*

Plants only contain several hundred miRNAs, which are largely conserved and limited in number as compared with small interfering RNAs (siRNAs), which are not conserved and numerous in number. Several endogenous siRNAs have been identified to play an important role in both ETI and PTI (Fig. 1) (Katiyar-Agarwal et al. 2006, 2007).

Arabidopsis nat-siRNAATGB2 was the first example of plant endogenous siRNAs that regulates gene expression in ETI responses (Fig. 1) (Katiyar-Agarwal et al. 2006). nat-siRNAATGB2 is induced specifically by *Pst* DC3000 carrying effector *avrRpt2*. It was generated by Dicer-like1 (DCL1) from the overlapping region of a natural antisense transcript (NATs) pair and targets the antisense *PPRL* transcript for degradation. The biogenesis of this siRNA was dependent on HYL1, HEN1, RDR6, SGS3, and NRPD1a. Accumulation of this siRNA depends on the cognate *R* gene *RPS2* and the downstream signaling component *NDR1* gene. Transgenic plants overexpressing the siRNA target gene *PPRL* showed more growth of *Pst* DC3000 (*avrRpt2*) and delayed hypersensitive response (HR), which indicates that *PPRL* plays a negative role in antibacterial defense in ETI.

Arabidopsis lsiRNA-1 (Katiyar-Agarwal et al. 2007) is one the novel class of endogenous siRNAs identified by Northern blot analysis during the search for more pathogen-induced small RNAs. This new class of siRNAs is 30–40-nt in length and is mainly induced by bacterial infection or specific growth conditions. Like nat-siRNAATGB2, AtlsiRNA-1 is also strongly and specifically induced by *Pst* DC3000 (*avrRpt2*), and its biogenesis pathway requires DCL1, HYL1, HEN1, HST1, AGO7, RDR6, NRPD1a, and NRPD1b. The target gene of AtlsiRNA-1 encodes a RNA-binding protein containing a RAP (RNA-binding domain abundant in Apicomplexans) domain. AtlsiRNA-1 employs a unique mechanism to repress *AtRAP* mRNA, which is by decapping and 5′–3′ decay mediated by an exoribonuclease XRN4. *atrap* mutant shows enhanced resistance to both virulent *Pst* and avirulent *Pst* (*avrRpt2*).

According to the widely accepted "gene-for-gene" theory (Flor 1956), successful plant disease resistance is triggered by the recognition of the pathogen effector proteins by the cognate R proteins and activates a series of resistance response events, including HRs, leading to local cell death and restrict the pathogen proliferation at the infection zone (Martin et al. 2003). To counteract the continuous evolution of pathogen effectors, plants have evolved many *R* genes. These *R* genes are generally clustered in the genome and encode proteins with conserved motifs, which are believed as a consequence of segmental chromosome duplication and rearrangements (Baumgarten et al. 2003; Meyers et al. 2003). In *Arabidopsis thaliana* ecotype Columbia, *RPP4* locus (first discovered in *Landsberg erecta* for recognition of *Peronospora parasitica* 5) comprises seven TIR-NBS-LRR class-*R* genes and is interspersed with three related and two unrelated genes (Noel et al. 1999; Yi and Richards 2007). Among this gene cluster, *RPP4* confers resistance to two races of *Hyaloperonospora parasitica*, while *SNC1* (for suppressor of *npr1-1*, constitutive 1) impairs resistance to both *P. syringe* and another race of *H. parasitica* (Stokes and Richards 2002; Zhang et al. 2003; Yang and Hua 2004), respectively. Yi et al. further extended Yang and Hua's findings that many paralogous *R* genes in this locus are positively regulated at the transcriptional level by SA amplification loop in *SNC1* overexpression plants (Yi and Richards 2007). Moreover, these *R* genes can be cosuppressed by siRNAs generated at this locus. On the one hand, activated *SNC1* can induce SA accumulation and defense responses that in turn inhibit cell growth, so RNA silencing might minimize the fitness cost associated with excessive *SNC1* expression. On the other hand, enhanced transcript levels of *SNC1* were observed in small RNA biogenesis-deficient mutants such as *dcl4*, *upf1*, and *ago1*, as well as in transgenic plants expressing P1/HC-Pro suppressor, which indicates that this locus is under the regulatory control of siRNAs.

3 Small RNA Pathway Components Play an Important Role in Plant Defense

3.1 *miRNA and siRNA Biogenesis Pathways*

miRNAs are derived from miRNA genes, which are transcribed by RNA polymerase II. The resulting single-stranded miRNA precursors could form stem-loop structure and be processed by DCL1 to generate miRNA /miRNA* duplex. DCL1 functions with two other proteins, HYL1 (HYPONASTIC LEAVES 1) and SE (SERRATE), which help ensure the dicing accuracy. HEN1 (HUA ENHANCER 1) methylates the small RNA duplex at the 3′ end (Yu et al. 2005), which is a crucial step to stabilize small RNAs. The matured miRNAs are incorporated predominantly into AGO1 to induce silencing of the target mRNA bearing fully or partly complementary sequences. Conversely, siRNAs are derived from perfectly paired region of double-stranded RNA (dsRNA) precursors. These regions could be the product of RNA-dependent RNA polymerase (RDR) or the overlapping regions of NATs. In plants, *trans*-acting siRNAs (ta-siRNAs), heterochromatic siRNAs (hc-siRNAs) or repeat-associated siRNAs (ra-siRNAs), nat-siRNAs, and lsiRNAs are four types of siRNAs identified so far with distinct biogenesis pathways (Katiyar-Agarwal and Jin 2010). During the arms race between host and pathogens, viruses and bacteria have evolved countermeasures to suppress host RNAi machinery. Here, we discuss the roles of small RNA pathway components in plant defense responses and silencing suppressors the pathogens evolved to suppress these RNAi pathway components.

3.2 *DCLs and Associated Proteins*

Arabidopsis encodes four DCLs; DCL1 is the major enzyme for processing miRNAs, and all the four DCLs are involved in siRNA formation. Some miRNAs, such as miR393, participate in PTI (Navarro et al. 2006). It has been shown that *dcl1-9* mutant is more susceptible to *Pst* DC3000 *hrcC*, and the transcriptional level of basal defense marker gene *WRKY30* was greatly reduced upon infection (Navarro et al. 2008). Moreover, PAMP-induced callose deposition was reduced to 50–70% in *Pst hrcC*-infected *dcl1-9* mutant (Li et al. 2010). These results confirmed the importance of DCL1 in PTI.

In antiviral immunity, DCL4 is the primary sensor and generates 21-nt siRNAs from several RNA and DNA viruses (Blevins et al. 2006; Deleris et al. 2006; Fusaro et al. 2006; Diaz-Pendon et al. 2007). In the absence of DCL4, 22- and 24-nucleotide-long virus-derived siRNAs are produced by DCL2 and DCL3, respectively. DCL1 may play an indirect role as a negative regulator of DCL4 (Qu et al. 2008). Although a number of studies indicate that *dcl2/4* double mutant is susceptible to RNA viruses (Bouche et al. 2006; Deleris et al. 2006; Diaz-Pendon et al. 2007), it has been found that SA-mediated resistance is sufficient to inhibit the CMV and TMV (*Tobacco mosaic virus*) infection in dcl2/3/4 triple mutant spraying with SA (Lewsey and Carr 2009).

Small RNA-binding proteins (DRBs) are cofactors of DCLs (Hiraguri et al. 2005; Nakazawa et al. 2007). HYL1, colocalized with DCL1 and SE in nuclear dicing bodies (Fang and Spector 2007; Fujioka et al. 2007; Song et al. 2007), is also an important component involved in antibacterial defense. *hyl1-1* mutant is susceptible to *Pst* (*avrRpt2*) and fails to accumulate nat-siRNAATGB2 and AtlsiRNA-1 (Katiyar-Agarwal et al. 2006, 2007). dsRNA-binding 4 (DRB4) contributes to antiviral defense by interacting with DCL4 (Qu et al. 2008). Pathogens have also evolved silencing suppressors to target various proteins of the RNAi pathways. Viral silencing suppressor P6 protein of *Cauliflower mosaic virus* (CaMV, family *Caulimoviridae*) inhibits the biogenesis of 21-nt siRNAs by physically binding with DRB4 protein, which subsequently interferes with the activity of the major plant antiviral silencing factor DCL4 (Haas et al. 2008).

HEN1 contributes to both antibacterial and antiviral defenses. *hen1-1* mutant compromised the resistance to *Pst* DC3000 *hrcC* (Navarro et al. 2008). Viral RNA of CMV accumulates five-fold more in *hen1-1* than in wild type (Boutet et al. 2003). Therefore, small RNA methylation and stabilization are crucial for plant defenses. P126 protein from TMV, a RNA replicase, binds to duplex viral siRNAs and inhibits HEN1-dependent methylation at the 3′ end, which destabilizes the viral siRNAs and attenuates the silencing efficiency (Vogler et al. 2007).

3.3 RNA-Dependent RNA Polymerases

Arabidopsis genome encodes six RDRs. RDR1 and RDR6 are considered to be involved in the antiviral defense. The activity of tobacco (*Nicotiana tabacum* cv. Xanthi, nn genotype) NtRDR1 was increased in TMV-infected or SA-treated plants. NtRDR1 antisense transgenic plants accumulated more TMV than wild-type control plants, indicating NtRDR1 plays a positive role in antiviral defense (Xie et al. 2001). Similarly, *Potato virus Y* (PVY) can also induce NtRDR1, and plants with reduced NtRDR1 transcripts accumulate higher PVY. The induction of another three defense-related genes (mitochondrial alternative oxidase, IVR, and ERF5) is also repressed in this plant (Rakhshandehroo et al. 2009). In *Arabidopsis*, both SA and compatible TMV can induce AtRDR1 gene expression. The authors suggest that the role of RDR1 in antiviral defense may not be through the RNAi pathway, but rather through the SA pathway (Yu et al. 2003). During the interaction between *Arabidopsis* and CMV-Δ2b (does not express the silencing suppressor protein 2b), viral siRNAs were mapped mostly to three viral RNA regions. Those mapped to the 5′ ends are associated with RDR1, whereas those mapped to the 3′ regions are associated with RDR6. Viral siRNAs were largely reduced in *rdr1/rdr6* double mutant. Thus, RDR1 and RDR6 work synergistically in *Arabidopsis* to amplify viral siRNAs, and RDR6 acts as a surrogate when RDR1 function is disrupted (Wang et al. 2010). However, mutated TuMV without silencing suppressor regained the ability to move systemically in *rdr1/ rdr6* double mutant plants (Garcia-Ruiz et al. 2010). The induction of RDR1 was not sufficient to prevent the systemic spreading of PVY (Rakhshandehroo et al. 2009).

Additional function of NtRDR1 was observed in *Nicotiana bethamiana*, which carries a nonfunctional endogenous NbRDR1 due to a 72-nt insertion (Yang et al. 2004). Transgenic *N. bethamiana* carrying tobacco NtRDR1 exhibits hypersusceptibility to *Plum pox potyvirus* (PPV) and several other viruses (Ying et al. 2010). The authors believe that transgenic NtRDR1 functions as a silencing suppressor to inhibit RDR6-dependent posttranscriptional silencing induced by sense transgenes (S-PTGS). *N. bethamiana* is one of the best hosts for many viruses. The silencing suppressor activity of NtRDR1 may explain the natural loss-of-function variant in *N. bethamiana* (Ying et al. 2010).

Some viruses have evolved suppressors that target RDRs directly or indirectly. For example, the V2 protein of *Tomato yellow leaf curl geminivirus* (TYLCV) is a suppressor that outcompetes host SGS3, a component associated with RDR6, for substrate dsRNA recognition at 5′ overhang and subsequently protects the viral dsRNAs from degradation (Glick et al. 2008; Elkashef and Ding 2009; Fukunaga and Doudna 2009).

3.4 Argonaute Proteins

Argonaute proteins are the effectors of RNAi machinery that associate with small RNAs and target complementary RNAs for transcriptional or posttranscriptional gene silencing (TGS). There are ten AGOs in *Arabidopsis*. AGO1 is involved in the biogenesis and function of most miRNAs. Two *ago1* mutants (*ago1-25* and *ago1-27*) showed attenuated flg22-induced basal defense responses, including reduced callose deposition, and reduced accumulation of *FRK1* and *WRKY29* transcripts. Pretreatment of flg22 induces PTI and subsequently inhibits bacterial growth, but this inhibitory effect was attenuated in both *ago1* mutants (Li et al. 2010). All of these results indicate that AGO1 is important for PTI against bacteriall infection. Bacteria have developed TTSS effectors, such as HopT1, to inhibit the activity of AGO1. In the SUC-SUL RNAi reporter line, expression of HopT1 inhibits the degradation of miRNA target transcripts, although the miRNA level remained the same. So, HopT1 can be considered as a bacteria-encoded suppressor of RNA silencing (BSRs) (Navarro et al. 2008).

AGO1 is also the target of many virus suppressors. Previous studies (Pazhouhandeh et al. 2006; Baumberger et al. 2007; Bortolamiol et al. 2007) have confirmed that P0 from beet Western yellow *polerovirus* (BWYV) contains an F-box-like domain and promotes proteolysis of AGO1. P0 does not interfere with preassembled miRNA/siRNA-containing RISC, but acts upstream of AGO1 loading and prevents the formation of small RNA-AGO1 complexes by AGO1 degradation (Csorba et al. 2010). P38 of Turnip crinkle virus (TCV) mimics host-encoded glycine/tryptophan (GW)-containing proteins and sequesters AGO1 from binding to other endogenous GW/WG proteins (Azevedo et al. 2010). Host GW/WG proteins bind to AGO proteins via the GW/WG "AGO-binding hooks" and are required for RISC assembly and activity (Jin and Zhu 2010). P1 protein from *Sweet potato mild mottle virus* (SPMMV, family *Potyviridae*) is another example of GW/WG containing VSR inhibits RISC activity by direct AGO1 binding. The three WG/GW motifs located at N-terminal part of P1 are efficient "Ago hooks" to inhibit functional RISC assembling. Thus, P38 and P1 proteins encoded by viruses are examples of how pathogens usurp and mimic host regulatory pathways. P25 protein of *Potato virus X* (PVX, family *Flexiviridae*) strongly interacts with *Arabidopsis* AGO1 through the degradation of AGO1 via the proteasome pathway (Chiu et al. 2010). Plants are less susceptible to PVX and Bamboo mosaic virus when treated with MG132, a proteasome inhibitor. P25 also interacts with AGO2, AGO3, and AGO4, but not with AGO5 and AGO9 by transient assay in *N. benthamiana* leaves. *Cucumber mosaic virus* (CMV, family *Bromoviridae*) encoded a nucleus-localized 2b protein, which can physically interact with both AGO1 and siRNAs of RISC and inhibit its slicing action (Diaz-Pendon et al. 2007; Goto et al. 2007). In addition, 2b was also found to interact with AGO4, and the 2b-AGO4 complex was mainly in the nucleus but absent from the nucleolus where the AGO4 and Pol IV complex reside (Gonzalez et al. 2010).

AGO4 and AGO6 are closely related AGO proteins that mediate mainly TGS by DNA methylation and/or chromatin modification (Vaistij et al. 2002; Huettel et al. 2007; Matzke et al. 2009). They predominantly associate with 24-nt siRNAs and silence mainly transposons and repeated sequences. Several studies indicated that viral, bacterial, and fungal infections alter the genomic methylation level and change the expression of many genes (Finnegan et al. 1998; Stokes and Richards 2002; Pavet et al. 2006). Geminiviruses are single-stranded DNA viruses that replicate in cell nuclei via double-stranded DNA intermediates that associated with host histone proteins (Pilartz and Jeske 1992, 2003). In transgenic plants carrying a geminivirus *Tomato leaf curl virus* (TLCV) gene C4, hypermethylation was observed at the asymmetric cytosines of the C4 transgene and its promoter region, and no C4 transcripts were detected (Bian et al. 2006). These results suggest that plants utilize DNA methylation as a defense mechanism against geminiviruses. Similar results were observed in two other geminiviruses, *Cabbage leaf curl virus* (CaLCuV) and *Beet curly top virus* (BCTV) (Raja et al. 2008). Asymmetric cytosine methylation and histone H3K9 methylation (Gene silencing marker) were reduced in RNA-directed DNA methylation (RdDM) pathway mutants, and AGO4 was also implicated in this process. These data suggest that the DNA genome of geminiviruses is silenced via RdDM pathway in plants. In addition to the antiviral function, AGO4 is also involved in defense response to *P. syringea* (Agorio and Vera 2007). *ago4-2* mutant is susceptible to both virulent *Pst* DC3000 and avirulent *Pst* carrying *avrRpm1* effector compare to the wild-type plants. However, whether this antibacterial function is through RdDM pathway is still not clear because other loss-of-function mutants in the RdDM pathway did not show compromised resistance, including genes operating upstream of AGO4, RDR2, and DCL3, and genes operating downstream of AGO4, such as DRD1, CMT3, DRM1, and DRM2.

AGO7 is preferentially associated with miR390, which triggers the production of ta-siRNAs (Montgomery et al. 2008). AGO7 is also required for the accumulation of AtlsiRNA-1, suggesting its role in *RPS2*-mediated ETI (Katiyar-Agarwal et al. 2007). *ago7* mutant (*zip-1*) attenuated the resistance to *Pst* DC3000 (*avrRpt2*), although it exhibits normal responses in PTI, including callose deposition, *FRK1* expression, seedling growth inhibition, and oxidative burst analysis (Li et al. 2010). Moreover, AGO1 and AGO7 have been found to contribute to viral RNA clearance (Qu et al. 2008).

Small RNA pathway components also play an important role in defense responses against multicellular eukaryotic pathogens, such as fungi. Inoculation of the *sgs* mutants (*sgs1-1*, *sgs2-1* and *sgs3-1*) with different pathogenic strains of the Vascular fungi *Verticillium* all resulted in a similar increased susceptibility phenotype (Ellendorff et al. 2009), but no altered defense was observed with the similar life-style vascular fungal pathogen *Fusarium oxysporum* f. sp. *raphani*. Further screening with other small RNA biogenesis mutants found that AGO7, DCL4, NRPD1a, RDR2, AGO1, HEN1, and HST1 all affect *Verticillium* defense. But those small RNA pathway components do not comply with one single RNA-silencing pathway among well-defined ones. This result suggests that plant defense against *Verticillium* may involve multiple small RNA-silencing pathways. Further in-depth analysis is needed to verify the specific roles of each component in the immune responses.

4 Conclusions and Future Prospects

Increasing evidence has demonstrated the importance of small RNAs and small RNA machinery in plant immune responses against various pathogens. They play essential roles in gene expression reprogramming and fine-tuning during host-microbial interaction. We predict that more pathogen-regulated host small RNAs and pathogen-derived silencing suppressors will be discovered. However, many important questions still remain. For example, what are the upstream signaling events that regulate small RNA biogenesis and metabolism in response to pathogen attack? What are the regulatory proteins that interact with small RNA pathway components in response to infection? Whether and how are these silencing signals transported? Can small RNAs and small RNA complex shuttle between hosts and pathogens? Increasing evidence suggests that pathogen infection alters host genomic chromatin structures. Whether and how long can these chromatin modifications maintain, and whether they can be transferred to the next generation? More exciting time will come when we will be able to answer these questions.

References

Agorio A, Vera P (2007) ARGONAUTE4 is required for resistance to *Pseudomonas syringae* in Arabidopsis. Plant Cell 19:3778–3790

Azevedo J, Garcia D, Pontier D et al (2010) Argonaute quenching and global changes in Dicer homeostasis caused by a pathogen-encoded GW repeat protein. Genes Dev 24:904–915

Baumberger N, Tsai CH, Lie M et al (2007) The Polerovirus silencing suppressor P0 targets ARGONAUTE proteins for degradation. Curr Biol 17:1609–1614
Baumgarten A, Cannon S, Spangler R et al (2003) Genome-level evolution of resistance genes in *Arabidopsis thaliana*. Genetics 165:309–319
Bent AF, Mackey D (2007) Elicitors, effectors, and R genes: the new paradigm and a lifetime supply of questions. Ann Rev Phytopathol 45:399–436
Bian XY, Rasheed MS, Seemanpillai MJ et al (2006) Analysis of silencing escape of tomato leaf curl virus: an evaluation of the role of DNA methylation. Mol Plant Microbe Interact 19:614–624
Blevins T, Rajeswaran R, Shivaprasad PV et al (2006) Four plant Dicers mediate viral small RNA biogenesis and DNA virus induced silencing. Nucleic Acids Res 34:6233–6246
Bortolamiol D, Pazhouhandeh M, Marrocco K et al (2007) The Polerovirus F box protein P0 targets ARGONAUTE1 to suppress RNA silencing. Curr Biol 17:1615–1621
Bouche N, Lauressergues D, Gasciolli V et al (2006) An antagonistic function for Arabidopsis DCL2 in development and a new function for DCL4 in generating viral siRNAs. EMBO J 25:3347–3356
Boutet S, Vazquez F, Liu J et al (2003) Arabidopsis HEN1: a genetic link between endogenous miRNA controlling development and siRNA controlling transgene silencing and virus resistance. Curr Biol 13:843–848
Chisholm ST, Coaker G, Day B et al (2006) Host-microbe interactions: shaping the evolution of the plant immune response. Cell 124:803–814
Chiu MH, Chen IH, Baulcombe DC et al (2010) The silencing suppressor P25 of Potato virus X interacts with Argonaute1 and mediates its degradation through the proteasome pathway. Mol Plant Pathol 11:641–649
Csorba T, Lozsa R, Hutvagner G et al (2010) Polerovirus protein P0 prevents the assembly of small RNA-containing RISC complexes and leads to degradation of ARGONAUTE1. Plant J 62:463–472
Deleris A, Gallego-Bartolome J, Bao J et al (2006) Hierarchical action and inhibition of plant Dicer-like proteins in antiviral defense. Science 313:68–71
Diaz-Pendon JA, Li F, Li WX et al (2007) Suppression of antiviral silencing by cucumber mosaic virus 2b protein in Arabidopsis is associated with drastically reduced accumulation of three classes of viral small interfering RNAs. Plant Cell 19:2053–2063
Dunoyer P, Himber C, Voinnet O (2006) Induction, suppression and requirement of RNA silencing pathways in virulent *Agrobacterium tumefaciens* infections. Nat Genet 38:258–263
Elkashef S, Ding SW (2009) Possible new RNA intermediate in RNA silencing. Nat Chem Biol 5:278–279
Ellendorff U, Fradin EF, de Jonge R et al (2009) RNA silencing is required for Arabidopsis defence against Verticillium wilt disease. J Exp Bot 60:591–602
Fahlgren N, Howell MD, Kasschau KD et al (2007) High-throughput sequencing of Arabidopsis microRNAs: evidence for frequent birth and death of MIRNA genes. PLoS ONE 2:e219
Fang Y, Spector DL (2007) Identification of nuclear dicing bodies containing proteins for microRNA biogenesis in living Arabidopsis plants. Curr Biol 17:818–823
Finnegan EJ, Genger RK, Peacock WJ et al (1998) DNA methylation in plants. Annu Rev Plant Physiol Plant Mol Biol 49:223–247
Flor HH (1956) The complementary genic systems in Flax and Flax Rust. Adv Genet 8:29–54
Fujioka Y, Utsumi M, Ohba Y et al (2007) Location of a possible miRNA processing site in SmD3/SmB nuclear bodies in Arabidopsis. Plant Cell Physiol 48:1243–1253
Fukunaga R, Doudna JA (2009) dsRNA with 5′ overhangs contributes to endogenous and antiviral RNA silencing pathways in plants. EMBO J 28:545–555
Fusaro AF, Matthew L, Smith NA et al (2006) RNA interference-inducing hairpin RNAs in plants act through the viral defence pathway. EMBO Rep 7:1168–1175
Garcia-Ruiz H, Takeda A, Chapman EJ et al (2010) Arabidopsis RNA-dependent RNA polymerases and dicer-like proteins in antiviral defense and small interfering RNA biogenesis during Turnip Mosaic Virus infection. Plant Cell 22:481–496

Glick E, Zrachya A, Levy Y et al (2008) Interaction with host SGS3 is required for suppression of RNA silencing by tomato yellow leaf curl virus V2 protein. Proc Natl Acad Sci USA 105:157–161

Gonzalez I, Martinez L, Rakitina DV et al (2010) Cucumber mosaic virus 2b protein subcellular targets and interactions: their significance to RNA silencing suppressor activity. Mol Plant Microbe Interact 23:294–303

Goto K, Kobori T, Kosaka Y et al (2007) Characterization of silencing suppressor 2b of cucumber mosaic virus based on examination of its small RNA-binding abilities. Plant Cell Physiol 48:1050–1060

Haas G, Azevedo J, Moissiard G et al (2008) Nuclear import of CaMV P6 is required for infection and suppression of the RNA silencing factor DRB4. EMBO J 27:2102–2112

Hiraguri A, Itoh R, Kondo N et al (2005) Specific interactions between Dicer-like proteins and HYL1/DRB-family dsRNA-binding proteins in Arabidopsis thaliana. Plant Mol Biol 57:173–188

Huettel B, Kanno T, Daxinger L et al (2007) RNA-directed DNA methylation mediated by DRD1 and Pol IVb: a versatile pathway for transcriptional gene silencing in plants. Biochim Biophys Acta 1769:358–374

Jagadeeswaran G, Saini A, Sunkar R (2009) Biotic and abiotic stress down-regulate miR398 expression in Arabidopsis. Planta 229:1009–1014

Jin H (2008) Endogenous small RNAs and antibacterial immunity in plants. FEBS Lett 582:2679–2684

Jin H, Zhu JK (2010) A viral suppressor protein inhibits host RNA silencing by hooking up with Argonautes. Genes Dev 24:853–856

Jones JD, Dangl JL (2006) The plant immune system. Nature 444:323–329

Katiyar-Agarwal S, Jin H (2010) Role of small RNAs in host-microbe interactions. Annu Rev Phytopathol 48:225–246

Katiyar-Agarwal S, Morgan R, Dahlbeck D et al (2006) A pathogen-inducible endogenous siRNA in plant immunity. Proc Natl Acad Sci USA 103:18002–18007

Katiyar-Agarwal S, Gao S, Vivian-Smith A et al (2007) A novel class of bacteria-induced small RNAs in Arabidopsis. Genes Dev 21:3123–3134

Lewsey MG, Carr JP (2009) Effects of DICER-like proteins 2, 3 and 4 on cucumber mosaic virus and tobacco mosaic virus infections in salicylic acid-treated plants. J Gen Virol 90:3010–3014

Li Y, Zhang Q, Zhang J et al (2010) Identification of microRNAs involved in pathogen-associated molecular pattern-triggered plant innate immunity. Plant Physiol 152:2222–2231

Martello G, Rosato A, Ferrari F et al (2010) A MicroRNA targeting dicer for metastasis control. Cell 141:1195–1207

Martin GB, Bogdanove AJ, Sessa G (2003) Understanding the functions of plant disease resistance proteins. Annu Rev Plant Biol 54:23–61

Matzke M, Kanno T, Daxinger L et al (2009) RNA-mediated chromatin-based silencing in plants. Curr Opin Cell Biol 21:367–376

Meyers BC, Kozik A, Griego A et al (2003) Genome-wide analysis of NBS-LRR-encoding genes in Arabidopsis. Plant Cell 15:809–834

Montgomery TA, Howell MD, Cuperus JT et al (2008) Specificity of ARGONAUTE7-miR390 interaction and dual functionality in TAS3 trans-acting siRNA formation. Cell 133:128–141

Nakazawa Y, Hiraguri A, Moriyama H et al (2007) The dsRNA-binding protein DRB4 interacts with the Dicer-like protein DCL4 in vivo and functions in the trans-acting siRNA pathway. Plant Mol Biol 63:777–785

Navarro L, Dunoyer P, Jay F et al (2006) A plant miRNA contributes to antibacterial resistance by repressing auxin signaling. Science 312:436–439

Navarro L, Jay F, Nomura K et al (2008) Suppression of the microRNA pathway by bacterial effector proteins. Science 321:964–967

Noel L, Moores TL, van Der Biezen EA et al (1999) Pronounced intraspecific haplotype divergence at the RPP5 complex disease resistance locus of Arabidopsis. Plant Cell 11:2099–2112

Padmanabhan C, Zhang X, Jin H (2009) Host small RNAs are big contributors to plant innate immunity. Curr Opin Plant Biol 12:465–472

Pavet V, Quintero C, Cecchini NM et al (2006) Arabidopsis displays centromeric DNA hypomethylation and cytological alterations of heterochromatin upon attack by *Pseudomonas syringae*. Mol Plant Microbe Interact 19:577–587

Pazhouhandeh M, Dieterle M, Marrocco K et al (2006) F-box-like domain in the polerovirus protein P0 is required for silencing suppressor function. Proc Natl Acad Sci USA 103:1994–1999

Pilartz M, Jeske H (1992) Abutilon mosaic geminivirus double-stranded DNA is packed into minichromosomes. Virology 189:800–802

Pilartz M, Jeske H (2003) Mapping of abutilon mosaic geminivirus minichromosomes. J Virol 77:10808–10818

Pruss GJ, Nester EW, Vance V (2008) Infiltration with *Agrobacterium tumefaciens* induces host defense and development-dependent responses in the infiltrated zone. Mol Plant Microbe Interact 21:1528–1538

Qu F, Ye X, Morris TJ (2008) Arabidopsis DRB4, AGO1, AGO7, and RDR6 participate in a DCL4-initiated antiviral RNA silencing pathway negatively regulated by DCL1. Proc Natl Acad Sci USA 105:14732–14737

Raja P, Sanville BC, Buchmann RC et al (2008) Viral genome methylation as an epigenetic defense against geminiviruses. J Virol 82:8997–9007

Rakhshandehroo F, Takeshita M, Squires J et al (2009) The influence of RNA-dependent RNA polymerase 1 on potato virus Y infection and on other antiviral response genes. Mol Plant Microbe Interact 22:1312–1318

Rhoades MW, Reinhart BJ, Lim LP et al (2002) Prediction of plant microRNA targets. Cell 110:513–520

Song L, Han MH, Lesicka J et al (2007) Arabidopsis primary microRNA processing proteins HYL1 and DCL1 define a nuclear body distinct from the Cajal body. Proc Natl Acad Sci USA 104:5437–5442

Stokes TL, Richards EJ (2002) Induced instability of two Arabidopsis constitutive pathogen-response alleles. Proc Natl Acad Sci USA 99:7792–7796

Sunkar R, Kapoor A, Zhu JK (2006) Posttranscriptional induction of two Cu/Zn superoxide dismutase genes in Arabidopsis is mediated by downregulation of miR398 and important for oxidative stress tolerance. Plant Cell 18:2051–2065

Vaistij FE, Jones L, Baulcombe DC (2002) Spreading of RNA targeting and DNA methylation in RNA silencing requires transcription of the target gene and a putative RNA-dependent RNA polymerase. Plant Cell 14:857–867

Vogler H, Akbergenov R, Shivaprasad PV et al (2007) Modification of small RNAs associated with suppression of RNA silencing by tobamovirus replicase protein. J Virol 81:10379–10388

Voinnet O (2008) Post-transcriptional RNA silencing in plant-microbe interactions: a touch of robustness and versatility. Curr Opin Plant Biol 11:464–470

Wang XB, Wu Q, Ito T et al (2010) RNAi-mediated viral immunity requires amplification of virus-derived siRNAs in *Arabidopsis thaliana*. Proc Natl Acad Sci USA 107:484–489

Xie Z, Fan B, Chen C et al (2001) An important role of an inducible RNA-dependent RNA polymerase in plant antiviral defense. Proc Natl Acad Sci USA 98:6516–6521

Yang S, Hua J (2004) A haplotype-specific Resistance gene regulated by BONZAI1 mediates temperature-dependent growth control in Arabidopsis. Plant Cell 16:1060–1071

Yang SJ, Carter SA, Cole AB et al (2004) A natural variant of a host RNA-dependent RNA polymerase is associated with increased susceptibility to viruses by *Nicotiana benthamiana*. Proc Natl Acad Sci USA 101:6297–6302

Yi H, Richards EJ (2007) A cluster of disease resistance genes in Arabidopsis is coordinately regulated by transcriptional activation and RNA silencing. Plant Cell 19:2929–2939

Ying XB, Dong L, Zhu H et al (2010) RNA-dependent RNA polymerase 1 from *Nicotiana tabacum* suppresses RNA silencing and enhances viral infection in *Nicotiana benthamiana*. Plant Cell 22:1358–1372

Yu D, Fan B, MacFarlane SA et al (2003) Analysis of the involvement of an inducible Arabidopsis RNA-dependent RNA polymerase in antiviral defense. Mol Plant Microbe Interact 16:206–216

Yu B, Yang Z, Li J et al (2005) Methylation as a crucial step in plant microRNA biogenesis. Science 307:932–935

Zhang Y, Goritschnig S, Dong X et al (2003) A gain-of-function mutation in a plant disease resistance gene leads to constitutive activation of downstream signal transduction pathways in suppressor of npr1-1, constitutive 1. Plant Cell 15:2636–2646

Antisense Transcription Associated with microRNA Target mRNAs

Christopher D. Rock, Qing-Jun Luo, and Fangqian Ou-Yang

Contents

Abstract Here, we review recent progress on understanding the biological significance and pervasiveness of transitivity (RNA silencing phenomena) associated with miRNAs in plants and comparisons with parallels in animals. The importance of RNA surveillance and splicing machineries in miRNA biogenesis as it relates to substrate availability for RNA interference (RNAi) is discussed. We summarize a meta-analysis of deep-sequencing datasets that reveals a novel case base substitution of miRNAs in *Arabidopsis* that results in perfect matches to their targets that spawn siRNAs. We discuss current models for RNAi triggers by miRNAs and the potential involvement of RNA editases, Dicers, and Argonautes. We conclude with some speculation on the role of transposon silencing in rice contributing to evolution of a class of miRNAs

C.D. Rock (✉), Q.-J. Luo, and F. Ou-Yang
Department of Biological Sciences, Texas Tech University, Lubbock, TX 79409-3131, USA
e-mail: chris.rock@ttu.edu

Q.-J. Luo
Present Address: Department of Pediatrics, Stanford University, Stanford, CA 94305-5208, USA

V.A. Erdmann and J. Barciszewski (eds.), *Non Coding RNAs in Plants*, RNA Technologies, DOI 10.1007/978-3-642-19454-2_3,

with predicted targets in 3′-UTRs, which may serve to assemble adaptive stress response networks and reveal a conserved mechanistic link between animal and plant miRNA function and transitivity.

Keywords *MIRNA* genes • RNA interference • Base substitution • miRNA evolution • Transposons • Interaction siRNA network

1 Introduction

Antisense transcription is a pervasive but poorly understood phenomenon associated with RNA interference (RNAi) and small RNAs (sRNAs) in plants and animals. Non coding loci, such as those producing miRNAs, P-element-Induced WImpy testes-interacting small RNAs (piRNAs), and small interfering RNAs (siRNAs), are key components of gene expression in eukaryotes, forming a regulatory network superimposed on the central dogma of molecular biology (Chapman and Carrington 2007; Voinnet 2009). It has been stated that sRNAs are the "dark matter" of biology (Ruvkun 2001); thus, ncRNAs may be the afterglow of a biological "Big Bang," at the core of how life originated through RNA evolution, a theory called "the RNA World" (Gilbert 1986). miRNAs are expressed through nucleolytic maturation of hairpin precursors transcribed by RNA Polymerase II or III (Borchert et al. 2006; Xie et al. 2005). siRNAs are derived either from endogenous transcripts that form perfect double-stranded RNA (dsRNA) structures, or from transcripts of transgenes, viral genomes, and protein-coding genes (and some miRNA targets). All three classes of sRNAs are involved in posttranscriptional gene regulation in plants, fungi and/or animals by acting as substrates for the ARGONAUTE (AGO)-containing RNA-Induced Silencing Complex (RISC), where the identity of the associated AGO or PIWI protein determines the functional output of the associated sRNA (transcriptional silencing, mRNA cleavage or translation repression) (Chapman and Carrington 2007; Voinnet 2009; Stefani and Slack 2008). The high sequence homology of plant miRNAs and siRNAs to their targets suggests the percentage of targeted plant genes (~1%) is much lower than mammals, yet both plant and animal miRNAs have epigenetic effects beyond cleavage and translational inhibition, raising the possibility that conserved mechanisms exist, notwithstanding the lack of conserved miRNA- and target sequences between kingdoms. There are two main questions to be addressed while approaching the subject of miRNAs and transitivity (the generation of secondary siRNAs by triggers): *how* are posttranscriptional gene silencing (PTGS) processes involved in miRNA activity, and perhaps more importantly, and *why*? The answer to the latter question is undoubtedly based on understanding the evolution of *MIRNA* genes and the molecules that make up the machineries of miRNA biogenesis, mRNA processing, and RNA action.

Nature constructs robust and precise systems (networks of proteins) from noisy and imperfect parts. All organisms are related by evolution, and therefore processes may be identical or highly similar between different organisms due to their shared descent, while variations on common themes represent evolutionary adaptations to diverse environments. The emerging evidence is that this is certainly the case with eukaryotic transcriptional gene silencing (TGS) and PTGS. miRNAs appear to have evolved from transposable elements (TEs), inverted duplications of target genes, and by diversification of pathogen and antiviral defense mechanisms, or even by random events. They might buffer mRNA populations against changes, in addition to promoting changes. There is emerging evidence that miRNAs and non coding RNAs are important for stress responses. Consistent with this notion, they are found as integral components of networks interfaced with feedback loops that monitor, report, and react to environmental and developmental changes. Yet, paradoxically, if miRNAs were to completely silence their own loci and/or that of their targets, the primary "normal" functions of miRNAs as homeostasis or fine-tuning effectors of growth and development would be lost. Some progress has been made recently into the first question of mechanisms and specificity (e.g., *which* miRNAs/templates serve as triggers, and *how*), but the latter (*why* do miRNAs and their targets undergo transitivity) is still outstanding. Several possibilities exist that may be informed by understanding *when* during development and *where* in the cell (or gene, or organism) transitive processes may occur.

It is becoming clear that sRNAs and antisense transcripts are not obscure biological curiosities, but fundamental to the integrity and cohesion of the genome, and a powerful influence on genome evolution that has been selected for during domestication of crops. However, understanding epigenetic control of gene expression in plants cannot be extrapolated from other eukaryotes. Higher plants have evolved during recent evolutionary time (<140 million years ago), possess kingdom-specific genes (e.g., coiled-coil zinc-finger domain SUPPRESSOR OF GENE SILENCING3 [SGS3], RNA Polymerases IV and V [Pol IV, Pol V]), and diverse RNA-dependent RNA Polymerases [RDRs]. They uniquely produce gametes from somatic tissues and can undergo parthenogenesis (apospory), and tolerate polyploidy and interspecific hybridization. They are also a huge repository of genetic diversity (>250,000 species), which makes them an excellent subject for understanding organic evolution, genome structure/function, and the physiological and trans-generational significance of non coding and sRNAs. Plant–microbe symbiotic development, gametogenesis, antibacterial-, viral-, and stress responses are dependent on natural-*cis*-antisense transcripts, but it is still under debate to what extent they are associated with sRNAs that couple exogenous signals to gene regulation (Navarro et al. 2008; Ron et al. 2010; Xu et al. 2008; Borsani et al. 2005; Ben Amor et al. 2009). Understanding ncRNAs may lead to new insights into fundamental processes such as tissue-specific, developmental, and environmental gene regulation, chromatin dynamics, and genome evolution.

This review focuses on the relationship between miRNAs and production of secondary sRNAs, presumably through antisense transcription and dsRNA, and presents some computational analyses of published datasets that can frame questions for future study. There are three fundamental aspects of these intersecting pathways that are yielding to mechanistic and structural studies: sRNA biogenesis catalyzed by DICER-LIKE and accessory factors, formation and sRNA loading of

RISCs, and targeting of complementary mRNAs. Several other chapters in this book intersect and buttress the approach we have taken, namely, that of studying the "noise" (secondary and modified sRNAs) associated with miRNAs. It is hoped that a collective view emerges that points the way toward an integrated understanding of ncRNAs, miRNAs, and RNAi and how variations in individual components affect function of overall RNA processing systems. The significance of this knowledge may be to provide cogent solutions to grand challenges such as development of value-added crops and gene therapies that efficiently respond to local and changing environments, and predicting the consequences of environmental changes to genes in terms of the related properties of efficiency, heterogeneity, and stochasticity inherent in all biological systems.

2 Original Observations of Plant miRNAs and Transitivity

Transcriptome profiling experiments have demonstrated the extensive presence of endogenous antisense transcripts both in plants and animals, but the mechanisms and significance of such transcriptional activities are still not clear. Transitivity, a prevalent phenomenon that drives transgene silencing, has been proposed to associate with such antisense transcription. Copy RNA synthesis may occur by primed or unprimed initiation, supported by the evidence that siRNAs spread both 5′ and 3′ along the target relative to the trigger in plants and *Neurospora* (Vaistij et al. 2002; Makeyev and Bamford 2002), while only in the antisense orientation in *Caenorhabditis elegans* (Pak and Fire 2007). There is biochemical evidence for both pathways (Tang 2005; Makeyev and Bamford 2002), and they probably overlap at some key point(s). The link between miRNAs and transitivity was not clearly revealed until Voinnet's group (Parizotto et al. 2004) observed that stringent mutations within miRNA target sequences can block cleavage, but may not entirely prevent transitivity through siRNAs. This suggests that miRNAs may have additional activities or determinants in posttranscriptional regulation that are independent of cleavage. Furthermore, miRNAs are found to guide the generation of *trans*-acting siRNAs (*ta*-siRNAs), a subclass of sRNAs, through antisense transcription associated with RDR6 and SGS3 (Vazquez et al. 2004; Allen et al. 2005; Axtell et al. 2006; Peragine et al. 2004). SGS3 acts as a homodimer (Elmayan et al. 2009) and stabilizes non coding *TAS1/2* cleavage products (Yoshikawa et al. 2005), possibly by binding dsRNA with a 5′ overhang in cooperation with RDR6 (Fukunaga and Doudna 2009). RDR6 and SGS3 physically interact and colocalize in cytoplasmic granules distinct from Processing-bodies (P-bodies, see below) (Kumakura et al. 2009).

Characterization of antisense transcripts and/or antisense sRNAs for miRNA targets, including *PLANTACYANIN*, *SPL3*, *SPL10*, *TIR1*, *HAP2C* and a clade of *PPR* genes (Lu et al. 2005; Wu and Poethig 2006; Axtell et al. 2006; Ronemus et al. 2006) further corroborate the link. By combining different techniques and analyses, including custom high-resolution tiling microarrays, semiquantitative reverse transcriptase real-time Polymerase Chain Reaction (qRT-PCR), and computational analysis of whole-genome tiling array and deep-sequencing sRNA data, we approached the question of antisense siRNAs and transcripts derived from miRNA target transcripts (Luo et al. 2009; Richardson et al. 2010). We observed that antisense transcription of miRNA targets in *Arabidopsis* can occur upstream of the miRNA binding site and that antisense transcripts near select miRNA target sites were elevated in *hua enhancer1* (*hen1-1*) and the *sgs3-14* mutants (Luo et al. 2009). Consistent with previous results, unambiguous antisense transcripts to several miRNA target genes depended on DICER-LIKE1 (DCL1), HYPONASTIC LEAVES1/DOUBLE-STRANDED RNA BINDING PROTEIN1 (HYL1/DRB1) and RDR6. However, we did not obtain direct evidence for an RDR6-dependent priming model of antisense transcription from miRNAs, because *rdr6-15* mutants can accumulate antisense transcripts associated with miRNA targets. Our results suggested that HEN1 and SGS3 might work in the same process/step to suppress synthesis/stability of miRNA target-associated antisense transcripts, which could serve as a link between miRNA and RNA silencing pathways, for example by competition between RNA surveillance/processing machineries and RNAi pathways for substrates such as intermediates/products in the miRNA pathway.

Interestingly, a recent report describes a similar observation in the moss *Physcomitrella patens* for miRNA target-associated antisense transcripts (Khraiwesh et al. 2010). Moss mutants without a *DCL1b* gene have normal miRNA levels (processed by DCL1a homologue), but cleavage of targets is abolished. Upstream and downstream antisense transcripts and siRNAs to miRNA targets are observed in wild type, similar to our results in *Arabidopsis*, but not in *DCL1b* knockouts (Khraiwesh et al. 2010). Furthermore, the target transcription rates are drastically reduced in mutants that accumulate miRNA:target-RNA duplexes and show spreading hypermethylation over the genes and their promoters, leading to silencing (Khraiwesh et al. 2010). These surprising results suggest that DCL1b functions via a pathway/mechanism other than canonical miRNA/siRNA biogenesis and programming of RISC, and serves as a specificity determinant for miRNA target transitivity and homeostasis. miRNA-mediated TGS has been described in animals (Kim et al. 2008a), albeit by different mechanisms involving miRNA target promoters. A *C. elegans* AGO family member, RDE1, functions as a scavenger protein by taking up sRNAs from many different sources, including the miRNA pathway, and triggering RNAi (Correa et al. 2010). It is intriguing to consider whether the antisense transcripts associated with higher plant miRNAs follow a similar path for their biogenesis as those in *P. patens* and *C. elegans*.

Along the line of reasoning that miRNAs and associated antisense transcripts may reveal important mechanisms of gene regulation similar to siRNAs, there are several evidences that hint at cross-talk/coupling between miRNAs and chromatin remodeling: dominant mutations in the miR166 binding sites of *PHABULOSA* (*PHB*) and *PHAVOLUTA* (*PHV*) cause decreased methyl-cytosine (in *cis*, especially CHG and CHH de novo establishment types over CpG maintenance type) of *PHB* and *PHV* (Bao et al. 2004), suggesting that miR166 interacts locally with nascent, newly processed *PHB* mRNA to alter chromatin downstream of the miRNA binding sites in *PHB* and *PHV* genes. It remains unknown whether cleavage of the target is required for this interaction, or whether transcription rates or tissue-specificity of *PHB* expression are altered by the miR166-mediated methylation. A recent report describes a DCL3-dependent class of novel 24 nt long miRNAs (lmiRNAs) in rice, which guides the methylation of about 80 nt flanking their own loci as well as target loci in *trans* via a DCL3- and AGO4-dependent pathway. When disrupted by RNAi, these *mirna* gene knockdowns resulted in some predicted targets being upregulated (Wu et al. 2010b), supporting a functional role for the epigenetic silencing of these miRNAs and/or targets. Since numerous miRNAs in *Arabidopsis* also can produce 24-nt lmiRNAs (Vazquez et al. 2008), it is plausible that *PHB/PHV* methylation is also mediated by this novel pathway.

We have recently uncovered a potential link between cryptic miR160 binding sites, antisense transcription, and 24 nt species of siRNAs in silencing of a clade of seven recently evolved *AUXIN RESPONSE FACTOR* genes (*ARF12/22*) (Rock, unpublished). These genes are expressed only in the female gametophyte and during embryogenesis, and have a retrotransposon-like simple sequence repeat in intron 2. These genes appear to be under positive selection, have a natural-*cis*-antisense transcript for *ARF14*, possess a candidate cryptic binding site for miR160, and spawn predominantly 24 nt species of siRNAs that are dependent on DNA cytosine methylation maintenance (*methyltransferase1*, mutant, *met1*) and establishment (DNA methyltransferase triple mutant, *drm1drm2cmt3*, *ddc*) pathways (Rock, unpublished). The cryptic miR160 sites in the candidate target sequences are partially complementary to miR160 and with a similar topology of mismatches as the noncleaved cryptic miR390 binding sites in *TAS3* that trigger production of *ta*-siRNAs sorted by AGO7 (Adenot et al. 2006; Axtell et al. 2006; Montgomery et al. 2008). There is an inverse relationship between *ARF14* antisense expression and *ARF14* sense strand expression during embryo development, which was regulated positively by RDR6 and negatively by DCL3. It was found that before ovule fertilization, *ARF14* sense expression was localized in punctate cytoplasmic foci of integumentary cells, whereas antisense *ARF12/22* expression was localized to the nucleus of these cells and to a callose-rich boundary between peripheral endosperm and the sporophyte endothelium which was altered in *rdr6-15* and *sgs3-11* mutants (Jiang and Rock, unpublished).

Why 24 nt species of siRNAs dominate the *ARF12/22* landscape is unknown, but recent results may provide a clue. Mutations in RDM12/IDN2 affect the levels of siRNAs from some of the RNA-directed DNA Methylation (RdDM) target loci, suggesting that RDM12/IDN2 could function with RDR2 or chromatin-remodeling effectors in RdDM and TGS, analogous to SGS3/RDR6 in PTGS (Zheng et al. 2010; Ausin et al. 2009). Base pairing between *ARF12/22* siRNAs and complementary nascent scaffold ncRNA transcripts produced by Pol II (e.g., *ARF14* antisense) or Pol V could generate dsRNAs with a 5′ overhang stabilized by SGS3 or SGS3-like RDM12/IDN2/homologues, including *At1g13790* and *At1g80790*, which are differentially expressed in the female gametophyte (Wuest et al. 2010). Recent reports on genetic analysis of miR160 and auxin in the developing embryo (Liu et al. 2007, 2010; Pagnussat et al. 2009) are consistent with a function of natural-*cis*-antisense siRNAs from the *ARF14* sense–antisense gene pair, which silence *ARF12/22* gene cluster expression in peripheral endosperm/endothelium and integument domains of the ovule.

3 RNA Processing and Antisense Transcripts

Once transcribed, mRNAs associate with a host of proteins throughout their lifetime. These mRNA-proteins (mRNPs) and heterogeneous nuclear ribonucleoproteins (hnRNPs) undergo a series of remodeling events that are influenced by and/or influence the translation and mRNA decay machinery. Nonsense-mediated RNA decay is a surveillance mechanism that detects and eliminates aberrant mRNAs whose expression would result in truncated proteins often deleterious to the organism (Chang et al. 2007). An important route for mRNA degradation produces uncapped mRNAs that are channeled into the exosome in the decay process initiated by decapping enzymes, endonucleases, and siRNAs and miRNAs. The exosome has many cofactors and is involved in a large number of pathways from 3′ processing of polyadenylated rRNA, small nuclear RNA (snRNA) and small nucleolar RNA (snoRNA), to decay of mRNAs and mRNA surveillance. Proper degradation of mRNA is crucial for the maintenance of cellular and organismal homeostasis. Degradation is also a function of assembly into P-bodies governed by the relative rates of translational repression, external stresses, and mRNA processive decay (Weber et al. 2008). Transcripts without a cap or a poly(A) tail might be substrates for the RNA silencing pathway where sRNAs could be generated from these "aberrant" RNA transcripts by antisense transcription (Luo and Chen 2007; Herr et al. 2006; Gregory et al. 2008).

A variety of mechanisms exist to prevent adventitious production of proteins from intergenic antisense transcripts, ranging from degradation within the nucleus to translational inhibition in the cytosol (Bickel and Morris 2006). mRNP homeostasis depends on rapid transitions between three functional states: (1) translated mRNPs in polysomes, (2) stored mRNPs, and (3) mRNPs under degradation. Massively parallel short-tag sequencing and microarray technologies have been used to study the dynamics and complexity of eukaryotic transcriptomes and epigenomes, giving new biological insights into the "active genome." For example, inhibition of the cellular RNA surveillance systems in *Arabidopsis* cells results in the accumulation (and suppression) of thousands of transcripts arising from annotated and unannotated regions of the genome (Gregory et al. 2008), underscoring the central position of regulated mRNA decay in physiology and development (Chekanova et al. 2007).

Specific examples of the links between RNA processing and miRNA biogenesis have been elucidated by transcriptome profiling of mutants of Zn-finger splicing factor *serrate* (*se*) and *aba-hypersensitive1/cap-binding protein80* (*abh1/cbp80*) showing deranged miRNA biogenesis (Laubinger et al. 2008; Gregory et al. 2008). A homologue of SERRATE, ARS2, has recently been shown as a component of the nuclear RNA cap-binding complex in mouse and to mediate both antiviral defense and developmental patterning in Drosophila (Sabin et al. 2009; Gruber et al. 2009), establishing that RNAi- and miRNA-dependent processes are deeply conserved between plants and animals. Similarly, ectopic overexpression in animals of the splicing factor SF2/ASF results in differential accumulation of many miRNAs (Wu et al. 2010a). Further capture and integration of additional types of genome-wide datasets will help to illuminate hidden features of the dynamic genomic landscape that are regulated by both genetic and epigenetic pathways.

In human cells, the addition of a 3′ terminal oligo U-tract on mRNAs or mRNA fragments can promote decapping and stabilization of the 3′ end of the RNA by binding the Lsm1-7 complex that ensures 5′-directional degradation (Song and Kiledjian 2007). This implies that the 3′ end of the 5′ fragment of miRNA target transcripts in plants could be stabilized by a similar mechanism and would have a longer half life than its 5′ end, thus increasing the probability for it to serve as a template for RNA silencing. SGS3 might be a plant-specific transporter/stabilizer of dsRNA products of miRNA targets, acting in opposition to the LSm-like complex and mRNA degradation by the 3′→5′ exosome (Chekanova et al. 2007), 5′→3′ exoribonuclease EXORIBONUCLEASE 4/ETHYLENE INSENSITIVE 5 (XRN4/EIN5) (Souret et al. 2004), and ABH1/CBP80 pathways (Gregory et al. 2008).

4 Models for Generation of Antisense Transcripts and siRNAs from miRNA Target Transcripts

The original hypothesis for the production of antisense transcripts and siRNAs from miRNA target mRNAs was that miRNAs could prime off the 3′ hydroxyl to generate antisense transcripts from their cognate sense transcripts, which in turn generate siRNAs for gene silencing (Peragine et al. 2004; Yu et al. 2005; Axtell et al. 2006; Ronemus et al. 2006). It had been speculated that methylation of miRNAs and siRNAs at the 3′-terminal hydroxyl group by HEN1 may serve to counteract the antisense transcription activity primed (in a formal sense) by unmethylated sRNAs (Yu et al. 2005). The lack of reduction of antisense transcripts in *mir164* knockout mutants also disfavors the priming model (Luo et al. 2009). Axtell et al. (2006) described a "two-hit" mechanism for transitivity of two classes miRNA target genes with one or two miRNA complementary sites. For the first class of miRNA targets, they are cleaved by two different miRNAs or *ta*-siRNAs at two complementary sites, including *TAS1-4* and *TAS3* in *P. patens*, and *ARF4* and many *PPR*s in Arabidopsis. Secondary siRNAs are generated between these two cleavage sites on target transcripts by RDR(s). The second class of miRNA targets has Arabidopsis *TAS3* as the only example. It carries one miR390 cleavage site at the 3′ end of the transcript, while a second, cryptic, miR390 binding site is on the transcript 5′ end, which can trigger siRNA production without cleavage. Thus, the two-hit model proposed for the generation of siRNAs from miRNA targets invokes two miRNA complementary sites within its transcript to confer transitivity from RDR actions. However, this model does not provide clues for the siRNA spawning from targets with only one miRNA cleavage site, for example *Arabidopsis TAS1a-c*, *TAS2*, *AGO1*, *TIR1* and some other transcripts. These target transcripts are cleaved by their cognate miRNAs on the 5′ end and generate downstream (toward the 3′ end of transcripts) siRNAs probably assisted by an unknown protein (Axtell et al. 2006).

Recently the Baulcombe (Chen et al. 2010) and Carrington (Cuperus et al. 2010) groups provided evidence for the association of 22-nt species of miRNAs with the generation of siRNAs from the above-mentioned targets and several others with one cleavage site. Strikingly, the cognate miRNAs or *ta*-siRNAs are preferentially 22-nt long, for example miR168, miR173, miR393, miR447, miR472, miR771, miR773, miR825*, miR828, miR856, and *ta*-siR2140. These sRNAs also exist as 21-nt forms in lower abundance which can guide the cleavage of target transcripts but lack the capability to trigger siRNA generation. 22-nt miRNAs are processed from hairpins with asymmetric bulges, which are supported by the solution structures of human miRNA precursors (Starega-Roslan et al. 2011; Cuperus et al. 2010). More importantly, 22-nt miRNAs or *ta*-siRNAs are loaded into the same AGO1 as their 21-nt species, which puts AGO1 squarely at the nexus of secondary siRNA-generating machinery. Therefore, Cuperus et al. (2010) proposed that the loading of 22-nt sRNAs into AGO1 may induce its tertiary structure change and recruit RDR6 and SGS3 to an AGO1-containing complex, which could route target transcripts into RNAi pathways. However, this model does not help to explain the transitivity associated with other miRNA targets because they are regulated by typical 21-nt miRNAs.

From current knowledge, we are trying to put forward a general model to fit with data from different angles. In our hypothesis, mature miRNA or siRNA duplexes, following Dicer cleavage, must be transferred from Dicer to AGO to form the active RISC, a process that leads to the selection of the guide (miRNA) strand and slicing/degradation of the passenger (miRNA*) strand. The human RISC loading complex and its components have recently been analyzed by negative-stain electron microscopy, providing a structural framework for testing how siRNA duplexes may be passed from the complex of Dicer and TAR-Binding Protein (TRBP) to Ago2 during RISC loading (reviewed in Kawamata and Tomari 2010). Taken together with the evidence in *Arabidopsis* that AGO1 can be found in the nucleus associated with Dicing-bodies (Fang and Spector 2007) and that AGOs interact with GW-repeat domain proteins associated with silencing (Karlowski et al. 2010; Azevedo et al. 2010; He et al. 2009), we speculate that miRNA:target and/or antisense associations could function as competitive inhibitors of miRNA action and/or triggers for siRNA production, similar to "target mimicry" for long-distance miR399 stress response in plants (Franco-Zorrilla et al. 2007) and "pseudogene miRNA decoys" and antisense transcript-mediated inhibition of miRNA function in animals (Faghihi et al. 2010; Poliseno et al. 2010).

5 Possible Biological Significance of miRNA Target-Associated Antisense Transcripts

The production of antisense transcripts and antisense sRNAs from miRNA targets probably induces a series of subsequent reactions in vivo. Antisense transcripts are prerequisites for formation of long dsRNA duplexes that may function in PTGS as hypothesized for natural antisense transcripts (Werner et al. 2009). This could result in the generation of secondary sRNAs and probable downregulation of transcripts with little homology to the primary sRNAs. This action would likely be restricted to some specific cell types or some extreme physiological conditions such that it would not affect the normal biological functions of the cognate genes in vivo. Our finding that not every miRNA target gene generates antisense transcripts or sRNAs (Luo et al. 2009) is in line with this notion.

The existence of antisense pathways generating RNA transcripts complementary to the sense strand of miRNA target mRNAs should be under stringent control for the majority of non-*TAS* miRNA targets. Because miRNAs are under strong selection pressure for their target mRNAs and act dominantly, their cell-specific expression must be tightly regulated. Therefore, transitivity may be under negative selective pressure because extensive amplification would compromise miRNA function. sRNAs can move through plasmodesmata and act noncell-autonomously in nearby cells, and RDR6 functions in transitive gene silencing in these neighbor cells too (Parizotto et al. 2004; Himber et al. 2003). The few neighboring cells adjacent to domains of cell-specific miRNA gene expression might be the source of antisense transcripts, which could also explain their typically low abundance signals. As previously suggested (Himber et al. 2003; Ronemus et al. 2006), coupled miRNA/siRNA mechanisms might function in tissues where the miRNA is not expressed to generate gradients of developmental effectors, e.g., in meristems and primordia, or to allow miRNA activity to be amplified where a limiting amount of miRNA may be present, e.g., in response to stress. Vaucheret and colleagues (Vaucheret 2009; Mallory and Vaucheret 2009; Vaucheret et al. 2006) have shown that minor perturbations of *MIRNA168* and/or its target *AGO1* expression by both miR168 and by transitive siRNAs generated after miR168-mediated cleavage leads to fine-tuned posttranscriptional adjustment of miR168 and AGO1 levels, thereby maintaining a proper balance of most other miRNAs and AGO1. This suggests that modulating the efficiency of assembling miRNA-programmed RISCs may be important in other contexts or require other determinants (see below). Another possible explanation for the lack of strong effects on miRNA target antisense transcript abundance in *hen1-1* and *sgs3-14* mutants is genetic redundancy, a hallmark of polyploid plant genomes. This hypothesis is congruent with phenotypes of *ago1*, *ago7*, *ago10/zll*, *dcl1*, *hyl1*, *rdr6*, and *rdm12/idn2* mutants that have only modestly altered miRNA or siRNAs and target gene abundances (Mallory et al. 2009; Peragine et al. 2004; Ausin et al. 2009), and the existence of parallel genetic pathways for miRNA activity defined by SERRATE, AS1, AS2, and ABH1 (Lobbes et al. 2006; Ueno et al. 2007; Yang et al. 2006; Gregory et al. 2008).

6 Speculation: Perfect Target-Match Small RNAs with One Mismatch to miRNAs as Potential Triggers for Antisense

sRNAs with the size of 20–24 nt are subject to posttranscriptional modifications, such as methylation, uridylation, adenylation, and base substitution (Ebhardt et al. 2009; Iida et al. 2009; de Hoon et al. 2010). We have endeavored to scrutinize the potential role of miRNA base substitution in promoting the generation of secondary siRNAs from miRNA target mRNAs and describe here some of our observations. Asymmetric bulges are intriguing because a similar topology of miRNA bulges in *MIRNA* foldbacks gives rise to 22 nt species shown to cause AGO1- and RDR6-dependent siRNA production from targets (Chen et al. 2010; Cuperus et al. 2010). Meta-analysis of publicly available sRNA deep-sequencing datasets showed that 20–22-nt one-base-mismatched sRNAs (OMM-sRNAs) are associated with mature miRNAs (Luo and Rock, submitted). There is biochemical (Mi et al. 2008) and genetic evidence for association of AGO1 activity (Vaucheret et al. 2006), HYL1/DRB1 miRNA strand selection (Eamens et al. 2009), and ABH1, SERRATE, DCL1, and HYL1-dependent miRNA biogenesis (Kim et al. 2008b; Laubinger et al. 2008; Fujioka et al. 2007; Fang and Spector 2007; Dong et al. 2008; Yang et al. 2006; Ronemus et al. 2006), concordant with our observed base substitutions.

Intriguingly, a class of sRNAs with single mismatch to mature miRNA sequences was found to match perfectly with one or more cognate miRNA targets. This class of candidate miRNAs includes family members of miR156/157, miR160, and miR172 (Table 1). These sRNAs perfectly matching to targets have a similar pattern of DCL1 dependence on abundance as their cognate miRNAs for those species where data exists (Table 1, data not shown). Remarkably, in *dcl2/3/4* triple mutants the normalized abundances of the edited miRNAs increase several fold for those species that have multiple reads in wild type, analogous to the AGO1- and DCL1 homeostasis effects on certain mature miRNA stabilities (Vaucheret et al. 2006) and HYL1/AGO1 effects on certain miRNA* abundances (Eamens et al. 2009).

The concordant expression patterns and abundance relationships (down in *dcl1*, up in *dcl234*) of these candidate miRNAs with their non-substituted miRNA counterparts support the working hypothesis that they are products or substrates of DCL1-processing, rather than generated by the PTGS pathway acting on miRNA targets to produce a single species of miRNA-like antisense sRNA. More intriguingly, these sRNAs were preferentially sequenced in high abundance from immunoprecipitated AGO1 (Table 1) (Mi et al. 2008). The AGO1 loading of these sRNA species suggests a functional significance in cellular conditions. Finally, the miRNA targets associated with these candidate substituted miRNAs are known to spawn antisense sRNAs (Table 1) (Luo et al. 2009), supporting a functional role of perfect match sRNAs in RNAi.

Based on these observations and previous reports of differential stabilities of these same miRNAs and miRNA* being dependent on miRNA biogenesis machinery and AGO1 (Kim et al. 2008b; Laubinger et al. 2008; Fujioka et al. 2007; Fang and Spector 2007; Dong et al. 2008; Yang et al. 2006; Eamens et al. 2009), we propose a working model (Fig. 1) whereby the edited miRNA hairpins affect the processing by DCL1 and/or HYL1-dependent loading or activity of AGO1. Using A→G substitution as an example, which is hypothesized to be catalyzed by Adenosine Deaminases Acting on RNA (ADARs) as in animals, we propose ADAR and ADAR-Like activities may compete with a DCL1:HYL1 complex for miRNA primary transcripts and/or precursors or miRNA:miRNA* duplexes in the nucleus. The base substitution may introduce a bulge at the edited site, which is then processed into edited miRNA:miRNA* duplex. Loading of such duplex into AGO1

Table 1 Small RNAs with one mismatch to miRNAs, in association with DICERs and ARGONAUTE1, have perfect match to miRNA targets that generate secondary siRNAs.

sRNA[a]	Avg Illumina[d]	Avg 454[e]	Col-0[f]	*dcl1-7*[h]	*dcl234*[i]	AGO1[k]	AGO2[k]	AGO4[k]	AGO5[k]	AGO7[l]	miRNA[m]	Target gene[n]/ α-sRNA[o]
TGACAGAAGAGAGaGAGCAC	28.6	3	12[g]	NA	77.5[j]	7.1/0	0	0	0	0	miR156abcdef	AT1G27370.1/ SPL10/4.40; AT5G43270.2/ SPL2/8.02; AT5G50670.1/ SPL/2.00
TTGACAGAAGAgAGaGAGCAC[b]	24.2	NA	9[g]	NA	96.5[j]	8.3/NA	0	0	0	0	miR156d, h/ miR157a, b, c	AT1G27370.1/ SPL10/4.40; AT5G43270.2/ SPL2/8.02
TGCCTGGCTCCCTGcATGCCA	13.6	6	21.3	26.0	94.5	7.1/12	0	0	0	0	miR160a, b, c	AT1G77850.1/ ARF17/0.48
AGAATCcTGATGATGCTGCAT[c]	9.8	18	22.3	0	29.5	7.7/220	0	1.4	3.2	0	miR172a, b	AT5G60120.1/ TOE2/0.95
AGAATCcTGATGATGCTGCAG	1.3	3	0.5	0	0	0.6/52	0	0	0	0	miR172c, d	AT4G36920.1/ AP2/4.26

The sRNA data sets from German et al. (2008) and Mi et al. (2008) were BLASTed to Arabidopsis TAIR9 annotated transcripts. The full-matched sRNAs were retained and BLASTed to miRNA hairpins. The abundance unit for sRNAs in this table is transcript per million (TPM)

[a]GEO accession number GSM280228 (German et al. 2008). The mismatched nucleotide for each sRNA with its cognate miRNAs is in lower case

[b]Two substituted positions are shown for this sRNA because it has one different substituted nucleotide each for different miRNAs, including miR156d, miR156h, miR157a, miR157b, and miR157c

[c]Also maps to non-expressed intergenic region on chromosome 3: 21,132,683. See http://mpss.udel.edu/at_sbs/

[d]Average normalized reads from all 49 sequenced sRNA libraries; combined read depth of 66,832,243

[e]Average normalized reads for four libraries from seedling, flower, silique, and leaves sequenced by 454/Roche pyrosequencing (Rajagopalan et al. 2006). Combined read depth of 717,842

[f]average normalized expression value (TPM) for Col-0 whole-aerial mock 7dpi samples (Garcia-Ruiz et al. 2010, GEO accession numbers: GSM506656, GSM506657, GSM506658 and GSM506682), except for those noted by [e]

[g]normalized expression value for Col-0 untreated wild type flower sample (SWT1, http://mpss.udel.edu/at_sbs/)

[h]average normalized expression value for dcl1-7 whole-aerial mock 7dpi samples (Garcia-Ruiz et al. 2010, GEO accession numbers: GSM506668, GSM506669, and GSM506670). NA: not available

[i]average normalized expression value for dcl234 whole-aerial mock 7dpi samples (Garcia-Ruiz et al. 2010, GEO accession numbers: GSM506680 and GSM506681), except for those noted by [j]

[j]average normalized expression value for two untreated dcl234 triple mutant flower samples (S234a and S234b, http://mpss.udel.edu/at_sbs/)

[k]normalized expression values from Mi et al. 2008, using Illumina Genome Analyzer (GEO accession number: GSM253622, GSM253623, GSM253624, and GSM253625). As a comparison, the abundance of OMM-sRNAs from immunoprecipitated AGO1 complex using 454 sequencing is listed after the slash "/" if available. See http://mpss.udel.edu/at/mpss_index.php

[l]normalized expression values from Montgomery et al. 2008 (GEO accession number: GSM304285)

[m]the corresponding miRNA for each sRNA is shown

[n]miRNA targets with full complementarity to each sRNA are shown by their AGI numbers and gene symbols, respectively

[o]the number of antisense siRNAs with perfect matches to the cDNA for each miRNA target was scored and normalized by dividing the length of each gene individually (antisense sRNA number/kb, data from German et al. 2008 (GEO accession number GSM280228)

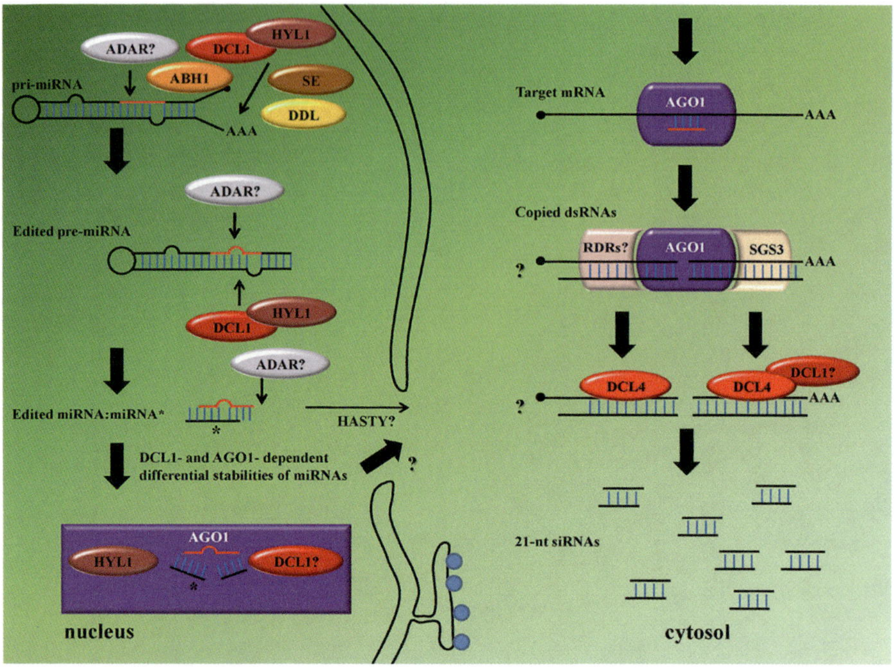

Fig. 1 Working model for base substitutions associated with miRNAs in Arabidopsis. In the processing of miRNA primary transcripts and/or precursors by DCL1, HYL1, ABH1, SE and/or DDL, ADAR may compete for the same substrates in the nucleus. Editing by hypothesized ADARs introduces a bulge on the editing site on the hairpin or miRNA:miRNA* duplex. miRNA loading into AGO(s) is assisted by HYL1 and/or DCL(s), whereby the miRNA* is preferentially degraded by the slicing activity of AGO(s). In some cases, base substitution on mature miRNA172, miR156/157, and miR160 results in perfect matches with cognate target mRNAs. Perfect complementarity might trigger a structural change of AGO1 when miRNA target mRNAs pair with miRNAs. This speculative conformation may induce dsRNA formation by recruiting SGS3 and RDR(s) to trigger copying the 5′ and 3′ halves of cleaved target transcripts. DCL4 and/or DCL1 would process these dsRNAs into 21-nt siRNAs in the cytoplasm

may need the assistance of HYL1 and/or DCL1, in which miRNA* is preferentially degraded by the slicing activity of AGO1. Editing may change the sequences to fully match with their cognate targets (see examples of miR156/157 and others in Table 1). Such perfect complementarity might trigger a structural change of AGO1 when miRNA target mRNAs pair with edited miRNAs. SGS3 and RDR(s) might recognize the structurally specific AGO1 and copy the 5′ and 3′ halves of cleaved target transcripts into dsRNAs which could be subsequently diced by DCL4 and/or DCL1 into 21-nt siRNAs. Thus, a molecular mechanism for generation of secondary siRNAs distinct from the 22-nt trigger models (Chen et al. 2010; Cuperus et al. 2010), which do not explain or address the cases of miR156/157, miR160, or miR172 targets we have described (Luo et al. 2009), is suggested by the OMM-miRNAs that are fully complementary to their targets, since perfectly dsRNA is a hallmark of PTGS (Dalmay et al. 2000).

Furthermore, miR168*, miR172* and miR408/408* are also substituted (data not shown) and accumulate to significant levels in *ago1* and *hyl1* mutants (Eamens et al. 2009), possibly due to base substitution effects on AGO loading efficiency, consistent with the model. Our model is not in conflict with the 22-nt trigger models and indeed may be underlying it and other phenomena associated with plant miRNAs such as differential stability (Eamens et al. 2009; Yang et al. 2006; Vaucheret et al. 2006), selective processing (Addo-Quaye et al. 2009; Bologna et al. 2009; Chekanova et al. 2007), and epigenetic modifications of target DNAs (Bao et al. 2004). Identification of the enzymatic activities for miRNA base substitution in plants should be informative for understanding the biological significance and evolution of RNA editing.

7 Speculation: Transposons – A Conserved Link Between Animal and Plant miRNA Function and Transitivity?

The first plant miRNA described (miR159/short RNA 40) was originally characterized as a TE-derived siRNA (Mette et al. 2002), and the functional relationships between these classes of sRNAs and their participation in epigenetic processes remains elusive. Zhang et al. (2009) discounted miR437 and six other conserved maize miRNAs on grounds that >50% of these miRNA family members mapped to annotated TEs. Interestingly, some published miRNAs are in antisense relationship to each other, most of which are homologous to TE sequences, for example *Arabidopsis* miR401 and miR855 (Richardson et al. 2010), and rice miR806, miR812, miR818, miR1884, and miR2123 (data not shown). We mapped rice smRNAs from three independent libraries (Xue et al. 2009; Heisel et al. 2008; Wu et al. 2009) to miRNA hairpins and could not find supportive evidence for 21 rice miRNAs annotated in miRBase based on the standard that >75% of reads must map to the miRNA/miRNA* duplex (Meyers et al. 2008). These discounted miRNAs are miR418, miR437, miR441, miR442, miR445, miR813, miR816, miR819, miR1319, miR1438, miR1440, miR1441, miR1866, miR2093, miR2096, miR2097, miR2099, miR2120, miR2121, miR2124, and miR2125. Wu et al. (2009) noted that for rice miR439, miR441-443, miR445, miR446, miR806-819, miR821, miR1319, miR1436-1442, and miR1847, scattered sRNAs are detected across both sense and antisense strands of the precursors, suggesting their origins may be from dsRNAs formed by sense and antisense transcripts of non-miRNA loci (Meyers et al. 2008).

The striking similarities between inverted repeat TEs and miRNAs, e.g., antisense transcription, have stimulated us to computationally characterize the TE-like miRNAs of rice for getting clues in relation to their evolution and functions. We found that nonconserved miRNAs (found only in rice) have significant sequence similarities with TEs. We customized miRNA prediction program miRanda V2.0 (Enright et al. 2003) and performed a comprehensive prediction for miRNA targets in *Arabidopsis thaliana* and *Oryza sativa* subsp. japonica genomes. Among the predicted targets of japonica rice miRNAs that met published criteria for target complementarity (Allen et al. 2005) (n=1,613, TIGR6.1), more than 25% were found in 5′ and 3′ untranslated regions (UTR) (Fig. 2a), whereas only a few cases (6% of total predicted targets) of *Arabidopsis* miRNA targets in 5′- or 3′-UTRs have been described (Fig. 2a) (Arteaga-Vazquez et al. 2006; Rhoades et al. 2002), including TE-like miR855/401 sense/antisense pair (Richardson et al. 2010).

Interestingly, the phenomenon of UTR-localized miRNA complementary sites was primarily attributable to nonconserved (recently described and presumably recently evolved) rice miRNA families (Fig. 2b). High ratios (>7:1) were found for the number of japonica rice 3′- and 5′-UTR regions targeted by nonconserved miRNAs compared to conserved (ancient) miRNA families. By contrast, UTR-localized miRNA complementary sites were not enriched as a percentage for recently discovered *Arabidopsis* miRNAs, suggesting different processes exist between rice and *Arabidopsis* that affect birth and death of recently evolved (probably lineage-specific) classes of miRNAs (Rajagopalan et al. 2006; Fahlgren et al. 2007). Thus, a large

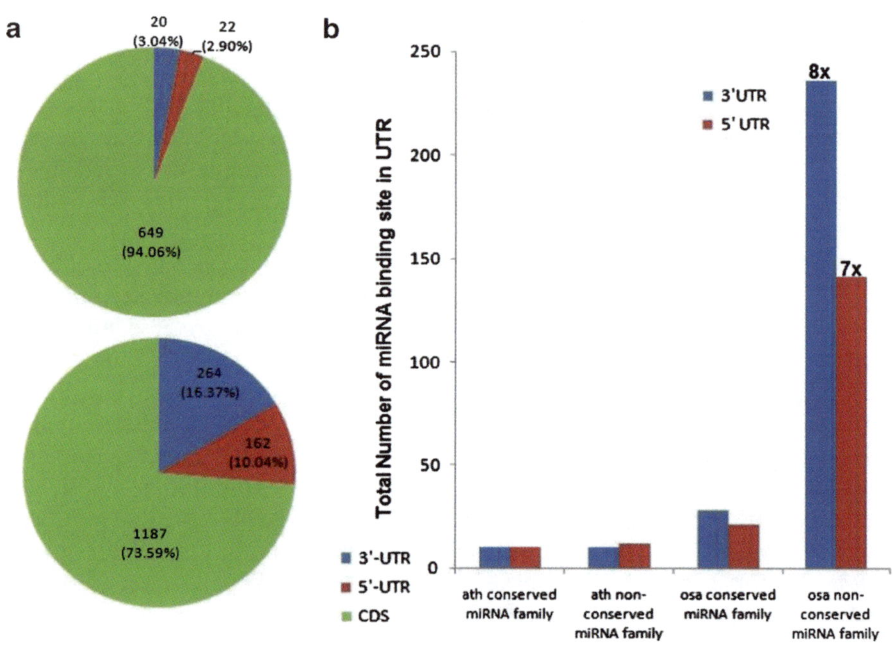

Fig. 2 Topology of predicted miRNA binding sites in target gene mRNA features in *Arabidopsis thaliana* and *Oryza sativa* ssp. japonica. (**a**) Summary of target sites' distribution in coding region (CDS, *green*), 3′-(*blue*) or 5′-(*red*) untranslated regions (UTRs). Numbers of genes and percentage (*in parentheses*) of different target features are indicated. (**b**) Comparison of miRNA binding sites in 3′- and 5′-UTRs between conserved and nonconserved miRNA families in *Arabidopsis* and rice. Conserved miRNAs are those existing in dicots and monocots. Nonconserved miRNAs are lineage-specific or found only in closely related species. Ath *Arabidopsis thaliana*; osa *Oryza sativa*

percentage of rice miRNAs possess a feature analogous to animal miRNAs (3′-UTR localization), implying the functionality of their complementary sites in 3′-UTR regions. This situation raises possibilities about the evolution of regulatory interactive siRNA network by TE-driven miRNA-target recognition in UTRs (MacLean et al. 2010). Consistent with this notion, 73% of rice 3′-UTR targets are repeat-related, and 31% are homologous to interspersed repeats (data not shown). The miRNA complementary sites in 3′-UTR targets mapped within the interspersed repeats in 82% of all cases, and 57% of their complementary sites bear the same subclass of TEs with their corresponding miRNAs and another 43% carry the same class of TEs, providing strong evidence that the repeat-associated miRNAs and their cognate 3′-UTR targets originated from the same TE. The most abundant class of associated repeats is DNA miniature inverted repeat transposable element (MITEs). We suggest that some miRNA in plants may evolve as proposed for animals – through the acquisition of miRNA binding sites mediated by the exaptation of TEs (Piriyapongsa and Jordan 2008; Piriyapongsa et al. 2007). PTGS mechanisms acting on MITEs in UTRs of rice genes could be agents of purifying selection for evolution of miRNA targets to counteract the presumed negative effects of a hairpin-containing mRNA, which would be diced to death. In this sense, the TE-associated sRNAs/miRNAs mapping to UTRs may represent intermediate molecular species linking TE silencing processes to miRNA target evolution. Rice gene topology may be more plastic than *Arabidopsis* and tolerate acquisition of novel sequences in non coding regions, in which case miRNA complementary site acquisition and selection/maintenance could become a driving force for the establishment of a regulatory network involving antisense transcripts and PTGS and/or translational inhibition.

Plant and animal miRNA targets are under purifying selection. In the case of 3′ UTR targets, the conservation of the miRNA complementary sites among family members is independent of protein functional constraints that could confound phylogenetic analysis. A canonical example of an ancient, deeply conserved miRNA targeting 3′ UTRs is miR169, which provides a reference for computational analysis of candidate 3′-UTR miRNA and TE-like target site evolution. Its complementary sites are conserved among *Arabidopsis* (Rhoades et al. 2002) and rice miRNA targets, where flanking sequences have diverged to show a clear target footprint (Fig. 3a). We also found significant similarity among the reverse complement sequence of miR1884 arm (as well as miR806/812/818/2123), *STOWAWAY2* TE, and its predicted target genes (Fig. 3b), supporting a functional constraint indicative of purifying selection maintaining the miR1884 and homologues' complementary sites. The qualitative nature of the miR1884 footprint compared to that of miR169 suggests that miR1884 and homologues are recently evolved, consistent with them only being described in rice (Zhu et al. 2008); however, we also find it is significantly homologous to *Sorghum bicolor* TE *candystripe1* ($E=5\times10^{-9}$; data not shown). Interestingly, *MIR1884* and homologous loci produce fewer antisense reads than does the most abundant case of *MIR808* (~130 vs. 2,660 reads/kbp of *MIRNA* hairpin; data not shown), consistent with a proposed miRNA activity.

By contrast, the reverse complement sequence of miR808 contains significant homology with its targets and *STOWAWAY1* TE sequence throughout the entire

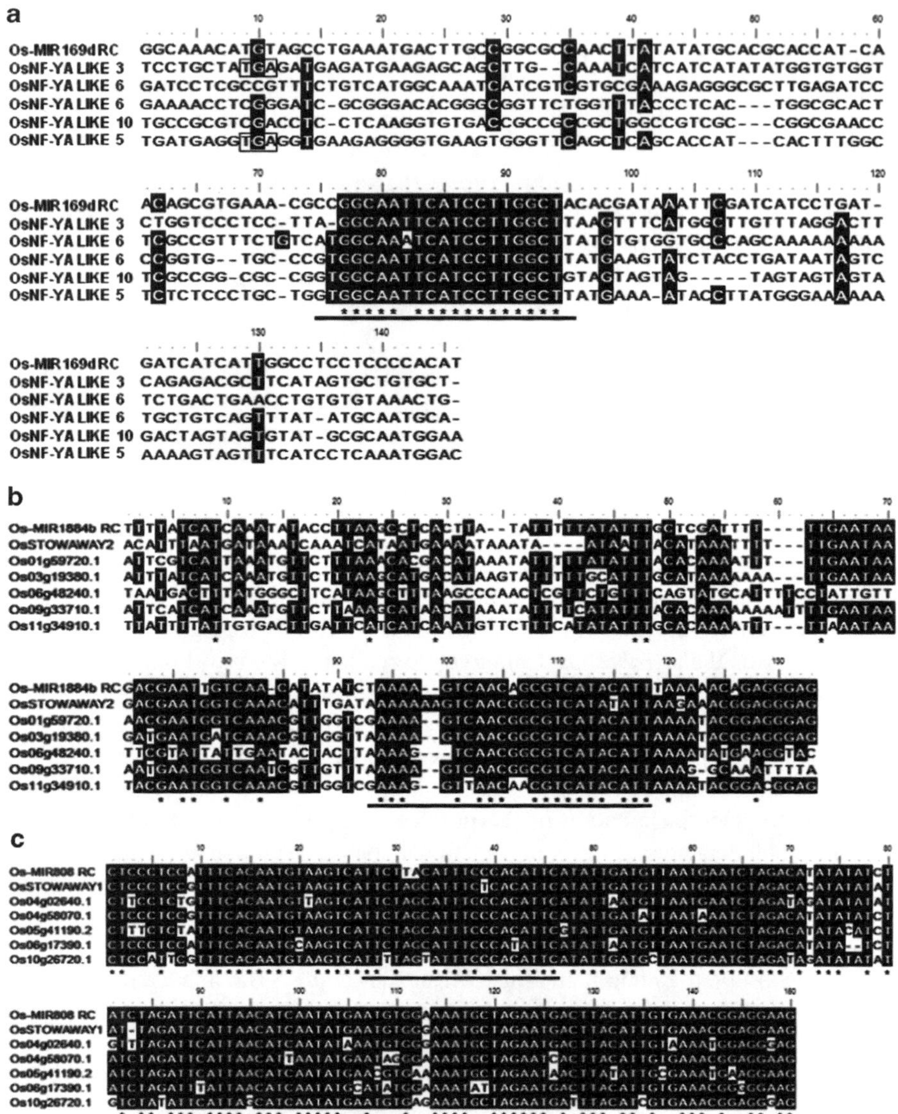

Fig. 3 Similarity between foldback arms and 3′UTR of target genes. Alignment of (**a**) reverse complement of MIR169 foldback arm sequences and corresponding target genes; (**b**) reverse complement of MIR1884 foldback arms, half of the transposable element (TE) sequence and corresponding target genes; (**c**) reverse complement of MIR808 precursor, entire sequence of TE and corresponding target genes. Upstream and downstream sequences flanking 3′-UTR miRNA complementary sites were extracted from plant reference genomes and made to equal length to the TE. For miR169 targets OsNF-YA-Like-3 and -5, the stop codon of the sequence is *boxed*. The consensus nucleotides (70% identity) are shaded with *black*. Conserved positions across all sequences are indicated by *asterisks*. miRNA complementary sites are *underlined*. Os01g59720.1, transposon protein; Os03g19380.1, ent-kaurene synthase; Os06g48240.1, helix-loop-helix DNA-binding domain containing protein; Os11g34910.1, respiratory burst oxidase; Os06g17390.1, GPI-anchored protein; Os10g26720.1, Os09g33710.1, Os04g02640.1, Os04g58070.1, and Os05g41190.2, unknown expressed proteins

length of the hairpin (Fig. 3c). Thus, the extremely large footprint beyond miRNA complementary binding sites in these 3′-UTRs suggests a possible recent origin from an antisense- and siRNA-producing TE intermediate for miR808 and its targets, amplified after domestication (~7,000 years ago) (Fuller et al. 2009). Our preliminary results are consistent with the notion that there is an evolutionary trade-off in which the benefit of TE silencing through 3′-UTRs imposes a fitness cost via deleterious effects on gene expression, unless they diverge and evolve into functional miRNA targets.

In conclusion, we propose that in rice MITE subfamilies can transpose into UTRs of protein-coding transcripts, which lead initially to the formation of hairpin structure reminiscent of the pre-miRNA sequence processed by Dicer to yield siRNAs and transitive production of antisense transcripts. TE-derived miRNAs can bind to their UTR targets and affect gene expression posttranscriptionally. This model is consistent with the genome defense mechanism against TE insertions, where the function of siRNAs is to silence invasive DNA but can evolve under purifying selection by the miRNA biogenesis and PTGS machineries into young miRNAs that target unrelated genes. The capture of interaction for TE-derived miRNA-like sequences and their complementary sites in UTRs by miRNA and/or RNAi pathways may reinforce the functional constraint between them. At the same time, manipulation of TEs in a complex genome with strong significance of repetitive sequences, like that of rice, will amplify the cascade of such newly evolved miRNA:target regulatory nodes and form an interactive siRNA network, for which established examples have been observed in *Arabidopsis* genome (MacLean et al. 2010). The creation of sRNA/miRNA regulatory nodes by transposition and RNAi could provide adaptive value if placed under positive selection, for example in response to stress, or during gametogenesis when sRNAs are in full bloom and performing functions yet to be elucidated (Naito et al. 2009; Jiao and Deng 2007; Olmedo-Monfil et al. 2010; Stefani and Slack 2008).

Acknowledgments This work was supported by the National Institutes of Health grant GM077245 to C.D.R. The funders had no role in the study design, in the collection, analysis, or interpretation of data, or in the writing of the manuscript or decision to submit the manuscript for publication.

References

Addo-Quaye C, Snyder JA, Park YB et al (2009) Sliced microRNA targets and precise loop-first processing of *MIR319* hairpins revealed by analysis of the *Physcomitrella patens* degradome. RNA 15:2112–2121

Adenot X, Elmayan T, Lauressergues D et al (2006) DRB4-dependent *TAS3 trans*-acting siRNAs control leaf morphology through AGO7. Curr Biol 16:927–932

Allen E, Xie Z, Gustafson AM et al (2005) microRNA-directed phasing during *trans*-acting siRNA biogenesis in plants. Cell 121:207–221

Arteaga-Vazquez M, Caballero-Perez J, Vielle-Calzada JP (2006) A family of microRNAs present in plants and animals. Plant Cell 18:3355–3369

Ausin I, Mockler TC, Chory J et al (2009) IDN1 and IDN2 are required for de novo DNA methylation in *Arabidopsis thaliana*. Nat Struct Mol Biol 16:1325–1327
Axtell MJ, Jan C, Rajagopalan R et al (2006) A two-hit trigger for siRNA biogenesis in plants. Cell 127:565–577
Azevedo J, Garcia D, Pontier D et al (2010) Argonaute quenching and global changes in Dicer homeostasis caused by a pathogen-encoded GW repeat protein. Genes Dev 24:904–915
Bao N, Lye KW, Barton MK (2004) MicroRNA binding sites in Arabidopsis class IIIHD-ZIP mRNAs are required for methylation of the template chromosome. Dev Cell 7:653–662
Ben Amor B, Wirth S, Merchan F et al (2009) Novel long non-protein coding RNAs involved in Arabidopsis differentiation and stress responses. Genome Res 19:57–69
Bickel KS, Morris DR (2006) Silencing the transcriptome's dark matter: mechanisms for suppressing translation of intergenic transcripts. Mol Cell 22:309–316
Bologna NG, Mateos JL, Bresso EG et al (2009) A loop-to-base processing mechanism underlies the biogenesis of plant microRNAs miR319 and miR159. EMBO J 28:3646–3656
Borchert GM, Lanier W, Davidson BL (2006) RNA polymerase III transcribes human microRNAs. Nat Struct Mol Biol 13:1097–1101
Borsani O, Zhu JH, Verslues PE et al (2005) Endogenous siRNAs derived from a pair of natural *cis*-antisense transcripts regulate salt tolerance in Arabidopsis. Cell 123:1279–1291
Chang YF, Imam JS, Wilkinson ME (2007) The nonsense-mediated decay RNA surveillance pathway. Annu Rev Biochem 76:51–74
Chapman EJ, Carrington JC (2007) Specialization and evolution of endogenous small RNA pathways. Nat Rev Genet 8:884–896
Chekanova JA, Gregory BD, Reverdatto SV et al (2007) Genome-wide high-resolution mapping of exosome substrates reveals hidden features in the Arabidopsis transcriptome. Cell 131:1340–1353
Chen H-M, Chen L-T, Patel K et al (2010) 22-nucleotide RNAs trigger secondary siRNA biogenesis in plants. Proc Natl Acad Sci USA 107:15269–15274
Correa RL, Steiner FA, Berezikov E et al (2010) MicroRNA-directed siRNA biogenesis in *Caenorhabditis elegans*. PLoS Genet 6:e1000903
Cuperus JT, Carbonell A, Fahlgren N et al (2010) Unique functionality of 22-nt miRNAs in triggering RDR6-dependent siRNA biogenesis from target transcripts in Arabidopsis. Nat Struct Mol Biol 17:997–1003
Dalmay T, Hamilton A, Rudd S et al (2000) An RNA-dependent RNA polymerase gene in Arabidopsis is required for posttranscriptional gene silencing mediated by a transgene but not by a virus. Cell 101:543–553
de Hoon MJL, Taft RJ, Hashimoto T et al (2010) Cross-mapping and the identification of editing sites in mature microRNAs in high-throughput sequencing libraries. Genome Res 20:257–264
Dong Z, Han MH, Fedoroff N (2008) The RNA-binding proteins HYL1 and SE promote accurate in vitro processing of pri-miRNA by DCL1. Proc Natl Acad Sci USA 105:9970–9975
Eamens AL, Smith NA, Curtin SJ et al (2009) The *Arabidopsis thaliana* double-stranded RNA binding protein DRB1 directs guide strand selection from microRNA duplexes. RNA 15:2219–2235
Ebhardt HA, Tsang HH, Dai DC, Liu Y, Bostan B, Fahlman RP (2009) Meta-analysis of small RNA-sequencing errors reveals ubiquitous post-transcriptional RNA modifications. Nucl Acids Res 37:2461–2470
Elmayan T, Adenot X, Gissot L et al (2009) A neomorphic *sgs3* allele stabilizing miRNA cleavage products reveals that SGS3 acts as a homodimer. FEBS J 276:835–844
Enright AJ, John B, Gaul U et al (2003) MicroRNA targets in Drosophila. Genome Biol 5:R1
Faghihi MA, Zhang M, Huang J et al (2010) Evidence for natural antisense transcript-mediated inhibition of microRNA function. Genome Biol 11:R56
Fahlgren N, Howell MD, Kasschau KD et al (2007) High-throughput sequencing of Arabidopsis microRNAs: evidence for frequent birth and death of *MIRNA* genes. PLoS ONE 2:e219
Fang Y, Spector DL (2007) Identification of nuclear dicing bodies containing proteins for microRNA biogenesis in living Arabidopsis plants. Curr Biol 17:818–823
Franco-Zorrilla JM, Valli A, Todesco M et al (2007) Target mimicry provides a new mechanism for regulation of microRNA activity. Nat Genet 39:1033–1037

Fujioka Y, Utsumi M, Ohba Y et al (2007) Location of a possible miRNA processing site in SmD3/SmB nuclear bodies in Arabidopsis. Plant Cell Physiol 48:1243–1253
Fukunaga R, Doudna JA (2009) dsRNA with 5′ overhangs contributes to endogenous and antiviral RNA silencing pathways in plants. EMBO J 28:545–555
Fuller DQ, Qin L, Zheng Y et al (2009) The domestication process and domestication rate in rice: spikelet bases from the lower Yangtze. Science 323:1607–1610
Garcia-Ruiz H, Takeda A, Chapman EJ et al (2010) Arabidopsis RNA-dependent RNA polymerases and Dicer-Like proteins in antiviral defense and small interfering RNA biogenesis during Turnip Mosaic Virus infection. Plant Cell 22:481–496
German MA, Pillay M, Jeong DH et al (2008) Global identification of microRNA-target RNA pairs by parallel analysis of RNA ends. Nat Biotechnol 26:941–946
Gilbert W (1986) Origin of life: the RNA world. Nature 319:618
Gregory BD, O'Malley RC, Lister R et al (2008) A link between RNA metabolism and silencing affecting Arabidopsis development. Dev Cell 14:854–866
Gruber JJ, Zatechka DS, Sabin LR et al (2009) Ars2 links the nuclear cap-binding complex to RNA interference and cell proliferation. Cell 138:328–339
He X-J, Hsu Y-F, Zhu S et al (2009) An effector of RNA-directed DNA methylation in Arabidopsis is an ARGONAUTE 4- and RNA-binding protein. Cell 137:498–508
Heisel SE, Zhang YJ, Allen E et al (2008) Characterization of unique small RNA populations from rice grain. PLoS ONE 3:e2871
Herr AJ, Molnar A, Jones A et al (2006) Defective RNA processing enhances RNA silencing and influences flowering of Arabidopsis. Proc Natl Acad Sci USA 103:14994–15001
Himber C, Dunoyer P, Moissiard G et al (2003) Transitivity-dependent and -independent cell-to-cell movement of RNA silencing. EMBO J 22:4523–4533
Iida K, Jin HL, Zhu JK (2009) Bioinformatics analysis suggests base modifications of tRNAs and miRNAs in *Arabidopsis thaliana*. BMC Genomics 10:155
Jiao Y, Deng XW (2007) A genome-wide transcriptional activity survey of rice transposable element-related genes. Genome Biol 8:R28
Karlowski WM, Zielezinski A, Carrere J et al (2010) Genome-wide computational identification of WG/GW Argonaute-binding proteins in Arabidopsis. Nucl Acids Res 38:4231–4245
Kawamata T, Tomari Y (2010) Making RISC. Trends Biochem Sci 35:368–376
Khraiwesh B, Arif MA, Seumel GI et al (2010) Transcriptional control of gene expression by microRNAs. Cell 140:111–122
Kim DH, Saetrom P, Snove O et al (2008a) MicroRNA-directed transcriptional gene silencing in mammalian cells. Proc Natl Acad Sci USA 105:16230–16235
Kim S, Yang JY, Xu J et al (2008b) Two cap-binding proteins CBP20 and CBP80 are involved in processing primary microRNAs. Plant Cell Physiol 49:1634–1644
Kumakura N, Takeda A, Fujioka Y et al (2009) SGS3 and RDR6 interact and colocalize in cytoplasmic SGS3/RDR6-bodies. FEBS Lett 583:1261–1266
Laubinger S, Sachsenberg T, Zeller G et al (2008) Dual roles of the nuclear cap-binding complex and SERRATE in pre-mRNA splicing and microRNA processing in *Arabidopsis thaliana*. Proc Natl Acad Sci USA 105:8795–8800
Liu PP, Montgomery TA, Fahlgren N et al (2007) Repression of AUXIN RESPONSE FACTOR10 by microRNA160 is critical for seed germination and post-germination stages. Plant J 52:133–146
Liu XD, Huang J, Wang Y et al (2010) The role of *floral organs in carpels*, an Arabidopsis loss-of-function mutation in microRNA160a, in organogenesis and the mechanism regulating its expression. Plant J 62:416–428
Lobbes D, Rallapalli G, Schmidt DD et al (2006) SERRATE: a new player on the plant microRNA scene. EMBO Rep 7:1052–1058
Lu C, Tej SS, Luo S et al (2005) Elucidation of the small RNA component of the transcriptome. Science 309:1567–1569
Luo Z, Chen Z (2007) Improperly terminated, unpolyadenylated mRNA of sense transgenes is targeted by RDR6-mediated RNA silencing in Arabidopsis. Plant Cell 19:943–958

Luo Q-J, Samanta MP, Koksal F et al (2009) Evidence for antisense transcription associated with microRNA target mRNAs in Arabidopsis. PLoS Genet 5:e1000457

MacLean D, Elina N, Havecker ER et al (2010) Evidence for large complex networks of plant short silencing RNAs. PLoS ONE 5:e9901

Makeyev EV, Bamford DH (2002) Cellular RNA-dependent RNA polymerase involved in post-transcriptional gene silencing has two distinct activity modes. Mol Cell 10:1417–1427

Mallory AC, Vaucheret H (2009) ARGONAUTE 1 homeostasis invokes the coordinate action of the microRNA and siRNA pathways. EMBO Rep 10:521–526

Mallory AC, Hinze A, Tucker MR et al (2009) Redundant and specific roles of the ARGONAUTE proteins AGO1 and ZLL in development and small RNA-directed gene silencing. PLoS Genet 5:e1000646

Mette MF, van der Winden J, Matzke M et al (2002) Short RNAs can identify new candidate transposable element families in Arabidopsis. Plant Physiol 130:6–9

Meyers BC, Axtell MJ, Bartel B, Bartel DP, Baulcombe D, Bowman JL, Cao X, Carrington JC, Chen X, Green PJ, Griffiths-Jones S, Jacobsen SE, Mallory AC, Martienssen RA, Poethig RS, Qi YJ, Vaucheret H, Voinnet O, Watanabe Y, Weigel D, Zhui JK (2008) Criteria for annotation of plant microRNAs. Plant Cell 20:3186–3190

Mi SJ, Cai T, Hu YG, Chen Y, Hodges E, Ni FR, Wu L, Li S, Zhou H, Long CZ, Chen S, Hannon GJ, Qi YJ (2008) Sorting of small RNAs into Arabidopsis Argonaute complexes is directed by the 5′ terminal nucleotide. Cell 133:116–127

Montgomery TA, Howell MD, Cuperus JT, Li DW, Hansen JE, Alexander AL, Chapman EJ, Fahlgren N, Allen E, Carrington JC (2008) Specificity of ARGONAUTE7-miR390 interaction and dual functionality in *TAS3 trans*-acting siRNA formation. Cell 133:128–141

Naito K, Zhang F, Tsukiyama T, Saito H, Hancock CN, Richardson AO, Okumoto Y, Tanisaka T, Wessler SR (2009) Unexpected consequences of a sudden and massive transposon amplification on rice gene expression. Nature 461:1130–1134

Navarro L, Jay F, Nomura K, He SY, Voinnet O (2008) Suppression of the microRNA pathway by bacterial effector proteins. Science 321:964–967

Olmedo-Monfil V, Duran-Figueroa N, Arteaga-Vazquez M, Demesa-Arevalo E, Autran D, Grimanelli D, Slotkin RK, Martienssen RA, Vielle-Calzada JP (2010) Control of female gamete formation by a small RNA pathway in Arabidopsis. Nature 464:628–632

Pagnussat GC, Alandete-Saez M, Bowman JL et al (2009) Auxin-dependent patterning and gamete specification in the Arabidopsis female gametophyte. Science 324:1684–1689

Pak J, Fire A (2007) Distinct populations of primary and secondary effectors during RNAi in *C. elegans*. Science 315:241–244

Parizotto EA, Dunoyer P, Rahm N et al (2004) *In vivo* investigation of the transcription, processing, endonucleolytic activity, and functional relevance of the spatial distribution of a plant miRNA. Genes Dev 18:2237–2242

Peragine A, Yoshikawa M, Wu G et al (2004) SGS3 and SGS2/SDE1/RDR6 are required for juvenile development and the production of *trans*-acting siRNAs in Arabidopsis. Genes Dev 18:2368–2379

Piriyapongsa J, Jordan IK (2008) Dual coding of siRNAs and miRNAs by plant transposable elements. RNA 14:814–821

Piriyapongsa J, Marino-Ramirez L, Jordan IK (2007) Origin and evolution of human microRNAs from transposable elements. Genetics 176:1323–1337

Poliseno L, Salmena L, Zhang J et al (2010) A coding-independent function of gene and pseudogene mRNAs regulates tumour biology. Nature 465:1033–1038

Rajagopalan R, Vaucheret H, Trejo J et al (2006) A diverse and evolutionarily fluid set of microRNAs in *Arabidopsis thaliana*. Genes Dev 20:3407–3425

Rhoades MW, Reinhart BJ, Lim LP et al (2002) Prediction of plant microRNA targets. Cell 110:513–520

Richardson CR, Luo Q-J, Gontcharova V et al (2010) Analysis of antisense expression by whole genome tiling microarrays and siRNAs suggests mis-annotation of Arabidopsis orphan protein-coding genes. PLoS ONE 5:e10710

Ron M, Saez MA, Williams LE et al (2010) Proper regulation of a sperm-specific *cis*-nat-siRNA is essential for double fertilization in Arabidopsis. Genes Dev 24:1010–1021

Ronemus M, Vaughn MW, Martienssen RA (2006) MicroRNA-targeted and small interfering RNA-mediated mRNA degradation is regulated by Argonaute, Dicer, and RNA-dependent RNA polymerase in Arabidopsis. Plant Cell 18:1559–1574

Ruvkun G (2001) Glimpses of a tiny RNA world. Science 294:797–799

Sabin LR, Zhou R, Gruber JJ et al (2009) Ars2 regulates both miRNA- and siRNA-dependent silencing and suppresses RNA virus infection in Drosophila. Cell 138:340–351

Song MG, Kiledjian M (2007) 3′ Terminal oligo U-tract-mediated stimulation of decapping. RNA 13:2356–2365

Souret FF, Kastenmayer JP, Green PJ (2004) AtXRN4 degrades mRNA in Arabidopsis and its substrates include selected miRNA targets. Mol Cell 15:173–183

Starega-Roslan J, Krol J, Koscianska E et al (2011) Structural basis of microRNA length variety. Nucl Acids Res 39:257–268

Stefani G, Slack FJ (2008) Small non-coding RNAs in animal development. Nat Rev Mol Cell Biol 9:219–230

Tang G (2005) siRNA and miRNA: an insight into RISCs. Trends Biochem Sci 30:106–114

Ueno Y, Ishikawa T, Watanabe K et al (2007) Histone deacetylases and ASYMMETRIC LEAVES2 are involved in the establishment of polarity in leaves of Arabidopsis. Plant Cell 19:445–457

Vaistij FE, Jones L, Baulcombe DC (2002) Spreading of RNA targeting and DNA methylation in RNA silencing requires transcription of the target gene and a putative RNA-dependent RNA polymerase. Plant Cell 14:857–867

Vaucheret H (2009) AGO1 homeostasis involves differential production of 21-nt and 22-nt miR168 species by *MIR168a* and *MIR168b*. PLoS ONE 4:e6442

Vaucheret H, Mallory AC, Bartel DP (2006) AGO1 homeostasis entails coexpression of *MIR168* and AGO1 and preferential stabilization of miR168 by AGO1. Mol Cell 22:129–136

Vazquez F, Vaucheret H, Rajagopalan R et al (2004) Endogenous *trans*-acting siRNAs regulate the accumulation of Arabidopsis mRNAs. Mol Cell 16:69–79

Vazquez F, Blevins T, Ailhas J et al (2008) Evolution of Arabidopsis *MIR* genes generates novel microRNA classes. Nucl Acids Res 36:6429–6438

Voinnet O (2009) Origin, biogenesis, and activity of plant microRNAs. Cell 136:669–687

Weber C, Nover L, Fauth M (2008) Plant stress granules and mRNA processing bodies are distinct from heat stress granules. Plant J 56:517–530

Werner A, Carlile M, Swan D (2009) What do natural antisense transcripts regulate? RNA Biol 6:43–48

Wu G, Poethig RS (2006) Temporal regulation of shoot development in *Arabidopsis thaliana* by miR156 and its target SPL3. Development 133:3539–3547

Wu L, Zhang QQ, Zhou HY et al (2009) Rice microRNA effector complexes and targets. Plant Cell 21:3421–3435

Wu H, Sun SY, Tu K et al (2010a) A splicing-independent function of SF2/ASF in microRNA processing. Mol Cell 38:67–77

Wu L, Zhou H, Zhang Q et al (2010b) DNA methylation mediated by a microRNA pathway. Mol Cell 38:465–475

Wuest SE, Vijverberg K, Schmidt A et al (2010) Arabidopsis female gametophyte gene expression map reveals similarities between plant and animal gametes. Curr Biol 20:506–512

Xie ZX, Allen E, Fahlgren N et al (2005) Expression of Arabidopsis *MIRNA* genes. Plant Physiol 138:2145–2154

Xu P, Chen F, Mannas JP et al (2008) Virus infection improves drought tolerance. New Phytol 180:911–921

Xue L-J, Zhang J-J, Xue H-W (2009) Characterization and expression profiles of miRNAs in rice seeds. Nucl Acids Res 37:916–930

Yang L, Liu ZQ, Lu F et al (2006) SERRATE is a novel nuclear regulator in primary microRNA processing in Arabidopsis. Plant J 47:841–850

Yoshikawa M, Peragine A, Park MY et al (2005) A pathway for the biogenesis of *trans*-acting siRNAs in Arabidopsis. Genes Dev 19:2164–2175

Yu B, Yang ZY, Li JJ et al (2005) Methylation as a crucial step in plant microRNA biogenesis. Science 307:932–935

Zhang LF, Chia JM, Kumari S et al (2009) A genome-wide characterization of microRNA genes in maize. PLoS Genet 5:e1000716

Zheng Z, Xing Y, He X-J et al (2010) An SGS3-like protein functions in RNA-directed DNA methylation and transcriptional gene silencing in Arabidopsis. Plant J 62:92–99

Zhu QH, Spriggs A, Matthew L et al (2008) A diverse set of microRNAs and microRNA-like small RNAs in developing rice grains. Genome Res 18:1456–1465

Post-Transcriptional Modifications of Plant Small RNAs

H. Alexander Ebhardt

Contents

Abstract Generally, small RNAs are generated template dependent by RNA polymerases or RNA dependent RNA polymerases with the transcription template being either DNA or RNA. Following transcription, small RNAs in plants undergo processing by Dicer-like RNase III type endonucleases to define their mature 5′ and 3′ termini. Part of the maturation process of small RNAs are post-transcriptional modifications either on the ribose or base moiety of the nucleotide. Additionally, deletion and/or addition of nucleotides by RNA-specific ribonucleotidyl transferases are also plausible (RNA 13:1834–1849, 2007). In this chapter, ubiquitous and sequence specific post-transcriptional modifications of plant small RNAs are discussed as well as their detection through biochemical means or analysis of DNA sequencing data of small RNAs.

Keywords 2′-*O*-methyl • 3′ Uridylation • ARGONAUTE complex association • Indel • MicroRNA • Post-transcriptional modification • RNA editing • Silencing RNA • Small RNA

H.A. Ebhardt (✉)
Institute of Molecular Systems Biology, Eidgenössische Technische Hochschule (ETH) Zürich, Wolfgang-Pauli-Str. 16, CH-8093 Zürich, Switzerland
e-mail: ebhardt@imsb.biol.ethz.ch

V.A. Erdmann and J. Barciszewski (eds.), *Non Coding RNAs in Plants*, RNA Technologies, DOI 10.1007/978-3-642-19454-2_4,

1 HEN1 Acts as Methyltransferase on Small RNAs

hen1 has been identified during a genetic screen exhibiting pleiotropic phenotypes during most stages of *Arabidopsis thaliana* development (Chen et al. 2002). Mechanistic insights into the broad affects of *HEN1* were provided when *hen1* was linked to microRNA biogenesis (Park et al. 2002). To pin point the exact mechanism of *hen1*'s contribution to microRNA biogenesis, *hen1* was purified as fusion construct (fused to glutathione *S*-transferase), added in vitro synthesized MIR173/ MIR173* duplex with 5′-phosphate and 3′-hydroxyl as well as *S*-adenosyl-L-[*methyl*-^{14}C] methionine. The reaction end point was resolved by gel electrophoresis, the ^{14}C label visualized by autoradiography and showed a *hen1* dependent incorporation of the ^{14}C label into the RNA. Neither a mutating the *S*-adenosyl-L-methionine binding pocket of *hen1* nor the fusion protein glutathione *S*-transferase were able to label the microRNA duplex. Through a series of experiments the authors conclude the methyl moiety was attached to the 3′ terminus of the microRNAs tested (Yu et al. 2005).

At the same time, Ebhardt and colleagues were investigating small RNAs from virally infected tobacco plants expressing a strong RNA silencing suppressor HC-Pro in a four point comparison study (HC-Pro$^-$, ± infection with Y-Satellite and its cucumber mosaic helper virus [CMV], HC-Pro$^+$ ± infection with Y-Satellite and CMV) (Wang et al. 2004). To their surprise, there was a discrepancy in HC-Pro$^+$ virally infected tobacco plants. On one hand, the 5′ ^{32}P labeling showed a bimodal distribution of 21/22 and 24 nt long small RNAs and on the other hand the histogram of the small RNA cloning data showed predominantly only 21/22 nt RNA species. To investigate this discrepancy, the authors chose a sodium periodate assay to probe for 2′ and 3′ hydroxyl moieties on the 3′ terminus of small RNAs. If both 2′ and 3′ hydroxyls are unmodified, the periodate treatment will result in a labile dialdehyde, which can be cleaved off in a β-elimination step (Alefelder et al. 1998; Hutvagner et al. 2001). Removing the 3′ terminal nucleoside causes the small RNAs to migrate approximately 2 nts faster than untreated RNA in a denaturing gel electrophoresis due to the remaining phosphate on the neo-3′ terminus. Treating all small RNAs from the four point comparison study with periodate revealed that only the 21/22 nt small RNAs from HC-Pro$^+$ plants infected with Y-Satellite and CMV were sensitive to periodate, while all other small RNAs were unaffected. Thus, the authors concluded that the 3′ termini of endogenous and exogenous plant small RNAs are ubiquitously modified (Ebhardt et al. 2005).

To investigate further the periodate sensitivity of 21/22 nt small RNAs in HC-Pro$^+$ plants infected with Y-Satellite and CMV, a streptavidin gel shift assay was developed to isolate from the pool of small RNAs specific endogenous and exogenous small RNAs (Ebhardt et al. 2005, 2011; Ebhardt and Unrau 2009). In the streptavidin gel shift assay, a biotinylated DNA antisense to the small RNA sequence of interest is hybridized with total 5′ radiolabeled plant small RNAs, followed by incubation with streptavidin and separation on a native polyacrylamide gel. Due to the steptavidin bound to biotin, the radiolabeled RNA will be supershifted and can be excised from the gel for subsequent biochemical analysis (Ebhardt and Unrau 2009). Using this approach, it was determined that exogenous viral small RNAs were not sensitive to periodate treatment in virally infected plant but were sensitive to periodate in virally infected HC-Pro$^+$ plants. However, microRNAs from either HC-Pro$^+$ or HC-Pro$^-$ plants were not sensitive to periodate. Other research showed that HC-Pro expression does affect endogenous small RNAs (Kasschau et al. 2003). To satisfactory explain both observations one should consider that Kasschau specifically chose plants expressing very high levels of HC-Pro (Kasschau et al. 2003). Thus, an all-encompassing explanation would include the concentration dependent effects of HC-Pro: at very high concentrations HC-Pro affects all small RNAs in the plant, while low to medium expression levels of HC-Pro only affect exogenous small RNAs.

In conclusion, the discrepancy between bimodal distribution of 21/22 and 24 nt small RNA of 5′ radiolabeled RNA vs. monomodal length distribution in the cloning data in HC-Pro$^+$ infected plants is due to the fact that exogenous 21/22 nt small RNAs were unmethylated on their 3′ terminus. In a mixture of 2′-*O*-methylated endogenous and unmethylated exogenous small RNAs, the latter population is a preferential substrate of T4 RNA ligase. As the 3′ adaptor ligation is typically the first step in a small RNA cloning protocol, it already introduced a bias (Ebhardt et al. 2005).

Having resolved the discrepancy between 5′ end labeling and length distribution of cloned small RNAs in virally infected HC-Pro$^+$ plants, it was argued that the modification on the 3′ terminus of small RNAs is a methyl moiety in concert with results already published (Yu et al. 2005). As all other small RNAs in the four point comparison study were insensitive to sodium periodate treatment, it was concluded that the 3′ terminal methyl moiety is ubiquitous to all plant small RNAs, a fact independently confirmed (Li et al. 2005). The 3′ terminal methyl moiety is a 2′-*O*-methyl (Ebhardt et al. 2005; Yang et al. 2006), a post-transcriptional modification often found in RNA (Omer et al. 2000, 2002).

In summary, the blockage of the 3′ terminus can be tested by treating small RNAs with sodium periodate followed by β-elimination. The traditional β-elimination step was simplified, removing the labile dialdehyde in formamide at 99°C (Ebhardt et al. 2005, 2011). Determining the nature of the modification requires either genetic data, such as *hen1* knock-out, or detailed biochemical analysis. In the latter case, a particular small RNA can be isolated from total plant small RNA material using streptavidin gel shift assay, followed by complete digestion of RNA by RNase H and A. The nucleotide fragments can be separated on a reverse phase column and elution times compared to 2′-*O*-methyl and 3′-*O*-methyl standard nucleotides (Yang et al. 2006). Indirect evidence of 3′ terminal 2′-*O*-methyl modification can be obtained through cloning of plant small RNAs if one of the subpopulations of small RNAs is differentially affected in a carefully controlled experiment. Why are small RNAs in plants methylated on their 3′ terminus (Ebhardt et al. 2005; Yu et al. 2005)?

One reason for protecting the 3′ terminus of small RNAs is to prevent endogenous or exogenous RNA dependent RNA polymerases (RDR or RdRp, respectively) to use the small RNA as initiation primer for template dependent amplification of RNA. Although microRNAs can be responsible for generating secondary silencing

RNAs to amplify the effect of the primary microRNA, e.g., *trans*-silencing RNAs, it is of essence for the primary microRNA to hybridize to the correct target. If off-set in length or sequence, the primary microRNA has the strong likelihood to miss the intended target, thus loosing the ability to silence the intended target altogether (Montgomery et al. 2008). Exemplifying the importance of primary silencing RNAs targeting the correct target mRNA are strong RNA silencing suppressor HC-Pro$^+$ tobacco plants infected with Y-Satellite and CMV, which show no phenotypic disease signs, despite high levels of viral RNA present in the plant (Wang et al. 2004). As HC-Pro prevents primary viral silencing RNAs from being methylated, the exogenous small RNAs serve as primers for RNA dependent RNA polymerases. The unintentionally extended small RNAs are processed by dicer-like RNase III out of register (Ebhardt and Unrau 2009), loaded into an ARGONAUTE containing RNA silencing complex but fail to direct the slicing complex to the correct endogenous mRNA. Ergo, high levels of viral small RNAs, without a phenotypic effect.

2 3′ Uridylation of Small RNAs

Another reason for small RNAs to terminate with a 2′-*O*-methyl is the extension by primer independent RNA-specific ribonucleotidyl transferases. The 2′-*O*-methyl group on the 3′ terminus of plant small RNAs protects microRNAs and silencing RNAs from 3′-end uridylation an observation made using two distinct methods (Li et al. 2005). First, two microRNAs were directionally cloned from wild type and *hen1* mutant *A. thaliana* using a 5′ ligation primer with a complementary sequence, in this case complementary to MIR173 and MIR167. Thus, the resulting pool of amplified and cloned small RNAs is enriched for these two microRNAs and variations of the 3′ terminus can easily be detected. Comparing MIR173 from wild type to *hen1* clearly shows that 36% of MIR173 isolates from *hen1* have additional uracil (U) attached to their 3′ termini whereas no 3′ uridylation events were observable in wild type MIR173 isolates. For MIR167 the uridylation frequency is even higher with 55% of MIR167 isolates in *hen1*. Following these observations, primer extension assay supplied with only α-[^{32}P] dATP detects degenerative 3′ termini in small RNAs. Using primer extension, degenerative 3′ termini in *hen1* mutants were shown for microRNAs, endogenous 24 nt silencing RNAs and *trans*-acting silencing RNAs. Li and colleagues conclude that the reduced levels of small RNAs in *hen1* mutants are due to the 3′ uridylation and subsequent recognition of the U tail by 3′–5′ exonuclease(s) similar to the observed U tails in microRNA cleaved mRNAs (Shen and Goodman 2004).

Besides the destabilizing nature of 3′ uridylation, in root and flower tissues of wild type *A. thaliana*, stable 3′ uridylated microRNAs were discovered (Ebhardt et al. 2009). Especially in flower tissue, some 3′ uridylated microRNAs were also observed to have their 5′ most nucleotide removed, a modification termed −1+UU (Ebhardt et al. 2009). Small RNA F42411 was of particular interest, as the −1+UU modified small RNA was almost four times as abundant as its apparent microRNA ath-MIR408. Thus, it is less likely that ath-MIR408−1+UU is a degradation product but rather is biologically significant. Due to the removal of the 5′ most nucleotide, the authors wondered whether the ARGONAUTE affector complex association was affected (Mi et al. 2008). And indeed, while ath-MIR408 resides in AGO2, ath-MIR408−1+UU resides in AGO1. Searching the Mi and colleagues dataset of AGO1, AOG2, AGO4 and AGO5 co-immunoprecipitated small RNAs for the pattern of −1+UU microRNAs revealed other examples, such as ath-MIR822 whereby the −1+UU modified small RNAs were found equally in AGO1 and AGO4 affector complex, while ath-MIR822 was found quantitatively only in AGO1 (Ebhardt et al. 2009).

3 Differential Length MicroRNAs

The −1+UU post-transcriptional modification raises the question whether this is a unique modification or if differential length microRNAs are common. To address this question, various data sets were queried from high throughput DNA sequencing of small RNAs in an in silico Northern extending the mature microRNA sequence found in MiRBase by one, two or three nucleotides on the 5′ terminus according to the microRNA's own hairpin sequence (Ebhardt et al. 2010). Unexpectedly, a fifth of all recorded microRNAs in MiRBase v14 (Griffiths-Jones et al. 2008) were extended by one or two nucleotides according to their own hairpin sequence. Some of the 5′ extended microRNAs were more abundant than the respective unextended parent sequence, e.g., ath-MIR156h was found on average 67% of the time in its extended +1 form across seven deep sequencing data sets. In cases where the +1 or +2 5′ extended microRNA was as abundant as the parental microRNA itself, a shift from AGO1 to AGO5 was observable (e.g., ath-MIR156h and ath-MIR775). This computational analysis clearly shows that differential length microRNAs are stable isoforms. However, bioinformatics analysis does not elucidate how these differential length microRNAs are generated, e.g., post-transcriptional modification of microRNAs (Martin and Keller 2007), differential processing of precursor microRNA hairpins by dicer-like RNase III enzymes or if the longer species of microRNAs are in fact *trans*-acting silencing RNAs. The latter possibility and the question of which microRNAs under which circumstances are generating *trans*-acting silencing RNAs is actively being investigated (Chen et al. 2010; Cuperus et al. 2010). Further, it was shown that microRNA isoforms (extended on the 3′ terminus) play distinct roles in plant development under environmental stress conditions play distinct roles in plant development (Vaucheret 2009).

4 RNA Editing

With the advent of pyrosequencing, large amounts of small RNA sequencing data were suddenly available. Deep sequencing enables cataloging small RNA inventory (Gustafson et al. 2005), studying evolutionary conserved small RNAs (Dolgosheina et al. 2008; Rajagopalan et al. 2006) or small RNA alterations under various environmental conditions (Borsani et al. 2005; Jones-Rhoades and Bartel 2004).

While most researchers were concerned with annotating small RNA sequences matching perfectly to their respective reference genomes, others were noticing sizable amounts of small RNA sequences that could not be mapped to the genome of origin without allowing at least one mismatch. There are plenty of technical reasons why there could be mismatches in the small RNA sequence: for one, there are three enzymes required to generate sequenceable DNA from small RNA, namely T4 RNA ligase, reverse transcriptase und DNA polymerase. Most of these enzymes are of viral origin and therefore have a high error rate. Also, the sequencing methodology itself has an error rate, in the case of pyrosequencing an error rate of 3% was determined by resequencing *Mycoplasma genitalium* (Margulies et al. 2005). However, a more detailed examination of enzymatic or technical sequencing errors reveals that insertions and deletions (indels) are the main source for error. In terms of substitution errors, in which one base is apparently replaced by another one, various enzymes have differential preferences. For example, a variant of Moloney Murine Leukemia Virus Reverse Transcriptase (MLV-RT) is often used to convert small RNA sequences into DNA. The error rate of MLV-RT is reported to be 1 in 15,000 of which two thirds are indels and one third substitutions. Of the substitutions, G–A are rare, while C–A and T–G are prevalent (Potter et al. 2003). However, when querying 193024 non-perfectly mapped small RNAs from *A. thaliana* to their genome of origin, A–G, C–U, G–A and T–C substitutions were prevalent while C–A and T–G were less common. These observations suggest that MLV-RT might not be the sole contributor to "sequencing errors." Thus, it was hypothesized that "sequencing errors" are not just the result of technical limitations but carry biologically significant information (Ebhardt et al. 2009). Especially in flower and root tissue, several significant RNA editing events were observed, e.g., ath-MIR399a C–U at position 3 or ath-MIR156 A–G at position 16. These editing events are tissue specific and can be rationalized with the activity of (de-)aminases acting on RNA (Bass 2002; Dance et al. 2001). If these edited microRNAs target alternative mRNAs when compared to their respective unedited microRNA is still an outstanding question.

5 Closing

The biological significance of post-transcriptional modification of plant small RNAs has been demonstrated. Thus far, post-transcriptional modifications of plant small RNAs have been identified due to tracing of an enzyme activity affecting small RNAs in plants (*hen1*) or careful evaluation of DNA sequencing data of small RNAs. Latter approach is limited by the use of three independent enzymes to generate cDNA and canonical nucleotides used in the process. It remains to be seen if all post-transcriptional modifications of microRNAs and other approximately 20–25 nt RNAs have been discovered yet.

Acknowledgments HAE is supported by a Natural Sciences and Engineering Research Council of Canada (NSERC) Post-doctoral Fellowship. HAE would also like to thank the reviewers for critically reading the manuscript.

References

Alefelder S, Patel BK, Eckstein F (1998) Incorporation of terminal phosphorothioates into oligonucleotides. Nucleic Acids Res 26:4983–4988

Bass BL (2002) RNA editing by adenosine deaminases that act on RNA. Annu Rev Biochem 71:817–846

Borsani O, Zhu J, Verslues PE et al (2005) Endogenous siRNAs derived from a pair of natural cis-antisense transcripts regulate salt tolerance in Arabidopsis. Cell 123:1279–1291

Chen X, Liu J, Cheng Y et al (2002) HEN1 functions pleiotropically in Arabidopsis development and acts in C function in the flower. Development 129:1085–1094

Chen HM, Chen LT, Patel K et al (2010) 22-Nucleotide RNAs trigger secondary siRNA biogenesis in plants. Proc Natl Acad Sci USA 107:15269–15274

Cuperus JT, Carbonell A, Fahlgren N et al (2010) Unique functionality of 22-nt miRNAs in triggering RDR6-dependent siRNA biogenesis from target transcripts in Arabidopsis. Nat Struct Mol Biol 17:997–1003

Dance GS, Beemiller P, Yang Y et al (2001) Identification of the yeast cytidine deaminase CDD1 as an orphan C→U RNA editase. Nucleic Acids Res 29:1772–1780

Dolgosheina EV, Morin RD, Aksay G et al (2008) Conifers have a unique small RNA silencing signature. RNA 14:1508–1515

Ebhardt HA, Unrau PJ (2009) Characterizing multiple exogenous and endogenous small RNA populations in parallel with subfemtomolar sensitivity using a streptavidin gel-shift assay. RNA 15:724–731

Ebhardt HA, Thi EP, Wang MB et al (2005) Extensive 3′ modification of plant small RNAs is modulated by helper component-proteinase expression. Proc Natl Acad Sci USA 102:13398–13403

Ebhardt HA, Tsang HH, Dai DC et al (2009) Meta-analysis of small RNA-sequencing errors reveals ubiquitous post-transcriptional RNA modifications. Nucleic Acids Res 37:2461–2470

Ebhardt HA, Fedynak A, Fahlman RP (2010) Naturally occurring variations in sequence length creates microRNA isoforms that differ in argonaute effector complex specificity. Silence 1:12

Ebhardt HA, Ovando OM, Unrau PJ (2011) Isolation and biochemical analysis of plant small RNAs. John M. Watson, Ming-Bo Wang (eds.), Antiviral resistance in plants. Humana Press (Springer)

Griffiths-Jones S, Saini HK, van Dongen S et al (2008) miRBase: tools for microRNA genomics. Nucleic Acids Res 36:D154–D158

Gustafson AM, Allen E, Givan S et al (2005) ASRP: the Arabidopsis small RNA project database. Nucleic Acids Res 33:D637–D640

Hutvagner G, McLachlan J, Pasquinelli AE et al (2001) A cellular function for the RNA-interference enzyme Dicer in the maturation of the let-7 small temporal RNA. Science 293:834–838

Jones-Rhoades MW, Bartel DP (2004) Computational identification of plant microRNAs and their targets, including a stress-induced miRNA. Mol Cell 14:787–799

Kasschau KD, Xie Z, Allen E et al (2003) P1/HC-Pro, a viral suppressor of RNA silencing, interferes with Arabidopsis development and miRNA function. Dev Cell 4:205–217

Li J, Yang Z, Yu B et al (2005) Methylation protects miRNAs and siRNAs from a 3′-end uridylation activity in Arabidopsis. Curr Biol 15:1501–1507

Margulies M, Egholm M, Altman WE et al (2005) Genome sequencing in microfabricated high-density picolitre reactors. Nature 437:376–380

Martin G, Keller W (2007) RNA-specific ribonucleotidyl transferases. RNA 13:1834–1849

Mi S, Cai T, Hu Y et al (2008) Sorting of small RNAs into Arabidopsis argonaute complexes is directed by the 5′ terminal nucleotide. Cell 133:116–127

Montgomery TA, Howell MD, Cuperus JT et al (2008) Specificity of ARGONAUTE7-miR390 interaction and dual functionality in TAS3 trans-acting siRNA formation. Cell 133:128–141

Omer AD, Lowe TM, Russell AG et al (2000) Homologs of small nucleolar RNAs in Archaea. Science 288:517–522

Omer AD, Ziesche S, Ebhardt H et al (2002) In vitro reconstitution and activity of a C/D box methylation guide ribonucleoprotein complex. Proc Natl Acad Sci USA 99:5289–5294

Park W, Li J, Song R et al (2002) CARPEL FACTORY, a Dicer homolog, and HEN1, a novel protein, act in microRNA metabolism in *Arabidopsis thaliana*. Curr Biol 12:1484–1495

Potter J, Zheng W, Lee J (2003) Thermal stability and cDNA synthesis capability of SuperScript™ III reverse transcriptase. Focus 25:19–24

Rajagopalan R, Vaucheret H, Trejo J et al (2006) A diverse and evolutionarily fluid set of microRNAs in *Arabidopsis thaliana*. Genes Dev 20:3407–3425

Shen B, Goodman HM (2004) Uridine addition after microRNA-directed cleavage. Science 306:997

Vaucheret H (2009) AGO1 homeostasis involves differential production of 21-nt and 22-nt miR168 species by MIR168a and MIR168b. PLoS ONE 4:e6442

Wang MB, Bian XY, Wu LM et al (2004) On the role of RNA silencing in the pathogenicity and evolution of viroids and viral satellites. Proc Natl Acad Sci USA 101:3275–3280

Yang Z, Ebright YW, Yu B et al (2006) HEN1 recognizes 21-24 nt small RNA duplexes and deposits a methyl group onto the 2′ OH of the 3′ terminal nucleotide. Nucleic Acids Res 34:667–675

Yu B, Yang Z, Li J et al (2005) Methylation as a crucial step in plant microRNA biogenesis. Science 307:932–935

Non Coding RNAs and Gene Silencing in Grape

Andrea Carra, Giorgio Gambino, Simona Urso, and Giuseppe Nervo

Contents

Abstract Grapevine (*Vitis vinifera* L.) is a worldwide fruit crop of primary economic interest for berry consumption and winemaking. The molecular basis of grape berry ripening has been partially elucidated with the isolation and functional characterization of transcription factors which regulate sugar accumulation and secondary metabolism. After the recent publication of the complete sequence draft of two grapevine genotypes, a set of small non coding RNAs has been isolated by Sanger and high-throughput sequencing of small RNA libraries. These include conserved and grapevine-specific microRNAs as well as other small RNAs potentially involved in berry ripening. Small non coding RNAs are effectors of silencing pathways that underlie transgene silencing phenomena observed in several experiments of *Agrobacterium*-mediated transformation

A. Carra (✉) and G. Nervo
Research Unit for Wood Production Outside Forests, CRA, Strada Frassineto 35, 15033 Casale Monferrato, Italy
e-mail: andrea.m.carra@gmail.com

G. Gambino
Plant Virology Institute, Grugliasco Unit, CNR, Via L. da Vinci 44, 10095 Grugliasco, Italy

S. Urso
CRA-GPG Genomic Research Centre, Via S. Protaso 302, 29017 Fiorenzuola d'Arda, Italy

V.A. Erdmann and J. Barciszewski (eds.), *Non Coding RNAs in Plants*,
RNA Technologies, DOI 10.1007/978-3-642-19454-2_5,

of grapevine. The knowledge of the silencing mechanisms in grapevine promises to facilitate the development of transient systems for gene functional studies.

Keywords *Agrobacterium* • Berry • Fruit • Grapevine • miRNA • Ripening • Silencing

1 Background

Grapevine (*Vitis vinifera* L.) is a worldwide fruit crop domesticated and grown for thousands of years for berry consumption and winemaking. Ripening grape berries accumulate large quantities of sugars and secondary metabolites, mainly polyphenols, which are key determinants of wine quality. Grape secondary metabolites, including flavonoids, tannins, and stilbenes, have also gained considerable interest as nutraceuticals.

Grape berry ripening is tightly regulated by environmental and developmental factors that, even after more than 10 years of studies at the molecular level, are only partially understood. A transcriptional network has been partially traced with the isolation and functional characterization of regulatory proteins that control the expression of genes encoding key enzymes of secondary metabolism pathways and sugar transporters. The regulation of genes encoding stress- and defense-related proteins, which also characterize the ripening berries, remains by contrast less known. The expression of ripening-related genes in grape is promoted by the establishment of a correct hormonal balance, which is characterized by reduced auxin (Davies et al. 1997) and increased abscisic acid (ABA) and brassinosteroids (Jeong et al. 2004; Symons et al. 2006).

A major breakthrough in grapevine research has been the publication of the complete sequence draft of two grapevine genotypes, the almost homozygous inbred line PN40024 (Jaillon et al. 2007) and the largely heterozygous 'Pinot noir' clone ENTAV115 (Velasco et al. 2007). In the context of these works, sets of conserved miRNAs have been predicted with bioinformatic tools. At the same time, the availability of the complete genome sequence greatly facilitated the isolation of nonconserved miRNAs and of other small RNAs with potential regulatory roles, opening the way to the investigation of gene regulation by RNA silencing in grape berry ripening.

2 High-Throughput Sequencing of Grapevine Small RNAs

To address the scope of small RNA-mediated regulation in grapevine, small RNA libraries generated from leaves of the inbred line PN40024 (Mica et al. 2010) and from leaves, tendrils, inflorescence, and young berries of the 'Pinot noir' clone ENTAV115 (Pantaleo et al. 2010) were sequenced with the Illumina Solexa technology. In the heterozygous clone ENTAV115, differently from most plant species examined to date, 21-nt small RNAs were cloned more frequently than 24-nt small RNAs. Read count was confirmed by gel staining, which evidenced two bands of approximately equal intensity corresponding to 24 and 21-nt RNA fractions, compared to *Arabidopsis*, tomato (*Solanum lycopersicum*), and *Nicotiana benthamiana* for which only 24-nt fractions were visible. Grapevine 24-nt small RNAs were, as expected, less redundant than the 21-nt small RNAs and mapped more frequently to introns than to intergenic regions of the genome, although this observation may be partially due the preliminary status of grape genome annotation. A relative abundance of 21-nt small RNAs was noted previously in other woody perennial plants: *Pinus cordata* (Morin et al. 2008) and *Populus balsamifera* (Barakat et al. 2007). Woody plants may have less 24-nt sRNAs because heterochromatin would be transcriptionally silenced over years, reducing the need for small RNA-mediated transcriptional silencing initiation (Pantaleo et al. 2010). As high-throughput sequencing of small RNAs from the homozygous line PN40024 do not confirm the preponderance of 21-nt sRNA (Mica et al. 2010), the abundance of 21-nt sRNA in the clone ENTAV115 may also be associated with heterozygosity. However, line PN40024 has been obtained with repeated self-fertilization, which may have resulted in additional need of 24-nt sRNAs for silencing of heterochromatic loci (Jullien and Berger 2010).

3 MicroRNAs in Grapevine

3.1 Conserved miRNAs in Grapevine

Traditional (Sanger) and high-throughput sequencing of small RNA libraries from different grapevine organs and tissues led to the experimental confirmation of 24 miRNA families widely conserved among plant species, and of an additional set of 26 miRNA families previously known in at least one plant species (Carra et al. 2009; Mica et al. 2010; Pantaleo et al. 2010). Pantaleo et al. (2010) profiled the expression of grapevine-conserved miRNAs in leaves, tendrils, inflorescence, and young immature berries by northern blot and read count from small RNA libraries prepared independently from these organs. Most grapevine-conserved miRNAs were more expressed in inflorescence, which had previously been observed in other plant species and associated with the primary roles played by miRNAs in development. In tendrils, which are modified inflorescences adapted for climbing, expression of conserved miRNAs was, by contrast, very low or undetectable, suggesting that a more relaxed control of gene expression by miRNAs may be important for tendril development.

Microarray analysis of miRNA expression in leaf, root, and inflorescence also identified a set of conserved miRNAs with higher expression in the root, which included miR397, miR398, and miR408 (Mica et al. 2010). In *Arabidopsis* and poplar, these miRNAs target copper proteins such as plantacyanin, laccases, and superoxide dismutase, which are involved in stress response and lignification (Sunkar and Zhu 2004; Lu et al. 2005; Jones-Rhoades et al. 2006; Sunkar et al. 2006). Thus, miR397, miR398, and miR408 may be important to regulate the high degree of lignification and stress tolerance, which are characteristics of grapevine roots.

Comparing to vegetative tissues or young berries, less data are available on the expression of miRNAs in ripening berries. Among ten conserved miRNAs isolated by Sanger sequencing from a grape berry small RNA library, the expression of miR160, miR164, and miR167, which are known negative regulators of auxin signal transduction (Mallory et al. 2004, 2005; Guo et al. 2005), was measured by northern blot in ripening berries and in vegetative tissues (Carra et al. 2009). Although auxin was shown to antagonize ripening, these miRNAs were found barely expressed in berries. Accordingly, a predicted miR164 target transcript encoding a NAC transcription factor had previously been shown to increase during maturation (Carra et al. 2009; Deluc et al. 2007). These findings suggest that miR160, miR164, and miR167 are not required for the repression of auxin signals in grape berries.

Mica et al. (2010) analyzed by microarray the expression of conserved miRNAs in berries of the grapevine black cultivar 'Corvina.' A few miRNAs were found to be differentially expressed in ripening berries, including miR169, miR395, and miR535, which were upregulated, and miR172 and miR396, which were downregulated. These miRNAs can, thus, be considered as candidate regulators of berry ripening. In *Arabidopsis*, miR169 is downregulated by water stress and targets the transcription factor NFYA5, which is involved in drought stress response (Li et al. 2008), and miR396 is upregulated by osmotic stress and targets a set of GRF transcription factors associated with cell expansion (Jones-Rhoades and Bartel 2004; Liu et al. 2008). Thus, miR169 and miR396 may be involved in the maintenance of water homeostasis in berries, which is a critical aspect of the ripening process. *Arabidopsis* miR395 is expressed in low sulfate conditions and targets several genes encoding proteins involved in sulfur metabolism, including sulfate transporters and ATP sulfurylases (Jones-Rhoades and Bartel 2004; Kawashima et al. 2009). Although sulfate metabolism is poorly known in grape, several enzymes and transporters in the flavonoid pathway are known to require sulfur. Thus, miR395 induction at the onset of berry ripening may help to adjust the sulfur levels optimal for secondary metabolism.

3.2 Grapevine Nonconserved miRNAs

Sanger sequencing of a small RNA library from grape berries allowed the identification of one nonconserved miRNA and four putative nonconserved miRNAs that did not meet the current criteria for miRNA annotation, as their star strands were not isolated (Meyers et al. 2008; Carra et al. 2009). A putative miRNA was later confirmed by high-throughput sequencing, along with the identification of additional 20 nonconserved miRNAs and 21 miRNA candidates (Pantaleo et al. 2010). Among the targets of grapevine nonconserved miRNAs and putative miRNAs, genes with predicted functions in defense and stress responses appear to be predominant. Using bioinformatic tools, the miRNA id47 and the putative miRNAs id97 and id113 were predicted to target four genes encoding TIR-NB-LRR domain proteins, nine genes encoding CC-NB-LRR proteins, and six genes encoding CC-NB-LRR proteins, respectively. For each of these miRNAs, one target gene was validated by 5′ RACE (Carra et al. 2009). By degradome analysis, Pantaleo et al. (2010) identified targets for 8 out of 21 nonconserved grapevine miRNAs. Also this analysis evidenced a relative abundance of target genes encoding proteins involved in defense, including three genes encoding proline-rich proteins, which have been recently associated with disease resistance in rice (Fukuoka et al. 2009), and six genes encoding NB-LRR proteins. Among nonconserved miRNAs that target R-like genes, id47, id97, and id113 are expressed in berries, and this suggests that they may play some role in maturation. Targeting of NB-LRR genes by miRNAs had previously been reported from other plants including poplar (Lu et al. 2005), *Arabidopsis* (Fahlgren et al. 2007), and loblolly pine (Lu et al. 2007), but the biological significance of this control has not been elucidated yet. The activity of plant R proteins is thought to be controlled principally at the posttranslational level, rather than at the transcriptional level (Moffett et al. 2002). Perhaps miRNAs adjust the expression level of the R proteins in different tissues to keep it balanced with that of host adaptor proteins. An alternative function may be the silencing of R genes that are no more useful, having lost their microbial counterpart in the evolutionary race with the emergence of new pathogenic strains.

4 Other Endogenous Small RNAs in Grapevine

Except for above-mentioned observations on the relative paucity of 24-nt small RNAs and their frequent mapping on intronic regions (Pantaleo et al. 2010), grapevine small RNAs not ascribable to the miRNA class were not further analyzed in high-throughput sequencing data. Such an analysis was attempted with Sanger sequencing data on a selection of small RNAs which were cloned in either multiple copies, or matched regions of predicted gene transcripts separated by 21-nt resembling a phased pattern, or matched genes belonging to the same family and clustered in a restricted region of the genome (Carra et al. 2009).

Two 21-nt small RNAs, named id4 and id65, were mapped in sense and antisense orientation, respectively, to the coding region of the grapevine gene GSVIVT00023692001, a predicted orthologue of the *Arabidopsis* cytokinin synthase AtIPT3. Expression of id65 was specifically detected in mature berries, and 5′ RACE revealed a complex pattern of 21-nt phased degradation fragments from GSVIVT00023692001 transcript. Furthermore, expression of GSVIVT00023692001 was drastically reduced in mature berries. These results suggest that a grapevine cytokinin synthase gene is targeted by id65 and the targeting results in double strand formation and phased cleavage by a DICER-LIKE protein to silence the cytokinin synthase gene in mature grape berries. The role of cytokinin in grape berry ripening is not known. Accumulation data available indicate a sharp peak of concentration at the onset of ripening followed by decrease to almost undetectable levels (Alleweldt et al. 1975); thus, id65 may be involved in the repression of cytokinin biosynthesis after this peak. As cytokinins are known antagonists of senescence-like physiological processes that present similarities to those associated with ripening, the repression of cytokinin biosynthesis may be important to allow grape maturation.

There are other, less characterized, small RNAs that match genes potentially involved in processes associated with ripening. For instance, four 21-nt small RNAs were mapped to three genes encoding BURP-domain proteins similar to *Arabidopsis* RD22 which is induced by ABA, drought and osmotic stress (Yamaguchi-Shinozaki and Shinozaki 1993; Carra et al. 2009). The RD22-like genes matched by the siRNAs are part of a cluster of 11 related genes spanning approximately 76 kb on grapevine chromosome 4. Information on these small RNAs is very preliminary, as they could not be detected by northern and attempts to validate targeting of a RD22-like gene transcript, matched in antisense orientation, evidenced an unusual cleavage site between the first and the second nucleotide of the complementary region, counting from the 5′ of the small RNA. As high ABA and osmotic stress characterize mature grape berries, RD22-like genes may be involved in ripening and small RNAs may participate in their regulation. Alternatively, posttranscriptional control of grapevine RD22-like genes may contribute to limit the effects of the excessive expansion of their family with a mechanism that was described in *Arabidopsis* for PPR genes (Howell et al. 2007). Finally, a 23-nt small RNA, cloned three times from a berry small RNA library, bears some interest for matching the coding sequence of VvMYBA1 (Carra et al. unpublished results), a key regulator of anthocyanin biosynthesis in grapevine that is inactivated by transposon insertion in white grapes (Kobayashi et al. 2002, 2004).

5 Silencing and Gene Expression Systems in Grapevine

5.1 Agrobacterium-Mediated Transformation and Gene Silencing in Grapevine

Small RNAs are typically associated with gene silencing at the posttranscriptional and transcriptional levels. Silencing evolved to protect plants from transposons and pathogenic nucleic acids and as a mechanism for the regulation of gene expression. Besides naturally occurring, and well before the elucidation of its molecular basis, silencing was involuntarily triggered in transgenic plants, resulting in inactivation of transgenes. Two types of events contribute to transgene silencing. The first is the position within the genome into which the T-DNA is integrated: T-DNA integration into transcriptionally silent regions (heterochromatic areas) or into highly repetitive and methylated DNA sequences has been correlated with reduced expression of the transgenes. A second event type is related to the configuration of the integrated T-DNAs: multiple T-DNA linked to each other in complex structures can integrate at one locus, inducing transgene silencing (Gelvin 2003).

Genetic transformation of grapevine was initially focused on virus resistance based on pathogen-derived resistance (PDR) (Sanford and Johnston 1985), in which resistance to a virus is engineered in transgenic plants through the expression of a segment of the virus genome. Among the many viral diseases affecting grapevine, one of the most harmful and widespread is fanleaf degeneration. The soilborne nepovirus *Grapevine fanleaf virus* (GFLV) causes fanleaf degeneration, characterized by substantial yield losses, low fruit quality, and a progressive decline of infected vines, which can eventually lead to plant mortality. Different research teams inserted the coat protein (CP) gene of GFLV (in sense, antisense orientation, or truncated forms) in several cultivars and rootstocks (Mauro et al. 1995; Gambino et al. 2005; Maghuly et al. 2006). Evaluation of GFLV resistance in these transgenic grapevines has been attempted in natural conditions of infection and by graft inoculation. Resistance to GFLV has been reported only in some lines of rootstocks expressing the CP gene (Mauro et al. 1995) after a 3-year trial in a naturally infected vineyard in France (Vigne et al. 2004).

As reported above, T-DNA integration in grapevine as well as other plants is a complicated process involving many unknown factors that frequently influence the transgene expression. Although T-DNA transfer was assumed to only involve DNA sequences between the right border (RB) and the left border (LB), evidence from earlier work shows that vector backbone sequences (DNA sequences from the transformation vector outside the T-DNA) may also be occasionally transferred to the plant (Ooms et al. 1982). Transgenic grapevines containing GFLV-CP gene in sense and antisense orientation (Gambino et al. 2005; Maghuly et al. 2006) exhibit multicopy transgene insertions in complex arrangements and different levels of mRNA expression that are not correlated with transgene copy number (Gambino et al. 2009). In these grapevines, vector backbone sequences were integrated in 28.6% of the transgenic plants, and multiple T-DNAs frequently integrated at the same position, resulting in the formation of tandem and inverted repeats (IRs). T-DNAs integrated generally into transcriptionally competent regions; therefore, the position within the grapevine genome into which the T-DNA integrated was not responsible for the transgene expression (Gambino et al. 2009). The transgene silencing in these lines seems to be associated with integration of multiple transgene copies in tandem or IRs at a single locus, and with the transfer of vector backbone sequences. In these transgenic grapevines, the correlation between accumulation of siRNAs, transgene methylation, and RNA silencing could not be confirmed in all lines (Gambino et al. 2010). In three silenced lines containing T-DNAs in complex arrangements, high levels of transgene methylation were observed in both symmetrical and asymmetrical (at lower levels) cytosines in the 35S promoter, GFLV-CP, and T7 terminator sequences. However, in other silenced lines low levels of DNA methylation were observed, and in addition, in the GFLV-CP transcript no siRNAs could be detected. It is possible that RNA signaling molecules were responsible for the RNA silencing, even if siRNAs were below the detection level. Scions of transgenic grapevines were wedge-grafted onto nontransgenic GFLV-infected rootstock and then cultivated in greenhouse conditions. Transgenic grapevines showing RNA silencing were unable to contrast the virus spread. siRNAs of 21–22 nt were detected in transgenic and nontransgenic grapevines following GFLV infection, but they were unable to block virus replication indicating that GFLV could circumvent this silencing pathway. This susceptibility to GFLV could be due to the high viral inoculum and to the constant viral pressure from the rootstock applied to relatively young and small plants: under these conditions the transgenic grapevines may be unable to suppress GFLV replication (Gambino et al. 2010). Though numerous studies demonstrated virus resistance mediated by the mechanism of RNA silencing (Lindbo and Dougherty 2005), these transgenic grapevines seem to be unable to suppress GFLV replication. In this case, the silencing process, although effective against the transgene, may be incomplete, leaving a small pool of viral transcripts that may be sufficient for GFLV replication.

The absence of siRNAs or the inability to identify siRNAs in transgenic silenced grapevines has been observed also in constructs that should lead to silencing with high efficiency as transgenes containing hairpin RNA (hpRNA) structures (Jardak-Jamoussi et al. 2009; Winterhagen et al. 2009). Jardak-Jamoussi et al. (2009) reported the development of IR constructs carrying fragments of a conserved region of the GFLV movement protein (MP). *N. benthamiana* plants transformed with these constructs exhibited different responses to viral inoculation, varying from resistance to retarded infection, recovery, and susceptibility. In transgenic resistant plants, siRNAs were not detected before the GFLV inoculation. The authors speculate that unfavorable environmental conditions, such as low temperature (Szittya et al. 2003) or some physiological stages (Missiou et al. 2004), could hinder siRNA accumulation. siRNAs could be detected only in resistant lines upon inoculation by the GFLV. This would indicate that prior to GFLV inoculation siRNA amount in resistant lines was not enough to be detected and that replication of GFLV seem to be essential to induce and to amplify the RNA degradation system. However, it is essential to confirm these observations in transgenic grapevine, the natural host of the virus. Winterhagen et al. (2009) transformed *N. benthamiana* with a defective interfering construct containing GFLV sequences and showed that siRNAs of transgenic origin was not found in silenced plants cultured in the greenhouse. Therefore, it is suggested that a low, undetectable amount of transgene-derived siRNA may be sufficient to establish efficient silencing. The reported results showed a significant and complex relationship between GFLV and the RNA silencing process in grapevine and herbaceous hosts.

5.2 Transient Expression Systems and Gene Silencing in Grapevine

Although currently there are still few publications on the topic, many studies of functional genomics are underway to investigate the role of important genes in grapevine biology. The gene transfer technology may provide an important contribution to this field, also considering that some tools (for example a large set of mutants) are available for model plants, but not for grapevine. As examined above, gene transfer for functional studies can be achieved by stable genetic transformation via *Agrobacterium*. For instance, grapevine plants expressing a alcohol dehydrogenase (Adh) cDNA under the constitutive 35S promoter displayed a lower sucrose content, a higher degree of polymerization of proanthocyanidins, and a generally increased content of volatile compounds, mainly in carotenoid- and shikimate-derived volatiles (Tesniere et al. 2006). However, *Agrobacterium*-mediated stable transformation of grapevine is hindered by low-efficiency, long regeneration time, and, most importantly, by many years of juvenile phase before fructification, which made practically not feasible the observation of transgenic fruits. These problems can be partially solved by transient expression systems as virus-induced gene silencing (VIGS), particle bombardment, or *Agrobacterium*-mediated transient transformation.

A prerequisite for VIGS is the availability of suitable viral vectors. In grapevine, the *Grapevine virus A* (GVA), a pathogen closely associated with the economically important rugose wood disease, is the main candidate to be used as a VIGS vector. The use of GVA-derived vectors for silencing the endogenous phytoene desaturase (PDS) gene in *N. benthamiana* has been recently demonstrated (Haviv et al. 2006; Muruganantham et al. 2009). In grapevine, an *Agrobacterium*-mediated method was used to inoculate roots of in vitro plants with GVA-derived vectors, resulting in efficient silencing of the endogenous PDS gene (Muruganantham et al. 2009). VIGS has also been used to introduce viral genes for PDR. Brumin et al. (2009), in an attempt to develop resistance to GVA, made a minireplicon that expressed the products required for virus replication, but that did not express the movement and CPs and the suppressor of RNA silencing, p10 (Zhou et al. 2006). The GVA minireplicon induced efficient gene silencing in agroinfiltrated leaves and transgenic *N. benthamiana* plants. These plants exhibited virus resistance specifically targeted against GVA, and this resistance was transmissible to nontransgenic scions grafted onto transgenic rootstocks probably through transgene-derived siRNA molecules. In addition, the authors showed that the plants were susceptible to *Grapevine virus B* (GVB), a GVA closely related virus. GVB infection resulted in increased accumulation of the transgene minireplicon RNA, which suggests suppression of the transgene-specific PTGS, probably through a suppressor of silencing from GVB. Accordingly, when aiming to induce virus resistance based on transgene-specific PTGS, it is necessary to take into account the possibility of infection by other viruses, which might cause suppression of the induced transgene-specific PTGS.

Particle bombardment has been used in different grapevine tissues such as undifferentiated callus and embryogenic cell suspensions (Kikkert et al. 2004; Vidal et al. 2006) to obtain stable transgenic plants after regeneration from transformed tissues. Biolistic-mediated gene transfer was used also for transient assays. Bogs et al. (2007) and Deluc et al. (2008) used a cell suspension from a 'Chardonnay' petiole callus culture to investigate the function of transcription factors that activate different structural genes of the flavonoid pathway.

Agrobacterium-mediated transient assay has become one of the preferred strategies for the functional characterization of genes; in particular, leaf agroinfiltration represents a simple and not invasive technique. Two recent reports have demonstrated the feasibility of the transient transformation assay applied to grape leaves. Santos-Rosa et al. (2008) achieved acceptable levels of β-glucuronidase (GUS) and green fluorescent protein (GFP) expression using vacuum infiltration of *Agrobacterium* on leaves obtained from in vitro grown grapevine plantlets. Agroinfiltration was then used to investigate the defensive role of stilbenes against

Plasmopara viticola by transient overexpression of the stilbene synthase gene in grape leaves before infection with the fungal pathogen. A protocol for efficient transient transformation of selected grapevine cultivars by combining different genotypes, *Agrobacterium* strains, and physiological conditions and by using a syringe without needle has recently been developed (Zottini et al. 2008). In this work, the combination of fluorescent marker tags and confocal microscope analyses allowed identification of the subcellular localization of transgene products and showed the opportunity to convert a transiently transformed leaf tissue into a stably transformed cell line.

Acknowledgments The authors thank Giampiero Valè (CRA-GPG Genomic Research Centre, Fiorenzuola d'Arda, Italy) for critical reading of the manuscript.

References

Alleweldt G, Duering H, Waits G (1975) Untersuchungen zum mechanismus der zuckereinlagerung in die wachsenden weinbeeren. Angew Bot 49:65–73

Barakat A, Wall K, Leebens-Mack J et al (2007) Large-scale identification of microRNAs from a basal eudicot (Eschscholzia californica) and conservation in flowering plants. Plant J 51:991–1003

Bogs J, Jaffe F, Takos A et al (2007) The grapevine transcription factor VvMYBPA1 regulates proanthocyanidin synthesis during fruit development. Plant Physiol 143:1347–1361

Brumin M, Stukalov S, Haviv S et al (2009) Post-transcriptional gene silencing and virus resistance in *Nicotiana benthamiana* expressing a Grapevine virus A minireplicon. Transgenic Res 18:331–345

Carra A, Mica E, Gambino G et al (2009) Cloning and characterization of small noncoding RNAs from grape. Plant J 59:750–763

Davies C, Boss PK, Robinson SP (1997) Treatment of grape berries, a nonclimacteric fruit with a synthetic auxin, retards ripening and alters the expression of developmentally regulated genes. Plant Physiol 115:1155–1161

Deluc LG, Grimplet J, Wheatley MD et al (2007) Transcriptomic and metabolite analyses of Cabernet Sauvignon grape berry development. BMC Genomics 8:429

Deluc L, Bogs J, Walker AR et al (2008) The transcription factor VvMYB5 contributes to the regulation of anthocyanin and proanthocyanidin biosynthesis in developing grape berries. Plant Physiol 147:2041–2053

Fahlgren N, Howell MD, Kasschau KD et al (2007) High-throughput sequencing of Arabidopsis microRNAs: evidence for frequent birth and death of MIRNA genes. PLoS ONE 2:e219

Fukuoka S, Saka N, Koga H et al (2009) Loss of function of a proline containing protein confers durable disease resistance in rice. Science 325:998–1001

Gambino G, Gribaudo I, Leopold S et al (2005) Molecular characterization of grapevine plants transformed with GFLV resistance genes: I. Plant Cell Rep 24:655–662

Gambino G, Chitarra W, Maghuly F et al (2009) Characterization of T-DNA insertions in transgenic grapevines obtained by Agrobacterium-mediated transformation. Mol Breed 24:305–320

Gambino G, Perrone I, Carra A et al (2010) Transgene silencing in grapevines transformed with GFLV resistance genes: analysis of variable expression of transgene, siRNAs production and cytosine methylation. Transgenic Res 19:17–27

Gelvin SB (2003) Agrobacterium-mediated plant transformation: the biology behind the "gene-jockeying" tool. Microbiol Mol Biol Rev 67:16–37

Guo HS, Xie Q, Fei JF et al (2005) MicroRNA directs mRNA cleavage of the transcription factor NAC1 to downregulate auxin signal for Arabidopsis lateral root development. Plant Cell 17:1376–1386

Haviv S, Galiakparov N, Goszczynski DE et al (2006) Engineering the genome of Grapevine virus A into a vector for expression of proteins in herbaceous plants. J Virol Meth 132:227–231

Howell MD, Fahlgren N, Chapman EJ et al (2007) Genome-wide analysis of the RNA-DEPENDENT RNA POLYMERASE6/DICER-LIKE4 pathway in Arabidopsis reveals dependency on miRNA and tasiRNA-directed targeting. Plant Cell 19:926–942

Jaillon O, Aury JM, Noel B et al (2007) The grapevine genome sequence suggests ancestral hexaploidization in major angiosperm phyla. Nature 449:463–468

Jardak-Jamoussi R, Winterhagen P, Bouamama B et al (2009) Development and evaluation of a GFLV inverted repeat construct for genetic transformation of grapevine. Plant Cell Tiss Organ Cult 97:187–196

Jeong ST, Goto-Yamamoto N, Kobayashi S et al (2004) Effects of plant hormones and shading on the accumulation of anthocyanins and the expression of anthocyanin biosynthetic genes in grape berry skins. Plant Sci 167:247–252

Jones-Rhoades MW, Bartel DP (2004) Computational identification of plant microRNAs and their targets, including a stress induced miRNA. Mol Cell 14:787–799

Jones-Rhoades MW, Bartel DP, Bartel B (2006) MicroRNAs and their regulatory roles in plants. Annu Rev Plant Biol 57:19–53

Jullien PE, Berger F (2010) DNA methylation reprogramming during plant sexual reproduction? Trends Genet 26:394–399

Kawashima CG, Yoshimoto N, Maruyama-Nakashita A et al (2009) Sulphur starvation induces the expression of microRNA-395 and one of its target genes but in different cell types. Plant J 57:313–321

Kikkert JR, Vidal JR, Reisch BI (2004) Stable plant transformation via particle bombardment/biolistics. In: Peña L (ed) Methods in molecular biology. Transgenic plants: methods and protocols, vol 286. Humana Press, Totowa, pp 61–67

Kobayashi S, Ishimaru M, Hiraoka K et al (2002) Myb-related genes of the Kyoho grape (Vitis labruscana) regulate anthocyanin biosynthesis. Planta 215:924–933

Kobayashi S, Goto-Yamamoto N, Hirochika H (2004) Retrotransposon-induced mutations in grape skin color. Science 304:982

Li WX, Oono Y, Zhu J et al (2008) The Arabidopsis NFYA5 transcription factor is regulated transcriptionally and posttranscriptionally to promote drought resistance. Plant Cell 20:2238–2251

Lindbo JA, Dougherty WG (2005) Plant pathology and RNAi: a brief history. Annu Rev Phytopathol 43:191–204

Liu H-H, Tian X, Li Y-J et al (2008) Microarray-based analysis of stress-regulated microRNAs in *Arabidopsis thaliana*. RNA 14:836–843

Lu S, Sun YH, Shi R et al (2005) Novel and mechanical stress–responsive microRNAs in Populus trichocarpa that are absent from Arabidopsis. Plant Cell 17:2186–2203

Lu S, Sun YH, Amerson H et al (2007) MicroRNAs in loblolly pine (*Pinus taeda* L.) and their association with fusiform rust gall development. Plant J 51:1077–1098

Maghuly F, Leopold S, Da Câmara MA et al (2006) Molecular characterization of grapevine plants transformed with GFLV resistance genes: II. Plant Cell Rep 25:546–553

Mallory AC, Dugas DV, Bartel DP et al (2004) MicroRNA regulation of NAC-domain targets is required for proper formation and separation of adjacent embryonic, vegetative, and floral organs. Curr Biol 14:1035–1046

Mallory AC, Bartel DP, Bartel B (2005) MicroRNA-directed regulation of Arabidopsis AUXIN RESPONSE FACTOR17 is essential for proper development and modulates expression of early auxin response genes. Plant Cell 17:1360–1375

Mauro MC, Toutain S, Walter B et al (1995) High efficiency regeneration of grapevine plants transformed with the GFLV coat protein gene. Plant Sci 112:97–106

Meyers BC, Axtell MJ, Bartel B et al (2008) Criteria for annotation of plant MicroRNAs. Plant Cell 20:3186–3190

Mica E, Piccolo V, Delledonne M et al (2010) High throughput approaches reveal splicing of primary microRNA transcripts and tissue specific expression of mature microRNAs in *Vitis vinifera*. BMC Genomics 11:109

Missiou A, Kalantidis K, Boutla A et al (2004) Generation of transgenic potato plants highly resistant to Potato virus Y (PVY) through RNA silencing. Mol Breed 14:185–197

Moffett P, Farnham G, Peart J et al (2002) Interaction between domains of a plant NBS-LRR protein in disease resistance-related cell death. EMBO J 21:4511–4519

Morin RD, Aksay G, Dolgosheina E et al (2008) Comparative analysis of the small RNA transcriptomes of *Pinus contorta* and *Oryza sativa*. Genome Res 18:571–584

Muruganantham M, Moskovitz Y, Haviv S et al (2009) Grapevine virus A-mediated gene silencing in *Nicotiana benthamiana* and *Vitis vinifera*. J Virol Meth 155:167–174

Ooms G, Bakker A, Molendijk L et al (1982) T-DNA organization in homogeneous and heterogeneous octopine-type crown gall tissues of *Nicotiana tabacum*. Cell 30:589–597

Pantaleo V, Szittya G, Moxon S et al (2010) Identification of grapevine microRNAs and their targets using high-throughput sequencing and degradome analysis. Plant J 62:960–976

Sanford JC, Johnston SA (1985) The concept of parasite-derived resistance: deriving resistance genes from the parasite's own genome. J Theor Biol 113:395–405

Santos-Rosa M, Poutaraud A, Merdinoglu D et al (2008) Development of a transient expression system in grapevine via agro-infiltration. Plant Cell Rep 27:1053–1063

Sunkar R, Zhu J-K (2004) Novel and stress-regulated micro RNAs and other small RNAs from Arabidopsis. Plant Cell 16:2001–2019

Sunkar R, Kapoor A, Zhu JK (2006) Posttranscriptional induction of two Cu/Zn superoxide dismutase genes in Arabidopsis is mediated by downregulation of miR398 and important for oxidative stress tolerance. Plant Cell 18:2051–2065

Symons GM, Davies C, Shavrukov Y et al (2006) Grapes on steroids. Brassinosteroids are involved in grape berry ripening. Plant Physiol 140:150–158

Szittya G, Silhavy D, Molnár A et al (2003) Low temperature inhibits RNA silencing-mediated defence by the control of siRNA generation. EMBO J 22:633–640

Tesniere C, Torregrosa L, Pradal M et al (2006) Effects of genetic manipulation of alcohol dehydrogenase levels on the response to stress and the synthesis of secondary metabolites in grapevine leaves. J Exp Bot 57:91–99

Velasco R, Zharkikh A, Troggio M et al (2007) A high quality draft consensus sequence of the genome of a heterozygous grapevine variety. PLoS ONE 2:e1326

Vidal J, Kikkert JR, Donzelli B et al (2006) Biolistic transformation of grapevine using minimal gene cassette technology. Plant Cell Rep 25:807–814

Vigne E, Komar V, Fuchs M (2004) Field safety assessment of recombination in transgenic grapevines expressing the coat protein gene of Grapevine fan leaf virus. Transgenic Res 13:165–179

Winterhagen P, Dubois C, Sinn M et al (2009) Gene silencing and virus resistance based on defective interfering constructs in transgenic *Nicotiana benthamiana* is not linked to accumulation of siRNA. Plant Physiol Biochem 47:739–742

Yamaguchi-Shinozaki K, Shinozaki K (1993) The plant hormone abscisic acid mediates the drought-induced expression but not the seed-specific expression of rd22, a gene responsive to dehydration stress in *Arabidopsis thaliana*. Mol Gen Genet 238:17–25

Zhou ZS, Dell'Orco M, Saldarelli P et al (2006) Identification of an RNA-silencing suppressor in the genome of Grapevine virus A. J Gen Virol 87:2387–2395

Zottini M, Barizza E, Costa A et al (2008) Agroinfiltration of grapevine leaves for fast transient assays of gene expression and for long term production of stable transformed cells. Plant Cell Rep 27:845–853

Genome-Wide Analysis of RNA Degradation in *Arabidopsis*

Yukio Kurihara and Motoaki Seki

Contents

Abstract During the past 10 years, a number of new findings have been made in RNA-mediated gene regulation and in regulation mechanisms of the RNAs using some advanced high-throughput technologies in the model plant, *Arabidopsis thaliana*. One of them is a genome-wide tiling array that allows us to receive the information of strand-specific transcriptome of not only protein-coding mRNAs but also long non coding transcripts. This chapter introduces several findings on plant nonsense-mediated decay (NMD), a well-known RNA regulatory machinery, and also summarizes the results of genome-wide analyses of RNA regulatory networks through NMD and NMD-related decay pathways, some of which have already been

M. Seki (✉)
Plant Genomic Network Research Team, RIKEN Plant Science Center,
1-7-22 Suehiro-cho, Tsurumi-ku, Yokohama-shi 230-0045, Japan
and
Kihara Institute for Biological Research, Yokohama City University,
641-12 Maioka-cho, Totsuka-ku, Yokohama 244-0813, Japan
e-mail: mseki@psc.riken.jp

Y. Kurihara
Plant Biology Laboratory, The Salk Institute for Biological Studies,
10010 North Torrey Pines Road, La Jolla, CA 92037, USA

V.A. Erdmann and J. Barciszewski (eds.), *Non Coding RNAs in Plants*,
RNA Technologies, DOI 10.1007/978-3-642-19454-2_6,

examined using the microarrays such as the tiling array. Their analyses revealed that unnecessary RNA species including long non coding junk RNAs are downregulated by the RNA decay machineries such as NMD. This chapter will help understand the existence of the transcriptional hidden layer and how genome-wide transcriptome is being constructed in plants.

Keywords NMD • Tiling array • RNA decay

1 Whole Genome Tiling Array Technology

Microarray technology has been used to identify the transcriptional profiles in several specific conditions in plants as well as other organisms. Most of the microarrays (exon arrays) have probes specific to annotated protein-coding mRNAs, but no probes specific to unannotated transcripts such as non coding RNAs (Fig. 1a). On the contrary, whole-genome tiling arrays have probes corresponding to both the annotated protein-coding mRNAs and the unannotated transcripts. For example, Affymetrix *Arabidopsis* tiling arrays (1.0F and 1.0R) are comprised of 25 nucleotides (nt)-long probes covering both Watson and Crick strands of whole genome with 10-nt gaps between two probes (Fig. 1a) (Zhang et al. 2006; Matsui et al. 2008).

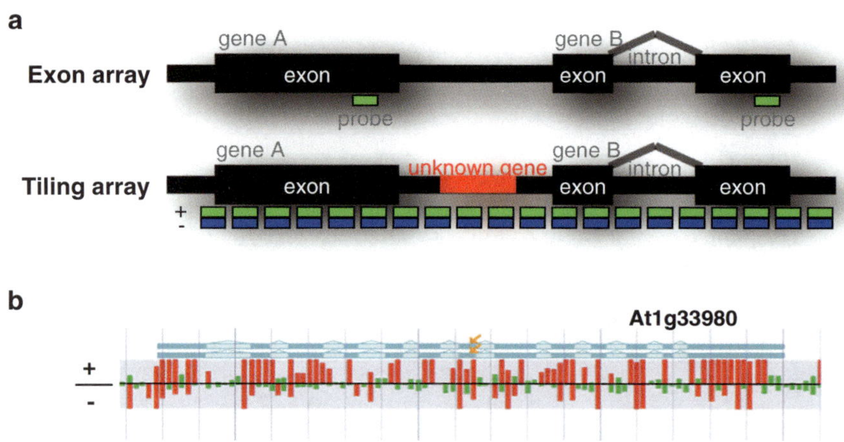

Fig. 1 (**a**) Scheme of exon array and tiling array. Exon arrays equip only probes specific to short strand-specific sequence of protein-coding genes. On the contrary, tiling arrays (1.0F and 1.0R) are comprised of 25 nucleotides (nt)-long probes covering both Watson and Crick strands of whole genome with 10-nt intervals between two probes. This architecture allows us to detect expression of the unannotated genome locus. (**b**) Example of the OmicBrowse (Toyoda et al. 2007) output of a gene (At1g33980 encodes UPF3 protein). The *blue* and *light-blue* regions are exons and introns, respectively, based on TAIR annotation (http://www.arabidopsis.org/). The *red* and *green* bars indicate relative signal intensity of probes (*red* >400, *green* <400). This illustration shows that signal intensities are correlated with exon–intron structure of the gene on the Watson strand (+)

The tiling arrays can monitor accumulation of strand-specific transcripts arisen from both Watson and Crick strands of genome and detect non coding RNAs including microRNA precursors and mRNA-like non coding RNAs (mlncRNAs), as well as protein-coding mRNAs that can be detected by exon arrays. It is also possible to check the exon–intron structure of the transcripts roughly (Fig. 1b). That is, continuous relatively higher signals on the transcript provide the rough position of exons, whereas continuous relatively lower signals provide the rough position of introns. These advantages, in combination with a genetic approach using lots of *Arabidopsis* mutants, permit us to investigate the hidden layer of the transcriptome that is regulated by RNA decay mechanisms.

2 Downregulation of Long Non Coding Transcripts by the NMD Pathway

2.1 NMD and RNA Decay Pathways

The nonsense-mediated mRNA decay (NMD) is a well-known mRNA surveillance mechanism conserved among eukaryotic organisms (Maquat 2004). NMD specially eliminates aberrant mRNAs with a premature termination codon (PTC) from cells (Fig. 2). Such transcripts may happen by genomic mutations, loading errors by RNA polymerase, inefficient splicing of pre-mRNA, and so on. For example, the majority of alternative spliced mRNAs carry PTC and may be subjected to NMD in *Arabidopsis* (Filichkin et al. 2010).

The NMD mechanism has been well characterized in yeast and animal systems. UPF1-3 proteins form the core complex on the mRNA, and then the UPF complex and ribosomes cooperatively search for any PTCs on the mRNA. If the mRNA is recognized as aberrant mRNA with PTC, it would be subjected to NMD pathway. In this way, eukaryotes have a quality-control mechanism to avoid production of truncated proteins translated from such aberrant mRNAs with PTC.

In the case of plant NMD, not only aberrant mRNAs with PTC but also mRNAs with long 3′UTR (more than 300 nt) tend to be NMD targets (Kertesz et al. 2006; Hori and Watanabe 2007). In addition, upstream ORF (uORF), which is located in the upstream of main ORF, also can be a trigger of NMD (Nyiko et al. 2009; Saul et al. 2009). In this case, a 50-amino-acid long uORF efficiently can trigger NMD (Nyiko et al. 2009).

A previous analysis has shown that ratios of PTC+/PTC– of some mRNAs are relatively higher in *Arabidopsis upf1* and *upf3* mutants than those in wild type, indicating that AtUPF1 and AtUPF3 are involved in NMD (Hori and Watanabe 2005; Yoine et al. 2006; Arciga-Reyes et al. 2006). It has been reported that aberrant mRNAs and AtUPF2 and AtUPF3 are enriched in the nucleolus, whereas AtUPF1 is dominantly localized in cytoplasm (Kim et al. 2009). Thus, it is possible that AtUPF2 as well as AtUPF1 and AtUPF3 function in NMD. It was also described

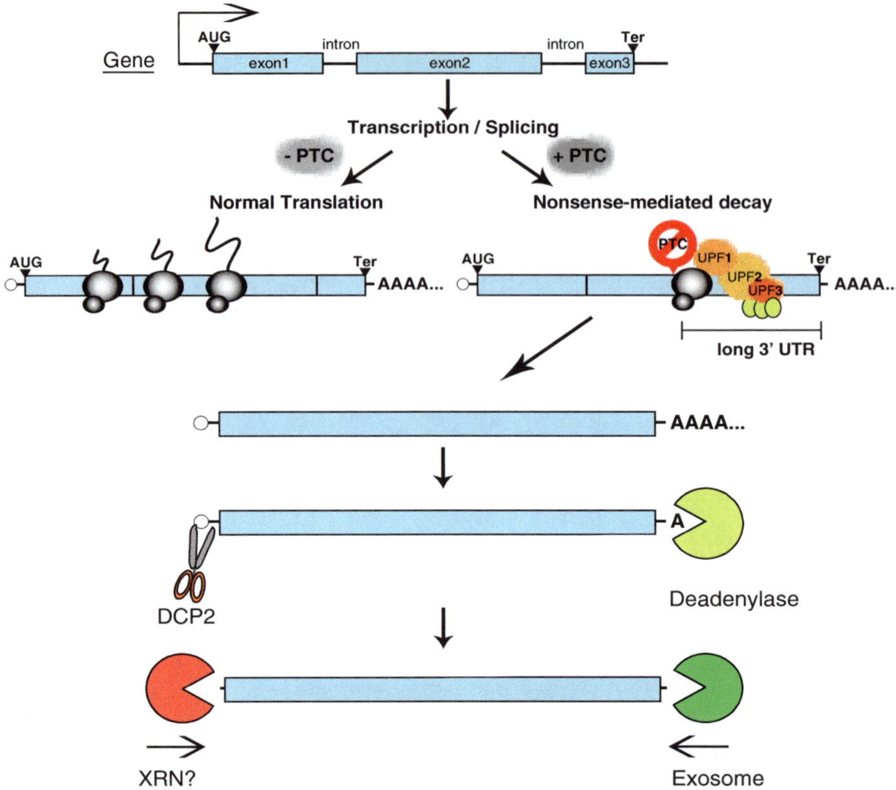

Fig. 2 Illustration of NMD pathway and NMD-related RNA decay machineries. Premature termination codon (PTC) on the aberrant mRNA is recognized by the UPF1-3 complex and ribosome, and the aberrant mRNA is sent to the downstream decay pathways. In general, the aberrant mRNA is deadenylated by CCR4-NOT complex, decapped by DCP2, and degraded in 5′–3′ direction by the exoribonuclease XRN? and/or in 3′–5′ direction by the exoribonuclease complex, exosome

that other factors except UPF proteins are involved in NMD in *Arabidopsis*. Animal SMG7 protein and the yeast ortholog are a key factor in NMD. AtSMG7 plays an important role in plant NMD. Ratios of PTC+/PTC− are relatively higher in the *smg7-1* mutant as the case of *upf* mutants (Riehs et al. 2008). ELF9, an RNA-binding protein, binds *SOC1* transcripts and reduces accumulation of the partially spliced *SOC1* transcripts with PTC, indicating that the *SOC1* transcript is a direct target of ELF9 and that ELF9 is involved in NMD (Song et al. 2009).

Aberrant mRNAs with NMD trigger are degraded by a combination of some RNA decay pathways. In yeast and mammals, the aberrant mRNAs are decapped by DCP1/DCP2 decapping complex and degraded from the free 5′ end by cytoplasmic 5′–3′ exoribonuclease, Xrn1 (He and Jacobson 2001; Lejeune and Maquat 2003).

Alternatively, the aberrant mRNAs are also degraded from 3′ end through the functions of deadenylase and 3′–5′ exonuclease complex, "exosome" (Lejeune and Maquat 2003; Mitchell and Tollervey 2003).

2.2 Tiling Array Analyses in *upf* Mutants

Genome-wide mapping of full-length cDNA clones revealed the existence of a lot of non-protein-coding transcripts harboring poly(A) tails (Seki et al. 2002). Most of the non-protein-coding RNAs annotated as AGI code do not have any main ORFs encoding long polypeptides, but instead have some short ORFs, which could not encode general proteins, and relatively longer 3′UTRs as an example of Fig. 3a shows (Fig. 3a, b). This architecture is very similar to those of NMD target mRNAs because the long 3′UTR is sometimes one of the triggers of NMD (Kertesz et al. 2006; Hori and Watanabe 2007).

That possibility is partly checked by whole-genome tiling array analysis in *upf1-1* and *upf3-1* mutants (Kurihara et al. 2009). The analysis estimated, through statistical filters, that at least 237 in *upf1-1* and 167 in *upf3-1* transcripts annotated as AGI code were upregulated more than 1.8-fold compared with wild type. Expectedly, among them, 31 in *upf1-1* and 25 in *upf3-1* transcripts were classified into mlncRNAs (Fig. 3b). About 80% of them are classified into natural antisense transcripts that arise from antisense strands of other genes based on TAIR8 annotation. All upregulated mlncRNAs annotated as an AGI code have relatively longer 3′UTRs (average 1,250 nt) probably sufficient to trigger NMD, if the lengths are calculated from the 5′-closest termination codons of the short ORFs to poly(A) sites of the TAIR gene model.

Of the upregulated AGI transcripts in *upf* mutants, about 80% were classified into protein-coding mRNAs. It was estimated that more than 50% of upregulated protein-coding transcripts carry one or more uORFs in front of the main ORFs, which is consistent with the previous notion that uORF is a trigger of NMD (Nyiko et al. 2009; Saul et al. 2009).

One advantage of the tiling array is to detect and analyze the accumulation of unannotated transcriptional units (TUs) which are putative non coding transcripts with poly(A) tails, such as antisense transcripts and intergenic transcripts. The tiling array analysis, using the ARTADE computational program to predict nonannotated TUs and their gene structures (Toyoda and Shinozaki 2005), revealed that at least 77 in *upf1-1* and 59 in *upf3-1* unannotated TUs were upregulated more than 1.8-fold. As in the case of AGI-tagged mlncRNAs, about 70% of the upregulated TUs were classified into natural antisense transcripts.

Although it is unknown why such antisense transcripts arise, one may imagine that double-stranded RNA paired with sense and antisense transcripts is a source of natural antisense transcript siRNA (nat-siRNA) (Borsani et al. 2005). However, this may be unlikely because previous informatics analysis indicated that small RNAs in public database are not enriched in the locus where sense and antisense

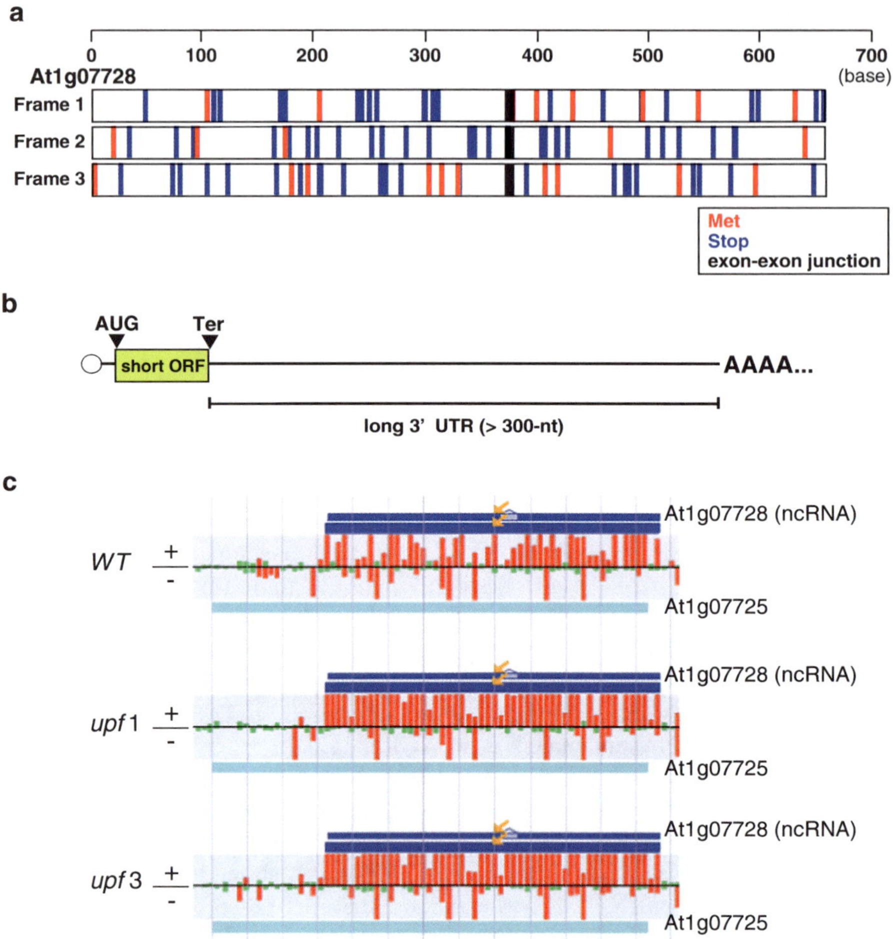

Fig. 3 (**a**) Example of mRNA-like non coding RNA (mlncRNA, At1g07728). The RNA sequence of this transcript is referred to a cDNA clone, RAFL09-83-M03 (Seki et al. 2002). Three possible frames are shown. *Red*, *blue*, and *black* bars indicate the positions of Met codon, termination codon, and exon–exon junction, respectively. (**b**) Characterization of most of mlncRNAs. They do not have any main ORFs, but have short ORFs and relatively longer 3′UTRs that are a trigger of NMD as an example of (**a**). (**c**) Output of tiling array result of At1g07728 locus on OmicBrowse (Toyoda et al. 2007). Accumulation of the transcript was increased in *upf1* and *upf3* mutants. The *blue* and *light-blue* regions are exons and introns, respectively. The *red* and *green* bars indicate relative signal intensity of probes (*red* >400, *green* <400). The tiling array data can be viewed at http://omicspace.riken.jp/gps/group/psca3 (Kurihara et al. 2009)

transcripts arise (Henz et al. 2007). This needs to be confirmed by examining a small RNA population in *upf* mutants in the future.

Importantly, accumulation of lots of mlncRNAs with poly(A) tail is upregulated in *upf* mutants (Kurihara et al. 2009). It is assumed that such non coding transcripts downregulated by NMD are transcriptional noise. This means that one of the significant roles of NMD is downregulation of the non coding transcripts as well as downregulation of aberrant mRNAs with NMD triggers.

3 Genome-Wide Analyses in Other RNA Decay Pathways and Their Relationships with NMD

3.1 Exosome-Mediated Decay

The exosome is a conserved 3′–5′ exoribonuclease complex involved in the degradation and maturation of a wide variety of RNAs (Belostotsky 2009). The doughnut-shaped core of exosome consists of ten common components, Rrp4, Rrp40–46, Mtr3, and Csl4 proteins, which are essential factors for viability in yeast. Six of them, Rrp41, Rrp42, Rrp43, Rrp45, Rrp46, and Mtr3, contain RNase PH domains, and three of them, Rrp4, Rrp40, and Csl4, contain S1 RNA binding motif. Although the core of yeast and human exosomes are catalytically inactive due to amino-acid replacements (instead, the catalytic activity is dependent on the factors, Rrp44/Dis3 (RNase II/R) and Rrp6 (RNase D), associated with exosome), it was shown that *Arabidopsis* Rrp41 subunit is catalytically active (Chekanova et al. 2000). In general, 3′ end of the mRNAs destined to be eliminated from the cell is deadenylated by CCR4-NOT complex and degraded by the exosome.

A previous study has shown that knockouts of *Arabidopsis Rrp4* and *Rrp41* were embryonic lethal, but the knockout plant of *Csl4* showed a phenotype similar to that of the wild-type plant, indicating that Csl4 is not essential for viability in *Arabidopsis* (Chekanova et al. 2007). The same study also identified the exosome substrates using tiling array analysis in *csl4-2* knockout mutant and the inducible RNAi mutants of the exosome-components *RRP4* and *RRP41* ($rrp4^{iRNAi}$ and $rrp41^{iRNAi}$). They include some kinds of structural RNAs such as small nuclear RNAs (snRNAs), small nucleolar RNAs (snoRNAs) and tRNAs, mRNAs (205 and 266 mRNAs in $rrp4^{iRNAi}$ and $rrp41^{iRNAi}$, respectively), a subset of mRNAs with 3′ extended end, and 5′ part of specific microRNA precursors. In addition, importantly, the exosome substrates include a large class of uncharacterized ncRNAs. Many of them are mapped to repetitive elements and small RNA-generating loci (210 and 156 ncRNAs, respectively) where genomic DNA is often methylated, indicating that exosome has a general role in quality control of siRNA biogenesis precursors (Chekanova et al. 2007). The other remarkable matter on exosome knockdown is upregulation of a distinct subclass of ncRNAs that are colinear with the 5′ ends (first exon) of known protein-coding transcripts. The authors suggested that their origin is distinct from that of the overlapping main TUs. However, it is unknown why they are transcriptionally active. Probably, they should be transcriptional noise as same as mlncRNAs upregulated in *upf* mutants. Thus, the exosome complex itself is also involved in quality control of large kinds of RNA species elusively including the noisy ncRNAs.

The analysis of the relationship between *upf* mutants and exosome knockdown revealed that the overlap of the upregulated transcripts is unexpectedly very narrow (six AGI transcripts) (Kurihara et al. 2009), although each analysis methodology was different. This result is pretty interesting because some but not all NMD target mRNAs should be 3′–5′ degraded through exosome function at least in yeast and animals. One possible reason is that 5′–3′ exoribonuclease activity in *Arabidopsis* may play a more powerful role in rapid degradation of NMD targets than exosome, or play redundantly with exosome.

3.2 Decapping Complex and Exoribonucleases

Decapping of mRNA is an inevitable process in the 5′–3′ mRNA decay and is executed by an well-characterized decapping complex. The complex in *Arabidopsis* consists of at least three proteins, VARICOSE, DCP1, and DCP2, which are essential for postembryonic development (Xu et al. 2006; Iwasaki et al. 2007). The in vitro experiments showed that, of these three proteins, DCP2 is a key enzyme possessing the pyrophosphatase activity for removing the cap structure, whereas DCP1 interacts with and stimulates DCP2 activity. The decapping complex is localized in several cytoplasmic granules, called processing bodies (P-bodies). DCP5, a homolog of human RNA-associated protein 55, is also localized in P-body and is required for its formation, efficient decapping reaction, and translational repression (Xu and Chua 2009).

It was showed that decay of some mRNAs was prolonged under transcription inhibitor in *dcp2* (*tdt-1*) knockdown and *varicose* (*vcs-7*) mutants compared with wild type, although some decay still occurred probably due to alternative exosome activity (Goeres et al. 2007). Those studies showed that 142 mRNAs were upregulated more than fivefold in *tdt-1* mutants using a microarray and that at least two of them possessed a cap structure. However, in a study on the relationship between decapping and NMD, we could not find any upregulated mRNAs overlapping between *tdt-1* and *upf* mutants (unpublished).

Recent report has shown that *SUPPRESSOR OF VARICOSE* (*SOV*) encoding an RNase II protein in L*er* ecotype can partially suppress severe *varicose* phenotypes of Col-0 background (Zhang et al. 2010). The SOV structure is similar to Rrp44/Dis3, an exosome-associated enzyme, which possesses a PIN endoribonuclease domain as well as RNase II domain and executes exosome-mediated RNA decay instead of inactive core complex in yeast and humans (Lebreton et al. 2008; Schaeffer et al. 2009). However, SOV does not possess any PIN domains, and SOV-CFP fusion protein showed cytoplasmic localization pattern, whereas *At*Rrp44-CFP was localized at nucleus. The above results indicate that novel RNA decay pathway redundantly functions with the decapping machinery at least in plants.

After decapping, the free 5′ end of mRNA is degraded by a specific exoribonuclease, Xrn1 in yeast. The *Arabidopsis* genome encodes three orthologs (XRN2–4) of nuclear exoribonuclease, Rat1 of yeast (Kastenmayer and Green 2000), but no ortholog of Xrn1. Especially, *XRN4* was identified as *ein5*, an ethylene-insensitive mutant, and then is required for downregulation of the mRNAs encoding the F-box proteins EBF1 and EBF2, which target EIN3 protein for degradation (Olmedo et al. 2006).

While XRN2 and XRN3, of which both possess a nuclear localization signal, are localized in the nucleus, XRN4, which does not possess any nuclear localization signals, is localized exclusively at the P-body with decapping complex in cytoplasm (Kastenmayer and Green 2000; Weber et al. 2008). The localization patterns showed that XRN2 and XRN3 have the yeast Rat1 function and XRN4 has the yeast Xrn1 function. Consistent with the prediction, XRN2 and XRN3 redundantly or independently play a role in maturation of ribosomal RNA as Rat1 does in yeast (Zakrzewska-Placzek et al. 2010). By contrast, XRN4 has a role in degrading (some but not all) 3′ products of miRNA-mediated mRNA cleavage (Souret et al. 2004; Gregory et al. 2008; German et al. 2008). In addition, all three XRN proteins act as endogenous RNA silencing suppressors (Gazzani et al. 2004; Gy et al. 2007).

Given the localization pattern of XRNs, XRN4 may be involved in degradation of NMD targets, because NMD occurs in the cytoplasm. However, tiling array analysis showed little overlap of upregulated transcripts of *upf* mutants and *xrn4* mutant (Gregory et al. 2008; Kurihara et al. unpublished data). So now, it is unknown which XRN enzyme is involved in the 5′–3′ decay during NMD in *Arabidopsis*.

4 Conclusions

In this chapter, we summarize the present picture of the plant RNA metabolism, featuring subjects around NMD machinery. One of the most fundamental topics in molecular biology is to understand how RNA molecules are degraded, since RNA itself and its behavior are involved in several biological processes. Especially, since plants do not move by themselves, they can sophisticatedly change their transcriptional profile in accordance with the environmental condition, in which several RNA decay machineries should be active. However, several noisy RNAs may be transcribed under a stressful condition (Matsui et al. 2008; Filichkin et al. 2010). Their behavior sometimes may be toxic. Therefore, it is important to reveal the basic mechanism to suppress their accumulation. In this context, the noisy ncRNAs such as NMD-targeted mlncRNAs have been reported to be transcribed but immediately degraded by several decay pathways in plants. Additional vigorous efforts and progressive technologies will be necessary to reveal the whole view of plant RNA metabolism in the future.

Acknowledgments This work was supported by a grant from RIKEN Plant Science Center (to M.S.) and by Grants-in-Aid for Scientific Research on Kiban (C) (no. 21570056) and on Priority Areas (no. 21027033) of the Ministry of Education, Culture, Sports, Science, and Technology of Japan (to M.S.). Y.K. was supported by a JSPS Postdoctoral Fellowship for Research Abroad.

References

Arciga-Reyes L, Wootton L, Kieffer M et al (2006) UPF1 is required for nonsense-mediated mRNA decay (NMD) and RNA*i in Arabidopsis*. Plant J 47:480–489

Belostotsky D (2009) Exosome complex and pervasive transcription in eukaryotic genomes. Curr Opin Cell Biol 21:352–358

Borsani O, Zhu J, Verslues PE et al (2005) Endogenous siRNAs derived from a pair of natural *cis*-antisense transcripts regulate salt tolerance in *Arabidopsis*. Cell 123:1279–1291

Chekanova JA, Shaw RJ, Wills MA et al (2000) Poly(A) tail-dependent exonuclease AtRrp41p from *Arabidopsis thaliana* rescues 5.8 S rRNA processing and mRNA decay defects of the yeast *ski6* mutant and is found in an exosome-sized complex in plant and yeast cells. J Biol Chem 275:33158–33166

Chekanova JA, Gregory BD, Reverdatto SV et al (2007) Genome-wide high-resolution mapping of exosome substrates reveals hidden features in the *Arabidopsis* transcriptome. Cell 131:1340–1353

Filichkin SA, Priest HD, Givan SA et al (2010) Genome-wide mapping of alternative splicing in *Arabidopsis thaliana*. Genome Res 20:45–58

Gazzani S, Lawrenson T, Woodward C et al (2004) A link between mRNA turnover and RNA interference in *Arabidopsis*. Science 206:1046–1048

German MA, Pillay M, Jeong D et al (2008) Global identification of microRNA-target RNA pairs by parallel analysis of RNA ends. Nat Biotechnol 26:941–946

Goeres DC, Van Norman JM, Zhang W et al (2007) Components of the *Arabidopsis* mRNA decapping complex are required for early seedling development. Plant Cell 19:1549–1564

Gregory BD, O'Malley RC, Lister R et al (2008) A link between RNA metabolism and silencing affecting *Arabidopsis* development. Dev Cell 14:854–866

Gy I, Gasciolli V, Lauressergues D et al (2007) *Arabidopsis* FIERY1, XRN2, and XRN3 are endogenous RNA silencing suppressors. Plant Cell 19:3451–3461

He F, Jacobson A (2001) Upf1p, Nmd2p, and Upf3p regulate the decapping and exonucleolytic degradation of both nonsense-containing mRNAs and wild-type mRNAs. Mol Cell Biol 21:1515–1530

Henz SR, Cumbie JS, Kasschau KD et al (2007) Distinct expression patterns of natural antisense transcripts in *Arabidopsis*. Plant Physiol 144:1247–1255

Hori K, Watanabe Y (2005) UPF3 suppresses aberrant spliced mRNA in *Arabidopsis*. Plant J 43: 530–540

Hori K, Watanabe Y (2007) Context analysis of termination codons in mRNA that are recognized by plant NMD. Plant Cell Physiol 48:1072–1078

Iwasaki S, Takeda A, Motose H et al (2007) Characterization of Arabidopsis decapping proteins AtDCP1 and AtDCP2, which are essential for post-embryonic development. FEBS Lett 581:2455–2459

Kastenmayer JP, Green PJ (2000) Novel features of the XRN-family in *Arabidopsis*: evidence that AtXRN4, one of several orthologs of nuclear Xrn2p/Rat1p, functions in the cytoplasm. Proc Natl Acad Sci USA 97:13985–13990

Kertesz S, Kerenyi Z, Merai Z et al (2006) Both introns and long 3′-UTRs operate *cis*-acting elements to trigger nonsense-mediated decay in plants. Nucleic Acids Res 34:6147–6157

Kim SH, Koroleva OA, Lewandowska D et al (2009) Aberrant mRNA transcripts and the nonsense-mediated decay proteins UPF2 and UPF3 are enriched in the *Arabidopsis* nucleolus. Plant Cell 21:2045–2057

Kurihara Y, Matsui A, Kawashima M et al (2009) Genome-wide suppression of aberrant mRNA-like noncoding RNAs by NMD in *Arabidopsis*. Proc Natl Acad Sci USA 106:2453–2458

Lebreton A, Tomecki R, Dziembowski A et al (2008) Endonucleolytic RNA cleavage by a eukaryotic exosome. Nature 456:993–996

Lejeune F, Maquat LE (2003) Nonsense-mediated mRNA decay in mammalian cells involves decapping, deadenylating, and exonucleolytic activities. Mol Cell 12:675–687

Maquat LE (2004) Nonsense-mediated mRNA decay: splicing, translation and mRNP dynamics. Nat Rev Mol Cell Biol 5:89–99

Matsui A, Ishida J, Morosawa T et al (2008) Arabidopsis transcriptome analysis under drought, cold, high-salinity and ABA treatment conditions using a tiling array. Plant Cell Physiol 49:1135–1149

Mitchell P, Tollervey D (2003) An NMD pathway in yeast involving accelerated degradation and exosome-mediated 3′→5′ degradation. Mol Cell 11:1405–1413

Nyiko T, Sonkoly B, Merai Z et al (2009) Plant upstream ORFs can trigger nonsense-mediated mRNA decay in a size-dependent manner. Plant Mol Biol 71:367–378

Olmedo G, Guo H, Gregory BD et al (2006) *ETHYLENE-INSENSITIVE5* encodes a 5′→3′ exoribonuclease required for regulation of the EIN3-targeting F-box proteins EBF1/2. Proc Natl Acad Sci USA 103:13286–13293

Riehs N, Akimcheva S, Puizina J et al (2008) *Arabidopsis* SMG7 protein is required for exit from meiosis. J Cell Sci 121:2208–2216

Saul H, Elharrar E, Gaash R et al (2009) The upstream open reading frame of the *Arabidopsis AtMHX* gene has a strong impact on transcript accumulation through the nonsense-mediated mRNA decay pathway. Plant J 60:1031–1042

Schaeffer D, Tsanova B, Barbas A et al (2009) The exosome contains domains with specific endoribonuclease, exoribonuclease and cytoplasmic mRNA decay activities. Nat Struct Mol Biol 16:56–62

Seki M, Narusaka M, Kamiya A et al (2002) Functional annotation of a full-length *Arabidopsis* cDNA collection. Science 296:141–145

Song H, Song J, Cho J et al (2009) The RNA binding protein ELF9 directly reduces *SUPPRESSOR OF OVEREXPRESSION OF CO1* transcript levels in *Arabidopsis*, possibly via nonsense-mediated mRNA decay. Plant Cell 21:1195–1211

Souret F, Kastenmayer JP, Green PJ (2004) AtXRN4 degrades mRNA in *Arabidopsis* and its substrates include selected miRNA targets. Mol Cell 15:173–183

Toyoda T, Shinozaki K (2005) Tiling array-driven elucidation of transcriptional structures based on maximum-likelihood and Markov models. Plant J 43:611–621

Toyoda T, Mochizuki Y, Player K et al (2007) OmicBrowse: a browser of multidimensional omics annotations. Bioinformatics 23:524–526

Weber C, Nover L, Fauth M (2008) Plant stress granules and mRNA processing bodies are distinct from heat stress granules. Plant J 56:517–530

Xu J, Chua N (2009) *Arabidopsis* decapping 5 is required for mRNA decapping, P-body formation, and translational repression during postembryonic development. Plant Cell 21:3270–3279

Xu J, Yang J, Niu Q et al (2006) *Arabidopsis* DCP2, DCP1, and VARICOSE form a decapping complex required for postembryonic development. Plant Cell 18:3386–3398

Yoine M, Ohto M, Onai K et al (2006) The *lba1* mutation of UPF1 RNA helicase involved in nonsense-mediated mRNA decay causes pleiotropic phenotypic changes and altered sugar signaling in *Arabidopsis*. Plant J 47:49–62

Zakrzewska-Placzek M, Souret FF, Sobczyk GJ et al (2010) *Arabidopsis thaliana* XRN2 is required for primary cleavage in the pre-ribosomal RNA. Nucleic Acids Res 38:4482–4502

Zhang X, Yazaki J, Sundaresan A et al (2006) Genome-wide high-resolution mapping and functional analysis of DNA methylation in *Arabidopsis*. Cell 126:1189–1201

Zhang W, Murphy C, Sieburth LE (2010) Conserved RNaseII domain protein functions in cytoplasmic mRNA decay and suppresses *Arabidopsis* decapping mutant phenotypes. Proc Natl Acad Sci USA 107:15981–15985

Small RNA Inheritance in Hybrids and Allopolyploids

Jie Lu and Z. Jeffrey Chen

Contents

Abstract Small RNAs, including microRNAs (miRNAs), small interfering RNAs (siRNAs), and *trans*-acting siRNAs (tasiRNAs), control gene expression and epigenetic regulation. Although the physiological and developmental roles of miRNAs and siRNAs have been extensively studied, their roles in morphological diversity among closely related species and in interspecific hybrids and allopolyploids are poorly understood. Here, we discussed recent findings of small RNA regulation with an emphasis on hybrids, interspecific hybrids, and allopolyploids. Divergence between siRNAs and inheritance of these siRNAs through maternal or paternal genome during gametogenesis may exert *trans*-acting effects on transposable elements and

Z.J. Chen (✉) and J.Lu
Section of Molecular Cell and Developmental Biology, Center for Computational Biology and Bioinformatics, The University of Texas at Austin, Austin, TX 78712, USA
and
Institute for Cellular and Molecular Biology, The University of Texas at Austin, One University Station, A-4800, Austin, TX 78712, USA
e-mail: zjchen@mail.utexas.edu

V.A. Erdmann and J. Barciszewski (eds.), *Non Coding RNAs in Plants*,
RNA Technologies, DOI 10.1007/978-3-642-19454-2_7,

on genes that are important to genomic stability and phenotypic variation. Moreover, expression changes in miRNAs and tasiRNAs between related species or parents may play a role in target gene regulation that is important to growth and development in hybrids and allopolyploids. Inheritance of small RNAs through maternal or paternal genome in interspecific hybrids and allopolyploids are reminiscent of parent-of-origin effects of small RNA regulation on offspring in heterozygous organisms including humans.

Keywords Epigenetics • Evolution • Gene expression • Polyploidy • Seed development • Small RNA

1 Introduction

Hybrids and polyploids (whole-genome duplication) are very common in many plants and some animals (Otto 2007; Leitch and Leitch 2008). Hybridization between different strains, varieties, and species often generate increased levels of heterozygosity and hybrid vigor. Hybrids between species and occasionally between genera followed by chromosome doubling can lead to allopolyploids that contain duplicate genomes originating in different species. As a result, stable allopolyploids provide permanent fixation of heterozygosity and hybrid vigor. Hybrid and allopolyploid plants often grow more vigorously and have better fitness and superior traits than the parents, which may facilitate natural selection and crop domestication. Indeed, some crops including maize and sorghum grow as hybrids, and many crops including wheat, cotton, and canola are allopolyploids.

The molecular mechanisms for phenotypic and morphological changes in hybrids and newly formed allopolyploids are poorly understood (Wendel 2000; Comai 2005; Chen 2007; Soltis and Soltis 2009). There is compelling evidence to suggest that consistent with the phenotypic changes that occur after hybridization, gene expression levels also change rapidly and dynamically in response to the "genomic shock" (McClintock 1984). In the last decade, much progress has been made to elucidate epigenetic mechanisms for changes in gene expression, including DNA methylation, histone modifications, and small RNAs (Lee and Chen 2001; Osborn et al. 2003; Salmon et al. 2005; Chen 2007; Chen et al. 2008; Martienssen 2010). Small RNAs are 20–24 nucleotide RNA molecules that play crucial roles in posttranscriptional regulation of gene expression in plants and animals. In this review, we discussed the recent findings concerning the expression and divergence of small RNAs in different species and predicted the potential roles of small RNAs in regulating genome stability and gene expression changes in hybrids and allopolyploids. A better understanding of small RNA regulation in hybrids and allopolyploids will provide novel strategies for genetic manipulation and conventional breeding to meet the growing demand of food, feed, fiber, and industrial raw materials.

2 Small RNAs in Plants

Plants produce several different classes of small RNAs from endogenous genomic loci, including microRNA (miRNA), small interfering small RNA (siRNA), tasiRNA, repeat-associated siRNA (rasiRNA), and natural antisense siRNA (nat-siRNA) (Baulcombe 2004; Vazquez 2006; Chapman and Carrington 2007; Chen 2009).

miRNA pathway is a posttranscriptional regulatory mechanism conserved in animal and plant kingdoms (Ruvkun 2001; Ambros 2004; Bartel 2009; Chen 2009). In animals, miRNAs bind to target mRNA through imperfect Waston–Crick base-pairing, which are usually located at the 3′ untranslated regions (UTRs), leading to a general mechanism for translational repression (Ambros 2004; Fabian et al. 2010; Siomi and Siomi 2010). In plants, however, miRNAs predominately mediate degradation of target mRNAs via perfect or near-perfect complementary sequences (Llave et al. 2002; Vaucheret 2006; Voinnet 2009). Plant miRNAs and siRNAs also inhibit mRNA translation (Chen 2004; Brodersen et al. 2008). miRNA precursors are transcribed from miRNA loci by RNA polymerase II and form hairpin RNA structures that are subsequently processed into mature miRNAs for degrading target mRNAs. Some non coding mRNAs targeted by miRNAs produce abundant secondary siRNAs around the cleavage sites in phased positions. These siRNAs act *in trans* to mediate the cleavage of mRNAs encoding other proteins and, thus, are denoted as tasiRNAs (Yoshikawa et al. 2005).

In addition to miRNAs and tasiRNAs, plants produce a plethora of siRNAs from transposable elements and pericentromeric repeats. Biogenesis of siRNAs from transposons requires the function of RNA polymerase IV (Pol IV), and, thus, is denoted as Pol IV siRNA (p4-siRNA). p4-siRNAs are usually 24-nt long and suppress transposon activities by mediating RNA-dependent DNA methylation (RdDM) and chromatin modification (Herr et al. 2005; Onodera et al. 2005; Chapman and Carrington 2007).

3 Conservation and Divergence of miRNAs Between Species

The mature miRNA sequences are generally conserved in animal or plant kingdom (Ambros 2004; Niwa and Slack 2007; Chen 2009), and a few are even conserved between animal and plant kingdoms (Arteaga-Vazquez et al. 2006). In *Arabidopsis*, both mature miRNA sequences and their target binding sites have very low levels of nucleotide variation and divergence compared to their flanking sequences, indicating strong purifying selection on the pairing of complementary sequences. By contrast, sequences flanking the mature miRNAs exhibit normal levels of polymorphism (Rhoades et al. 2002).

Some miRNA families, known as canonical miRNAs, are conserved across higher and lower plant species, while others are species specific (noncanonical miRNAs).

For example, among the first set of 16 miRNAs identified in *Arabidopsis*, eight have orthologs in rice (Reinhart et al. 2002). The microarray analysis of 23 miRNA families in nine species identified the expression of 11 in a gymnosperm (*Pinus resinosa*) and eight in a fern (*Ceratopteris thalictroides*), suggesting that many canonical plant miRNAs have remained essentially unchanged since before the emergence of flowering plants (Axtell and Bartel 2005). Canonical miRNAs usually have multiple copies in the genome and express at higher levels with explicit regulatory functions in development (Rajagopalan et al. 2006; Axtell et al. 2007; Fahlgren et al. 2007; Ma et al. 2010). On the contrary, noncanonical miRNAs are often species specific, encoded by single loci, and expressed at low levels, and sometime rely on DCL4 instead of DCL1 for processing. Among 199 annotated miRNA loci in miRBase, 26 miRNA families (encoded by 99 miRNA loci) are annotated in one or more non-Brassicaceae species, while 104 miRNA families (encoded by 106 loci) are only annotated in Brassicaceae species (Ma et al. 2010), suggesting that species specific miRNAs are usually single-copy genes. Their functions are more obscure, and they may exert regulatory effects in a manner that is quite different from canonical miRNAs. One example is *Arabidopsis thaliana* miR838, which is derived from a hairpin within the intron of DCL1 mRNA (Rajagopalan et al. 2006). miR838 is predicted to maintain DCL1 homeostasis through processing by miRNA biogenesis machinery, which produces a pool of truncated, nonfunctional DCL1 mRNAs when the DCL1 protein levels are high. Alternatively, many homologs of noncanonical miRNAs simply degenerate and do not have relevant functions in closely related species, suggesting that these "young" miRNAs are species specific and function only in one species, but not in others (Ma et al. 2010).

Despite the high level of sequence conservation of miRNAs in plants, the expression levels of many miRNAs are not necessarily conserved among different species. Using miRNA microarray and deep sequencing, Ha et al. (2009a, b) found that expression levels of miRNAs and tasiRNAs were highly variable in *A. thaliana* and its close relative *Arabidopsis arenosa* (Ha et al. 2009b). In leaves, among the 85 miRNAs and 23 tasiRNAs on the microarrays, 22 miRNAs (~26%) and 6 tasiRNAs (~25%) were expressed differently between these two closely related species. For example, miR163, a recently evolved 24-nt miRNA, was highly expressed in *A. thaliana* leaves, but almost undetectable in *A. arenosa*. Interestingly, the homolog of miR163 locus in *A. arenosa* generates a 23-nt mature miRNA that can be detected in inflorescences at a level of 30-fold lower than that in *A. thaliana*, suggesting rapid changes of sequence and expression of recently evolved miRNAs in closely related species.

4 Nonadditive Expression of miRNAs in Hybrids and Allotetraploids

Nonadditive gene expression is defined as deviation of the expression level of a gene in a hybrid from the mid-parent value (MPV), which is the sum of two parental alleles (null hypothesis: 1+1=2). In *Arabidopsis*, more than 15% genes are differentially expressed between *A. thaliana* and *A. arenosa*, among which ~68% are nonadditively expressed in the allotetraploids, consistent with the nonadditive phenotypic changes (Wang et al. 2006). Similar to nonadditive gene expression, Ha et al. (2009a, b) found that ~50% miRNAs and tasiRNAs are differentially expressed between *A. thaliana* and *A. arenosa* using miRNA microarrays, and ~56% differentially miRNAs are non-additively expressed in the allotetraploids of these two species (Ha et al. 2009b). Interestingly, for both gene and miRNA expression, those that are expressed higher in *A. thaliana* than in *A. arenosa* are suppressed in allotetraploids, suggesting expression dominance of *A. arenosa* miRNA genes over that of *A. thaliana* miRNA genes. It is possible that the combination of diverged progenitors' alleles leads to *cis*- and *trans*-effects on miRNA gene expression and their biogenesis genes in interspecific hybrids and new allopolyploids. For miRNAs, the posttranscriptional level includes the processing of miRNA precursors and the transportation and modification of mature miRNAs. Therefore, the nonadditive expression of miRNA biogenesis genes can lead to the nonadditive expression of miRNAs. Indeed, *DCL1* and *AGO1*, two key components of the miRNA biogenesis pathway in plants, displayed nonadditive expression levels in allotetraploids. By contrast, the genes responsible for 24-nt siRNA production showed additive expression in allotetraploids (Ha et al. 2009b).

Bateson–Dobzhansky–Muller model predicts that hybrid incompatibilities are caused by interactions between genes that have functionally diverged in respective hybridizing species (Bateson 1909; Dobzhansky 1936; Muller 1942). In *Arabidopsis*, small RNA biogenesis machinery is composed of proteins belonging to multiple gene families. For example, there are four Dicer-Like genes, six RDR genes, and ten Argonaute genes that are responsible for generating different classes of small RNAs. In *Drosophila*, genes related to RNAi involved in antiviral function (DCR2, R2D2, and AGO2) evolve significantly faster than paralogous genes with "house-keeping" functions (Obbard et al. 2006). It is very likely that independently evolved small RNA biogenesis genes in *A. arenosa* have different active sites or established interactions with a different set of proteins. When these two sets of machineries are brought together in interspecific hybrids or allotetraploids, the processing efficiency may be compromised by their divergent functions and interactions. Indeed, high-throughput sequencing analysis showed that more miRNAs were downregulated in F_7 *Arabidopsis* allotetraploids than in either of the parents (Ha et al. 2009a). The overall siRNA density in the F_1 was significantly lower than in F_7 and natural allotetraploid *Arabidopsis suecica*, suggesting the functional incompatibility of small RNA machineries immediately after hybridization, which is restored in the later generations.

The sequence divergence of miRNA precursors may also affect miRNA processing in interspecific hybrids or allotetraploids. A genome-wide comparison of small RNAs from *A. thaliana* and its related species *Arabidopsis lyrata* suggests that less conserved miRNAs are highly divergent in miRNA hairpin structures and miRNA processing precision (Ma et al. 2010). Although plant miRNAs are defined by one specific sequence from the stem-loop precursor, imprecisely processed products are observed for less-conserved miRNAs. Therefore, it will be interesting to closely examine the processing accuracy of miRNAs and recently evolved miRNAs in interspecific hybrids and allotetraploids.

Nonadditive expression of miRNA targets can be caused by nonadditive expression of miRNAs. Indeed, the expression levels of miRNAs are negatively correlated with those of nonadditively expressed miRNA target genes in the allotetraploids (Ha et al. 2009b). Interestingly, over 50% miRNA targets are among the nonadditively expressed genes. The enriched targeting of miRNAs for nonadditively expressed genes in *Arabidopsis* allotetraploids may be resulted from divergence between target site sequences. The evolutionarily conserved miRNA families are usually conserved in target complementary sites across plant species (McConnell et al. 2001; Rhoades et al. 2002; Ha et al. 2008). However, the targets of miRNAs with sequence divergence and imprecise processing tend to be variable between closely related species in *Arabidopsis* (Ma et al. 2010). Even for conserved miRNAs that have identical target complementary sites in different species, preferential targeting has been observed in the allotetraploids probably because the accessibility of target sites is affected by local RNA secondary structures (Long et al. 2007; Ha et al. 2009b).

5 Conservation and Divergence of p4-siRNAs Between Species

Unlike miRNAs, siRNAs are not conserved in plants or animals and even among closely related species. For example, plants have a specific class of siRNAs, derived from the p4-siRNA pathway that does not exist in animals. In animals, piwi-RNAs (piRNA) are functionally similar to p4-siRNAs in plants. piRNAs (24–30 nt) are derived predominantly from transposons and other repetitive sequences through a Dicer-independent mechanism (Lau et al. 2006; Vagin et al. 2006; Hartig et al. 2007). They are associated with a subset of Argonaute proteins Piwi and MiWi2 (Aravin et al. 2006, 2007; Carmell et al. 2007; Hartig et al. 2007). In *Drosophila*, the most abundant piRNAs originate from the antisense strand of transposons and preferentially interact with the Argonaute proteins Piwi and Aubergine (Aub), whereas sense-strand piRNAs associate with Argonaute 3 (Ago3) (Aravin et al. 2007; Brennecke et al. 2007; Hartig et al. 2007). Piwi, Aub, and Ago3 bind with piRNAs and trigger the cleavage of target RNAs to repress transposons (Saito et al. 2006; Vagin et al. 2006; Brennecke et al. 2007). Piwi proteins are required for male and female fertility in *Drosophila* (Lin and Spradling 1997). Interestingly, piRNAs originate from disproportionally few master regulator loci in *Drosophila*, and 81% of piRNAs are derived from 142 discrete genomic locations comprising only 3.5% of the *Drosophila* genome (Brennecke et al. 2007). piRNAs from these master regulator loci can target retrotransposons elsewhere in the genome for cleavage. For example, master regulator locus *flamenco* spanning 180 kb on X chromosome produces abundant piRNAs to suppress *gypsy*, *Idefix*, and *ZAM* retroelements (Desset et al. 2003; Brennecke et al. 2007). Importantly, *flamenco* is conserved in terms of abundance and strand orientation in closely related *Drosophila* species (*D. yakuba* and *D. erecta*) (Brennecke et al. 2007).

In *Arabidopsis*, discrete genomic loci for p4-siRNA production have also been identified (Zhang et al. 2007). One example is *INVERTED REPEAT 71* (*IR71*), a large inverted repeat that can form a hairpin structure by its Crick strand and generate siRNAs requiring all four Dicer-like proteins. Unlike in *Drosophila*, however, no conserved p4-siRNA hot spot was found between *A. thaliana* and *A. lyrata*, probably because the analysis was limited to the conserved genomic segments (syntenic regions) between two species, whereas rapidly evolved pericentromeric regions may have been excluded from the analysis (Ma et al. 2010). Pericentromeres are rich in transposons, retroelements, 5S rDNA arrays, and pseudogenes, and, thus, are predisposed to generate p4-siRNAs. Unlike euchromatic genomic segments, where *A. thaliana* and its closely related species are often collinear, pericentromeric regions are highly divergent between the closely related species. For example, in *A. thaliana*, retroelements and transposons comprise ~59% (380 kb) of 643-kp peri-CEN3 region, and ~53% (364 kb) of the 686-kb peri-CEN5 region. By contrast, mobile elements are less abundant in the other three closely related species, comprising ~27%, ~16%, and ~11% of the pericentromeres in *A. arenosa*, *C. rubella*, and *O. pumila*, respectively. The data suggest that *A. thaliana* has undergone recent and significant expansions of its pericentromeres (Hall et al. 2006). Therefore, p4-siRNAs derived from these highly diverged regions could be present in one species but absent in its related species, which may serve as species barriers.

6 p4-siRNAs in Interspecific Hybrids

Interestingly, p4-siRNAs are expressed only from maternal chromosomes in developing seeds (Mosher et al. 2009). By contrast, both alleles from maternal and paternal origins are detectable during vegetative growth, indicating that imprinting of p4-siRNAs is limited to endosperm development. A burst of p4-siRNA expression was observed at 5 days after pollination (DAP) in developing endosperm, but not in embryo, which is reminiscent of the enrichment of rasiRNAs in male gametes (Slotkin et al. 2009). In *Arabidopsis* pollen, TEs are unexpectedly reactivated and transpose only in the pollen vegetative nucleus (VN), which accompanies the sperm cells but does not provide DNA to the fertilized zygote. VN may contribute siRNAs to sperm cells and reprogram the transposon silencing in next generation. As a functional analogy, endosperm, a plant terminal organ that nourishes embryo, may also produce siRNAs to suppress transposon movement in embryo.

7 Derepression of Transposons During Interspecific Hybridization

Barbara McClintock predicted that transposons could be derepressed and mobilized in response to "genomic shock" (McClintock 1984). Interspecific hybridization between *A. thaliana* and *A. arenosa* tetraploid results in a high level of seed lethality, and the normally silenced heterochromatic element *ATHILA* was expressed from the paternal, but not the maternal, origin (Josefsson et al. 2006). Another example of transposon reactivation is found in three hybrid species of sunflowers, *Helianthus anomalus*, *H. deserticola*, and *H. paradoxus*, which are the progenies of ancient hybridization between *H. annuus* and *H. petiolaris* (Rieseberg et al. 2003). Strikingly, the genomes of these hybrid species are at least ~50% larger than that of their parents despite the same chromosome numbers (Ungerer et al. 2006). The expansion of genome size in hybrid species is largely explained by the proliferation of Ty3/gypsy-like long terminal repeat (LTR) sequences in hybrids, suggesting the derepression of these elements by interspecific hybridization.

It is tempting to speculate that sequence divergence or higher copy numbers of transposons in paternal genome can escape the suppression from the siRNAs produced by maternal genome. Recent studies on *Drosophila* hybrid dysgenesis have provided strong evidence for the crucial role of small RNAs in transposon mobilization in hybrids. Hybrid dysgenesis has been characterized in many *Drosophila* species (Engels and Preston 1979; Kidwell 1981; Bingham et al. 1982). For example, in *Drosophila melanogaster*, the progeny of crosses between wild-caught males and laboratory-strain females are sterile, whereas the genetically identical progeny of the reciprocal cross remain fertile (Picard 1976; Kidwell 1977). This was attributable to the mobilization of P-element or I-element transposons, which are present in wild-caught flies but absent from laboratory strains, leading to defects in gametogenesis (Pelisson 1981; Rubin et al. 1982; Kidwell 1983; Bucheton et al. 1984; Castro and Carareto 2004; Chambeyron and Bucheton 2005). In germ-line cells, piRNAs epigenetically repress the mobilization of transposons, which are crucial to normal gametogenesis (Brennecke et al. 2008). These piRNAs are maternally deposited into oocytes of the daughters. Laboratory-strain females lacking P-element and I-element are not able to deposit enough piRNAs to their daughters' oocytes so that the transposons from paternal chromosomes are mobilized and disrupt female gametogenesis. Similarly, the maternal inheritance of p4-siRNA in *Arabidopsis* endosperm may suggest a role of these p4-siRNAs in hybrid incompatibility.

8 siRNAs in Nucleolar Dominance and Genomic Imprinting

In addition to repressing transposable elements and heterochromatic repeats, recent studies have revealed an important role for siRNAs in gene expression in interspecific hybrids. Nucleolar dominance is an epigenetic phenomenon in which the rRNA genes of one progenitor are silenced in interspecific hybrids of plants or animals, independent of the parent-of-origin effects (Reeder 1985; Pikaard 2000). The siRNA biogenesis genes RDR2 and DCL3 are required for the silencing of rRNA genes from *A. thaliana* genome in natural *Arabidopsis* allotetraploid *A. suecica*. Knockdown of RDR2 and DCL3 in *A. suecica* disrupts the nucleolar dominance and restores the expression of rRNA genes from the *A. thaliana* loci (Preuss et al. 2008). Intergenic spacers of rRNA genes generate 24-nt siRNAs from both strands, which probably specify the *de novo* DNA methylation patterns of corresponding rDNA loci (Finigan and Martienssen 2008). It remains to be determined why and how the siRNAs only target *A. thaliana* rRNA loci without affecting *A. arenosa* loci.

Genomic imprinting is another epigenetic phenomenon that refers to the expression of one parental allele of a gene, which is dependent on parent of origin, in contrast to nucleolar dominance. Tandem repeats adjacent to the coding regions are a common feature of known imprinted genes in *Arabidopsis* (Gehring et al. 2006; Kinoshita et al. 2007; Villar et al. 2009). Studies on maternally imprinted *FWA* and paternally imprinted *PHE1* showed that the tandem repeats near the genes are enriched in p4-siRNAs and are necessary for the silencing of the transgene alleles (Chan et al. 2006; Kinoshita et al. 2007; Villar et al. 2009). These findings in nucleolar dominance and genomic imprinting collectively suggest a role for siRNAs in reprogramming gene expression patterns in heterozygous organisms, hybrids, and allopolyploids.

The modes of action for siRNAs on gene expression may not be limited to the transposons or tandem repeats in the upstream or downstream of a gene. In *Arabidopsis* genome, it is estimated that ~7.8% expressed genes contained a region with close similarity to a known transposon sequence. It is likely for the p4-siRNAs derived from a transposon family to target a protein-coding gene in *trans* (Lockton and Gaut 2009). However, in *Arabidopsis*, mutants deficient in p4-siRNA biogenesis, such as *nrpd1a*, *rdr2*, and *dcl3*, do not have obvious developmental abnormality (Mosher et al. 2009), suggesting that transposon insertions into functionally important genes are selected against. Alternatively, *Arabidopsis* genome has a relatively low amount of transposons (~10%) (Arabidopsis Genome Initiative 2000), and most transposons are immobile (Tsukahara et al. 2009). Indeed, maize has a much higher proportion of transposons (60–80%) in the genome (SanMiguel et al. 1996; Messing et al. 2004), and the mutation of *RDR2* causes dramatic changes in gene expression and shoot apical meristem morphology, suggesting a role of p4-siRNAs in overall gene expression regulation (Jia et al. 2009).

9 Roles for miRNAs and p4-siRNAs in Hybrid Formation and Development

Different biogenesis and functions of miRNAs, tasiRNAs, and p4-siRNAs render them different roles in the formation and development of interspecific hybrids and allopolyploids. miRNAs and tasiRNAs are involved in cell patterning, organ identity, and development timing (Aukerman and Sakai 2003; Palatnik et al. 2003; Chen 2004; Mallory et al. 2004a, b). Maternal inheritance of p4-siRNAs may affect genome stability and inheritance. We propose a model to explain how miRNAs and p4-siRNAs are reprogrammed to mediate genomic incompatibility and phenotypic variations in interspecific hybrids and allopolyploids (Fig. 1). Sequences and functions of miRNAs and tasiRNAs are relatively conserved in closely related species. Different species may gain new expression patterns as a consequence of divergence in regulatory *cis*-elements or emergence of species-specific *trans*-acting factors. The *cis* and *trans*

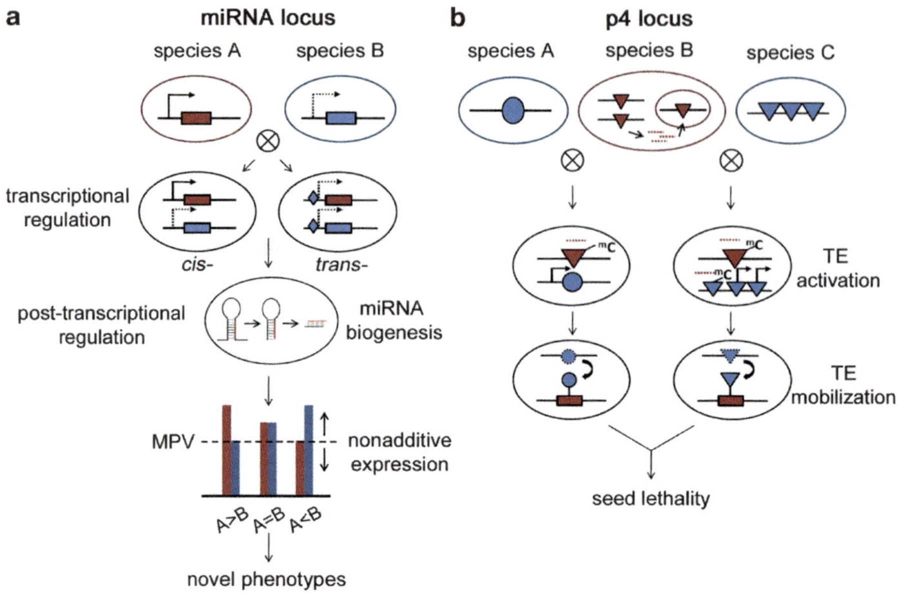

Fig. 1 A model for small RNA regulation in interspecific hybrids and allopolyploids. (**a**) Divergent miRNA loci from two different species are subject to both transcriptional and posttranscriptional regulation. After hybridization, the *cis*- and *trans*- elements regulate the transcription of miRNA genes from both species, which may result in nonadditive expression of miRNA precursors. During miRNA precursor processing, the nonadditive expression of the genes encoding miRNA biogenesis enzymes may lead to the nonadditive accumulation of mature miRNAs. The functional divergence of miRNA processing factors can also cause the incompatibility of small RNA biogenesis machineries in hybrids, accountable for the overall reduction of small RNAs in the newly formed *Arabidopsis* allotetraploids. Nonadditive expression of miRNAs is associated with the nonadditive expression of some miRNA target genes in hybrids and allopolyploids, which may lead to the reprogramming of downstream genes and mediate the emergence of novel phenotypes in newly formed hybrids and allopolyploids. *Oval*: female gamete (*red*) from species A and male gamete (*blue*) from species B; *diamond*: *trans*-element from species B; *square*: miRNA locus from species A (*red*) and species B (*blue*); *bold arrow*: enhanced gene expression; *dashed arrow*: weakened gene expression. (**b**) P4-siRNA loci are highly divergent in DNA sequences and copy numbers among related species. Upon hybridization, maternally derived p4-siRNAs in endosperm may suppress the transposon activities in embryo through RdDM. The transposons escaped from the targeting of maternal siRNAs due to the sequence divergence or increased copy numbers in paternal genome can be reactivated and transposed to the embryogenesis genes, leading to postzygotic sterility and seed abortion. *Oval*: female gamete (*red*) from species B or male gamete (*blue*) from species A or C; *red dashed line*: maternally derived siRNAs from species B; *triangles*: p4-siRNA loci in maternal (*red*) genome from species B or in paternal (*blue*) genome from species C; *circle*: diverged p4-siRNA sequences in paternal genome from species A; ^{m}C: cytosine methylation; *square*: embryogenesis genes

regulatory effects on miRNA and tasiRNA loci result in transcriptional activation or suppression of precursors in interspecific hybrids or allopolyploids. In addition to transcriptional regulation, posttranscriptional regulation may also affect biogenesis and functions of miRNAs and tasiRNAs. The complex biogenesis pathways are responsible for accumulation of mature miRNAs and tasiRNAs. Functional divergence and nonadditive expression of RNAi biogenesis genes between species and in interspecific hybrids or allopolyploids may directly affect miRNA and tasiRNA abundance. Both transcriptional and posttranscriptional regulation can cause nonadditive accumulation of miRNAs and tasiRNAs in interspecific hybrids and allopolyploids. Many miRNA and tasiRNA targets encode transcription factors that are important for growth and development. For example, miR156 and miR172 act sequentially to control the flowering time in *Arabidopsis* (Wu et al. 2009). miR164 is associated with cellular senescence and aging (Kim et al. 2009). The nonadditive expression of these miRNAs and their targets may provide insights into late flowering and increased fitness that are often observed in allopolyploids.

p4-siRNAs are likely to mediate interspecies compatibility. Heterochromatic transposons evolve faster than euchromatic genes and diverge rapidly between strains, varieties, and species. p4-siRNA originating from heterochromatic regions may serve as species barriers during hybrid formation. Maternal inheritance of p4-siRNAs suggests that copy number and sequence divergence between transposons in paternal genome are important to hybrid compatibility between two species. One model suggests that maternal p4-siRNAs in central cells are provided for the zygotes to suppress transposon activities (Slotkin et al. 2009). The two maternal nuclei in a central cell undergo genome-wide demethylation, but this does not occur in the egg cell or sperm. These siRNAs guide *de novo* methylation to the corresponding sequences that are consequently silenced. If the paternal genome has additional copies of a transposon or transposons that are divergent or absent in the maternal genome, this will lead to the incompatibility of maternal and paternal siRNAs in offspring. As a result, some transposons will be mobilized into embryogenesis genes or random genomic locations, leading to genomic instability and possibly postzygotic lethality, as shown in many interspecific hybrids and early generations of allopolyploids. Collectively, small RNAs are associated with gene expression changes as well as epigenetic reprogramming, which shape the physiological and morphological renovation of interspecies hybrids and allopolyploids. Although the data in general hybrids are not well documented, siRNAs and miRNAs may also offer similar mechanisms for changes in seed fertility, development, and growth vigor in hybrids.

10 Future Perspectives

Small RNAs play important roles in epigenetic regulation of gene expression. miRNAs control developmental programs and morphological traits, while siRNAs maintain genome stability and affect the expression of genes that are often associated with transposons. Interspecific hybridization brings together two genomes with diverged sequences and expression levels of small RNAs in the same cell. As a result, nonadditive expression occurs in rDNA loci and protein-coding genes including small RNAs. Nonadditive expression of small RNAs including miRNAs is regulated at transcriptional and posttranscriptional levels. Many questions remain to be addressed.

What are the relative contributions of *cis*-regulatory elements in miRNA promoters and *trans*-acting factors to the nonadditive expression of miRNAs? Do homologs of small RNA biogenesis genes in the closely related species, *A. thaliana* and *A. arenosa*, have similar or different functions and interacting factors that modulate transcriptional and posttranscriptional regulation? How do nonadditively expressed miRNAs affect developmental and morphological variations, including those in leaf shape, plant stature, flowering time, and senescence in interspecific hybrids and allopolyploids? How do maternally inherited p4-siRNAs affect genomic stability and transposon activity in response to the "genomic shock" How many p4-siRNA loci are divergent in closely related species, leading to species-specific p4-siRNA loci? Do the copy numbers of transposon families vary in closely related species and affect siRNA expression levels? How do divergent p4-siRNAs affect postzygotic lethality in interspecific hybrids? Can we ameliorate the hybrid seed lethality by expressing paternal genome-specific p4-siRNA loci in female gametophytes? Although our work is focused on small RNAs, similar mechanisms may also account for the inheritance of other non coding RNAs in interspecific hybrids and allopolyploids. A better understanding of small RNA regulation in hybrids and allopolyploids will help us effectively select the best combinations of parents for producing hybrids and allopolyploid plants and manipulate small RNA expression to overcome species barriers and produce "super hybrids" for food, feed, fiber, and fuels.

Acknowledgment The work is supported by the grants from the National Science Foundation Plant Genome Research Program (DBI0733857) and National Institutes of Health (GM067015), USA to ZJC.

References

Ambros V (2004) The functions of animal microRNAs. Nature 431:350–355

Arabidopsis Genome Initiative (2000) Analysis of the genome sequence of the flowering plant *Arabidopsis thaliana*. Nature 408:796–815

Aravin A, Gaidatzis D, Pfeffer S et al (2006) A novel class of small RNAs bind to MILI protein in mouse testes. Nature 442:203–207

Aravin AA, Hannon GJ, Brennecke J (2007) The Piwi-piRNA pathway provides an adaptive defense in the transposon arms race. Science 318:761–764

Arteaga-Vazquez M, Caballero-Perez J, Vielle-Calzada JP (2006) A family of microRNAs present in plants and animals. Plant Cell 18:3355–3369

Aukerman MJ, Sakai H (2003) Regulation of flowering time and floral organ identity by a MicroRNA and its APETALA2-like target genes. Plant Cell 15:2730–2741

Axtell MJ, Bartel DP (2005) Antiquity of microRNAs and their targets in land plants. Plant Cell 17:1658–1673

Axtell MJ, Snyder JA, Bartel DP (2007) Common functions for diverse small RNAs of land plants. Plant Cell 19:1750–1769

Bartel DP (2009) MicroRNAs: target recognition and regulatory functions. Cell 136:215–233

Bateson W (1909) Heredity and variation in modern lights. In: Seward AC (ed) Darwin and modern science. Cambridge University Press, Cambridge, pp 85–101

Baulcombe D (2004) RNA silencing in plants. Nature 431:356–363

Bingham PM, Kidwell MG, Rubin GM (1982) The molecular basis of P-M hybrid dysgenesis: the role of the P element, a P-strain-specific transposon family. Cell 29:995–1004

Brennecke J, Aravin AA, Stark A et al (2007) Discrete small RNA-generating loci as master regulators of transposon activity in Drosophila. Cell 128:1089–1103

Brennecke J, Malone CD, Aravin AA et al (2008) An epigenetic role for maternally inherited piRNAs in transposon silencing. Science 322:1387–1392

Brodersen P, Sakvarelidze-Achard L, Bruun-Rasmussen M et al (2008) Widespread translational inhibition by plant miRNAs and siRNAs. Science 320:1185–1190

Bucheton A, Paro R, Sang HM et al (1984) The molecular basis of I-R hybrid dysgenesis in *Drosophila melanogaster*: identification, cloning, and properties of the I factor. Cell 38:153–163

Carmell MA, Girard A, van de Kant HJ et al (2007) MIWI2 is essential for spermatogenesis and repression of transposons in the mouse male germline. Dev Cell 12:503–514

Castro JP, Carareto CM (2004) *Drosophila melanogaster* P transposable elements: mechanisms of transposition and regulation. Genetica 121:107–118

Chambeyron S, Bucheton A (2005) I elements in Drosophila: in vivo retrotransposition and regulation. Cytogenet Genome Res 110:215–222

Chan SW, Zhang X, Bernatavichute YV et al (2006) Two-step recruitment of RNA-directed DNA methylation to tandem repeats. PLoS Biol 4:e363

Chapman EJ, Carrington JC (2007) Specialization and evolution of endogenous small RNA pathways. Nat Rev Genet 8:884–896

Chen X (2004) A microRNA as a translational repressor of APETALA2 in *Arabidopsis* flower development. Science 303:2022–2025

Chen ZJ (2007) Genetic and epigenetic mechanisms for gene expression and phenotypic variation in plant polyploids. Annu Rev Plant Biol 58:377–406

Chen X (2009) Small RNAs and their roles in plant development. Annu Rev Cell Dev Biol 25:21–44

Chen M, Ha M, Lackey E et al (2008) RNAi of met1 reduces DNA methylation and induces genome-specific changes in gene expression and centromeric small RNA accumulation in Arabidopsis allopolyploids. Genetics 178:1845–1858

Comai L (2005) The advantages and disadvantages of being polyploid. Nat Rev Gen 6(11):836–846

Desset S, Meignin C, Dastugue B et al (2003) COM, a heterochromatic locus governing the control of independent endogenous retroviruses from *Drosophila melanogaster*. Genetics 164:501–509

Dobzhansky T (1936) Studies on hybrid sterility. II. Localization of sterility factors in Drosophila Pseudoobscura hybrids. Genetics 21:113–135

Engels WR, Preston CR (1979) Hybrid dysgenesis in *Drosophila melanogaster*: the biology of female and male sterility. Genetics 92:161–174

Fabian MR, Sonenberg N, Filipowicz W (2010) Regulation of mRNA translation and stability by microRNAs. Annu Rev Biochem 79:351–379

Fahlgren N, Howell MD, Kasschau KD et al (2007) High-throughput sequencing of Arabidopsis microRNAs: evidence for frequent birth and death of miRNA genes. PLoS ONE 2:e219

Finigan P, Martienssen RA (2008) Nucleolar dominance and DNA methylation directed by small interfering RNA. Mol Cell 32:753–754

Gehring M, Huh JH, Hsieh TF et al (2006) DEMETER DNA glycosylase establishes MEDEA polycomb gene self-imprinting by allele-specific demethylation. Cell 124:495–506

Ha M, Pang M, Agarwal V et al (2008) Interspecies regulation of microRNAs and their targets. Biochim Biophys Acta 1779:735–742

Ha M, Kim ED, Chen ZJ (2009a) Duplicate genes increase expression diversity in closely related species and allopolyploids. Proc Natl Acad Sci USA 106:2295–2300

Ha M, Lu J, Tian L et al (2009b) Small RNAs serve as a genetic buffer against genomic shock in Arabidopsis interspecific hybrids and allopolyploids. Proc Natl Acad Sci USA 106(42): 17835–17840

Hall AE, Kettler GC, Preuss D (2006) Dynamic evolution at pericentromeres. Genome Res 16: 355–364

Hartig JV, Tomari Y, Forstemann K (2007) piRNAs – the ancient hunters of genome invaders. Genes Dev 21:1707–1713
Herr AJ, Jensen MB, Dalmay T et al (2005) RNA polymerase IV directs silencing of endogenous DNA. Science 308:118–120
Jia Y, Lisch DR, Ohtsu K et al (2009) Loss of RNA-dependent RNA polymerase 2 (RDR2) function causes widespread and unexpected changes in the expression of transposons, genes, and 24-nt small RNAs. PLoS Genet 5:e1000737
Josefsson C, Dilkes B, Comai L (2006) Parent-dependent loss of gene silencing during interspecies hybridization. Curr Biol 16:1322–1328
Kidwell MG (1977) Reciprocal differences in female recombination associated with hybrid dysgenesis in *Drosophila melanogaster*. Genet Res 30:77–88
Kidwell MG (1981) Hybrid dysgenesis in *Drosophila melanogaster*: the genetics of cytotype determination in a neutral strain. Genetics 98:275–290
Kidwell MG (1983) Hybrid dysgenesis in DROSOPHILA MELANOGASTER: factors affecting chromosomal contamination in the P-M system. Genetics 104:317–341
Kim JH, Woo HR, Kim J et al (2009) Trifurcate feed-forward regulation of age-dependent cell death involving miR164 in Arabidopsis. Science 323:1053–1057
Kinoshita Y, Saze H, Kinoshita T et al (2007) Control of FWA gene silencing in *Arabidopsis thaliana* by SINE-related direct repeats. Plant J 49:38–45
Lau NC, Seto AG, Kim J et al (2006) Characterization of the piRNA complex from rat testes. Science 313:363–367
Lee HS, Chen ZJ (2001) Protein-coding genes are epigenetically regulated in *Arabidopsis* polyploids. Proc Natl Acad Sci USA 98:6753–6758
Leitch AR, Leitch IJ (2008) Genomic plasticity and the diversity of polyploid plants. Science 320:481–483
Lin H, Spradling AC (1997) A novel group of pumilio mutations affects the asymmetric division of germline stem cells in the Drosophila ovary. Development 124:2463–2476
Llave C, Xie Z, Kasschau KD et al (2002) Cleavage of Scarecrow-like mRNA targets directed by a class of Arabidopsis miRNA. Science 297:2053–2056
Lockton S, Gaut BS (2009) The contribution of transposable elements to expressed coding sequence in *Arabidopsis thaliana*. J Mol Evol 68:80–89
Long D, Lee R, Williams P et al (2007) Potent effect of target structure on microRNA function. Nat Struct Mol Biol 14:287–294
Ma Z, Coruh C, Axtell MJ (2010) *Arabidopsis lyrata* small RNAs: transient MIRNA and small interfering RNA loci within the Arabidopsis genus. Plant Cell 22:1090–1103
Mallory AC, Dugas DV, Bartel DP et al (2004a) MicroRNA regulation of NAC-domain targets is required for proper formation and separation of adjacent embryonic, vegetative, and floral organs. Curr Biol 14:1035–1046
Mallory AC, Reinhart BJ, Jones-Rhoades MW et al (2004b) MicroRNA control of PHABULOSA in leaf development: importance of pairing to the microRNA 5′ region. EMBO J 23:3356–3364
Martienssen RA (2010) Heterochromatin, small RNA and post-fertilization dysgenesis in allopolyploid and interploid hybrids of Arabidopsis. New Phytol 186:46–53
McClintock B (1984) The significance of responses of the genome to challenge. Science 226:792–801
McConnell JR, Emery J, Eshed Y et al (2001) Role of PHABULOSA and PHAVOLUTA in determining radial patterning in shoots. Nature 411:709–713
Messing J, Bharti AK, Karlowski WM et al (2004) Sequence composition and genome organization of maize. Proc Natl Acad Sci USA 101:14349–14354
Mosher RA, Melnyk CW, Kelly KA et al (2009) Uniparental expression of PolIV-dependent siRNAs in developing endosperm of Arabidopsis. Nature 460:283–286
Muller HJ (1942) Isolating mechanisms, evolution and temperature. Biol Symp 6:71–125
Niwa R, Slack FJ (2007) The evolution of animal microRNA function. Curr Opin Genet Dev 17:145–150
Obbard DJ, Jiggins FM, Halligan DL et al (2006) Natural selection drives extremely rapid evolution in antiviral RNAi genes. Curr Biol 16:580–585

Onodera Y, Haag JR, Ream T et al (2005) Plant nuclear RNA polymerase IV mediates siRNA and DNA methylation-dependent heterochromatin formation. Cell 120:613–622

Osborn TC, Pires JC, Birchler JA et al (2003) Understanding mechanisms of novel gene expression in polyploids. Trends Genet 19:141–147

Otto SP (2007) The evolutionary consequences of polyploidy. Cell 131:452–462

Palatnik JF, Allen E, Wu X et al (2003) Control of leaf morphogenesis by microRNAs. Nature 425:257–263

Pelisson A (1981) The I–R system of hybrid dysgenesis in *Drosophila melanogaster*: are I factor insertions responsible for the mutator effect of the I–R interaction? Mol Gen Genet 183:123–129

Picard G (1976) Non-mendelian female sterility in *Drosophila melanogaster*: hereditary transmission of I factor. Genetics 83:107–123

Pikaard CS (2000) The epigenetics of nucleolar dominance. Trends Genet 16:495–500

Preuss SB, Costa-Nunes P, Tucker S et al (2008) Multimegabase silencing in nucleolar dominance involves siRNA-directed DNA methylation and specific methylcytosine-binding proteins. Mol Cell 32:673–684

Rajagopalan R, Vaucheret H, Trejo J et al (2006) A diverse and evolutionarily fluid set of microRNAs in *Arabidopsis thaliana*. Genes Dev 20:3407–3425

Reeder RH (1985) Mechanisms of nucleolar dominance in animals and plants. J Cell Biol 101:2013–2016

Reinhart BJ, Weinstein EG, Rhoades MW et al (2002) MicroRNAs in plants. Genes Dev 16:1616–1626

Rhoades MW, Reinhart BJ, Lim LP et al (2002) Prediction of plant microRNA targets. Cell 110:513–520

Rieseberg LH, Raymond O, Rosenthal DM et al (2003) Major ecological transitions in wild sunflowers facilitated by hybridization. Science 301:1211–1216

Rubin GM, Kidwell MG, Bingham PM (1982) The molecular basis of P-M hybrid dysgenesis: the nature of induced mutations. Cell 29:987–994

Ruvkun G (2001) Glimpses of a tiny RNA world. Science 294:797–799

Saito K, Nishida KM, Mori T et al (2006) Specific association of Piwi with rasiRNAs derived from retrotransposon and heterochromatic regions in the Drosophila genome. Genes Dev 20:2214–2222

Salmon A, Ainouche ML, Wendel JF (2005) Genetic and epigenetic consequences of recent hybridization and polyploidy in Spartina (Poaceae). Mol Ecol 14:1163–1175

SanMiguel P, Tikhonov A, Jin Y-K et al (1996) Nested retrotransposons in the intergenic regions of the maize genome. Science 274:765–768

Siomi H, Siomi MC (2010) Posttranscriptional regulation of microRNA biogenesis in animals. Mol Cell 38:323–332

Slotkin RK, Vaughn M, Borges F et al (2009) Epigenetic reprogramming and small RNA silencing of transposable elements in pollen. Cell 136:461–472

Soltis PS, Soltis DE (2009) The role of hybridization in plant speciation. Annu Rev Plant Biol 60:561–588

Tsukahara S, Kobayashi A, Kawabe A et al (2009) Bursts of retrotransposition reproduced in Arabidopsis. Nature 461:423–426

Ungerer MC, Strakosh SC, Zhen Y (2006) Genome expansion in three hybrid sunflower species is associated with retrotransposon proliferation. Curr Biol 16:R872–R873

Vagin VV, Sigova A, Li C et al (2006) A distinct small RNA pathway silences selfish genetic elements in the germline. Science 313:320–324

Vaucheret H (2006) Post-transcriptional small RNA pathways in plants: mechanisms and regulations. Genes Dev 20:759–771

Vazquez F (2006) Arabidopsis endogenous small RNAs: highways and byways. Trends Plant Sci 11:460–468

Villar CB, Erilova A, Makarevich G et al (2009) Control of PHERES1 imprinting in Arabidopsis by direct tandem repeats. Mol Plant 2:654–660

Voinnet O (2009) Origin, biogenesis, and activity of plant microRNAs. Cell 136:669–687

Wang J, Tian L, Lee HS et al (2006) Genomewide nonadditive gene regulation in *Arabidopsis* allotetraploids. Genetics 172:507–517

Wendel JF (2000) Genome evolution in polyploids. Plant Mol Biol 42:225–249

Wu G, Park MY, Conway SR et al (2009) The sequential action of miR156 and miR172 regulates developmental timing in Arabidopsis. Cell 138:750–759

Yoshikawa M, Peragine A, Park MY et al (2005) A pathway for the biogenesis of trans-acting siRNAs in Arabidopsis. Genes Dev 19:2164–2175

Zhang X, Henderson IR, Lu C et al (2007) Role of RNA polymerase IV in plant small RNA metabolism. Proc Natl Acad Sci USA 104:4536–4541

Phased Small RNAs in Rice

Lewis H. Bowman, Cameron Johnson, Gail Pruss, and Vicki Vance

Contents

Abstract Small RNAs have emerged as important regulators of gene expression in eukaryotic organisms. Plant small RNAs are cleaved from longer fully or partially double-stranded RNA by Dicer. In most small RNA pathways, populations of overlapping small RNAs are produced. However, in some pathways double-stranded RNA molecules having unique start sites for cleavage are generated, and processive cleavage of these double-stranded RNAs generates a population of non-overlapping, phased small RNAs. Rice contains at least 40-fold more phased small RNA-generating

L.H. Bowman (✉), G. Pruss, and V. Vance
Department of Biological Sciences, University of South Carolina, Columbia, SC 29208, USA
e-mail: lbowman@mailbox.sc.edu; pruss@biol.sc.edu; vvance@mailbox.sc.edu

C. Johnson
Section of Plant Biology, College of Biological Sciences,
University of California Davis, Davis, CA 95616, USA
e-mail: camjohnson@ucdavis.edu

V.A. Erdmann and J. Barciszewski (eds.), *Non Coding RNAs in Plants*,
RNA Technologies, DOI 10.1007/978-3-642-19454-2_8,

loci than Arabidopsis, and mutations in genes involved in the biogenesis of phased small RNAs have much more severe phenotypes in rice and maize than in Arabidopsis. These considerations suggest that pathways producing phased small RNAs play a more important role in monocots than in Arabidopsis. To put this new development in context here we review the biogenesis and functions of phased small RNAs in rice.

Keywords Inflorescence development • MicroRNA • Phased small RNAs • Rice • *Trans*-acting si-RNA

1 Introduction to Phased Small RNAs

The discovery of endogenous small RNAs and the recognition that these small RNA molecules are important regulators of gene expression in eukaryotes have revolutionized our understanding of the roles of RNA in eukaryotic organisms. Development, genome stability, and response to both biotic and abiotic stress are just some of the processes in which regulation by endogenous small RNAs has been shown to play an important role (for recent reviews see Eamens et al. 2008; Girard and Hannon 2008; Mosher and Baulcombe 2008; Poethig 2009; Ramachandran and Chen 2008; Voinnet 2009). Most recently, the advent of new deep-sequencing technologies has enabled the identification of novel small RNA generating loci and promises to help further reveal the spectrum of endogenous small RNAs and their functions in eukaryotic organisms (Kircher and Kelso 2010).

Most small RNAs are 21–24 nucleotides (nt) in length and are produced by Dicer ribonuclease cleavage of double-stranded (ds) RNA. The dsRNA can arise from perfectly or partially self-complementary sequence in single-stranded RNA, from regions of overlapping transcription, or from synthesis of a second strand by RNA-dependent RNA polymerase (RDR). The two major types of small RNA are micro (mi) RNAs and short interfering (si) RNAs. miRNAs derive from unique genetic loci that produce no protein product, and the sequences of many miRNAs are conserved across species. In contrast to miRNAs, siRNAs can be produced from any gene and, in general, are not characterized by unique sequences. There is a class of siRNAs that are more uniquely defined, however, and these are the *trans*-acting (ta) siRNAs in plants, which were the first examples of phased small RNAs to be discovered (reviewed in Allen and Howell 2010; Vaucheret 2005; Voinnet 2008).

Phased small RNAs are generated by processive cleavage by Dicer starting from a unique site on the dsRNA. A unique origin for the first cleavage is essential for generating a phased set, as cleavage from random starting sites would produce overlapping populations of small RNAs that would not be in phase. So far, two general mechanisms have been identified for generating a unique origin for Dicer cleavage. Structural features of a partially self-complementary transcript can direct Dicer to cleave at a specific site (Fig. 1a). This is the mechanism used to specify cleavage of a miRNA from its precursor transcript. A unique origin can also be set by small

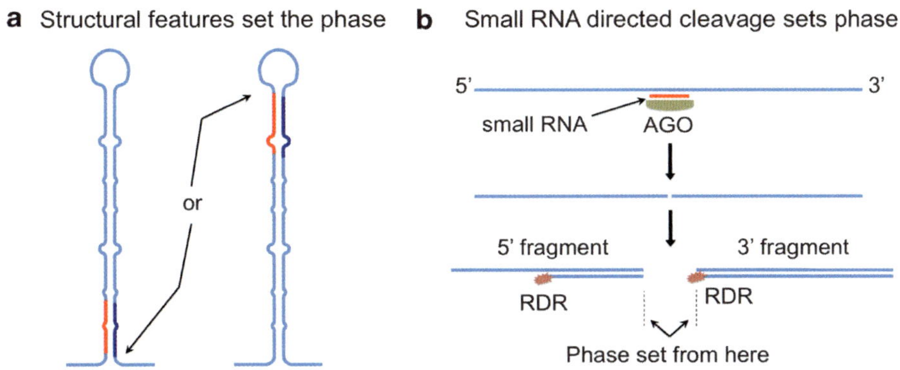

Fig. 1 Setting the phase of phased small RNAs. (**a**) The phase of small RNAs derived from hairpin RNAs is set by Dicer, which recognizes structural features of the hairpin RNA. The initial cleavage can be proximal or distal to the loop. (**b**) The phase of small RNAs for pathways employing RDR to generate dsRNA is set by a small RNA-directed cleavage of the precursor

RNA-directed cleavage of single-stranded (ss) RNA, generating a 5′ or 3′-fragment that can be used as a template for RDR to generate dsRNA having a unique end (Fig. 1b). This is the mechanism used to set the phase of ta-siRNAs. The small RNA that directs the cleavage in the case of ta-siRNAs is a miRNA, and Argonaute proteins are the enzymes that carry out small RNA-directed cleavage of ssRNA.

Phased small RNAs were initially thought to comprise only a minor component of endogenous small RNAs, as the first classes of phased small RNAs discovered had only a small number of members (Howell et al. 2007). However, our high-throughput sequencing of small RNA populations from different tissues in rice has identified a large number of previously unreported groups or clusters of phased 21 and 24-nt small RNAs that are preferentially expressed in the inflorescence of rice plants (Johnson et al. 2009). This finding raises the interesting possibility that phased small RNAs might play a much larger and more important role in plants than previously thought. In the remainder of this review, we discuss the classes of phased endogenous small RNAs in plants, focusing on those expressed in rice.

2 Phased Small RNAs Derived from Fold-Back Structures in RNA

2.1 Phased Small RNAs Arising from miRNA Precursors

miRNAs are produced from precursor transcripts that have regions of self complementarity and can adopt a partially double-stranded structure (Figs. 1a and 2a). Many miRNA precursors give rise not only to their specific miRNA but also to other small RNAs known as miRNA-like RNAs (Fahlgren et al. 2007; Kurihara and Watanabe 2004; Rajagopalan et al. 2006; Zhang et al. 2010). In many cases, these miRNA-like RNAs are in phase, and the Dicer cleavage that produces the miRNA sets the phasing (Fig. 2a). Most miRNA precursors are cleaved at the base of the miRNA hairpin, and this cleavage is dependent on a 15-nt imperfect stem at the base of the precursor (Cuperus et al. 2010b; Mateos et al. 2010; Werner et al. 2010; Song et al. 2010; Kurihara and Watanabe 2004; Vaucheret et al. 2006). However, for some miRNAs, such as miRNA and miRNA, features of the loop direct cleavage of the

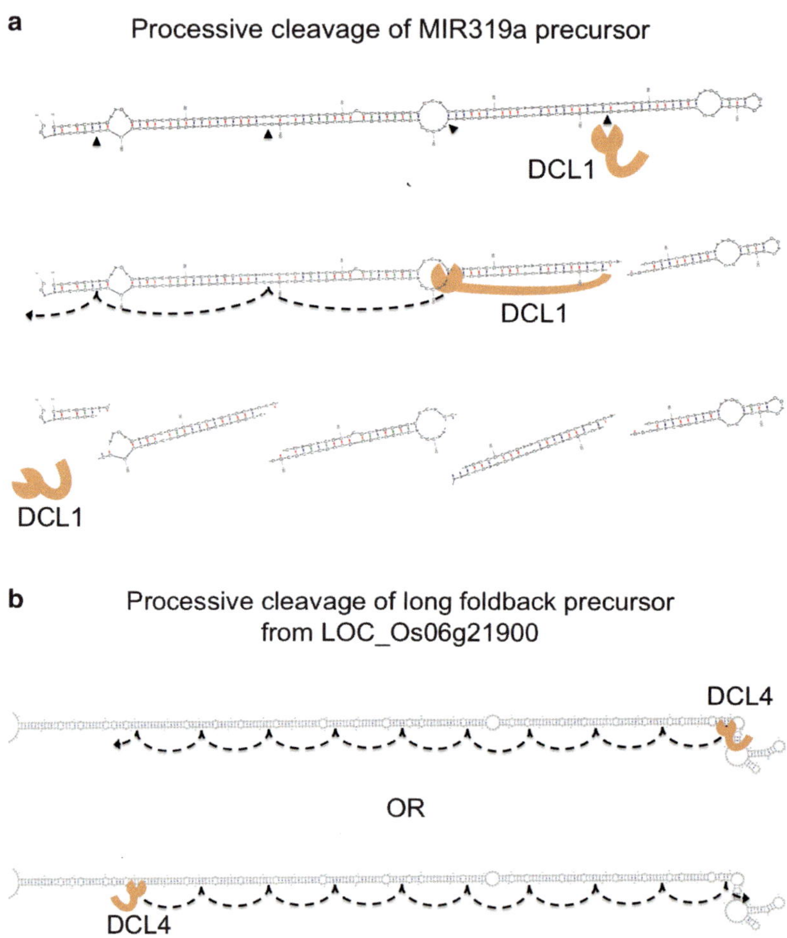

Fig. 2 Generation of small RNAs from hairpin RNAs. (**a**) The generation of miRNA-like RNAs from the miR319a precursor is shown. The initial loop proximal cleavage of the precursor is followed by processive cleavage by DCL1. (**b**) The generation of phased small RNAs from a hairpin derived from a region overlapping Os06g21900 is shown. It is not known if the initial cleavage of the hairpin is loop proximal or loop distal

miRNA precursor (Fig. 2a) (Addo-Quaye et al. 2009; Bologna et al. 2009; Song et al. 2010). The second Dicer cleavage liberates the miRNA as a duplex with its complementary strand. If the double-stranded region of the miRNA precursor extends beyond the miRNA, Dicer may continue cleaving the precursor to produce a set of phased small RNAs. In *Arabidopsis*, 35 miRNA-like RNAs have been identified, and these small RNAs differ from miRNAs with respect to their 5′ terminal nucleotide (Zhang et al. 2010). The predominant 5′ nucleotide of miRNAs is U. By contrast, only about half of the miRNA-like RNAs start with 5′U, while the other half start with 5′A (Zhang et al. 2010).

Several lines of evidence suggest that miRNA-like RNAs have some biological importance (Zhang et al. 2010). The generation of miRNA-like RNAs from the precursors of miR159, miRNA, and miR319 is conserved in different plant species, and in the case of miR159 and miR319, individual miRNA-like RNAs are conserved. Furthermore, most of the miRNA-like RNAs identified to date associate with an ARGONAUTE protein. However, analysis of degradome libraries has identified 3′ fragments that are consistent with cleavages directed by only one of the conserved and five of the nonconserved currently known miRNA-like RNAs (see Zhang et al. 2010). This result suggests that most of the miRNA-like RNAs are either nonfunctional or act in some manner other than directing RNA cleavage. Alternatively, the degradome analysis may not be sensitive enough to detect cleavages directed by most miRNA-like RNAs. Further experimentation, including mutational analysis of predicted miRNA-like RNA target sites, is required to determine if miRNA-like RNAs are of physiological or developmental importance. Interestingly, more rice than *Arabidopsis* miRNA precursors, 75 vs. 19, give rise to miRNA-like RNAs (Zhang et al. 2010). It is not known if this difference reflects an increased importance of phased small RNA in rice compared to *Arabidopsis*.

2.2 Phased Small RNAs Arising from Long Fold-Back Structures

Two loci have been identified that produce phased small RNAs in rice grains (Heisel et al. 2008; Zhu et al. 2008). RNA transcripts from these loci form long hairpin structures that are cleaved by Dicer to produce phased small RNAs. The phased small RNAs derived from these loci are extremely abundant and together account for 20–30% of the 21-nt small RNAs expressed in rice grains. One of these long hairpin structures is derived from exon 2 and 3 of locus Os06g21900 and consists of a set of 18 phased small RNAs generated by ten Dicer cleavages of the long hairpin. The phased small RNAs derived from this locus are almost exclusively expressed in rice grains. However, a very small number of reads were detected for some of these phased small RNAs in libraries derived from mixed tissues (Johnson et al. 2007). These small RNAs are not evolutionarily conserved, as northern blot hybridization techniques failed to detect them in various maize, soybean, and *Arabidopsis* tissues (Heisel et al. 2008). Bioinformatic analysis predicted RNA targets for eight of the 18 phased small RNAs; however, no RNA targets could be predicted for the three most abundant of the 18 phased small RNAs (Zhu et al. 2008). 5′ RACE analysis validated two of the predicted targets: Os04g43210, which encodes a putative myo-inositol transporter, and Os12g17310, which encodes a putative myosin heavy-chain protein (Zhu et al. 2008). These results indicate that at least some of the Os06g21900-derived phased small RNAs are functional and direct target cleavage. However, it is also possible that some of these small RNAs regulate gene expression by other means, such as at the level of transcription, RNA processing, or translation. Whether downregulation of Os04g43210 and Os12g17310 by these phased small RNAs is important in grain development has not yet been determined.

Abundant phased small RNAs are also derived from Os12g42390, a gene that was originally annotated as miR436 (Sunkar et al. 2005), but is no longer considered a miRNA locus. A set of 24 phased small RNAs from this locus is generated by 13 Dicer cleavages. Unlike the small RNAs from Os06g21900, which are exclusively expressed in rice grains, the small RNAs derived from Os12g42390 are expressed in a wide range of tissues (Johnson et al. 2007; Nobuta et al. 2007). In our experiments, their expression in mature leaves was greater than in root apices, shoot apices, or inflorescences. However, the expression in all these tissues is much less than in rice grains. There are no validated targets for the phased small RNAs derived from this locus; therefore, the function, if any, of these small RNAs is unknown.

Some features of the biogenesis pathway for rice phased small RNAs arising from long fold-back structures are known. Current evidence indicates that the long fold-back portions of the transcripts from both Os06g21900 and Os12g42390 are cleaved by DICER-LIKE (DCL) 4 to produce the phased small RNAs (Fig. 2b). Accumulation of small RNAs derived from Os12g42390 is significantly reduced in a Dicer mutant (Liu et al. 2007), and the accumulation of phased small RNAs derived from a transgenic Os06g21900 locus expressed in *Arabidopsis* is reduced in a DCL4 mutant, but not in a Dicer mutant (Heisel et al. 2008). It is not known how the phasing is set for these two fold-back structures, although it is likely that structural features of the hairpin direct the initial phase setting cleavage. Many of the sequence features that specify the initial cleavage of miRNA precursors by DCL1 have been identified, but neither the Os06g21900 nor the Os12g42390 long hairpin has these features. However, DCL4 likely has different structural determinants for cleavage than DCL1. Whatever these structural determinants are, they appear to be conserved because *Arabidopsis* DCL4 correctly processed the Os06g21900 long fold-back.

3 Biogenesis and Function of ta-siRNAs

3.1 Biogenesis of TAS1, TAS2, and TAS4 ta-siRNAs

Arabidopsis has four ta-siRNA gene families: ta-siRNA, ta-siRNA, ta-siRNA, and ta-siRNA (for recent reviews see Allen and Howell 2010; Vaucheret 2005; Voinnet 2008). The TAS1, TAS2, and TAS4 families do not occur in monocots. However,

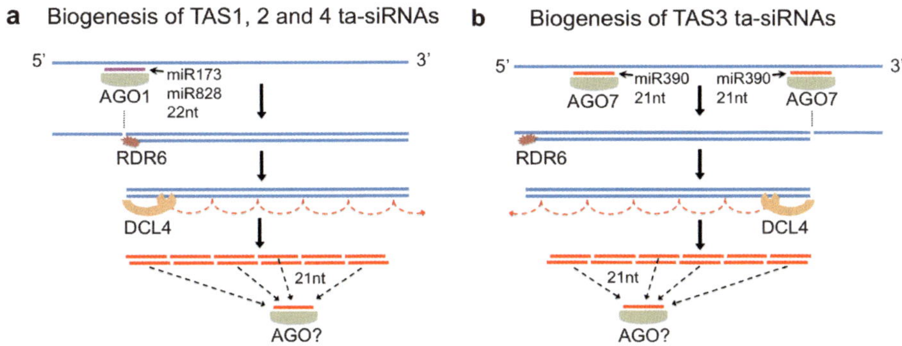

Fig. 3 The biogenesis of ta-siRNAs. Panel (**a**) shows the biogenesis pathways for TAS1, TAS2, and TAS4 ta-siRNAs, and (**b**) shows the biogenesis pathway for TAS3 ta-siRNAs

biogenesis of the phased inflorescence small RNAs in rice (see below) appears to be similar or identical to the *Arabidopsis* TAS1, TAS2, and TAS4 ta-siRNA biogenesis pathway. In ta-siRNA biogenesis, a long non coding RNA is cleaved by a miRNA to set the phasing (Fig. 1b). In the case of the TAS1, TAS2, and TAS4 pathway, the 3′ cleavage product is copied by RNA-dependent RNA polymerase (RDR) to produce dsRNA that is cleaved by DCL4 to generate the ta-siRNAs (Fig. 3a). The phase-setting miRNAs in this pathway, miR173 and miR828, are 22 nt and associate with ARGONAUTE. Twenty-one nucleotide versions of these miRNAs are not effective at ta-siRNA generation, showing that it is important for the TAS1, TAS2, and TAS4 pathway that the size of the phase setting miRNA be 22 nt (Chen et al. 2010; Cuperus et al. 2010a). To explain this requirement for a 22-nt miRNA, it has been proposed that AGO1 can assume two different conformations depending on whether it is bound to a 21- or 22-nt small RNA. In this model, the 22-nt conformation would function to recruit RDR6 and thereby promote synthesis of the dsRNA substrate needed for production of the ta-siRNAs (Chen et al. 2010; Cuperus et al. 2010a).

3.2 Biogenesis of TAS3 ta-siRNAs

The TAS3 family is conserved throughout the plant kingdom, and there are at least four TAS3 loci in rice. Current evidence suggests that the TAS3 ta-siRNA biogenesis pathway is similar in *Arabidopsis* and rice (Fig. 3b). The TAS3 pathway is different from that of TAS1, TAS2, and TAS4 in several respects. First, the miRNA (miR390) that sets the phasing in the TAS3 pathway is 21 nt, not 22 nt as in the TAS1, 2, and 4 pathway. Second, ARGONAUTE is associated with the phase-setting miRNA in the TAS3 pathway instead of AGO1 as in the TAS1, 2, and 4 pathway. Lastly, there are two miRNA binding sites flanking the TAS3 ta-siRNA producing regions instead of one as in the case of TAS1, TAS2, and TAS4. miR390 binds at both of these sites, but does not direct cleavage of the TAS3 precursor at the 5′ site because there are mismatches between miRNA and the TAS3 precursor at positions critical for cleavage. However, the 5′ miR390 binding site is essential for the generation of ta-siRNAs and cannot be replaced by a binding site for a miRNA that associates with AGO1. By contrast, the 3′ miR390 binding site can be replaced by other miRNA binding sites; it is only necessary that the complementary miRNA direct cleavage at this site. In the TAS3 pathway, the 5′ cleavage fragment is made double stranded by RDR6 and cleaved by DCL4 to produce the phased ta-siRNAs (Fig. 3b).

In *Arabidopsis*, the functional TAS3 ta-siRNAs target auxin response factor (ARF) mRNA. These ta-siRNAs (called ta-siRNA-ARFs) regulate juvenile to adult phase transition as well as growth of leaves and lateral roots. The ta-siRNA-ARFs are also essential for the establishment of leaf polarity, and the intercellular movement of ta-siRNA-ARFs from the adaxial surface to the abaxial surface generates a gradient of ta-siRNA-ARFs, which is important for this process (Chitwood et al. 2009). Analysis of rice partial loss of function mutations in the genes involved in TAS3 ta-siRNA biogenesis suggests that ta-siRNA-ARFs also regulate phase transitions and polarity in monocots (Itoh et al. 2008; Satoh et al. 2003; Toriba et al. 2010). In addition, the overexpression of rice ta-siRNA-ARFs in *Arabidopsis* and rice affected the establishment of polarity and phase transitions in both, suggesting a conserved function for these ta-siRNA-ARFs (Wang et al. 2010).

Analysis of rice DCL4, RDR6, and AGO7 mutants suggests that TAS3 ta-siRNAs or other small RNAs that require these genes for their biogenesis play a more important developmental role in rice than in *Arabidopsis*. Mutations in DCL4, RDR6, and AGO7 have minimal effects on the phenotype of *Arabidopsis*, whereas rice plants defective in these genes display much more severe effects (Peragine et al. 2004; Xie et al. 2005; Yoshikawa et al. 2005). For example, *Arabidopsis* having loss-of-function mutations in DCL4, RDR6, or AGO7 have downward curled leaves and early juvenile to adult phase transition (Peragine et al. 2004; Xie et al. 2005; Yoshikawa et al. 2005). By contrast, DCL4 loss-of-function mutants in rice completely lack shoot apical meristems, while RDR6 and AGO7 loss-of-function mutations in rice result in the complete absence of an embryonic shoot (Itoh et al. 2000; Nagasaki et al. 2007). Similarly, mutations in two genes required for ta-siRNA biogenesis, SGS3 and AGO7, produce more severe phenotypes in maize than in *Arabidopsis* (Douglas et al. 2010; Nogueira et al. 2007).

There are three general nonexclusive explanations for the apparently increased role of these ta-siRNA biogenesis genes in monocot development as compared to *Arabidopsis*: (1) The targets of the TAS3 ta-siRNA-ARFs may play a more important role in monocot development than in *Arabidopsis*. (2) The rice TAS3 loci may produce ta-siRNAs that target more mRNAs than the corresponding loci in *Arabidopsis*. (3) Rice may have additional small RNA pathways that require the genes involved in the biogenesis of TAS3 ta-siRNAs.

4 Biogenesis and Function of Rice Phased Inflorescence Small RNAs

4.1 Number of Rice Loci Generating Phased Small RNAs

A spatial cluster analysis of small RNAs expressed in rice shoot apices, root apices, inflorescences, and mature leaves identified more than 800 loci that produce phased 21-nt small RNAs and more than 30 loci that produce phased 24-nt small RNAs (Johnson et al. 2009). The overwhelming majority of these clusters are preferentially expressed in inflorescence tissue derived from early to midstage 7, prior to gamete development (Itoh et al. 2005). It is reasonable to suggest that more phased clusters would be identified if small RNAs were sequenced to a greater depth. Regardless of the precise number of loci that produce clusters of phased small RNAs in rice, the number far exceeds that found in *Arabidopsis*. To date, only 18 loci in *Arabidopsis* have been found to produce phased small RNAs (Howell et al. 2007).

4.2 Characteristics of the Loci Producing Phased Inflorescence Small RNAs

The majority of the loci producing 21-nt phased inflorescence small RNAs are grouped into 31 larger regions, or superclusters, containing ten or more loci that are separated by less than 100 kbp. The 21-nt clusters are generally found in unannotated regions of the genome. Only about one fifth as many of these clusters as expected by chance overlap repeat or protein-coding loci. However, ~20% of the loci producing 21-nt phased inflorescence small RNAs could be aligned to another 21-nt cluster region. By contrast, the 24-nt clusters are evenly spread throughout the genome and overlap repeat loci more frequently than expected by chance.

4.3 Biogenesis of Phased Inflorescence Small RNAs

Comparison of the sequences of the clusters, including some flanking sequence, using MEME (Bailey et al. 2006) identified two 22-nt motifs. A 22-nt motif is located at one end of about 85% of the phased 21-nt clusters (705 of 831 clusters) and is offset exactly 12-nt from the cluster's phase frame. The other 22-nt motif is located at one end of about 74% of the phased 24-nt clusters (28 of 38 clusters) and is also offset by exactly 12-nt from the cluster's phase in 27 of these clusters. The offset of 12-nt is consistent with target cleavage after the 10th nucleotide when

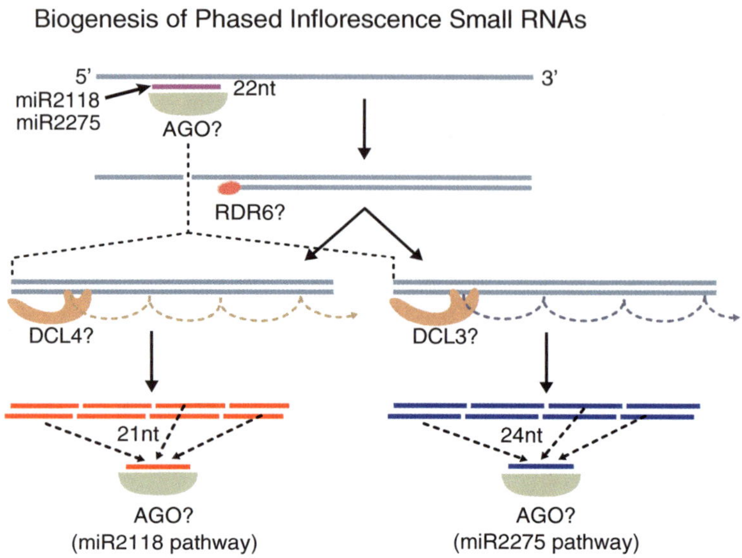

Fig. 4 The proposed pathway for biogenesis of the 21- and 24-nt rice phased inflorescence small RNAs. The RDR and DCL gene family members involved in the biogenesis of the 21- and 24-nt small RNAs have not yet been determined

measured from the 5'end of a targeting-miRNA. Examination of our small RNA libraries identified two miRNA families that correspond to these two motifs. The miR2118 family consists of 22-nt small RNAs that are complementary to the motif flanking the 21-nt clusters, and the miR2275 family is complementary to the motif flanking the 24-nt clusters. However, 125 of the 21-nt clusters do not have the miRNA motif, and 14 of the 24-nt clusters lack the miRNA motif. It is possible that the miR2118 or miR2275 motifs would have been detected in these clusters if the stringency of the match criteria had been reduced or the region searched for the motif had been extended. However, it is also possible that phasing within these clusters is set by other miRNAs or siRNAs and that these clusters might be regulated differently from the rest.

The observations that the motifs are found at only one end of the clusters and that the complementary miRNAs are 22 nt suggest that the biogenesis of these phased small RNAs is similar or identical to the TAS1, TAS2, and TAS4 pathway in *Arabidopsis*. Thus, a likely pathway for biogenesis of the rice phased inflorescence small RNAs entails miR2118 or miR2275 directed cleavage of a non coding RNA, followed by the generation of dsRNA by RDR6 in the case of the 21-nt clusters or, perhaps, RNA-dependent RNA polymerase (RDR) in the case of the 24-nt clusters. DCL4 or DCL3 cleavage of the dsRNA would then produce the phased 21-nt or 24-nt small RNAs, respectively (Fig. 4).

4.4 Functions of the Phased Inflorescence Small RNAs

One likely function for the phased inflorescence small RNAs is that they negatively regulate the expression of genes in *trans*. Because there are greater than 6,000 of these small RNAs, they have the potential to exert a very large influence on gene expression in the developing inflorescence. However, it is possible that the vast majority of the phased inflorescence small RNAs do not have a function, as only one or two small RNAs from each of the TAS loci have been shown to direct small RNA cleavage. Even if only one small RNA from each locus is functional, the phased small RNAs could still direct a massive change in gene expression. The 21-nt small RNAs are expected to act by directing RNA cleavage or inhibit mRNA translation, while the 24-nt small RNAs are expected to induce DNA methylation and chromatin modifications. Hundreds of potential targets of the 21-nt small RNAs were identified using stringent rules established for the targeting of *Arabidopsis* miRNAs. However, no Gene Ontology (GO) annotation class of genes was preferentially enriched in the predicted targets. The predicted targets of the 21-nt phased small RNAs have yet to be validated, and the role of these small RNAs in inflorescence development remains to be determined.

Rather than having an important function in regulating gene expression during inflorescence development, the phased small RNAs could simply be the consequence of the miRNA induced degradation of these transcripts. In *Arabidopsis*, phased small RNAs are generated from some members of a recently expanded clade of genes coding for pentatricopeptide repeat (PPR) proteins (Howell et al. 2007). These members are targeted by multiple miRNAs. Some of the phased small RNAs produced from the miRNA-targeted genes target other genes in the same clade, but not genes in other PPR clades. It was proposed that posttranscriptional silencing of the recently expanded PPR clade might mitigate any deleterious effects caused by the increased number of PPR genes (Howell et al. 2007). It is possible that a similar phenomenon underlies the generation of the 21-nt phased inflorescence small RNAs. In this model, the superclusters of the 21-nt cluster producing loci would be due to a recent expansion of these genes, and the expression of these or related genes would be harmful to the developing inflorescence. The 21-nt phased small RNAs would then serve to reduce the expression of the genes from which they were derived, as well as that of related genes.

References

Addo-Quaye C, Snyder JA, Park YB et al (2009) Sliced microRNA targets and precise loop-first processing of MIR319 hairpins revealed by analysis of the Physcomitrella patens degradome. RNA 15:2112–2121

Allen E, Howell MD (2010) miRNAs in the biogenesis of trans-acting siRNAs in higher plants. Semin Cell Dev Biol 21:798–804

Bailey TL, Williams N, Misleh C et al (2006) MEME: discovering and analyzing DNA and protein sequence motifs. Nucleic Acids Res 34:W369–W373

Bologna NG, Mateos JL, Bresso EG et al (2009) A loop-to-base processing mechanism underlies the biogenesis of plant microRNAs miR319 and miR159. EMBO J 28:3646–3656

Chen HM, Chen LT, Patel K et al (2010) 22-Nucleotide RNAs trigger secondary siRNA biogenesis in plants. Proc Natl Acad Sci USA 107:15269–15274

Chitwood DH, Nogueira FT, Howell MD et al (2009) Pattern formation via small RNA mobility. Genes Dev 23:549–554

Cuperus JT, Carbonell A, Fahlgren N et al (2010a) Unique functionality of 22-nt miRNAs in triggering RDR6-dependent siRNA biogenesis from target transcripts in Arabidopsis. Nat Struct Mol Biol 17:997–1003

Cuperus JT, Montgomery TA, Fahlgren N et al (2010b) Identification of MIR390a precursor processing-defective mutants in Arabidopsis by direct genome sequencing. Proc Natl Acad Sci USA 107:466–471

Douglas RN, Wiley D, Sarkar A et al (2010) Ragged seedling2 encodes an ARGONAUTE7-like protein required for mediolateral expansion, but not dorsiventrality, of maize leaves. Plant Cell 22:1441–1451

Eamens A, Wang MB, Smith NA et al (2008) RNA silencing in plants: yesterday, today, and tomorrow. Plant Physiol 147:456–468

Fahlgren N, Howell MD, Kasschau KD et al (2007) High-throughput sequencing of Arabidopsis microRNAs: evidence for frequent birth and death of MIRNA genes. PLoS ONE 2:e219

Girard A, Hannon GJ (2008) Conserved themes in small-RNA-mediated transposon control. Trends Cell Biol 18:136–148

Heisel SE, Zhang Y, Allen E et al (2008) Characterization of unique small RNA populations from rice grain. PLoS ONE 3:e2871

Howell MD, Fahlgren N, Chapman EJ et al (2007) Genome-wide analysis of the RNA-DEPENDENT RNA POLYMERASE6/DICER-LIKE4 pathway in Arabidopsis reveals dependency on miRNA- and tasiRNA-directed targeting. Plant Cell 19:926–942

Itoh JI, Kitano H, Matsuoka M et al (2000) Shoot organization genes regulate shoot apical meristem organization and the pattern of leaf primordium initiation in rice. Plant Cell 12: 2161–2174

Itoh J, Nonomura K, Ikeda K et al (2005) Rice plant development: from zygote to spikelet. Plant Cell Physiol 46:23–47

Itoh J, Sato Y, Nagato Y (2008) The SHOOT ORGANIZATION2 gene coordinates leaf domain development along the central-marginal axis in rice. Plant Cell Physiol 49:1226–1236

Johnson C, Bowman L, Adai AT et al (2007) CSRDB: a small RNA integrated database and browser resource for cereals. Nucleic Acids Res 35:D829–D833

Johnson C, Kasprzewska A, Tennessen K et al (2009) Clusters and superclusters of phased small RNAs in the developing inflorescence of rice. Genome Res 19:1429–1440

Kircher M, Kelso J (2010) High-throughput DNA sequencing-concepts and limitations. Bioessays 32:524–536

Kurihara Y, Watanabe Y (2004) Arabidopsis micro-RNA biogenesis through Dicer-like 1 protein functions. Proc Natl Acad Sci USA 101:12753–12758

Liu B, Chen Z, Song X et al (2007) *Oryza sativa* dicer-like4 reveals a key role for small interfering RNA silencing in plant development. Plant Cell 19:2705–2718

Mateos JL, Bologna NG, Chorostecki U et al (2010) Identification of microRNA processing determinants by random mutagenesis of Arabidopsis MIR172a precursor. Curr Biol 20:49–54

Mosher RA, Baulcombe DC (2008) Bacterial pathogens encode suppressors of RNA-mediated silencing. Genome Biol 17:237–241

Nagasaki H, Itoh J, Hayashi K et al (2007) The small interfering RNA production pathway is required for shoot meristem initiation in rice. Proc Natl Acad Sci USA 104:14867–14871

Nobuta K, Venu RC, Lu C et al (2007) An expression atlas of rice mRNAs and small RNAs. Nat Biotechnol 25:473–477

Nogueira FT, Madi S, Chitwood DH et al (2007) Two small regulatory RNAs establish opposing fates of a developmental axis. Genes Dev 21:750–755

Peragine A, Yoshikawa M, Wu G et al (2004) SGS3 and SGS2/SDE1/RDR6 are required for juvenile development and the production of trans-acting siRNAs in Arabidopsis. Genes Dev 18: 2368–2379

Poethig RS (2009) Small RNAs and developmental timing in plants. Curr Opin Genet Dev 19: 374–378

Rajagopalan R, Vaucheret H, Trejo J et al (2006) A diverse and evolutionarily fluid set of microRNAs in *Arabidopsis thaliana*. Genes Dev 20:3407–3425

Ramachandran V, Chen X (2008) Small RNA metabolism in Arabidopsis. Trends Plant Sci 13:368–374

Satoh N, Itoh J, Nagato Y (2003) The SHOOTLESS2 and SHOOTLESS1 genes are involved in both initiation and maintenance of the shoot apical meristem through regulating the number of indeterminate cells. Genetics 164:335–346

Song L, Axtell MJ, Fedoroff NV (2010) RNA secondary structural determinants of miRNA precursor processing in Arabidopsis. Curr Biol 20:37–41

Sunkar R, Girke T, Jain PK et al (2005) Cloning and characterization of microRNAs from rice. Plant Cell 17:1397–1411

Toriba T, Suzaki T, Yamaguchi T et al (2010) Distinct regulation of adaxial-abaxial polarity in anther patterning in rice. Plant Cell 22:1452–1462

Vaucheret H (2005) MicroRNA-dependent trans-acting siRNA production. Sci STKE 2005:pe43

Vaucheret H, Mallory AC, Bartel DP (2006) AGO1 homeostasis entails coexpression of MIR168 and AGO1 and preferential stabilization of miR168 by AGO1. Mol Cell 22:129–136

Voinnet O (2008) Use, tolerance and avoidance of amplified RNA silencing by plants. Trends Plant Sci 13:317–328

Voinnet O (2009) Origin, biogenesis, and activity of plant microRNAs. Cell 136:669–687

Wang J, Gao X, Li L et al (2010) Overexpression of Osta-siR2141 caused abnormal polarity establishment and retarded growth in rice. J Exp Bot 61:1885–1895

Werner S, Wollmann H, Schneeberger K et al (2010) Structure determinants for accurate processing of miR172a in *Arabidopsis thaliana*. Curr Biol 20:42–48

Xie Z, Allen E, Wilken A et al (2005) DICER-LIKE 4 functions in trans-acting small interfering RNA biogenesis and vegetative phase change in *Arabidopsis thaliana*. Proc Natl Acad Sci USA 102:12984–12989

Yoshikawa M, Peragine A, Park MY et al (2005) A pathway for the biogenesis of trans-acting siRNAs in Arabidopsis. Genes Dev 19:2164–2175

Zhang W, Gao S, Zhou X et al (2010) Multiple distinct small RNAs originate from the same microRNA precursors. Genome Biol 11:R81

Zhu QH, Spriggs A, Matthew L et al (2008) A diverse set of microRNAs and microRNA-like small RNAs in developing rice grains. Genome Res 18:1456–1465

Small RNA in Legumes

Ghazanfar Abbas Khan, Elodie Hudik, Céline Sorin, Caroline Hartmann, Martin Crespi, and Christine Lelandais-Brière

Contents

Abstract Legumes (Fabaceae) are major crops for food and fodder production worldwide. They have the ability to develop nitrogen-fixing root nodules in symbiosis with soil bacteria of the *Rhizobium* genus. Plant small RNA (sRNA) from 20 to 24 nt, either microRNA (miRNA) or short interfering RNA (siRNA), negatively regulate the expression of specific target genes, at transcriptional or posttranscriptional levels, and can control development, growth, and adaptation to environmental constraints. Since 2007, in silico analysis and deep sequencing of sRNA in legume model and crop species allowed to identify 167 novel miRNA families, in addition to the conserved ones. Although many miRNA differentially accumulate in organs, during symbiosis or in response to abiotic stresses, only few physiological roles

M. Crespi (✉), G.A. Khan, and E. Hudik
Institut des Sciences du Végétal, Centre National de la Recherche Scientifique (C.N.R.S.), 91198 Gif-sur-Yvette, Cedex, France
e-mail: Martin.Crespi@isv.cnrs-gif.fr

C. Sorin, C. Hartmann, and C. Lelandais-Brière
Institut des Sciences du Végétal, Centre National de la Recherche Scientifique (C.N.R.S.), 91198 Gif-sur-Yvette, Cedex, France
and
Université Paris Diderot-Paris 7, 5 rue Thomas Mann, 75205 Paris, Cedex 13, France

V.A. Erdmann and J. Barciszewski (eds.), *Non Coding RNAs in Plants*, RNA Technologies, DOI 10.1007/978-3-642-19454-2_9,

could be assigned as yet to these miRNA. In *Lotus japonicus*, the role of *trans*-acting siRNA on auxin-dependent regulation of leaf and flower development was shown. In *Medicago truncatula* and soybean, several miRNA were shown to act in the regulation of symbiotic interactions. The functions of other legume-specific siRNA, such as stress-responsive natural-antisense siRNA or repeat-associated siRNA involved in heterochromatic silencing, or miRNA remain largely unexplored.

Keywords Abiotic stress • Development • Legume • microRNA • Nodulation • Symbiosis • Transacting siRNA

1 Introduction

Fabaceae, previously called Leguminosae, is a large and agronomically important plant family. They include 727 genera and around 20,000 species (Lewis et al. 2005), including major crops such as *Glycine max* (soybean), *Phaseolus vulgaris* (common bean), *Pisum sativum* (pea), *Cicer rietinum* (chickpea), *Medicago sativa* (alfalfa), and *Arachis hypogeae* (peanut). Contrarily, some legume species are weedy pests in different parts of the world, including *Cytisus scoparius* (broom), *Pueraria lobata* (kudzu), and a number of Lupinus species (Doyle and Luckow 2003). Legume crops are essential for human and animal nutrition because of their high seed protein content. In addition, their cultivation requires limited use of chemical fertilizers due to their capacity to acquire some nutrients, especially nitrogen, through symbiotic interactions with soil microbes. The symbiotic relationships are mutually beneficial for both partners: the plant provides carbon-based energy to the microbe in exchange for essential nutrients (Stacey et al. 2006). The symbiotic interaction with rhizobial bacteria results in the formation of root nodules, in which atmospheric nitrogen fixation by the bacteria and subsequent nitrate uptake by the host plant take place. The symbiotic nitrogen fixation process is of great significance in agricultural ecosystems, especially when the reduced availability of nitrate and phosphate is the major limiting factor for crop productivity. Moreover, in suitable environmental conditions, legumes can also establish symbiotic interactions with arbuscular mycorrhizal (AM) fungi. The AM symbiosis results in the formation of arbuscules, which are the sites for phosphorus nutrient exchange (Harrison 2005; Hause and Fester 2005; Parniske 2008). These mycorrhizal symbioses exist in a wide range of plants, in contrast to the rhizobial symbiosis, which is nearly exclusive to legumes.

Actually, three main legume species have been chosen for genomic studies throughout the world (Sato et al. 2010). *M. truncatula* has a small diploid genome with an estimated size of 500 Mb. It is self-fertile and has a short generation time of approximately 2 months. It produces a sufficient number of seeds per plant and is also amenable to genetic transformation. *Lotus japonicus* has a diploid genome of small size (about 470 Mb) and a short life cycle of about 2–3 months. Finally, the cultivated soybean (*G. max*) has an amphidiploid genome, approximately 1.1 Gb in size. It takes 3–4 months from sowing to harvesting. For these three species, the genomes are almost completely sequenced (*L. japonicus*: http://www.kazusa.or.jp/lotus/, *M. truncatula*: http://medicago.org/genome/, soybean: http://www.phytozome.

net/soybean), and genomic tools such as large cDNA/EST databases, microarrays, high-density linkage maps, and mutant libraries are available (Sato et al. 2010).

The recent discovery of small RNA (sRNA) as important components of gene regulation in plants has initiated studies about their role in legumes and particularly in the rhizobial symbiosis and nitrogen fixation processes. sRNA are short non coding RNA between 20 and 24 nucleotides (nt) in length. In plants, they can be divided into two major classes: microRNA (miRNA) and short-interfering RNA (siRNA). Both regulate negatively the expression of specific target genes at the transcriptional and posttranscriptional levels (Bartel 2004; Jones-Rhoades et al. 2006; Mallory and Vaucheret 2006). Some conserved targets of sRNA are key players in a number of developmental processes and in responses to environmental constraints (Chuck and O'Connor 2010; Sunkar et al. 2007; Sunkar 2010). The biogenesis and action of miRNA are now well documented especially in the model plant *Arabidopsis thaliana* (Voinnet 2009). miRNA genes, mainly intergenic, are transcribed by RNA polymerase II. The initial transcript, called primary miRNA, forms a hairpin-like secondary structure that is subsequently processed by DCL1 and associated proteins. This hairpin is cleaved from the flanking regions, which generates a structure called the miRNA precursor (pre-miRNA). The pre-miRNA is then processed into a miRNA/miRNA* duplex, where the miRNA* corresponds to the imperfect complement of the miRNA in the folded stem. Both strands of the duplex are modified by 3′ methylation by Hua Enhancer 1 (HEN 1) before exportation from the nucleus to the cytoplasm by HASTY (Bollman et al. 2003; Chen 2005; Jones-Rhoades et al. 2006). The mature miRNA, 21 nt in size, is then loaded into the RNA-Induced Silencing protein Complex (RISC), where it determines the recognition of a target mRNA by base-pairing. This interaction directs the cleavage and/or the inhibition of translation of the target.

On the contrary, plant siRNA are divided into three subclasses: (1) *trans*-acting siRNA (tasiRNA), (2) natural antisense transcript-derived siRNA (natsiRNA), and (3) repeat associated siRNA (rasiRNA) also called heterochromatic associated siRNA. tasiRNA are created by the cleavage of a long non coding TAS transcript by a miRNA and subsequent synthesis of a long double stranded RNA (dsRNA) molecule through the action of RNA-dependent RNA Polymerase 6 (RDR6) associated to the SGS3 protein. These long dsRNA are further processed by DICER LIKE 4 (DCL4) into 21-nucleotide tasiRNA, which are by themselves able to target specific mRNA *in trans* (Allen et al. 2005; Gasciolli et al. 2005). NatsiRNA, also 21 nt in length, are formed from overlapping convergent genes forming dsRNA due to natural antisense transcription and are produced by DCL1 or DCL2 (Borsani et al. 2005; Katiyar-Agarwal et al. 2006). Finally, the 24-nt rasiRNA, involved in transcriptional repression, are derived from long dsRNA mainly corresponding to genomic repetitive sequences and are generated by DCL3 (Xie et al. 2004).

Since 2007, deep-sequencing technologies have generated large databases of legume sRNA (e.g., more than a million sRNA loci in soybean, Joshi et al. 2010) and revealed their large diversity encompassing the whole genome (Lelandais-Brière et al. 2009). The functions of most of the identified sRNA are still elusive, but this interesting area of research is advancing rapidly. In this chapter, we describe the recent data concerning legume sRNA and particularly their role in plant development, abiotic stress responses, and symbiotic interactions.

2 Identification of Legume Small RNA

Initial investigations of legume sRNA have been focused on conserved miRNA. A miRNA can be considered as conserved when a stem-fold structure of the precursor contains a mature miRNA sequence conserved in different plants, mainly *A. thaliana*, rice, and poplar (Ambros et al. 2003; Meyers et al. 2008). Large sets of conserved miRNA were first identified thanks to computational approaches that predict hairpin secondary structures (i.e., putative pre-miRNA) containing a mature miRNA sequence in its stem (Meyers et al. 2006, 2008). In soybean, 69 precursors corresponding to 33 miRNA families were identified from a total of 394,370 EST (Zhang et al. 2008), while Zhou et al. (2008) found 38 conserved miRNA genes among *M. truncatula* EST and genomic sequence databases. In a more extended analysis, Sunkar and Jagadeeswaran (2008) identified 682 miRNA in 155 diverse plants with the use of public databases, from which 19 conserved miRNA families were common to the three legume models. Unfortunately, due to the lack of complete genomic sequences in legumes, such computational approaches are greatly hindered and could not be used for a genome-wide search of novel legume miRNA.

Alternatively, sequencing of sRNA can serve to identify novel mi/siRNA, and the first reports of sRNA libraries concerned soybean (Subramanian et al. 2008; Wang et al. 2009a, b; Joshi et al. 2010) and *M. truncatula* (Szittya et al. 2008; Jagadeeswaran et al. 2009; Lelandais-Brière et al. 2009). These studies mainly focused on the identification of conserved and novel miRNA, and corresponding data are discussed downstream in this chapter. However, since 2009, sRNA datasets were also published for three other legume crops, peanut (Zhao et al. 2010), common bean (Arenas-Huertero et al. 2009), and *Glycine soja*, the wild ancestor of domesticated soybean (Chen et al. 2009). In these species, 13, 6, and 9 novel legume- or species-specific miRNA were identified, respectively. To date, a total of 509 miRNA genes are registered in the miRNA database miRBAse (Release 15: April 2010, http://www.mirbase.org/) for Fabaceae species with 23, 85, 13, 3, 375, 8, and 2 sequences from *A. hypogeae*, *G. max*, *G. soja*, *L. japonicus*, *M. truncatula*, *P. vulgaris*, and *Vigna unguiculata* (cowpea), respectively. Except miR170 and miR391, all conserved miRNA families have been found at least in one legume species. In addition, a total of 167 novel miRNA families, at least identified in one sequencing project of legume sRNA, are also registered in miRBAse. Nevertheless, a deeper comparative genomic analysis is still needed to clarify without ambiguity whether these miRNA are species- or legume-specific.

3 Conserved Small RNAs and Plant Development in Legumes

The role of sRNA, mostly miRNA and tasiRNA, in plant growth and development has been intensively studied especially in the model plant *A. thaliana* (Chuck and O'Connor 2010). In legumes, to date, only two reports described the involvement of sRNA in root, flower, and leaf development (Fig. 1, Boualem et al. 2008; Yan et al. 2010).

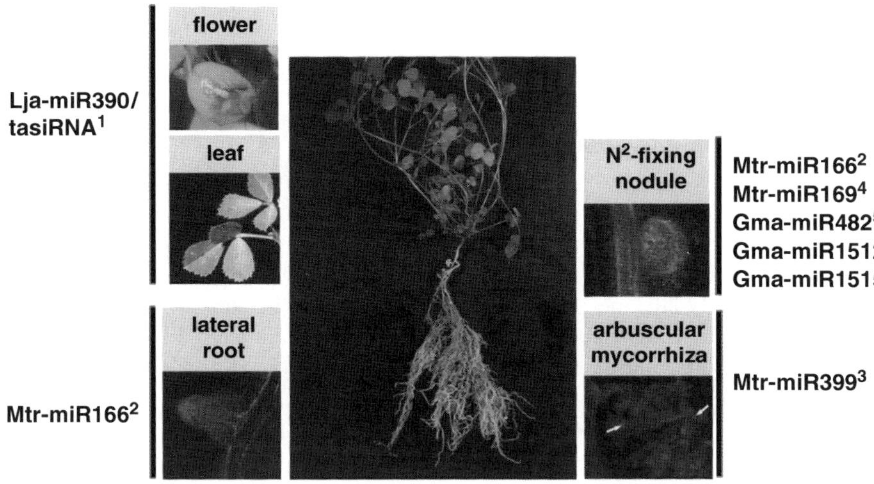

Fig. 1 Small RNA (sRNA) involved in legume development and symbiotic interactions. Only sRNAs, whose role was functionally demonstrated, have been represented. Gma *Glycine max* (soybean); Mtr *Medicago truncatula*; Lja *Lotus japonicus*. *1* Yan et al. (2010), *2* Boualem et al. (2008), *3* Branscheid et al. (2010), *4* Combier et al. (2006), *5* Li et al. (2010)

In *M. truncatula*, SBS sequencing (Illumina) of leaf sRNA allowed to obtain around four million reads, among which 619,175 distinct sRNA mapped to the available genome. Search for miRNA, according to the criteria proposed by Jones-Rhoades et al. (2006), identified 25 families of conserved miRNA and eight novel species-specific miRNA expressed in leaves (Szittya et al. 2008). The most abundant conserved miRNA in this library were miR156, miR159, and miR166, which target mRNA encoding transcription factors (TF) of the Squamosa-Promoter-Like binding (SPL), MYB, and homeodomain-leucine zipper III (HD-ZIP III) families, respectively. These data were concordant with previous reports on their role in aerial plant tissues. Indeed, the miR156/SPL couple regulates both the vegetative reproductive phase transition (Wang et al. 2009a, b) and leaf cell number in *A. thaliana* (Wang et al. 2008; Usami et al. 2009).

Leaves are produced on the periphery of the shoot apical meristem. These organs have a dorsoventral asymmetry, with the side toward the stem called adaxial and the side oriented away from the stem called abaxial. Both sides are morphologically and anatomically different.

The involvement of TAS3-derived *trans*-acting siRNA in leaf adaxial–abaxial polarity has been well described in maize and *A. thaliana* (Adenot et al. 2006; Garcia et al. 2006). In this species, the primary TAS3 non coding transcripts are initially cleaved by a miR390/AGO7 complex. Long dsRNA, produced through transcription of the resulting cleavage products by RDR6 and its SGS3 partner, are then sliced into phased siRNA called TAS3 tasiRNA. Some of these tasiRNA negatively regulate several Auxin Response Factors (ARF3 and ARF4 essentially), which are involved in proper auxin responses necessary for developmental timing and lateral organ patterning (Fahlgren et al. 2006; Marin et al. 2010). This regulatory pathway is well conserved in higher plants including legumes. Indeed, in *M. truncatula*, Jagadeeswaran et al. (2009) confirmed that Mtr-miR390 was able to direct the initial cleavage of a TAS3 transcript, resulting in the production of tasiRNA that target MtARF3 and MtARF4 homologues. Moreover, in *L. japonicus*, two loss of function mutants, called *reduced leaflet* (*rel1* and *rel3*), were identified as impaired in genes coding for SGS3 and AGO7 proteins (Yan et al. 2010). *L. japonicus* plants display classical compound leaves, with a rachis and five leaflets. Moreover, palisade mesophyll and xylem tissues are located on the adaxial side of the leaf, and spongy mesophyll cells and phloem on the abaxial side. In the *rel* mutants, leaflets were narrower than in wild type, and the two basal leaflets were completely lost. In addition, cells around the central vasculature were homogeneous, and the phloem surrounded xylem cells as a ring. Scanning electron microscopy (SEM) analysis revealed that the leaflets were completely covered with long and narrow epidermal cells, specific from the abaxial side of the wild-type plants. This suggested that, in *rel* mutants, leaflets were abaxialized and leaf polarity was disturbed. Furthermore, flowers of these mutants were reduced in size and infertile. Further analysis demonstrated that, in later stages of development, flowers presented defects. For example, certain petal primordia formed a nearly symmetric and trumpet-shaped structure containing small undifferentiated epidermal cells, indicating that the adaxial–abaxial patterning of floral organs was also disturbed. When *rel* mutants were complemented with the wild-type alleles of Lj-SGS3 and Lj-AGO7 respectively, a wild-type phenotype was recovered. This clearly indicates that tasiRNA and presumably those derived from TAS3 transcripts are involved in the adaxial–abaxial polarity of leaf and flower organs in *L. japonicus*, as shown in *A. thaliana*.

The conserved miRNA miR166 also regulates leaf polarity by restricting the expression of its HD-ZIP III targets to the adaxial side in maize and *A. thaliana* (Nogueira et al. 2006, 2007). In addition, this miRNA plays a crucial role in vascular bundle organization and lateral root development (Hawker and Bowman 2004; Prigge et al. 2005). Boualem et al. (2008) studied Mtr-miR166 and its conserved HD-ZIP targets in transgenic roots of *M. truncatula* obtained by *Agrobacterium rhizogenes*-mediated transformation. This protocol, which enables efficient production of composite plants with transformed roots and wild-type aerial organs, allows a rapid validation and functional analysis of genes in roots. These authors showed that Mtr-miR166 and its targets had similar spatial expression pattern in root vascular bundles and meristems. Overexpression of Mtr-miR166 led to a lower expression of its predicted targets and resulted in a significant decrease in the number of lateral roots. Additionally, the organization of root vascular bundles was highly perturbed in these transgenic roots, with an increased number of xylem poles, strongly diverging from the typical stele of dicot roots. This suggested that Mtr-miR166 regulation is crucial for root development in *M. truncatula*, certainly through proper differentiation of vasculature. Accordingly, an Ath-miR166 isoform has also been recently linked to the differentiation of vascular tissues and cortex–endoderm cell–cell communication in the primary root apex (Carlsbecker et al. 2010). In fact, mutation of the HD-ZIP target at the miR166 site yielded a defect in the phloem–xylem differentiation.

4 Legume microRNA and Abiotic Stresses

Because of their sessile nature, plants have to cope with various adverse environmental constraints, such as drought, salinity, temperature extremes, or mineral limitation in soil. They have evolved complex adaptive mechanisms that involve the regulation of thousands of genes at both transcriptional and posttranscriptional levels. Although many abiotic stress-responsive sRNA have been identified in plants, only very few have been experimentally associated to drought or salt tolerance mechanisms (Sunkar and Zhu 2004; Sunkar et al. 2007). In *A. thaliana*, miR169 is involved in drought tolerance via the repression of TFs of the nuclear factor Y-A family (NFY-A) (Li et al. 2008). Overexpression of another conserved miRNA, miR396, led to decreased salt tolerance in rice (Gao et al. 2010) but increased drought tolerance in *A. thaliana* (Liu et al. 2009). In legumes, some studies reported the identification of stress-responsive miRNA. However, there was no demonstration yet to show their real impact on stress tolerance.

Recently, Trindade et al. (2010) analyzed by northern blots the expression of ten conserved miRNA in response to water deficit in *M. truncatula*. Mtr-miR166 was slightly upregulated, whereas Mtr-miR169 was specifically downregulated in roots of water-deficit-treated plants, a regulation similar to that described in *A. thaliana*. Surprisingly, the target of Mtr-miR169, MtHAP2, was also downregulated in similar water-deficit conditions, suggesting MtHAP2 was not only regulated by Mtr-miR169. By contrast, in common bean, Pvu-miR169 accumulated in abscisic acid (ABA)-treated seedlings (Arenas-Huertero et al. 2009). These results are consistent with previous data showing that miR169, or perhaps different isoforms of miR169, behave differently in drought conditions. For example, in rice, osa-MIR169g, which contained a dehydration-responsive element (DRE) in its promoter region, accumulated upon water deficit (Zhao et al. 2007), while in *A. thaliana*, miR169a and miR169c isoforms were downregulated by drought (Li et al. 2008). Furthermore, it is possible that expression patterns in response to abiotic stress conditions may change in different tissues and organs (e.g., leaves vs. whole seedlings at different developmental stages) as global RNA preparations were generally used.

The conserved miRNA, miR398 and miR408, accumulated under water deficit in both *M. truncatula* and common bean (*P. vulgaris*). Both miRNA are proposed to be related to copper homeostasis (Abdel-Ghany and Pilon 2008; Ding and Zhu 2009). In *M. truncatula*, the cleavage of COXb5 and plantacyanin mRNA targets by Pvu-miR398 and Pvu-miR408 respectively was validated by RACE-PCR experiments. As expected, these miRNA and their targets were inversely regulated in conditions of water limitation (Trindade et al. 2010). Additionally, the *PvSPL7* gene, a TF that activates expression of several copper-related miRNA as well as genes encoding copper transporters and a copper chaperone in *A. thaliana* (Yamasaki et al. 2009) was upregulated in shoots in response to water deficit. Arenas-Huertero et al. (2009) also identified novel miRNA in bean that are regulated by different osmotic-related stresses such as cold, high salinity, drought, or ABA application. For instance, Pvu-miR1514a, Pvu-miR2118, and Pvu-miR2119 accumulation increased in response to drought and ABA treatment, but not salinity. In *M. truncatula*, 28 Mtr-miRNA, including five conserved, showed differential accumulation in sRNA libraries of root tips treated or not treated by salt (Lelandais-Brière et al. unpublished results).

Several conserved miRNA also play major roles in adaptation to limited or toxic concentrations of soil nutrients, such as sulfate, copper, and phosphate. For example, the expression of Ath-miR395, which targets both ATP sulfurylases and sulfate transporters, increased during sulfate starvation. Although *A. thaliana* plants overexpressing this miRNA accumulated more sulfate than the wild-type, they still displayed sulfur deficiency symptoms, indicating that it was not the main mechanism determining these symptoms (Liang et al. 2010). Ath-miR398, whose expression is highly responsive to copper levels, targets two Cu/Zn superoxide dismutases (CSD1 and CSD2), which are key regulators of zinc–copper homeostasis, and a chaperone CCS1 essential for copper delivery to the CSDs (Dugas and Bartel 2008; Beauclair et al. 2010). Because of its link with mycorrhizal symbiosis and as *A. thaliana* is unable to perform this symbiotic interaction, the role of miR399 in phosphorus signaling is discussed in this context. Additional miRNA, which accumulate differentially under various conditions of mineral and iron availability, have recently been identified thanks to miRNA macroarrays (Valdés-López et al. 2010; Zeng et al. 2010). For example, 57 and 10 miRNA were responsive to phosphorus deficiency in soybean and common bean, respectively, whereas four legume-specific miRNA (Pvu-miR1511, Gma-miR1513, Gma-miR1515 and Gma-miRNA1516) were strongly induced upon iron limitation in common bean. Finally, in this species, several miRNA were highly responsive to high toxic concentrations of manganese (Valdés-López et al. 2010). Legume miRNA could, thus, be crucial for a variety of responses to both limited and high, even toxic, concentrations of minerals.

5 Legume microRNA in Arbuscular Mycorrhizal Symbiosis

Phosphorus is a major component of many macromolecules and plays important roles in energy transfer, regulation of enzymatic reactions, and different metabolic pathways. It is one of the most-limiting macroelements for plant growth and is consequently essential in agriculture. It is taken up by roots as inorganic phosphate (Pi), one of the less available nutrients for the plant in the soil. To meet the plant phosphorus requirements, the majority of world agriculture, thus, depends on the addition of chemical fertilizers. In *A. thaliana*, it is well known that miR399 plays a key role in maintenance of Pi homeostasis. Ath-miR399 negatively regulates the ubiquitin conjugating enzyme, UBC24, a repressor of phosphate transporters via protein degradation. Under adequate phosphorus availability, miR399 does not accumulate, but upon phosphate starvation, the induction of miR399 leads to the repression of UBC24 expression and to the subsequent accumulation of phosphate-responses proteins, including phosphate transporters. Concordant with these data, *A. thaliana* miR399 overexpressing plants accumulated more phosphate in shoots than the wild type (Fujii et al. 2005; Chiou et al. 2006). In common bean, Valdés-Lopez et al. (2008) suggested that Pvu-miR399 was involved in Pi homeostasis and signaling through a mechanism similar to Arabidopsis (Bari et al. 2006; Doerner 2008). Indeed, these authors obtained transgenic roots silenced for PvDCL1, the main actor of miRNA biogenesis. Analysis of miR399 and Pi deficiency-related transcripts accumulation suggested that the repression of miRNA biogenesis was correlated with an enhanced expression of the miR399 target PvPHO2 and a lower expression of a subset of Pi starvation-responsive genes, including a high-affinity phosphate transporter PvPHT1. Finally, Pvu-miR399 should also serve as phloem-mobile long-distance signal communicating Pi status between shoots and roots (Liu et al. 2010).

To improve phosphate availability, many plants have developed mutualistic associations with AM fungi that expand their capacity to explore the soil volume surrounding the plant through a large network of fungal hyphae (Parniske 2008). The establishment of an AM symbiosis requires a complex developmental program of both plant and fungus. As shown in Fig. 1, miR399 has recently been shown to be involved in the AM symbiosis (Branscheid et al. 2010). Indeed, mycorrhizal roots of Pi-depleted *M. truncatula* and tobacco plants displayed higher miR399 and lower MtPHO2 levels than the nonmycorrhizal roots. In addition, increased expression of PHO2-dependent genes induced by P-starvation and higher Pi contents were reported. The authors concluded that increased miR399 levels in the mycorrhizal roots could limit *MtPHO2* expression, which otherwise would increase in response to symbiotic Pi uptake linking miR399 regulation to the AM symbiosis.

6 Identification of miRNA Regulated During the Nitrogen-Fixing Symbiosis

Legumes additionally form nitrogen-fixing nodules due to the symbiotic interaction with rhizobia as mentioned before. Symbiotic nodules are formed in plants grown under nitrogen-limiting conditions, but not when nitrogen levels are adequate. The establishment of the rhizobia–legume symbiosis is a complex process that has been well documented (Patriarca et al. 2004; Crespi and Frugier 2008). First, the plant host secretes flavonoids and related compounds from their roots. Sensing of these compounds by a potential symbiotic bacterium leads to the activation of *nod* genes, which are involved in the synthesis of bacterial lipochitooligosaccharidic signaling molecules called "Nod Factors." Nod Factor perception by the plant, involving a LysM receptor-like kinase, initiates epidermal infection and stimulates subcellular changes both in the root epidermis and in the cortex. Cell cycle reinitiation in the cortex leads to the formation of the nodule primordium. Consequently, bacteria invade the roots through infection threads formed by epidermal cells (Fournier et al. 2008). The bacteria are then released into the cytoplasm of the nodule primordia cells, where they differentiate into nitrogen-fixing bacteroids. In species forming indeterminate nodules (e.g., alfalfa, pea, and other temperate legumes), mature nitrogen-fixing nodules contain common regions from tip to base: the meristematic zone, the invasion zone, the fixation zone containing the bacteroids, and finally a senescence zone. In subtropical legumes such as *G. max* (soybean), *Phaseolus* (bean), or *Lotus*, determinate nodules are characterized by the lack of a persistent meristem. By contrast, indeterminate nodules maintain an active apical meristem all along the life of the nodule. Symbiotic nodule organogenesis and establishment of nitrogen fixation involve a variety of regulatory molecules including kinases (Grimsrud et al. 2010), TFs (Godiard et al. 2007; Brechenmacher et al. 2008), and other regulatory genes that are now well documented. Recent reports have showed that some of these key regulatory factors are regulated by miRNA (Simon et al. 2009).

To identify legume- or nodulation-specific miRNA, several sRNA libraries have been constructed from mature or developing nodules (Wang et al. 2009a, b; Lelandais-Brière et al. 2009; Joshi et al. 2010) or from roots inoculated with symbiotic rhizobia (Subramanian et al. 2008; Arenas-Huertero et al. 2009). In soybean, Subramanian et al. (2008) used the 454 technology (Roche) to sequence sRNA extracted from roots 3 h post inoculation (hpi) with the symbiotic bacterium *Bradyrhizobium japonicum*. Among the 354,000 reads obtained, they identified a total of 20 conserved miRNA families and 35 novel miRNA that belonged to 30 families (called Gma-miR1507 to 1536). Some miRNA were differentially regulated during the first 12 h post inoculation: miR159, miR393, and miR1508 were rapidly upregulated and maintained this level till 12 hpi. miR168 and miR172 showed a rapid and transient induction at 1 or 3 hpi, while miR160 and miR169 expression was repressed in response to rhizobial infection. These various and oscillating profiles of miRNA accumulation suggest specific roles of these miRNA in early nodulation stages. To extend the set of soybean sRNA, Joshi et al. (2010) constructed sRNA libraries from different organs including nodules of 1, 2, and 3 weeks. SBS sequencing followed by bioinformatical analyses allowed to obtain 1.2 million sequences that mapped on the Gmax1.01 release version genomic sequences and gene model predictions of Williams 82 soybean genome. This yielded 129 miRNA including 87 novel nonconserved ones. Small RNA libraries from mature nodules were constructed in both soybean (Wang et al. 2009a, b) and *M. truncatula* (Lelandais-Brière et al. 2009). Among 375 different sRNA sequences, Wang et al. (2009a, b) identified 32 miRNAs of which 20 were specific to soybean and four were new. A subset of these soybean-specific miRNA, such as GM172 and GM222, were specifically upregulated in nitrogen-fixing nodules. Lelandais-Brière et al. (2009) used 454 pyrosequencing (Roche) to sequence sRNA from *M. truncatula* roots tips and mature nodules. Among a total of 844,110 reads, 36 conserved and 100 putative novel miRNA were identified. Blastn search against genomic sequences of seven plant species including *L. japonicus* and soybean suggested that 15 of the novel miRNA were specific to *M. truncatula* and eight were legume-specific. These novel legume miRNA may play crucial roles in the nodulation process. According to northern analysis, several miRNA show higher levels in nodules than in roots and seedlings. We also noticed that some miRNA isoforms were only sequenced in mature nodules and could, thus, have specific roles in nitrogen-fixing symbiosis (Lelandais-Brière et al. 2009). This idea was reinforced by RACE-PCR experiments showing that, in addition to the conserved SPL target, a 20-nt isoform of Mtr-miR156 specifically cleaved a nonconserved mRNA target encoding a WD40 transducin-like protein in *M. truncatula* roots (Naya et al. 2010).

In situ hybridization experiments helped to localize ten Mtr-miRNA in different regions of the mature nodules of *M. truncatula* (Lelandais-Brière et al. 2009). The majority of them accumulated in the meristematic zone, which supports the hypothesis that these miRNA could be involved in stem cell renewal or meristem maintenance in mature indeterminate nodules. Spatial tissue-specific detection of legume-specific miRNA in both determinate and indeterminate nodules will be useful to better compare their role in these processes. Only one miRNA, a specific isoform of miR399, accumulated in the nitrogen-fixing zone. Wang et al. (2009a, b) also reported that only few miRNA were specifically induced (Gma-miR172, Gma-miR222) or repressed (Gma-miR1508 and Gma-miR1509) in mature nitrogen-fixing nodules. These miRNAs are, thus, good candidates for important regulators of late nodulation and nitrogen fixation.

The conserved Mtr-miR172 and Mtr-miR398 showed broad expression patterns in the nodule invasion zone. In *Arabidopsis*, miR172 targets APETALA2-related (AP2) TFs, a very large gene family involved in a variety of processes (Bowman et al. 1989; Aukerman and Sakai 2003; Würschum et al. 2006; Yant et al. 2010). Expression of this miRNA was induced during nodulation in soybean and common bean (Wang et al. 2009a, b; Valdés-López et al. 2010) and this was correlated with downregulation of an AP2 target in soybean nodules treated with toxic concentrations of manganese. In addition, AP2 mRNA levels responded to P-starvation in common bean nodules in correlation with the downregulation of PvmiR172 (Hernández et al. 2007, 2009). Hence, miR172 could, thus, be involved in different steps of the nodulation process and in nodule responses to various stresses activating AP2-dependent pathways in the invasion zone. On the contrary, in *A. thaliana*, miR398 was initially found to repress genes coding for proteins involved in copper homeostasis and detoxification of reactive oxygen species (ROS), including Cu/Zn superoxidases (CSD) (Sunkar et al. 2006). Protection against ROS could be anticipated to play an important role in *Rhizobium*–legume symbiosis, since the nitrogenase is extremely oxygen sensitive. Moreover, the activities of bacterial and plant superoxide dismutases seem to be crucial for symbiosis initiation (Santos et al. 2000; Rubio et al. 2001). In *L. japonicus*, immunolocalization experiments allowed to locate the different superoxide dismutase isoforms in young, mature, and old nodules (Rubio et al. 2007) and showed that the CSD isoforms were expressed in infected cells of young nodules and disappeared with nodule age. A likely function of these isoforms at early stages of infection could be to provide H_2O_2 for the cross-linking of the matrix glycoprotein of infection threads. Although additional studies are required to confirm that CSDs are regulated by miR398 in legumes, it is tempting to suppose that the accumulation of this miRNA in the infection zone could account for CSD repression in mature nodules. Other miRNA, potentially linked to plant–bacteria interactions, are induced during nodulation. For instance, miR393, the first miRNA linked to plant immune responses, displayed a clear accumulation in soybean roots 3–6 h after inoculation with *B. japonicum* (Subramanian et al. 2008). In *A. thaliana*, this miRNA regulates the auxin receptor F-Box protein TIR1, which is responsible for the degradation of ARF repressor proteins and was shown to be induced by the flagellin-derived bacterial PAMP, flg22 (Navarro et al. 2006). Gma-miR482, a novel soybean miRNA that targets NBS-LRR receptor kinases linked to disease resistance, was also early induced after *B. japonicum* inoculation (Wang et al. 2009a, b). Moreover, in *M. truncatula*, more than ten novel miRNA were predicted to target NBS-LRR resistance genes (Szittya et al. 2008; Lelandais-Brière et al. 2009).Together, these data strongly suggest that miRNAs are largely involved in the regulation of microbe infection and plant defense responses that occur but have to be modulated during symbiosis (Oldroyd and Downie 2008).

Hormones are critical for nodule initiation and development (reviewed by Ding and Oldroyd 2009). Moreover, recent reports have made clear that miRNA are crucial mediators of phytohormonal signaling and homeostasis especially auxin. For instance, miR164 is upregulated in response to auxin and targets the NAC1 TF involved in lateral root initiation in *A. thaliana* (Guo et al. 2005). This miRNA was slightly downregulated during *B. japonicum* infection of soybean roots (Subramanian et al. 2008). In addition, Ath-miR160 and miR167 regulate Auxin Response Factors (ARF10/16/17 for miR160 and ARF6/8 for miR167). These TFs repress or activate target genes involved in the modulation of free auxin levels (Guilfoyle and Hagen 2007). Both miRNA showed higher accumulation levels in *M. truncatula* and soybean mature nodules than in roots, seedlings, or leaves (Lelandais-Brière et al. 2009; Wang et al. 2009a, b). Moreover, Gma-miR160 accumulation was repressed during the early stages of soybean–rhizobia interaction. According to in situ hybridization studies, Mtr-MiR160 accumulated in all regions of the mature nodules, whereas Mtr-miR167 expression was restricted to the apical region and developing vascular tissues of the nodules (Lelandais-Brière et al. 2009). These miRNAs could, thus, play specific role in the fine regulation of auxin levels during nodule organogenesis and vascular differentiation.

7 Functional Analyses Revealed Specific Roles for miRNA in Rhizobial Symbiosis

Although many nodulation-responsive miRNA have been identified, only five have been functionally analysed and associated to this process (Fig. 1): Mtr-miR169 (Combier et al. 2006), Mtr-miR166 (Boualem et al. 2008), and three novel soybean miRNA (Li et al. 2010). The TF MtHAP2-1, regulated by Mtr-miR169, is involved in the redifferentiation of cortical cells into the nodule primordium. RNA interference (RNAi) experiments demonstrated that a reduction in MtHAP2-1 transcript levels significantly affected nodule development. The nodulation process was delayed, and nodule growth was arrested 8–10 days after inoculation. This arrested growth was associated to a loss of the nitrogen-fixing ability of nodules. Overexpression of Mtr-miR169 led to a similar nodulation phenotype correlated with a reduction in *MtHAP2-1* transcript levels due to miR169-mediated cleavage (Combier et al. 2006). In addition, miR169-resistant versions of MtHAP2-1 showed defects in meristem differentiation and final nodule size, suggesting that miR169-mediated degradation of MtHAP2-1 is required for nodule differentiation.

Overexpression of Mtr-miR166 led to a decreased number of lateral roots and nodules in *M. truncatula* (Boualem et al. 2008). This miRNA and its HD-ZIP III targets were strongly expressed in the nodule primordium and in the apical region and vascular bundles of mature nodules. This expression pattern strongly suggests a role of these TFs in nodule development, a process involving the formation of a new meristem as well as vascular tissue differentiation. The nodulation phenotype could also be linked to the severe defects in vascular bundle formation in root tissues.

Recently, Li et al. (2010) reported the functional analysis of some novel miRNA in soybean. Gma-miR482 targets Resistance (R) gene receptor kinases related to defense responses. Expression of this miRNA increased during the initial stages of nodule development with a peak at 6 days post inoculation (dpi), suggesting that this miRNA plays an important role during initial stages of nodule organogenesis. Its expression was not affected in the nonnodulating soybean mutant NOD49 and the supernodulating soybean mutant NTS382. NOD49 contains a mutation in one of the Nod Factor Receptor genes, NFR1 (Wan et al. 2008), while NTS382 contains a defect in the Nodule Autoregulation Receptor Kinase (NARK) (Kinkema and Gresshoff 2008). NARK activates a shoot-derived inhibition mechanism, which significantly reduces additional nodulation after initial interactions with rhizobia. In the absence of NARK-induced autoregulation, there was a significantly increased nodulation in the NTS382 mutant. This demonstrates that *B. japonicum*-induced expression of miR482 is not likely linked to NFR1 and GmNARK1 signaling. Misexpression of miR482, by using either a constitutive promoter or the *Rhizobium*-responsive pENOD40 promoter, resulted in an approximately twofold increase in nodule number, suggesting again that miRNA-directed modulation of plant defense responses might be crucial to the establishment of symbiosis (Simon et al. 2009). The novel miRNA Gma-miR1511 and Gma-miR1512 target transcripts encoding a phosphatase 2C and a calmodulin binding protein, respectively. Their expression profiles oscillated during the initial stages of nodule development. *B. japonicum*-induced expression of these miRNA showed strikingly opposite expression levels between wild-type and NOD49 roots. Their expression levels transiently increased ~6 hpi in wild-type roots but decreased in response to *B. japonicum* inoculation in NOD49 roots, suggesting a regulation by NFR1. Furthermore, the transient induction of miR1511 and miR1512 in response to *B. japonicum* was not observed in NTS382. Misexpression of miR1512 under control of the pENOD40 nodule specific promoter resulted in a higher nodule number. Finally, Gma-miR1515, which negatively regulates a Dicer-like protein encoding gene, showed unaltered expression during early stages of nodule development, but its expression was strongly impaired in nonnodulating mutants. Furthermore, ectopic overexpression of this miRNA also resulted in an increase in the number of nodules. Hence, Nod Factor regulates miRNA expression, and certain miRNA may participate in the regulatory mechanisms involved in nodule autoregulation.

8 Concluding Remarks

Hundreds of conserved and novel miRNA have been identified in model or cultivated legumes since 2007. Many miRNA regulated during symbiotic interactions and/or in response to abiotic stresses were detected in these plants, but functional analysis is only starting and further work is needed to understand their precise roles. In particular, comparative analyses of the symbiosis-related miRNA between legume species (e.g., forming determinate vs. indeterminate nodules) will be of great interest to understand their regulatory functions in the equilibrium between cell proliferation and cell differentiation in nodules. Although short-interfering RNA play important roles in development and adaptive responses to the environment in *A. thaliana*, the large diversity of legume siRNA obtained by deep sequencing has not yet been explored. Some of these siRNA may be involved in epigenetic control of gene expression patterns, and it will be of interest to identify mutants affected in the various si/miRNA biogenesis pathways. Future challenges will certainly be to identify and understand the role of the legume-specific mi/siRNA in three main agricultural traits: environmental stress and pathogen responses, nitrogen fixing and mycorrhizal symbiosis, and seed formation.

References

Abdel-Ghany SE, Pilon M (2008) MicroRNA-mediated systemic down-regulation of copper protein expression in response to low copper availability in Arabidopsis. J Biol Chem 283:15932–15945

Adenot X, Elmayan T, Lauressergues D et al (2006) DRB4-dependent TAS3 trans-acting siRNAs control leaf morphology through AGO7. Curr Biol 16:927–932

Allen E, Xie Z, Gustafson AM et al (2005) microRNA-directed phasing during trans-acting siRNA biogenesis in plants. Cell 121:207–221

Ambros V, Bartel B, Bartel DP et al (2003) A uniform system for microRNA annotation. RNA 9:277–279

Arenas-Huertero C, Pérez B, Rabanal F et al (2009) Conserved and novel miRNAs in the legume *Phaseolus vulgaris* in response to stress. Plant Mol Biol 70:385–401

Aukerman MJ, Sakai H (2003) Regulation of flowering time and floral organ identity by a MicroRNA and its APETALA2-like target genes. Plant Cell 15:2730–2741

Bari R, Datt Pant B, Stitt M et al (2006) PHO2, microRNA399, and PHR1 define a phosphate-signaling pathway in plants. Plant Physiol 141:988–999

Bartel DP (2004) MicroRNAs: genomics, biogenesis, mechanism, and function. Cell 116:281–297

Beauclair L, Yu A, Bouché N (2010) microRNA-directed cleavage and translational repression of the copper chaperone for superoxide dismutase mRNA in Arabidopsis. Plant J 62:454–462

Bollman KM, Aukerman MJ, Park MY et al (2003) HASTY, the Arabidopsis ortholog of exportin 5/MSN5, regulates phase change and morphogenesis. Development 130:1493–1504

Borsani O, Zhu J, Verslues PE et al (2005) Endogenous siRNAs derived from a pair of natural cis-antisense transcripts regulate salt tolerance in Arabidopsis. Cell 123:1279–1291

Boualem A, Laporte P, Jovanovic M et al (2008) MicroRNA166 controls root and nodule development in *Medicago truncatula*. Plant J 54:876–887

Bowman JL, Smyth DR, Meyerowitz EM (1989) Genes directing flower development in Arabidopsis. Plant Cell 1:37–52
Branscheid A, Sieh D, Pant BD et al (2010) Expression pattern suggests a role of MiR399 in the regulation of the cellular response to local Pi increase during arbuscular mycorrhizal symbiosis. Mol Plant Microbe Interact 23:915–926
Brechenmacher L, Kim MY, Benitez M et al (2008) Transcription profiling of soybean nodulation by *Bradyrhizobium japonicum*. Mol Plant Microbe Interact 21:631–645
Carlsbecker A, Lee JY, Roberts CJ et al (2010) Cell signalling by microRNA165/6 directs gene dose-dependent root cell fate. Nature 465:316–321
Chen X (2005) MicroRNA biogenesis and function in plants. FEBS Lett 579:5923–5931
Chen R, Hu Z, Zhang H (2009) Identification of microRNAs in wild soybean (*Glycine soja*). J Integr Plant Biol 51:1071–1079
Chiou TJ, Aung K, Lin SI et al (2006) Regulation of phosphate homeostasis by MicroRNA in Arabidopsis. Plant Cell 18:412–421
Chuck G, O'Connor D (2010) Small RNAs going the distance during plant development. Curr Opin Plant Biol 13:40–45
Combier JP, Frugier F, de Billy F et al (2006) MtHAP2-1 is a key transcriptional regulator of symbiotic nodule development regulated by microRNA169 in *Medicago truncatula*. Genes Dev 20:3084–3088
Crespi M, Frugier F (2008) De novo organ formation from differentiated cells: root nodule organogenesis. Sci Signal 1:re11
Ding Y, Oldroyd GE (2009) Positioning the nodule, the hormone dictum. Plant Signal Behav 4:89–93
Ding YF, Zhu C (2009) The role of microRNAs in copper and cadmium homeostasis. Biochem Biophys Res Commun 386:6–10
Doerner P (2008) Phosphate starvation signaling: a threesome controls systemic P(i) homeostasis. Curr Opin Plant Biol 11:536–540
Doyle JJ, Luckow MA (2003) The rest of the iceberg. Legume diversity and evolution in a phylogenetic context. Plant Physiol 131:900–910
Dugas DV, Bartel B (2008) Sucrose induction of Arabidopsis miR398 represses two Cu/Zn superoxide dismutases. Plant Mol Biol 67:403–417
Fahlgren N, Montgomery TA, Howell MD et al (2006) Regulation of AUXIN RESPONSE FACTOR3 by TAS3 ta-siRNA affects developmental timing and patterning in Arabidopsis. Curr Biol 16:939–944
Fournier J, Timmers AC, Sieberer BJ et al (2008) Mechanism of infection thread elongation in root hairs of *Medicago truncatula* and dynamic interplay with associated rhizobial colonization. Plant Physiol 148:1985–1995
Fujii H, Chiou TJ, Lin SI et al (2005) A miRNA involved in phosphate-starvation response in Arabidopsis. Curr Biol 15:2038–2043
Gao P, Bai X, Yang L et al (2010) Over-expression of osa-MIR396c decreases salt and alkali stress tolerance. Planta 231:991–1001
Garcia D, Collier SA, Byrne ME et al (2006) Specification of leaf polarity in Arabidopsis via the trans-acting siRNA pathway. Curr Biol 16:933–938
Gasciolli V, Mallory AC, Bartel DP et al (2005) Partially redundant functions of Arabidopsis DICER-like enzymes and a role for DCL4 in producing trans-acting siRNAs. Curr Biol 15:1494–1500
Godiard L, Niebel A, Micheli F et al (2007) Identification of new potential regulators of the *Medicago truncatula*-Sinorhizobium meliloti symbiosis using a large-scale suppression subtractive hybridization approach. Mol Plant Microbe Interact 20:321–332
Grimsrud PA, den Os D, Wenger CD et al (2010) Large-scale phosphoprotein analysis in *Medicago truncatula* roots provides insight into in vivo kinase activity in legumes. Plant Physiol 152:19–28
Guilfoyle TJ, Hagen G (2007) Auxin response factors. Curr Opin Plant Biol 10:453–460

Guo HS, Xie Q, Fei JF et al (2005) MicroRNA directs mRNA cleavage of the transcription factor NAC1 to downregulate auxin signals for arabidopsis lateral root development. Plant Cell 17:1376–1386

Harrison MJ (2005) Signaling in the arbuscular mycorrhizal symbiosis. Annu Rev Microbiol 59:19–42

Hause B, Fester T (2005) Molecular and cell biology of arbuscular mycorrhizal symbiosis. Planta 221:184–196

Hawker NP, Bowman JL (2004) Roles for Class III HD-Zip and KANADI genes in Arabidopsis root development. Plant Physiol 135:2261–2270

Hernández G, Ramírez M, Valdés-López O et al (2007) Phosphorus stress in common bean: root transcript and metabolic responses. Plant Physiol 144:752–767

Hernández G, Valdés-López O, Ramírez M et al (2009) Global changes in the transcript and metabolic profiles during symbiotic nitrogen fixation in phosphorus-stressed common bean plants. Plant Physiol 151:1221–1238

Jagadeeswaran G, Zheng Y, Li YF et al (2009) Cloning and characterization of small RNAs from *Medicago truncatula* reveals four novel legume-specific microRNA families. New Phytol 184:85–98

Jones-Rhoades MW, Bartel DP, Bartel B (2006) MicroRNAs and their regulatory roles in plants. Annu Rev Plant Biol 57:19–53

Joshi T, Yan Z, Libault M (2010) Prediction of novel miRNAs and associated target genes in *Glycine max*. BMC Bioinformatics 11(Suppl 1):S14

Katiyar-Agarwal S, Morgan R, Dahlbeck D et al (2006) A pathogen-inducible endogenous siRNA in plant immunity. Proc Natl Acad Sci USA 103:18002–18007

Kinkema M, Gresshoff PM (2008) Investigation of downstream signals of the soybean autoregulation of nodulation receptor kinase GmNARK. Mol Plant Microbe Interact 21:1337–1348

Lelandais-Brière C, Naya L, Sallet E et al (2009) Genome-wide *Medicago truncatula* small RNA analysis revealed novel microRNAs and isoforms differentially regulated in roots and nodules. Plant Cell 21:2780–2796

Lewis G, Schrire B, Mackinder B et al (2005) Legumes of the World. A remarkable new and profusely illustrated book detailing all 727 genera of legumes. Royal Botanic Gardens, Kew, England. pp. 392–431

Li WX, Oono Y, Zhu J et al (2008) The Arabidopsis NFYA5 transcription factor is regulated transcriptionally and posttranscriptionally to promote drought resistance. Plant Cell 20:2238–2251

Li H, Deng Y, Wu T et al (2010) Misexpression of miR482, miR1512, and miR1515 increases soybean nodulation. Plant Physiol 153:1759–1770

Liang G, Yang F, Yu D (2010) MicroRNA395 mediates regulation of sulfate accumulation and allocation in *Arabidopsis thaliana*. Plant J 62:1046–1057

Liu D, Song Y, Chen Z et al (2009) Ectopic expression of miR396 suppresses GRF target gene expression and alters leaf growth in Arabidopsis. Physiol Plant 136:223–236

Liu JQ, Allan DL, Vance CP (2010) Systemic signaling and local sensing of phosphate in common bean: cross-talk between photosynthate and MicroRNA399. Mol Plant 3:428–437

Mallory AC, Vaucheret H (2006) Functions of microRNAs and related small RNAs in plants. Nat Genet 38(Suppl):S31–S36

Marin E, Jouannet V, Herz A et al (2010) miR390, Arabidopsis TAS3 tasiRNAs, and their AUXIN RESPONSE FACTOR targets define an autoregulatory network quantitatively regulating lateral root growth. Plant Cell 22:1104–1117

Meyers BC, Souret FF, Lu C et al (2006) Sweating the small stuff: microRNA discovery in plants. Curr Opin Biotechnol 17:139–146

Meyers BC, Axtell MJ, Bartel B et al (2008) Criteria for annotation of plant MicroRNAs. Plant Cell 20:3186–3190

Navarro L, Dunoyer P, Jay F et al (2006) A plant miRNA contributes to antibacterial resistance by repressing auxin signaling. Science 312:436–439

Naya L, Khan GA, Sorin C et al (2010) Cleavage of a non-conserved target by a specific miR156 isoform in root apexes of *Medicago truncatula*. Plant Signal Behav 5:328–331

Nogueira FT, Sarkar AK, Chitwood DH et al (2006) Organ polarity in plants is specified through the opposing activity of two distinct small regulatory RNAs. Cold Spring Harb Symp Quant Biol 71:157–164

Nogueira FT, Madi S, Chitwood DH et al (2007) Two small regulatory RNAs establish opposing fates of a developmental axis. Genes Dev 21:750–755

Oldroyd GE, Downie JA (2008) Coordinating nodule morphogenesis with rhizobial infection in legumes. Annu Rev Plant Biol 59:519–546

Parniske M (2008) Arbuscular mycorrhiza: the mother of plant root endosymbioses. Nat Rev Microbiol 6:763–775

Patriarca EJ, Tatè R, Ferraioli S et al (2004) Organogenesis of legume root nodules. Int Rev Cytol 234:201–262

Prigge MJ, Otsuga D, Alonso JM et al (2005) Class III homeodomain-leucine zipper gene family members have overlapping, antagonistic, and distinct roles in Arabidopsis development. Plant Cell 17:61–76

Rubio MC, Ramos J, Webb KJ et al (2001) Expression studies of superoxide dismutases in nodules and leaves of transgenic alfalfa reveal abundance of iron-containing isozymes, posttranslational regulation, and compensation of isozyme activities. Mol Plant Microbe Interact 14:1178–1188

Rubio MC, Becana M, Sato S et al (2007) Characterization of genomic clones and expression analysis of the three types of superoxide dismutases during nodule development in *Lotus japonicus*. Mol Plant Microbe Interact 20:262–275

Santos R, Hérouart D, Puppo A et al (2000) Critical protective role of bacterial superoxide dismutase in rhizobium-legume symbiosis. Mol Microbiol 38:750–759

Sato S, Isobe S, Tabata S (2010) Structural analyses of the genomes in legumes. Curr Opin Plant Biol 13:146–152

Simon SA, Meyers BC, Sherrier DJ (2009) MicroRNAs in the rhizobia legume symbiosis. Plant Physiol 151:1002–1008

Stacey G, Libault M, Brechenmacher L et al (2006) Genetics and functional genomics of legume nodulation. Curr Opin Plant Biol 9:110–121

Subramanian S, Fu Y, Sunkar R et al (2008) Novel and nodulation-regulated microRNAs in soybean roots. BMC Genomics 9:160

Sunkar R (2010) MicroRNAs with macro-effects on plant stress responses. Semin Cell Dev Biol 21:88

Sunkar R, Jagadeeswaran G (2008) In silico identification of conserved microRNAs in large number of diverse plant species. BMC Plant Biol 8:37

Sunkar R, Zhu JK (2004) Novel and stress-regulated microRNAs and other small RNAs from Arabidopsis. Plant Cell 16:2001–2019

Sunkar R, Kapoor A, Zhu JK (2006) Posttranscriptional induction of two Cu/Zn superoxide dismutase genes in Arabidopsis is mediated by downregulation of miR398 and important for oxidative stress tolerance. Plant Cell 18:2051–2065

Sunkar R, Chinnusamy V, Zhu J et al (2007) Small RNAs as big players in plant abiotic stress responses and nutrient deprivation. Trends Plant Sci 12:301–309

Szittya G, Moxon S, Santos DM et al (2008) High-throughput sequencing of *Medicago truncatula* short RNAs identifies eight new miRNA families. BMC Genomics 9:593

Trindade I, Capitão C, Dalmay T et al (2010) miR398 and miR408 are up-regulated in response to water deficit in *Medicago truncatula*. Planta 231:705–716

Usami T, Horiguchi G, Yano S et al (2009) The more and smaller cells mutants of Arabidopsis thaliana identify novel roles for SQUAMOSA PROMOTER BINDING PROTEIN-LIKE genes in the control of heteroblasty. Development 136:955–964

Valdés-López O, Arenas-Huertero C, Ramírez M et al (2008) Essential role of MYB transcription factor: PvPHR1 and microRNA: PvmiR399 in phosphorus-deficiency signalling in common bean roots. Plant Cell Environ 31:1834–43

Valdés-López O, Yang SS, Aparicio-Fabre R et al (2010) MicroRNA expression profile in common bean (*Phaseolus vulgaris*) under nutrient deficiency stresses and manganese toxicity. New Phytol 187:805–818

Voinnet O (2009) Origin, biogenesis, and activity of plant microRNAs. Cell 136:669–687
Wan J, Zhang XC, Neece D et al (2008) A LysM receptor-like kinase plays a critical role in chitin signaling and fungal resistance in Arabidopsis. Plant Cell 20:471–481
Wang JW, Schwab R, Czech B et al (2008) Dual effects of miR156-targeted SPL genes and CYP78A5/KLUH on plastochron length and organ size in *Arabidopsis thaliana*. Plant Cell 20:1231–1243
Wang JW, Czech B, Weigel D (2009a) miR156-regulated SPL transcription factors define an endogenous flowering pathway in *Arabidopsis thaliana*. Cell 138:738–779
Wang Y, Li P, Cao X et al (2009b) Identification and expression analysis of miRNAs from nitrogen-fixing soybean nodules. Biochem Biophys Res Commun 378:799–803
Würschum T, Gross-Hardt R, Laux T (2006) APETALA2 regulates the stem cell niche in the Arabidopsis shoot meristem. Plant Cell 18:295–307
Xie Z, Johansen LK, Gustafson AM et al (2004) Genetic and functional diversification of small RNA pathways in plants. PLoS Biol 2:E104
Yamasaki H, Hayashi M, Fukazawa M et al (2009) SQUAMOSA Promoter Binding Protein-Like7 is a central regulator for copper homeostasis in Arabidopsis. Plant Cell 21:347–361
Yan J, Cai X, Luo J et al (2010) The REDUCED LEAFLET genes encode key components of the trans-acting small interfering RNA pathway and regulate compound leaf and flower development in *Lotus japonicus*. Plant Physiol 152:797–807
Yant L, Mathieu J, Dinh TT et al (2010) Orchestration of the floral transition and floral development in Arabidopsis by the bifunctional transcription factor APETALA2. Plant Cell 22:2156–2170
Zeng HQ, Zhu YY, Huang SQ et al (2010) Analysis of phosphorus-deficient responsive miRNAs and cis-elements from soybean (*Glycine max* L.). J Plant Physiol 167:1289–1297
Zhang B, Pan X, Stellwag EJ (2008) Identification of soybean microRNAs and their targets. Planta 229:161–182
Zhao B, Ge L, Liang R et al (2007) Members of miR-169 family are induced by high salinity and transiently inhibit the NF-YA transcription factor. BMC Mol Biol 10:29
Zhao CZ, Xia H, Frazier TP et al (2010) Deep sequencing identifies novel and conserved microRNAs in peanuts (*Arachis hypogaea* L.). BMC Plant Biol 10:3
Zhou ZS, Huang SQ, Yang ZM (2008) Bioinformatic identification and expression analysis of new microRNAs from *Medicago truncatula*. Biochem Biophys Res Commun 374:538–542

Physcomitrella patens Small RNA Pathways*

Muhammad Asif Arif, Isam Fattash, Basel Khraiwesh, and Wolfgang Frank

Contents

Abstract Small, non coding RNAs (sRNAs) are a distinct class of regulatory RNAs in plants and animals controlling a variety of biological processes. Given the great impact of sRNAs in biology, recent studies in model seed-plant species, particularly in *Arabidopsis thaliana*, focused on the identification, biogenesis and functional analysis of sRNAs. In seed plants, several classes of sRNAs with specific sizes and

---

* Muhammad Asif Arif, Isam Fattash and Basel Khraiwesh contributed equally to this work.

W. Frank (✉) and I. Fattash
Department of Plant Biotechnology, University of Freiburg, Schänzlestraße 1, 79104 Freiburg, Germany
e-mail: wolfgang.frank@biologie.uni-freiburg.de

M.A. Arif
Institute for Genetics, Heinrich-Heine-University Düsseldorf, Universitätsstraße 1, 40225 Düsseldorf, Germany

B. Khraiwesh
Department of Plant Systems Biology, Flanders Institute for Biotechnology (VIB)

Department of Plant Biotechnology and Genetics, Ghent University, Technologiepark 927, 9052 Gent, Belgium

V.A. Erdmann and J. Barciszewski (eds.), *Non Coding RNAs in Plants*, RNA Technologies, DOI 10.1007/978-3-642-19454-2_10,

dedicated functions have evolved through a series of pathways, namely, microRNAs (miRNAs), repeat-associated small interfering RNAs (ra-siRNAs), natural antisense transcript-derived small interfering RNAs (nat-siRNAs) and *trans*-acting small interfering RNAs (ta-siRNAs). In the last few years, the analysis of plant sRNA pathways has been extended to the bryophyte *Physcomitrella patens*, a non-flowering, non-vascular ancient land plant, that diverged from the lineage of seed plants approximately 450 million years ago. Based on a number of characteristic features and its phylogenetic key position in land-plant evolution, *P. patens* emerged as a plant model species to address basic as well as applied topics in plant biology. The analysis of *P. patens* sRNA pathways has been recently advanced by the deep sequencing of sRNA libraries, the release of the *P. patens* genome that allowed the mapping of sRNA producing loci and first molecular analyses of *P. patens* mutants with targeted disruption of genes encoding essential components of endogenous sRNA pathways. Even though the major sRNA pathways are evolutionarily conserved in *P. patens*, there are particular differences in the functional components of sRNA pathways and the biological function of sRNAs. These include a specific amplification of initial miRNA and ta-siRNA signals by the generation of transitive siRNAs, deviating functions and specificities of DICER-LIKE proteins and an epigenetic gene silencing pathway that is triggered by miRNAs. Further, the conservation of miRNA biogenesis in *P. patens* was used to establish specific gene silencing by the expression of artificial miRNAs suited for functional gene analysis by reverse genetics approaches. These findings underline that *P. patens* serves as a valuable model system to study the evolution, diversity and function of plant sRNAs. Here, we summarise the current knowledge on different sRNA biogenesis pathways, their biological relevance and the expression of artificial miRNAs in *P. patens*.

Keywords Development • DICER-LIKE • Gene silencing • *Physcomitrella patens* • Small non coding RNA

1 Introduction

Bryophytes (mosses) diverged from the lineage of seed plants approximately 450 million years ago (Kenrick and Crane 1997). Thus, the moss *Physcomitrella patens* occupies an important phylogenetic position that allows to study the evolution of biological mechanisms in land plants (Fig. 1a). Furthermore, *P. patens* is characterised by specific features, making it an attractive model system in plant biology. Most strikingly, *P. patens* exhibits a high frequency of homologous recombination, making it an ideal system for reverse genetics approaches by the simple generation of targeted gene knockout mutants (Schaefer and Zryd 1997). *P. patens* displays a phytohormone-independent regeneration capacity that allows an easy regeneration of transgenic lines from transfected protoplasts and a stable vegetative maintenance of cultures. In contrast to seed plants, the heterophasic and heteromorphic life cycle of mosses (Fig. 1b) is characterised by a predominant haploid generation (the gametophyte) and a reduced diploid generation (the sporophyte). The dominant haploid gametophytic generation

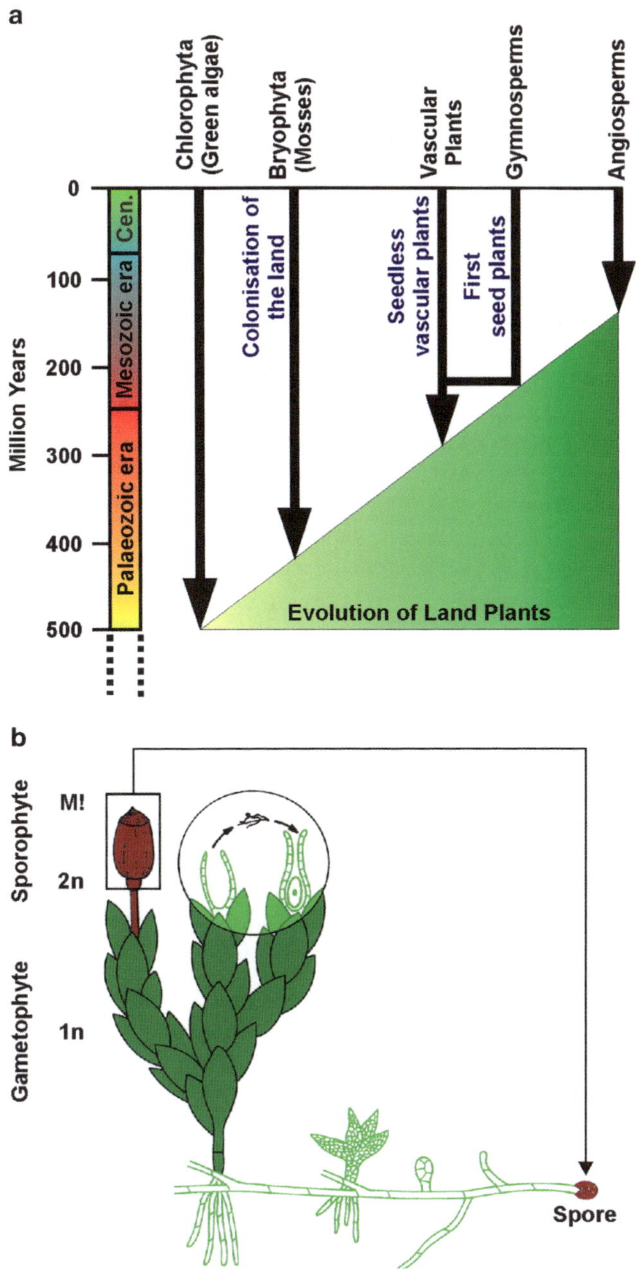

Fig. 1 (**a**) The evolution of land plants. Mosses diverged from the lineage of seed plants approximately 450 million years (MY) ago. Palaeozoic era: 542–251 MY; Mesozoic era: 251–65.5 MY; Cenozoic era (Cen.): 65.5 MY to date. (**b**) Life cycle of *Physcomitrella patens*: A haploid spore germinates and grows into protonema existing of chloronema and caulonema cells. Starting with a three-faced apical cell, the bud formation is initiated, which gives rise to the leafy adult gametophyte. In monoecious moss species, both sex organs (antheridia and archegonia) are present on one and the same plant. Fertilisation of the egg cells takes place in the presence of water. From the fertilised egg, the sporophyte grows out of the archegonia. The diploid sporophyte is highlighted by the surrounding rectangle. Within the spore capsule, the cells undergo meiosis (M!) and new spores are formed

enables the direct phenotypic analysis of loss-of-function mutations without time-consuming back-crossing experiments. Furthermore, the sequence of the 480 Mbp *P. patens* genome was released in 2008 (Rensing et al. 2008). The availability of the *P. patens* genome sequence is a valuable tool for reconstructing the evolution of plant genomes and further progress in the field of functional genomics. Since the discovery of sRNAs in animals (Lee et al. 1993) and the elucidation of their important regulatory role in a large variety of biological processes, sRNA biogenesis and biological functions of sRNAs have also been intensively studied in seed plants (Axtell et al. 2007; Baulcombe 2004; Khraiwesh et al. 2010; Lu et al. 2008; Vazquez 2006). These studies not only revealed the existence of conserved sRNA pathways in both animals and plants, e.g. the miRNA pathway (Bartel 2004; Gregory et al. 2004; Park et al. 2005), but also provided insight into specific sRNA pathways restricted to plants such as the ta-siRNA pathway (Allen et al. 2005; Peragine et al. 2004; Rajagopalan et al. 2006; Talmor-Neiman et al. 2006; Vazquez et al. 2004). Besides the comparison of animal and plant sRNA pathways, another challenge is the investigation of these pathways in different plant species to understand the complexity and particular specifications of sRNA regulation in plants. Furthermore, such analyses will be the basis to fully understand the invention of sRNA-mediated gene regulation during land-plant evolution. To address this point, in the last few years such studies have been applied to the moss *P. patens* combining sRNA sequencing approaches and the analysis of specific targeted *P. patens* knockout mutants defective in the processing of particular sRNAs (Axtell and Bartel 2005; Axtell et al. 2006; Cho et al. 2008; Fattash et al. 2007; Khraiwesh et al. 2010; Talmor-Neiman et al. 2006). Besides the analysis of *P. patens* sRNA pathways, molecular tools were developed exploiting the mode of action of sRNAs for the down-regulation of genes in reverse genetics applications. These approaches include the use of conventional inverted RNAi constructs (Bezanilla et al. 2003, 2005a) as well as the expression of highly specific artificial miRNAs (Khraiwesh et al. 2008).

2 Classes of Small RNA (sRNA)

Independent approaches combining traditional cloning, computational prediction and high-throughput sequencing of sRNA libraries have identified several classes of sRNAs in *P. patens* (Arazi et al. 2005; Axtell et al. 2006, 2007; Cho et al. 2008; Fattash et al. 2007) and revealed a high complexity in sRNA pathways comparable to higher plants. In land plants, three main classes of small regulatory RNAs are distinguished based on their specific biogenesis: small interfering RNAs (siRNAs), microRNAs (miRNAs) and *trans*-acting small interfering RNAs (ta-siRNAs). The biogenesis and functions of these small RNA classes are well characterised in the model plant *Arabidopsis thaliana*. In general, small RNAs are generated from at least partially dsRNA precursors by the action of ribonuclease III-like Dicer proteins (Bernstein et al. 2001; Vazquez 2006). The small RNA duplexes generated by Dicer activity have a characteristic 2-nucleotide overhang at the 3′ end due to an offset cutting of the Dicers. In plants, these 3′ overhangs are stabilised by 2′-*O*-methylation (Gan et al. 2006; Park et al. 2002; Yang et al. 2006; Yu et al. 2005). Only one strand of the processed sRNA duplex subsequently associates with an RNA-induced silencing complex (RISC) that scans for nucleic acids complementary to the loaded sRNA to execute its function (Hammond et al. 2000; Hutvagner and Simard 2008; Noma et al. 2004; Voinnet 2009). In plants, sRNAs act in gene silencing by different ways, namely, by mediating RNA slicing (Bartel 2004; Baumberger and Baulcombe 2005; Morel et al. 2002; Qi et al. 2005), translational repression (Brodersen et al. 2008; Chen 2004; Lanet et al. 2009) and histone modification and DNA methylation (Khraiwesh et al. 2010; Matzke and Matzke 2004; Schramke and Allshire 2004). The first two mechanisms control gene expression post-transcriptionally, whereas the latter affects gene expression at the transcriptional level.

2.1 miRNAs

2.1.1 Biogenesis of miRNAs

In the miRNA biogenesis pathway, primary miRNAs (pri-miRNAs) are transcribed from nuclear encoded genes by RNA polymerase II (Lee et al. 2004; Voinnet 2009) leading to precursor transcripts with a characteristic hairpin structure. In plants, the processing of these pri-miRNAs into pre-miRNAs is catalysed by Dicer proteins (plant Dicers are referred to as DICER-LIKE [DCL] proteins), which contrasts the situation in animals where the first processing step is catalysed by the RNase III enzyme Drosha residing in a multiprotein microprocessor complex (Denli et al. 2004; Gregory et al. 2004). Thereafter, in both plants and animals Dicer proteins in the nucleus recognise the pre-miRNA hairpin precursor and catalyse the release of a specific RNA duplex from the double-stranded stem region consisting of the mature miRNA and its complementary miRNA* (Gregory et al. 2004; Han et al. 2004; Kurihara and Watanabe 2004; Lee et al. 2002; Park et al. 2002; Reinhart et al. 2002) (Fig. 2a). In *A. thaliana*, the miRNA duplexes are transported to the cytoplasm by HASTY, the plant orthologue of the animal Exportin 5 (Bartel 2004; Baulcombe 2004; Bohnsack et al. 2004; Cai et al. 2004; Kim 2004; Park et al. 2005). In the cytoplasm, the miRNA is incorporated into an Argonaute (AGO) protein, the catalytic component of RISC, and guides RISC to bind to cognate target transcripts by sequence complementarity (Bartel 2004). In principle, plant DCL proteins have particular substrate specificities with respect to the dsRNA precursors from which they catalyse the maturation of sRNAs (Kurihara and Watanabe 2004; Park et al. 2002; Reinhart et al. 2002; Xie et al. 2004, 2005). Plant DCL proteins can be divided into four groups, each having one *A. thaliana* DCL homologue with a specific function in sRNA processing including AtDCL1 essential for miRNA biogenesis (Golden et al. 2002; Park et al. 2002). The *P. patens* genome also encodes four DCL proteins (Table 1) (Axtell et al. 2007), but the DCL repertoire differs from that in *A. thaliana*. *P. patens* encodes two proteins highly similar to AtDCL1 and two DCL proteins

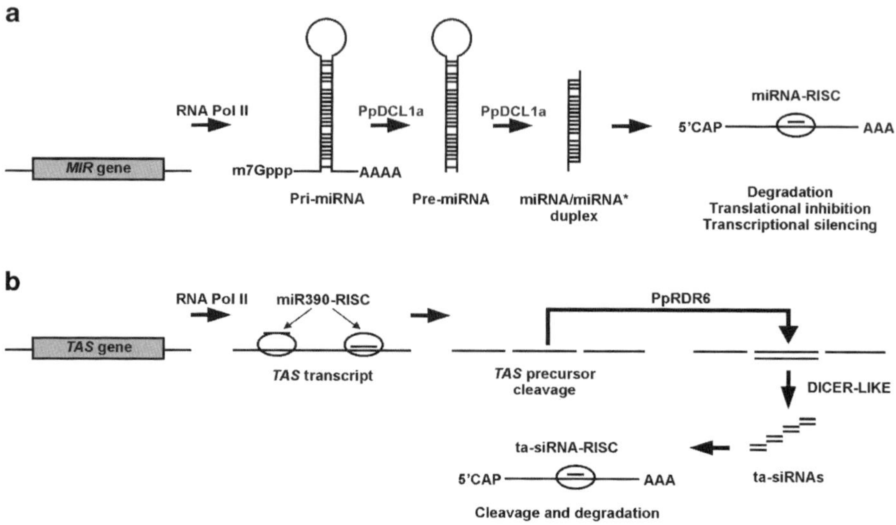

Fig. 2 (**a**) *P. patens* miRNA pathway. *MIR* genes are transcribed by RNA polymerase II into pri-miRNA transcripts and further processed into pre-miRNAs that have with a characteristic hairpin structure. From the stem of this miRNA precursor the miRNA/miRNA* duplex is excised by PpDCL1a. The mature miRNA is incorporated into miRNA-RISC and guides the complex to cognate target RNAs. (**b**) *P. patens* ta-siRNA pathway. *TAS* genes are transcribed by RNA polymerase II into *TAS* precursors harbouring two miR390 binding sites. After *TAS* precursor cleavage at both miR390 sites, the middle cleavage product is converted into dsRNA by PpRDR6 and subsequently processed into phased ta-siRNAs by a DICER-LIKE protein. ta-siRNAs are loaded into RISC where they act like miRNAs

homologous to AtDCL3 and AtDCL4, respectively, whereas an AtDCL2 homologue is missing (Table 1). Consequently, the *P. patens* proteins were termed PpDCL1a, PpDCL1b, PpDCL3 and PpDCL4. Given the high similarity of PpDCL1a and PpDCL1b to the *A. thaliana* AtDCL1 protein, it was unclear whether both proteins may have redundant functions in miRNA biogenesis. Experimental evidence for an essential function of PpDCL1a in miRNA biogenesis was obtained from the analysis of Δ*PpDCL1a* mutants (Khraiwesh et al. 2010). Δ*PpDCL1a* null mutants showed strongly reduced expression levels or complete lack of miRNAs and elevated steady-state transcript levels of cognate miRNA targets (Khraiwesh et al. 2010). *A. thaliana dcl1* mutants that lack miRNAs accompanied with increased miRNA target expression levels are embryo-lethal (Golden et al. 2002; Park et al. 2002). Similarly, Δ*PpDCL1a* mutants displayed severe developmental disorders affecting cell size and shape, retarded growth that was partially complemented by growth on medium supplemented with vitamins and an arrestment at the filamentous protonema stage, since these mutants failed to develop leafy gametophores (Fig. 3a). The lack of gametophores also causes sterility of Δ*PpDCL1a* mutants, since the gametophores bear the male and female sex organs (antheridia and archegonia). Based on the essential function of PpDCL1a in miRNA biogenesis, it was concluded that PpDCL1a is the functional equivalent to the AtDCL1 protein from *A. thaliana*.

Table 1 *P. patens* and *A. thaliana* homologues of proteins involved in sRNA pathways

Protein family	*P. patens* homologues	NCBI/gene model number	*E*-value, %identity	Molecular function	References
AtDCL1	PpDCL1a PpDCL1b	EF670436 DQ675601	0.0, 68% 0.0, 65%	miRNA biogenesis Indispensable for target cleavage	Khraiwesh et al. (2010), Kurihara and Watanabe (2004), Park et al. (2002), and Reinhart et al. (2002)
AtDCL2	n.i[a]	–	–	Generates endogenous siRNAs from a convergently transcribed and overlapping gene pairs Transitive silencing of transgenes Produces siRNAs from viruses in *Atdcl4* mutants	Borsani et al. (2005), Bouche et al. (2006), and Mlotshwa et al. (2008)
AtDCL3	PpDCL3	EF670437	1e-116, 32%	Generates siRNAs that guide chromatin modification in *P. patens* and *A. thaliana*	Cho et al. (2008) and Xie et al. (2004)
AtDCL4	PpDCL4[b]	EF670438	1e-124, 33%	Generates *trans*-acting siRNAs (ta-siRNAs)	Dunoyer et al. (2005), Gasciolli et al. (2005), Liu et al. (2007a), Xie et al. (2005), and Yoshikawa et al. (2005)
AtAGO1	PpAGO1a[b] PpAGO1b[b] PpAGO1c[b]	Phypa_205541 Phypa_158832 Phypa_141045	0.0, 78% 0.0, 77% 0.0, 75%	Associates with the majority of miRNAs to guide the cleavage of their targets	Axtell et al. (2007), Baumberger and Baulcombe (2005), Qi et al. (2005), and Vaucheret et al. (2004, 2006)
AtAGO2	n.i	–	–	Not analysed	
AtAGO3	n.i	–	–	Not analysed	

(continued)

Table 1 (continued)

Protein family	*P. patens* homologues	NCBI/gene model number	*E*-value, %identity	Molecular function	References
AtAGO4	PpAGO4[b]	Phypa_200513	1e-164, 38%	Involved in 24 nt siRNA-mediated gene silencing	Havecker et al. (2010), Pontes et al. (2006, 2009), Wierzbicki et al. (2008), and Zilberman et al. (2003, 2004)
AtAGO5	n.i	–	–	Not analysed	
AtAGO6	PpAGO6[b]	Phypa_117253	1e-152, 39%	Involved in 24 nt siRNA mediated DNA methylation	Havecker et al. (2010) and Zheng et al. (2007)
AtAGO7	n.i	–	–	Associates specifically with miR390 and directs cleavage of the *AtTAS3* precursor	Adenot et al. (2006), Allen et al. (2005), Montgomery et al. (2008), Peragine et al. (2004), Rajagopalan et al. (2006), and Vazquez et al. (2004)
AtAGO8	n.i	–	–	Not analysed	
AtAGO9	PpAGO9[b]	Phypa_134255	6.1e-160, 40%	Preferentially interacts with 24 nt siRNAs derived from transposable elements (TEs), required to silence TEs in female gametes and their accessory cells. Cell fate determination in the ovule	Havecker et al. (2010) and Olmedo-Monfil et al. (2010)
AtAGO10	n.i	–	–	Implicated in miRNA-directed translational inhibition and repression of miR165/166 levels	Brodersen et al. (2008), Liu et al. (2008), and Mallory et al. (2009)

AtRDR1	PpRDR1[b]	Phypa_219654	1.3e-204, 48%	Synthesis of long dsRNA from transgenes that can initiate different RNAi pathways	Sugiyama et al. (2005) and Vaistij et al. (2002)
AtRDR2	n.i	–	–	Biogenesis of 24 nt siRNAs from repeat loci involved in DNA methylation	Chan et al. (2004) and Kasschau et al. (2007)
AtRDR3a	n.i.	–	–		Pandey et al. (2008) and Zong et al. (2009)
AtRDR3b	PpRDR3b[b]	Phypa_169723	1.8e-96, 33%	Not analysed	
AtRDR3c	PpRDR3c[b]	Phypa_172848	6.1e-89, 31%		
AtRDR6	PpRDR6	Phypa_379	2.8e-226, 42%	Initiation and maintenance of dsRNA-induced RNAi in *A. thaliana* Conversion of *TAS* precursors into dsRNA in *P. patens* and *A. thaliana*	Dalmay et al. (2000), Elmayan et al. (2009), Peragine et al. (2004), and Talmor-Neiman et al. (2006)
AtHEN1	PpHEN1[b]	Phypa_148777	3e-56, 33%	Methylates miRNA-miRNA* duplex at the 3′ end	Park et al. (2002), Yang et al. (2006), and Yu et al. (2005)
AtHYL	PpHYL1[b]	Phypa_34761	7e-31, 50%	Interacts with AtDCL1 and confers stability to miRNA precursors	Fang and Spector (2007), Han et al. (2004), Qin et al. (2010), and Xie et al. (2004)
AtHASTY	PpHASTY1[b]	Phypa_137344	1.3e-228, 40%	Exports miRNA-miRNA* duplex to the cytoplasm	Park et al. (2005) and Peragine et al. (2004)
	PpHASTY2[b]	Phypa_151199	2.1e-173, 41%		
AtSE	PpSE1[b]	Phypa_133793	1.7e-92, 41%	Interacts with AtDCL1 and confers stability to miRNA precursors	Fang and Spector (2007), Han et al. (2004), Kurihara and Watanabe (2004), and Xie et al. (2004)
	PpSE2[b]	Phypa_124567	1.8e-70,41%		
	PpSE3[b]	Phypa_99415	3.3e-53, 35%		

(continued)

Table 1 (continued)

Protein family	*P. patens* homologues	NCBI/gene model number	*E*-value, %identity	Molecular function	References
AtSGS3	PpSGS3[b]	Phypa_448213	3.0e-71, 37%	Involved in the production of ta-siRNAs, through direct or indirect stabilisation of *TAS* cleavage products	Elmayan et al. (2009) and Peragine et al. (2004)
AtPol IV	PpPol IV[b]	Phypa_132119	1.3e-72, 49%	Required for the biogenesis of 24 nt siRNAs (with RDR2 and DCL3) that associate with AGO4 and direct DNA and histone modifications	Pontes et al. (2006, 2009) and Wierzbicki et al. (2008)
AtPol V	PpPol V[b]	Phypa_129844	1e-132, 70%	Generates transcripts from heterochromatic regions (with DRD1) that are discussed to bind siRNA-AGO4 complexes directing DNA and histone modifications	
AtDRM1 AtDRM2	PpDRM1[b] PpDRM2[b]	Phypa_148057 Phypa_133529	5e-92, 51% 6.3e-87, 47%	Involved in the siRNA-directed de novo DNA methylation and maintenance of DNA methylation at CHH sites	Cao et al. (2003), Cao and Jacobsen (2002), and Chan et al. (2004)
AtDRD1	PpDRD1[b]	Phypa_113504	1e-109, 35%	Cooperates with Pol V	Cao and Jacobsen (2002), Chan et al. (2006), Huettel et al. (2006, 2007), Law et al. (2010), Matzke et al. (2006), Pontes et al. (2009), and Wierzbicki et al. (2008)

AtSNF2	PpSNF2[b]	Phypa_211797	1.3e-187, 46%	Involved in the spreading of transgene silencing (with AtRDR2 and AtPol IV) and in the production of endogenous 24 nt siRNAs	Jeddeloh et al. (1999), Morel et al. (2000), Shaked et al. (2006), and Smith et al. (2007)
AtRDM12	PpRDM12[b]	Phypa_98999	1e-46, 26%	Involved in the de novo DNA methylation and siRNA-mediated maintenance of DNA methylation	Ausin et al. (2009) and Zheng et al. (2010)

Protein sequences from *A. thaliana* (TAIR; http://www.arabidopsis.org) were used for reciprocal BLASTP searches against the *P. patens* V1.2_proteins database (http://www.cosmoss.org)

DCL DICER LIKE; *AGO* ARGONAUTE; *RDR* RNA-DEPENDENT RNA POLYMERASE; *HEN1* HUA ENHANCER 1; *HYL1* HYPONASTIC LEAVES 1; *SE* SERRATE; *SGS3* SUPPRESSOR OF GENE SILENCING 3; *DRM* DOMAINS REARRANGED METHYLASE; *DRD1* DEFECTIVE IN RNA-DIRECTED DNA METHYLATION 1; *RDM12* RNA-DIRECTED DNA METHYLATION 12

[a]Not identified

[b]Unknown function in *P. patens*

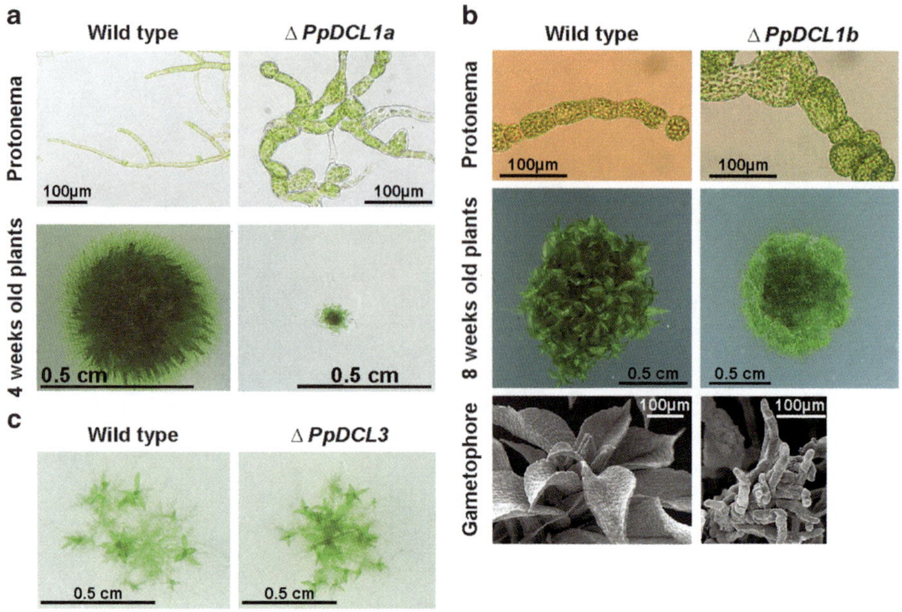

Fig. 3 *P. patens DCL* mutants. (**a**) Phenotypic comparison of *P. patens* wild type and a Δ*PpDCL1a* mutant line. Protonema tissue grown in liquid medium supplemented with glucose and vitamins and developing colonies on solid supplemented medium. (**b**) Phenotypic comparison of *P. patens* wild type and a Δ*PpDCL1b* mutant line. Protonema tissue grown in liquid minimal medium and developing colonies on solid minimal medium. (**c**) Protonema tissue of identical density from wild type and a Δ*PpDCL3* mutant was spotted onto solid minimal medium. Note the accelerated gametophore development in Δ*PpDCL3* mutants

2.1.2 *Physcomitrella patens* miRNA Repertoire

Since the discovery of first identified miRNAs, lin-4 and let-7 in *Caenorhabditis elegans* (Lagos-Quintana et al. 2001; Lau et al. 2001; Lee and Ambros 2001), enormous efforts have been directed towards miRNA identification in eukaryotic organisms. In plants, the discovery of miRNAs is still an ongoing process especially in many economical and evolutionary important species. One important factor in miRNA identification is the availability of a sequenced genome that facilitates the mapping of putative miRNAs to the genome and that predicts hairpin-like structures of the encoded *MIR* precursor transcripts. At the time of writing this review, 939, 730, 58 and 230 miRNAs had been identified in dicots, monocots, lycopods and bryophytes, respectively, as inferred from the current miRBase version 15 (www.mirbase.org) (Griffiths-Jones et al. 2006, 2008).

In *P. patens*, three major methods were applied for miRNA discovery: cloning and sequencing of individual sRNAs using traditional molecular methods, massive parallel sequencing of sRNA libraries and bioinformatic prediction followed by experimental validation (Arazi et al. 2005; Axtell et al. 2006, 2007; Fattash et al. 2007; Talmor-Neiman et al. 2006). Conserved miRNAs families between *P. patens* and other land plants were discovered using the microHARVESTER algorithm that predicts conserved miRNAs in genomic sequences (Fattash et al. 2007). This method is limited to the identification of conserved miRNAs and excludes the identification of species-specific miRNAs. The reported *P. patens* miRNAs were identified from wild-type plants covering major developmental stages (protonema, young gametophores, gametophores and sporophytes). The identification of miRNAs was restricted to plants that were cultivated under standard growth conditions, and thus miRNAs that may response to certain physiological conditions such as abiotic stress may have escaped identification. These analyses led to the identification of 230 different miRNAs in *P. patens*, which fall into 109 miRNA families. Interestingly, among these 109 families, 13 families are conserved between *P. patens*, monocotyledons and dicotyledons, while the 47 miRNAs families identified in the green alga *Chlamydomonas reinhardtii* are species-specific and are not related to land-plant miRNAs (Fig. 4). Similar to the existence of conserved miRNA families in animals (Sempere et al. 2006), 13 miRNA families are conserved in different land-plant lineages (Fig. 4) (Axtell and Bartel 2005; Axtell et al. 2007).

Based on the high complementarity between miRNAs and their target RNAs, various algorithms have been implemented to predict authentic miRNA targets in

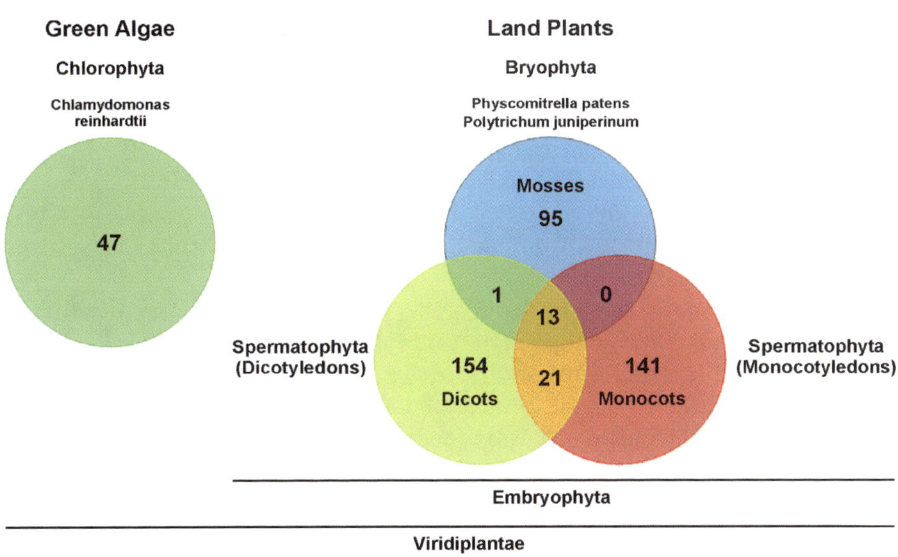

Fig. 4 Venn diagram comparing miRNA families in monocotyledons, dicotyledons, *P. patens* and *Chlamydomonas reinhardtii* based on miRBase version 15. *C. reinhardtii* miRNA families are species specific and do not show any similarity to miRNA families of land plants. Thirteen miRNA families (miR156, miR160, miR166, miR167, miR171, miR319, miR390, miR395, miR408, miR414, miR419, miR529 and miR535) are conserved between monocotyledons, dicotyledons and *P. patens*. A single miRNA family, miR477, is conserved between dicotyledons and *P. patens*

P. patens (Arazi et al. 2005; Axtell et al. 2007; Fattash et al. 2007; Talmor-Neiman et al. 2006). The first applied parameters predicted *P. patens* miRNA targets from assembled EST data (Nishiyama et al. 2003), allowing a maximum of three mismatches between the miRNA and its complementary binding site within the target RNA (Arazi et al. 2005). Later on, parameters developed by Schwab et al. (2005) were applied to predict miRNA targets from assembled EST data and raw WGS trace files (Fattash et al. 2007; Talmor-Neiman et al. 2006). Finally, a large number of miRNA and ta-siRNA targets were predicted from the *P. patens* genome assembly (Axtell et al. 2007) by applying modified prediction parameters (Allen et al. 2005) that allow a mismatch at position 10 or 11 between the sRNA and its potential target. Overall, the different parameters led to the prediction of 230 miRNA targets in *P. patens*. Target validation was initially performed by 5′RACE (Arazi et al. 2005; Axtell et al. 2007; Fattash et al. 2007; Talmor-Neiman et al. 2006). In contrast to the miRNA-induced cleavage mapping of single RNAs by 5′RACE, a newly developed technique of sequencing, the 5′ ends of uncapped mRNAs (degradome sequencing) allows for a genome-wide identification of miRNA-directed sliced RNA products (Addo-Quaye et al. 2008). The degradome sequencing was successfully applied to identify and validate miRNA targets in *P. patens* (Addo-Quaye et al. 2009). At the time of writing this review, 55 targets were confidently identified by degradome sequencing (Addo-Quaye et al. 2009) whereas 34 targets were validated by 5′RACE (Axtell et al. 2007; Fattash et al. 2007; Talmor-Neiman et al. 2006). Among these, 15 targets were independently identified by both methods resulting in a total number of 74 experimentally validated targets that are regulated by miRNAs belonging to 32 different miRNA families (Table 2). Compared to the large number of miRNAs and their predicted targets, the number of validated targets appears to be incomplete. Even though both methods involve amplification steps, it cannot be excluded that low-abundant cleavage products are below the detection limit. Furthermore, to our knowledge target validation was exclusively performed with RNA from protonema tissue. Since the expression of several miRNA is restricted to other tissues (Arazi et al. 2005; Fattash et al. 2007), it is likely that further targets will be validated using RNA from additional tissues.

These analyses (Addo-Quaye et al. 2009; Arazi et al. 2005; Axtell and Bartel 2005; Axtell et al. 2007) confirmed the first evidence that conserved miRNAs regulate conserved targets (Floyd and Bowman 2004), suggesting conserved functions of these miRNAs in the control of important biological processes, since mutations within the miRNA as well as within their targets were repressed during evolution. The conserved miRNA-target pairs include miR156 regulating *SBP* transcripts, miR166 mediating cleavage of class III *HD-ZIP* mRNAs, miR171 controlling mRNAs encoding GRAS domain transcription factors, miR408 regulating a transcript coding for a plastocyanin domain-containing protein and miR319 cleaving *MYB* transcription factor mRNAs (Table 2). Functional analysis of selected conserved miRNAs revealed their function in the control of specific developmental processes in seed plants. For example, in *A. thaliana*, overexpression of ath-miR156 extends the juvenile vegetative phase and delays flowering, whereas overexpression of the ath-miR156 targets *AtSPL3* and *AtSPL9* accelerates flowering

Table 2 Experimentally confirmed *P. patens* miRNA targets

P. patens miRNA	Target accession	Target annotation	Validated by	References
miR156abc[a]	Phypa_168927	SPL3 (SQUAMOSA PROMOTER BINDING PROTEIN-LIKE 3)	5′RACE and degradome	Arazi et al. (2005) and Axtell et al. (2007)
miR156abc[a]	Phypa_74968	SPL1 (SQUAMOSA PROMOTER BINDING PROTEIN-LIKE 1)	5′RACE and degradome	Axtell et al. (2007)
miR156abc[a]	Phypa_168928	SPL2 (SQUAMOSA PROMOTER BINDING PROTEIN-LIKE 2)	Degradome	Addo-Quaye et al. (2009)
miR160a[a]	Phypa_108888	ARF16 (AUXIN RESPONSE FACTOR 16)	Degradome	Addo-Quaye et al. (2009)
miR166jkl[a]	Phypa_116038	Class III HD-Zip protein HB12	5′RACE	Axtell et al. (2007)
miR166jkl[a]	Phypa_182184	Class III HD-Zip protein HB11	5′RACE	Axtell et al. (2007)
miR166jkl[a]	Phypa_183629	Class III HD-Zip protein HB13	5′RACE	Axtell et al. (2007)
miR166jkl[a]	Phypa_184087	Class III HD-Zip protein HB10	5′RACE	Axtell et al. (2007)
miR166jkl[a]	Phypa_192868	Class III HD-Zip protein HB14	5′RACE	Axtell et al. (2007)
miR171ab[a]	Phypa_143602	GRAS family transcription factor	5′RACE and degradome	Axtell et al. (2007) and Addo-Quaye et al. (2009)
miR171a[a]	Phypa_189182	GRAS family transcription factor	5′RACE and degradome	Axtell et al. (2007) and Addo-Quaye et al. (2009)
miR319ab	Phypa_165935	Cyclin family protein	5′RACE	Axtell et al. (2007)
miR319ab[a]	Phypa_99470	MYB family transcription factor	5′RACE	Axtell et al. (2007)
miR319ab[a]	Phypa_86194	MYB family transcription factor	5′RACE	Axtell et al. (2007)
miR390a	Phypa_81023	Unknown protein	Degradome	Addo-Quaye et al. (2009)
miR408b[a]	Phypa_49402	Plastocyanin-like domain-containing protein	5′RACE and degradome	Axtell et al. (2007) and Addo-Quaye et al. (2009)
miR408b	Phypa_207583	LPR2 (LOW PHOSPHATE ROOT2)	Degradome	Addo-Quaye et al. (2009)
miR477h	Phypa_233854	CONSTANS-like	5′RACE and degradome	Addo-Quaye et al. (2009) and Fattash et al. (2007)
miR477h	Phypa_230817	CONSTANS-like	Degradome	Addo-Quaye et al. (2009)
miR477h	Phypa_93769	Heat shock protein-related	Degradome	Addo-Quaye et al. (2009)
miR477h	Phypa_74633	Ribosomal protein L29	Degradome	Addo-Quaye et al. (2009)

(continued)

Table 2 (continued)

P. patens miRNA	Target accession	Target annotation	Validated by	References
miR477h	Phypa_79360	Heat shock protein-related	Degradome	Addo-Quaye et al. (2009)
miR477h	Phypa_171809	Basic helix–loop–helix (bHLH) transcription factor	5′RACE	Axtell et al. (2007)
miR477a-5p	Phypa_230817	CONSTANS-like	Degradome	Addo-Quaye et al. (2009)
miR529g	Phypa_65352	Member of the ERF (ethylene response factor) subfamily B-1 of ERF/AP2 transcription factor family	5′RACE	Axtell et al. (2007)
miR529g	Phypa_78785	ESR1 (ENHANCER OF SHOOT REGENERATION 1)	Degradome	Addo-Quaye et al. (2009)
miR529d	Phypa_74968	SPL1 (SQUAMOSA PROMOTER BINDING PROTEIN-LIKE 1)	Degradome	Addo-Quaye et al. (2009)
miR529d	Phypa_168928	SPL2 (SQUAMOSA PROMOTER BINDING PROTEIN-LIKE 2)	Degradome	Addo-Quaye et al. (2009)
miR534a	Phypa_121620	BOP2 (BLADE ON PETIOLE2)	5′RACE and degradome	Axtell et al. (2007) and Addo-Quaye et al. (2009)
miR534a	Phypa_119190	BOP2 (BLADE ON PETIOLE2)	5′RACE and degradome	Axtell et al. (2007) and Addo-Quaye et al. (2009)
miR536c	Phypa_170879	F-box family protein	5′RACE and degradome	Axtell et al. (2007) and Addo-Quaye et al. (2009)
miR537a	Phypa_16569	Hydroxyproline-rich glycoprotein family protein	5′RACE	Fattash et al. (2007)
miR538abc	Phypa_109598	SEP1 (SEPALLATA1)	5′RACE and degradome	Axtell et al. (2007) and Addo-Quaye et al. (2009)
miR538abc	Phypa_83991	SHP1 (SHATTERPROOF 1)	5′RACE and degradome	Axtell et al. (2007) and Addo-Quaye et al. (2009)
miR538abc	Phypa_94754	SHP1 (SHATTERPROOF 1)	5′RACE and degradome	Axtell et al. (2007) and Addo-Quaye et al. (2009)

miR898a	Phypa_135864	Protein kinase family protein	Degradome	Addo-Quaye et al. (2009)
miR902cdebfg(5′)	Phypa_170794	Basic helix–loop–helix (bHLH) transcription factor	5′RACE	Axtell et al. (2007)
miR902cdebfg(5′)	Phypa_172669	Basic helix–loop–helix (bHLH) transcription factor	5′RACE	Axtell et al. (2007)
miR902jk-5p	Phypa_166219	AMS (ABORTED MICROSPORES)	Degradome	Addo-Quaye et al. (2009)
miR902b-5p	Phypa_170794	TT8 (TRANSPARENT TESTA 8)	Degradome	Addo-Quaye et al. (2009)
miR902b-5p	Phypa_172669	Basic helix–loop–helix (bHLH) transcription factor	Degradome	Addo-Quaye et al. (2009)
miR902b-5p	Phypa_233706	NAI1; DNA binding/transcription factor	Degradome	Addo-Quaye et al. (2009)
miR902bj-5p	Phypa_158675	Unknown protein	Degradome	Addo-Quaye et al. (2009)
miR904	Phypa_205541	AGO1a (ARGONAUTE 1a)	5′RACE	Axtell et al. (2007)
miR904	Phypa_158832	AGO1b (ARGONAUTE 1b)	5′RACE	Axtell et al. (2007)
miR904	Phypa_141045	AGO1c (ARGONAUTE 1c)	5′RACE	Axtell et al. (2007)
miR1023b-5p	Phypa_159483	STOP1 (SENSITIVE TO PROTON RHIZOTOXICITY 1)	Degradome	Addo-Quaye et al. (2009)
miR1023a-5p	Phypa_218544	STOP1 (SENSITIVE TO PROTON RHIZOTOXICITY 1)	Degradome	Addo-Quaye et al. (2009)
miR1027a	Phypa_85651	Unknown protein	Degradome	Addo-Quaye et al. (2009)
miR1028c-3p	Phypa_202708	Protein arginine *N*-methyltransferase family protein	Degradome	Addo-Quaye et al. (2009)
miR1028b-5p	Phypa_87991	Unknown protein	Degradome	Addo-Quaye et al. (2009)
miR1029	Phypa_162837	DRE-binding transcription factor	5′RACE and degradome	Axtell et al. (2007) and Addo-Quaye et al. (2009)
miR1038-5p	Phypa_163703	HD2A (HISTONE DEACETYLASE 2A)	Degradome	Addo-Quaye et al. (2009)
miR1038-5p	Phypa_158891	HD2A (HISTONE DEACETYLASE 2A)	Degradome	Addo-Quaye et al. (2009)
miR1039-5p	Phypa_119635	Vesicle-associated membrane family protein	Degradome	Addo-Quaye et al. (2009)
miR1043-3p	Phypa_216982	WLIM1; transcription factor	Degradome	Addo-Quaye et al. (2009)

(continued)

Table 2 (continued)

P. patens miRNA	Target accession	Target annotation	Validated by	References
miR1043-3p	Phypa_183406	ADH2 (ALCOHOL DEHYDROGENASE 2)	Degradome	Addo-Quaye et al. (2009)
miR1049	Phypa_170179	Unknown protein	Degradome	Addo-Quaye et al. (2009)
miR1049	Phypa_171661	EMB975 (EMBRYO DEFECTIVE 975)	Degradome	Addo-Quaye et al. (2009)
miR1065	Phypa_161460	Zinc finger (C2H2 type) protein (WIP4)	Degradome	Addo-Quaye et al. (2009)
miR1073-5p	Phypa_63006	CSD2 (COPPER/ZINC SUPEROXIDE DISMUTASE 2)	Degradome	Addo-Quaye et al. (2009)
miR1073-5p	Phypa_85183	CSD2 (COPPER/ZINC SUPEROXIDE DISMUTASE 2)	Degradome	Addo-Quaye et al. (2009)
miR1073-5p	Phypa_107257	L-ascorbate oxidase	Degradome	Addo-Quaye et al. (2009)
miR1073-5p	Phypa_225061	L-ascorbate oxidase/copper ion binding	Degradome	Addo-Quaye et al. (2009)
miR1073-5p	Phypa_182545	Glyoxal oxidase-related	Degradome	Addo-Quaye et al. (2009)
miR1078	Phypa_87833	Ankyrin protein kinase	Degradome	Addo-Quaye et al. (2009)
miR1211	Phypa_204321	MOT1 (MOLYBDATE TRANSPORTER 1)	Degradome	Addo-Quaye et al. (2009)
miR1216	Phypa_65041	Unknown protein	Degradome	Addo-Quaye et al. (2009)
miR1216	Phypa_158683	Unknown protein	Degradome	Addo-Quaye et al. (2009)
miR1216	Phypa_64945	GIL1 (GRAVITROPIC IN THE LIGHT)	Degradome	Addo-Quaye et al. (2009)
miR1219a	Phypa_203442	Auxin-responsive factor, putative (ARF1)	5′RACE	Axtell et al. (2007)
miR1219a	Phypa_224167	Auxin-responsive factor, putative (ARF1)	5′RACE	Axtell et al. (2007)
miR1026ab	Phypa_209063	Basic helix–loop–helix (bHLH) transcription factor	5′RACE	Khraiwesh et al. (2010)
miR1222b	Phypa_162039	Unknown protein	Degradome	Addo-Quaye et al. (2009)

[a]Conserved miRNAs and conserved targets between *P. patens* and seed plants

(Gandikota et al. 2007; Wang et al. 2008, 2009; Wu and Poethig 2006; Yamaguchi et al. 2009) Three *A. thaliana* mRNAs encoding auxin response transcription factors (*AtARF10*, *AtARF16* and *AtARF17*) are negatively regulated by ath-miR160 to control development (Liu et al. 2007b; Mallory et al. 2005; Wang et al. 2005). Overexpression of *A. thaliana* ath-miR171 causes pleiotropic developmental defects including a reduced number of cauline and rosette leaves and reduced shoot branching (Song et al. 2010). miR165/166 controls class III *HD-ZIP* transcription factor mRNAs involved in the determination of the adaxial/abaxial leaf axis (Emery et al. 2003; Juarez et al. 2004; Timmermans et al. 2004; Yao et al. 2009) and vascular development (Kim et al. 2005). Given the evolutionary conservation of these miRNAs, it is tempting to speculate that the homologous conserved *P. patens* miRNAs control the development of analogous morphological structures in the *P. patens* gametophyte and first functional analyses are required to address this interesting topic of the evolution of miRNAs in plants. Furthermore, several moss-specific miRNAs target mRNAs encoding transcription factors that are associated with developmental control. For example, miR534 directs the cleavage of transcripts encoding ankyrin repeat containing proteins including homologues of the *A. thaliana* genes *BLADE ON PETIOLE 1* and *2*, miR538 regulates mRNAs of the *MADS* box transcription factor family and miR902 controls several mRNAs encoding basic helix–loop–helix transcription factors (Addo-Quaye et al. 2009; Arazi et al. 2005; Axtell et al. 2007).

2.2 *Trans-Acting Small Interfering RNAs (ta-siRNA)*

Ta-siRNAs are plant-specific sRNAs that originate from nuclear-encoded *TAS* transcripts that are cleaved by miRNAs, converted into double-stranded RNA and processed into phased siRNA duplexes by Dicer proteins. In *A. thaliana* several components of the ta-siRNA biogenesis pathway have been identified including RDR6, SGS3, DCL4 and DRB4 (Adenot et al. 2006; Allen et al. 2005; Gasciolli et al. 2005; Peragine et al. 2004; Vazquez et al. 2004; Xie et al. 2005; Yoshikawa et al. 2005). In *A. thaliana*, four *TAS* gene families (*TAS1-4*) have been identified. ta-siRNA production from *TAS1* and *TAS2* requires miR173-directed cleavage; miR390-dependent cleavage is required for *TAS3*, and miR828 is assigned to *TAS4*. *TAS2* and *TAS4* are encoded by single genes, while *TAS1* and *TAS3* each harbour three members (Allen et al. 2005; Peragine et al. 2004; Rajagopalan et al. 2006; Vazquez et al. 2004; Yoshikawa et al. 2005). In *A. thaliana*, most miRNAs are incorporated into AGO1-containing RISC to direct cleavage of their targets, whereas miR390 interacts specifically with AGO7. The resulting miR390–AGO7 complex has a specific role in the phased processing of ta-siRNAs from *TAS3* precursors (Montgomery et al. 2008). In contrast to *A. thaliana*, *P. patens* only encodes the *TAS3* family with four members (*TAS3a-d*) each harbouring dual miR390 sites. In *P. patens*, all *TAS3* precursors are cleaved at both miR390 binding sites, triggering subsequent conversion of the middle cleavage product into dsRNA by PpRDR6

(Fig. 2b) (Axtell et al. 2006; Talmor-Neiman et al. 2006). Like in *A. thaliana*, the rice genome also encodes three *TAS3* precursors (*TAS3a-c*), and all *TAS3* family members from both species harbour dual miR390 binding sites. In *A. thaliana* and rice, however, miR390-directed cleavage only occurs at the 3′ miR390 binding site exhibiting high sequence complementarity to miR390. Consequently, ta-siRNAs are only generated from the cleavage product located 5′ to the cleaved miR390 site (Axtell et al. 2006; Lu et al. 2008). In *A. thaliana* and rice, phased processing of ta-siRNAs from the dsRNA *TAS* cleavage products is catalysed by DCL4 proteins (Dunoyer et al. 2005; Liu et al. 2007a; Xie et al. 2005). Even though functional studies on three *P. patens DCL* knockout mutants (Δ*PpDCL1a*, Δ*PpDCL1b* and Δ*PpDCL3*) have been recently published (Cho et al. 2008; Khraiwesh et al. 2010), experimental evidence for an essential role of PpDCL4 in the processing of ta-siRNAs from *TAS* precursors is still missing. However, based on the presence of this DCL4 homologue in *P. patens* (Table 1), it is tempting to speculate that PpDCL4 is the functional equivalent to the *A. thaliana* and rice DCL4 proteins. Furthermore, the molecular analysis of the three existing Δ*PpDCL* knockout mutants did not indicate a role for these PpDCL proteins in the phased processing of ta-siRNAs. The absence of ta-siRNAs in the Δ*PpDCL1a* mutant can be explained by the lack of miR390 that is required to initiate the ta-siRNA pathway by the cleavage of *TAS* precursors (Khraiwesh et al. 2010) (Fig. 2b). So far, only the PpRDR6 protein has been shown to be essential for ta-siRNA production in *P. patens*. Like the *A. thaliana rdr6* mutant, targeted Δ*PpRDR6* null mutants lack ta-siRNAs due to the requirement of RDR6 for the conversion of *TAS* cleavage products into dsRNA (Peragine et al. 2004; Talmor-Neiman et al. 2006). Furthermore, it was suggested that the rice RDR6 is also required for ta-siRNA production (Nagasaki et al. 2007; Toriba et al. 2010). *A. thaliana rdr6* mutants show mild phenotypic deviations with elongated and slightly downward curled rosette leaves and an accelerated transition to the adult phase (Peragine et al. 2004). *P. patens* Δ*PpRDR6* mutants do not show altered morphology but, similar to the *A. thaliana* mutants, exhibit an accelerated juvenile to adult gametophyte transition (Talmor-Neiman et al. 2006). By contrast, strong rice *rdr6* mutants (*OsSHL2*) lack the shoot apical meristem (Nagasaki et al. 2007; Satoh et al. 2003), whereas a weak mutation leads to defects in the adaxial-abaxial patterning of floral organs (Toriba et al. 2010), indicating a broader impact of ta-siRNA-controlled gene expression on developmental programmes.

Like for miRNAs, ta-siRNAs targets can be predicted based on the sequence complementarity between a ta-siRNA and its binding motif present in an RNA. In *A. thaliana*, *TAS3* ta-siRNAs regulate several *AUXIN RESPONSE FACTOR* (*ARF*) mRNAs including *ARF3* and *ARF4* (Allen et al. 2005; Williams et al. 2005). ta-siRNA regulation of these transcripts is required for the determination of leaf polarity and proper timing of vegetative shoot development (Adenot et al. 2006; Fahlgren et al. 2006; Garcia et al. 2006; Xie et al. 2005). Likewise, rice *TAS3*-derived ta-siRNAs control *ARF2/3* belonging to the same gene family (Lu et al. 2008). Strikingly, *P. patens* ta-siRNAs, which are solely generated from *TAS3* precursors, also target *ARF* transcripts (Phypa_203442, Phypa_224167), even though the sequences of ta-siRNAs are not conserved between moss and seed plants (Axtell et al. 2007). These observations indicate that at least *ARF* targeting *TAS3* ta-siRNA function is conserved between *A. thaliana*, rice and *P. patens*, irrespective of the varying ta-siRNA sequences. The conserved regulation by *TAS3* ta-siRNAs can be explained by the divergence of a common ancestor or the convergent evolution of *TAS3* precursors in mosses and seed plants.

Besides *ARF* transcripts, *P. patens* ta-siRNAs also regulate an mRNA encoding an AP2/EREPB transcription factor (Phypa_65352) (Talmor-Neiman et al. 2006). *A. thaliana TAS1* and *TAS2* family ta-siRNAs target transcripts encoding PPR (pentatricopeptide repeat) proteins, whereas *MYB* transcription factor transcripts are controlled by *TAS4* ta-siRNAs (Allen et al. 2005; Peragine et al. 2004; Rajagopalan et al. 2006; Vazquez et al. 2004). In addition to their role in mediating RNA target cleavage, *A. thaliana* ta-siRNAs may also function in the nucleus to control mRNA splicing. A binding motif for a *TAS1a*-derived ta-siRNA was identified within an intron of a pre-mRNA encoding a FAD binding domain-containing protein (*At2g46740*), and elevated levels of unspliced *At2g46740* mRNA were observed in ta-siRNA-deficient mutants (Vazquez et al. 2004). Evidence for a similar function of *P. patens* ta-siRNAs is still missing.

2.3 Biogenesis of Short Interfering RNAs (siRNAs)

siRNAs are generated from perfectly double-stranded RNA that can originate from different sources such as RNA transcribed from inverted repeats or convergent transcription of adjacent genes in the genome (Axtell 2009; Dunoyer et al. 2010). The dsRNA is cleaved into 21–24 nt siRNAs by DCLr proteins, and the size of the released siRNAs depends on the specific catalytic activity of the respective DCL protein. Usually, dsRNA is cleaved by multiple DCL proteins, thereby generating siRNA classes with different sizes (Axtell 2009). Like miRNAs, siRNAs are loaded into AGO protein containing RISC that guide target regulation by base pairing in which the mode of RISC action largely depends on the AGO protein (Baumberger and Baulcombe 2005; MacRae et al. 2008). First evidence on the generation of siRNAs in *P. patens* was obtained from the expression of inverted repeat *GUS* and *GFP* RNAi constructs that caused silencing of GUS and GFP signals in *P. patens* lines with a constitutive *GUS* and *GFP* expression (Bezanilla et al. 2003, 2005b). The total endogenous sRNA population of flowering plants is characterised by two distinct peaks at 21 and 24 nt (Morin et al. 2008; Nobuta et al. 2010; Rajagopalan et al. 2006). The 21 nt size fraction largely constitutes miRNAs and ta-siRNAs, while 24 nt sRNAs are mainly generated from intergenic and repetitive genomic regions (Morin et al. 2008; Rajagopalan et al. 2006). In *A. thaliana*, these 24 nt sRNAs are primarily generated by AtDCL3 (Xie et al. 2004). The *P. patens* top 100 non-miRNA and non-ta-siRNA producing regions fall into two distinct categories and were classified on the basis of their abundance and size (Cho et al. 2008). One class is dominated by

21 nt RNAs, whereas a second class comprises a mixture of 21–24 nt RNAs in a strikingly consistent ratio. Accordingly, the loci generating these two types of sRNAs were annotated as *Pp21SR* (21 nt small RNA) and *Pp23SR* (21, 23, and 24 nt small RNA) loci, respectively (Cho et al. 2008). The *Pp23SR* loci generally span larger genomic regions between 5 and 50 kb in length (median 11.9 kb), while the *Pp21SR* cover loci 100–1,000 nt in length (median 247.5 bp). *Pp21SR* loci mainly generate single-stranded sRNA precursors because the cloned sRNAs from these loci share the same orientation. The *Pp23SR* loci are templates for the production of long dsRNA precursors that are processed into siRNAs as inferred from the sense and antisense polarity of siRNAs derived from these loci. Most of the *Pp23SR* loci overlap with regions similar to LTR-retrotransposons and helitron elements, and 22–24 nt siRNA accumulation from these loci requires PpDCL3, whereas the accumulation of 21 nt siRNAs from these loci was unaffected in Δ*PpDCL3* mutants (Cho et al. 2008). Furthermore, *Pp23SR* loci are characterised by dense cytosine-methylation and the depletion of the 22–24 nt siRNAs in Δ*PpDCL3* mutants caused a derepression of LTR retrotransposon-associated reverse transcriptases pointing to an epigenetic control of these elements by the specific set of 22–24 nt siRNAs. Thus, the 22–24 nt sRNAs from *Pp23SR* loci are analogous in function to AtDCL3 generated 24 nt siRNAs and are involved in the repression of transposons (Cho et al. 2008). Repetitive siRNAs also control *P. patens* development, since Δ*PpDCL3* mutants show an accelerated gametophore formation (Fig. 3c) (Cho et al. 2008). The biological function of 21 nt siRNAs produced from *Pp21SR* and *Pp23SR* loci, however, remains unknown.

2.4 Secondary siRNAs

Initial sRNA silencing signals can be amplified by the generation and action of secondary siRNAs, a phenomenon known as transitivity (Axtell et al. 2006; Howell et al. 2007; Moissiard et al. 2007). Primary sRNAs that bind to their cognate target RNAs do not only trigger RNA cleavage and subsequent RNA destruction but they may also serve as primers for RdRP activity, which extends local RNA double strands and generates templates for the production of secondary siRNAs by Dicer action (Mlotshwa et al. 2008; Moissiard et al. 2007). In plants, transitivity is not common, but does occur occasionally for miRNA targets and, furthermore, is able to spread into upstream and downstream regions of the initial sRNA trigger, whereas in animals spreading of the initial signal occurs only upstream of the trigger (Alder et al. 2003; Axtell et al. 2006; Howell et al. 2007; Luo et al. 2009; Mlotshwa et al. 2008; Moissiard et al. 2007; Nishikura 2001).

Studies in *C. elegans* showed that the silencing generated from a small amount of dsRNA can become persistent and strong by the generation of secondary siRNAs. *C. elegans* secondary siRNAs possess a 5′ triphosphate group, and they almost exclusively have an antisense orientation, indicating that secondary siRNAs are synthesised de novo by RdRP in a primer-independent manner (Pak and Fire 2007; Sijen et al. 2007). In *P. patens*, the mapping of miRNA-mediated RNA cleavage

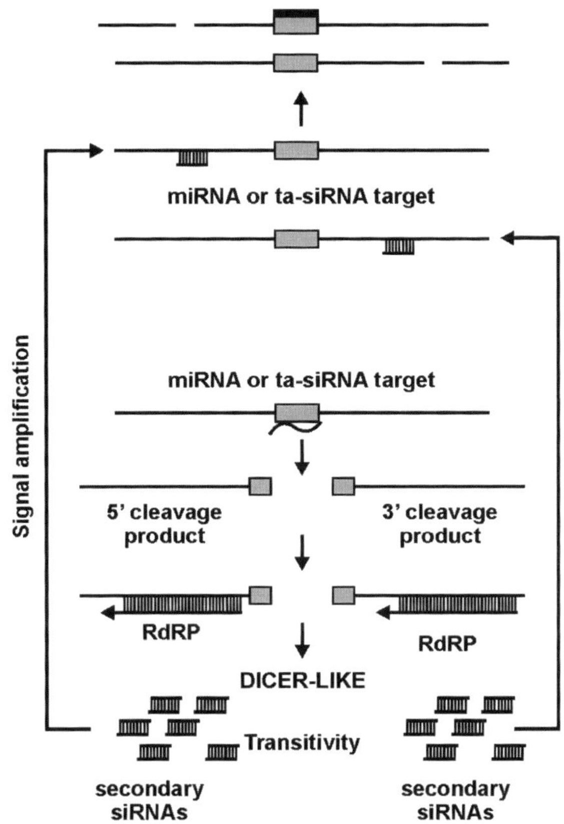

Fig. 5 Scheme for the generation of secondary, transitive siRNAs in *P. patens*. dsRNA is synthesised from cleaved miRNA or ta-siRNA targets by RdRP and processed into secondary siRNAs that mediate cleavage of the target RNA upstream and downstream of the miRNA/ ta-siRNA recognition motif, resulting in an amplification of the initial sRNA trigger. *Black line*, mRNA; *grey box*, miRNA/ta-siRNA binding site; *curved line*, miRNA/ ta-siRNA

sites by 5′RACE yielded in the amplification of additional products. Besides the expected miRNA-directed RNA cleavage products, the additional degradation products could be generated by the action of secondary siRNAs that guide cleavage of the miRNA targets at additional sites (Khraiwesh et al. 2010). It was demonstrated that the generation of these secondary siRNAs involves RdRP activity to generate dsRNA, since cDNA could be synthesised from the antisense strand of miRNA target RNAs and secondary siRNAs were detected in sense and antisense orientation (Khraiwesh et al. 2010). Furthermore, the generation of secondary siRNAs spreads into upstream and downstream regions relative to the miRNA binding site, causing transitivity of the initial miRNA trigger (Fig. 5) (Khraiwesh et al. 2010). The generation of secondary siRNAs was also observed after ta-siRNA-mediated cleavage of the ta-siRNA target *PpEREBP/AP2*. Thus, biogenesis of transitive siRNAs in *P. patens* differs from biogenesis of secondary siRNAs in *C. elegans*, as these occur in antisense polarity only due to an unprimed de novo synthesis by RdRP (Pak and Fire 2007; Sijen et al. 2007). Furthermore, transitivity in *P. patens* only occurs after miRNA-mediated cleavage of target RNAs, since Δ*PpDCL1b* mutants defective in miRNA-directed target cleavage do not generate transitive siRNAs (Khraiwesh et al. 2010). Just recently, it has become evident that secondary siRNA biogenesis in *A. thaliana* is triggered by miRNAs and ta-siRNAs with a size of 22 nt, rather than by the more typical 21 nt miRNAs and ta-siRNAs. Subsequently, eight *A. thaliana* miRNAs (miR168, miR173, miR393, miR447, miR472, miR473, miR828 and miR856) and one ta-siRNA (ta-siR2140) were identified, which are known triggers of siRNA production (Chen et al. 2010). Since secondary siRNAs in *P. patens* also derive from RNAs that are targeted by 21 nt miRNAs (miR160, miR166) and a 21 nt ta-siRNA (ta-siRNA 6(+)), further studies are required to address the differences and specificities in secondary siRNA production in mosses and seed plants.

3 miRNA-Mediated Epigenetic Gene Silencing

Besides the posttranscriptional control of RNA targets, endogenous siRNAs were shown to trigger epigenetic modifications at cognate target loci and to be associated with RNA-directed DNA methylation (RdDM) and chromatin remodelling (Dunoyer et al. 2010; Gao et al. 2010; Hamilton et al. 2002; Kanno et al. 2010; Xie et al. 2004; Zilberman et al. 2003). In plants, dsRNAs that contain sequences that are homologous to promoter regions can trigger promoter methylation via RdDM and transcriptional gene silencing (Matzke and Birchler 2005; Melquist and Bender 2003). Usually, RdDM does not spread substantially into adjacent sequences and is largely confined to the region of RNA-DNA sequence homology (Aufsatz et al. 2002; Hall et al. 2002). Besides the epigenetic silencing of LTR retrotransposons by PpDCL3-dependent 22–24 nt siRNAs (Cho et al. 2008), further evidence for the existence of sRNA-mediated epigenetic gene silencing in *P. patens* was obtained from the analysis of Δ*PpDCL1b* mutants (Khraiwesh et al. 2010). Similar to Δ*PpDCL1a* mutants, which are deficient in miRNA biogenesis, Δ*PpDCL1b* mutant lines showed developmental disorders throughout all stages of development (Fig. 3b) including abnormalities in cell division, cell size, cell shape and growth polarity, and they developed only a small number of gametophores, which in addition were malformed (Khraiwesh et al. 2010). Loss of *PpDCL1b* did not affect miRNA biogenesis, since miRNAs accumulated to equal amounts as in the wild type. However, a novel function for plant DCL proteins in miRNA-directed target cleavage was inferred for PpDCL1b because miRNA-triggered cleavage of miRNA target RNAs was abolished in the Δ*PpDCL1b* mutant lines. It is unlikely that PpDCL1b is directly involved in target cleavage since AGO proteins catalyse RNA-directed target cleavage in RISC (MacRae et al. 2008). Since animal Dicers were shown to be components of RISC-loading complexes (RLC) (Liu et al. 2003; MacRae et al. 2008; Tabara et al. 2002; Tomari and Zamore 2005), it was proposed that PpDCL1b has a similar function and acts in loading miRNAs into RISC. Given the abolished cleavage of miRNA targets in the Δ*PpDCL1b* mutants, the steady-state transcript levels of miRNA targets were expected to be elevated. Conversely, all analysed miRNA targets had strongly reduced expression levels in Δ*PpDCL1b* mutants. Subsequently, it was shown that the reduced transcript levels were due to cytosine methylation at CpG residues of the cognate miRNA target loci causing transcriptional silencing of these loci. Furthermore, DNA methylation was not restricted to the region of the encoded miRNA binding site, but spread into upstream and downstream regions including introns and promoter regions (Khraiwesh et al. 2010). In two miRNA target genes, *PpHB10* and *PpC3HDZIP1,* the miRNA binding motif is disrupted by an intron, making it unlikely that that DNA methylation is initiated by the formation of an miRNA:DNA hybrid. Instead, indirect evidence for the presence of miRNA:mRNA duplexes in the Δ*PpDCL1b* mutants led to the hypothesis that these duplexes interact with an RNA-induced silencing complex (RITS) and guide the duplex to the cognate genomic regions to initiate DNA methylation. Moreover, it was proposed that DNA methylation of miRNA target genes in the Δ*PpDCL1b* mutants is triggered by a high miRNA:target RNA ratio due to the abolished target cleavage. This model was substantiated by the expression of an artificial miRNA (amiRNA; see also below) targeting a control gene, *PpGNT1*, that is usually not under miRNA control. Methylation of the *PpGNT1* locus correlated with the expression of the *PpGNT1*-amiRNA, since it was only detected in transgenic lines with high *PpGNT1*-amiRNA expression levels. The dependence of DNA methylation on miRNA levels was also shown for an abscisic acid (ABA) responsive *PpbHLH*-miR1026 regulon in *P. patens* wild type. ABA application caused an increase of miR1026 and a decrease of its *PpbHLH* target RNA. Strikingly, besides miR1026-mediated *PpbHLH* cleavage, ABA application triggered methylation of CpG sites at the *PpbHLH* locus (Khraiwesh et al. 2010).

It was speculated that the miRNA:RNA duplexes are recognised by a RITS complex, but the involvement of such a complex in the silencing of miRNA targets in *P. patens* remains to be shown. In the fission yeast *Schizosaccharomyces pombe*, a RITS complex that associates with sRNAs and contains AGO1 (the fission yeast AGO homologue), a chromodomain protein (Chp1), and the Tas3 protein was shown to function in sRNA-directed transcriptional gene silencing (Verdel et al. 2004). This complex binds to nascent transcript and recruits an RdRP containing complex (RDRC; Rdp1, Cid12 and Hrr1) that may increase siRNA production. siRNAs bound to the RITS complex together with nascent transcripts lead to recruitment of the CLRC complex (Clr4-Rik1-Cul4), promoting H3K9 methylation and heterochromatin formation (Bayne et al. 2010; Sugiyama et al. 2005). However, clear homologues of the proteins identified in yeast RITS seem to be lacking in *P. patens* as inferred from the current *P. patens* gene models (Genome version 1.2). In *A. thaliana*, a nucleolar complex is involved in the siRNA-directed silencing of endogenous repeat regions (Pontes et al. 2009; Wierzbicki et al. 2008). Twenty-four nucleotide siRNAs generated from these regions by RNA polymerase IV (Pol IV), RDR2 and DCL3 associate with AGO4. In parallel, the nuclear RNA Pol V together with DRD1 generates transcripts from these loci. Similar to the yeast RITS siRNA-AGO4 complexes are discussed to bind to Pol V transcripts guiding the de novo cytosine methyltransferase DRM2 and histone modifying complexes to the target loci. Homologues of these proteins are also present in *P. patens* (Table 1) that might be promising candidates for functional analyses to obtain further mechanistic insight into *P. patens* RdDM pathways.

4 *Physcomitrella patens* Homologues of Small RNA Pathways Components

To date, the functional analysis of sRNA pathway components in *P. patens* is limited to *PpDCL1a*, *PpDCL1b*, *PpDCL3* and *PpRDR6*. In order to obtain a comprehensive view on the presence of *P. patens* homologues, we used *A. thaliana* proteins shown to act in different sRNA pathways as queries for reciprocal BLASTP searches in the *A. thaliana* TAIR database (http://www.arabidopsis.org) and the *P. patens* V1.2 protein database (http://www.cosmoss.org). This analysis identified *P. patens* homologues of all protein families involved in sRNA pathways in *A. thaliana* (Table 1), indicating their wide conservation over evolutionary time and, furthermore, pointing to large functional overlaps in different plant taxa. However, the size of certain protein families involved in sRNA pathways can differ between *P. patens* and seed plants. For example, the *P. patens* AGO family comprises six members, whereas ten members are present in *A. thaliana* (Axtell et al. 2007; Morel et al. 2002). These ten AGO proteins are clustered into three clades: AtAGO1, AtAGO5 and AtAGO10 within the first clade, AtAGO2, AtAGO3 and AtAGO7 within the second clade and AtAGO4, AtAGO6, AtAGO8 and AtAGO9 within the third clade (Vaucheret 2008). *P. patens* encodes three homologues of AtAGO1, the core component of miRNA-RISC, whereas homologues of *A. thaliana* AGO2, AGO3, AGO5, AGO7, AGO8 and AGO10 are missing (Table 1). Thus, besides a large overlap of sRNA-related proteins, there are particular differences in the protein repertoire that may cause deviating functions of sRNA pathways in *P. patens* and seed plants.

5 miRNA-Dependent Autoregulatory Feeback Control of sRNA Pathway Components

miRNAs are also involved in the negative feedback control of transcripts encoding catalytic sRNA pathway components. For example, in *A. thaliana*, miR162 regulates the *AtDCL1* mRNA that encodes the essential enzyme of miRNA biogenesis, suggesting an autoregulation of the complete miRNA pathway (Xie et al. 2003). Another feedback control may affect the maturation of the *AtDCL1* pre-mRNA. Intron 14 of the *AtDCL1* gene harbours a miR828 precursor sequence (Rajagopalan et al. 2006). Thus, processing of the miR828 by AtDCL1 could compete with the splicing of the *AtDCL1* pre-mRNA to control functional *AtDCL1* mRNA levels. So far, a miRNA-mediated regulation of *PpDCL1a* or *PpDCL1b* has not been identified yet. However, intron 7 of *PpDCL1a*, which is essential for miRNA biogenesis, harbours an miR1047 precursor, reminiscent to the mirtron present in *AtDCL1* (Axtell et al. 2007). Whether the proposed competition of miRNA processing and mRNA splicing are critical for *AtDCL1* and *PpDCL1a* homeostasis still awaits experimental evidence. Another conserved miRNA-mediated feedback was reported for *AGO1* mRNAs in *A. thaliana* and *P. patens*. In *A. thaliana*, the single *AtAGO1* mRNA is targeted by miR168, whereas miR904 targets three *PpAGO1* homologues (*PpAGO1a-c*) (Axtell et al. 2007; Vaucheret et al. 2004, 2006), presenting negative feedback loops, since AGO1 proteins catalyse the cleavage of miRNA targets (Vaucheret et al. 2004, 2006). Perturbation of this control loop by the expression of a miR168-resistant *AtAGO1* mRNA led to elevated *AtAGO1* transcript levels and affected development, indicating the biological relevance of this negative control loop. In addition, these lines also showed an increased abundance of miRNA target RNAs, suggesting that elevated AtAGO1 levels interfere with proper miRNA-RISC activity (Vaucheret et al. 2004). It has to be shown whether the miR904-mediated control of the *P. patens PpAGO1a-c* homologues has a similar function in the maintenance of miRNA-RISC activity. Further, a functional analysis of the three *PpAGO1* homologues will show whether they act redundantly and are functionally equivalent to the single *A. thaliana AtAGO1* or they exhibit diverse functions. The sequences of miRNAs controlling *DCL1* mRNAs (mirtrons ath-miR838 and ppt-miR1047) and *AGO1* transcripts (ath-miR168 and ppt-miR904) are not conserved between both species pointing to a convergent evolution of these control pathways.

6 Expression of Artificial miRNAs in *P. patens*

In plants and animals, the miRNA sequence within a miRNA precursor can be exchanged without affecting the processing of the miRNA, as long as the number of matches and mismatches in the fold-back structure remains unaltered (Parizotto et al. 2004; Zeng et al. 2002). This allows to modify miRNA sequences and to create artificial miRNAs (amiRNAs) that are able to target any gene of interest and to knockdown its expression at the post-transcriptional level. This method was successfully applied in different seed plants (Alvarez et al. 2006; Niu et al. 2006; Schwab et al. 2006; Warthmann et al. 2008) and subsequently adapted for specific gene knockdown in *P. patens* (Khraiwesh et al. 2008). amiRNAs targeting the *P. patens* genes *PpFtsZ2-1* and *PpGNT1*, respectively, were engineered within the *A. thaliana* miR319a precursor backbone. Upon constitutive expression of the modified precursor, both amiRNAs were precisely processed and caused cleavage and effective knockdown of the corresponding transcripts. amiRNA-mediated silencing of *PpFtsZ2-1*, which is indispensable for plastid division, caused the formation of macrochloroplasts and, thus, was sufficient to phenocopy Δ*PpFtsZ2-1* null mutants (Strepp et al. 1998). Furthermore, the *PpFtsZ2-1*-amiRNA was highly specific, since it did not affect the expression of a closely related *PpFtsZ* homologue.

Similar to natural *P. patens* miRNAs, the expression of amiRNAs caused transitivity by the generation of secondary siRNAs (Khraiwesh et al. 2008). The expression of amiRNAs in *P. patens* complements the generation of targeted knockout mutants and may have particular advantages. For example, amiRNAs can be designed to target several related genes, or they can be expressed by spatio-temporal specific or inducible promoters. The use of highly specific amiRNAs may also displace conventional inverted repeat-based RNAi constructs, since the latter produce a diverse set of siRNAs that may affect off-targets and were found to be occasionally unstable (Bezanilla et al. 2005b).

In *A. thaliana*, a similar approach was reported relying on the expression of artificial ta-siRNAs (ata-siRNAs) by engineering the *TAS1c* locus to silence the *FAD2* gene (de la Luz Gutierrez-Nava et al. 2008). The generation of ata-siRNAs has not been reported for *P. patens*, but the expression of ata-siRNAs appears to be feasible, since all *P. patens TAS3* precursors generate a specific set of phased ta-siRNAs that could be modified to target several transcripts simultaneously.

7 Conclusions and Outlook

In the last few years, considerable progress in the understanding of *P. patens* sRNA pathways has been made by high-throughput sRNA and "degradome" sequencing, as well as the functional analysis of essential components of sRNA biogenesis. In future, a combination of these techniques and the inclusion of gene expression profiling using an available whole-genome 90K microarray can be applied to the available *P. patens* mutants with perturbed sRNA pathways. The information obtained from such analyses will add to a comprehensive understanding of sRNA pathways on a genome-wide scale.

The functional analysis of sRNA components is currently limited to Δ*PpDCL1a*, Δ*PpDCL1b*, Δ*PpDCL3* and Δ*RDR6* mutants. However, it became evident that *P. patens* sRNA pathways have specific features that differ from sRNA pathways in seed plants. Since all identified components of seed-plant sRNA pathways have homologues in *P. patens*, their analysis by the generation of targeted knockout mutants may reveal further conserved as well as deviating functions.

Long-term studies of Δ*PpDCL3* mutants that lack a particular siRNA class associated with the silencing of LTR retrotransposons can provide an insight into the relevance of this siRNA-mediated control in the maintenance of genome stability.

So far, functional studies on single *P. patens* miRNAs by miRNA overexpression, miRNA target mimicry or the generation of miRNA-resistant lines by altering miRNA binding sites (which is feasible in *P. patens* without introducing additional gene copies) are completely missing. The analysis of deeply conserved miRNAs together with their conserved targets will clarify whether these miRNAs control homologous and/or analogous processes in mosses and seed plants. This also applies for *P. patens* ta-siRNAs that vary in sequence, but regulate conserved targets in seed plants and *P. patens*. Furthermore, it is desirable to start comprehensive studies of all

P. patens miRNAs and ta-siRNAs to elucidate their biological function. Many of them regulate transcripts encoding proteins with a regulatory function pointing to an important role in the control of diverse biological processes.

Acknowledgements We gratefully acknowledge financial support to our work on small RNA pathways by the Landesstiftung Baden-Württemberg (P-LS-RNS/40 to W.F.), the German Federal Ministry of Education and Research (BMBF; FRISYS 0313921 to W.F.), the German Research Foundation (DFG; FR 1677/3-1 to W.F.) and the German Academic Exchange Service (DAAD; Ph.D. fellowships to M.A.A. and I.F.).

References

Addo-Quaye C, Eshoo TW, Bartel DP et al (2008) Endogenous siRNA and miRNA targets identified by sequencing of the Arabidopsis degradome. Curr Biol 18:758–762

Addo-Quaye C, Snyder JA, Park YB et al (2009) Sliced microRNA targets and precise loop-first processing of MIR319 hairpins revealed by analysis of the *Physcomitrella patens* degradome. RNA 15:2112–2121

Adenot X, Elmayan T, Lauressergues D et al (2006) DRB4-dependent TAS3 trans-acting siRNAs control leaf morphology through AGO7. Curr Biol 16:927–932

Alder MN, Dames S, Gaudet J et al (2003) Gene silencing in *Caenorhabditis elegans* by transitive RNA interference. RNA 9:25–32

Allen E, Xie Z, Gustafson AM et al (2005) microRNA-directed phasing during trans-acting siRNA biogenesis in plants. Cell 121:207–221

Alvarez JP, Pekker I, Goldshmidt A et al (2006) Endogenous and synthetic microRNAs stimulate simultaneous, efficient, and localized regulation of multiple targets in diverse species. Plant Cell 18:1134–1151

Arazi T, Talmor-Neiman M, Stav R et al (2005) Cloning and characterization of micro-RNAs from moss. Plant J 43:837–848

Aufsätz W, Mette MF, van der Winden J et al (2002) RNA-directed DNA methylation in Arabidopsis. Proc Natl Acad Sci USA 99(Suppl 4):16499–16506

Ausin I, Mockler TC, Chory J et al (2009) IDN1 and IDN2 are required for de novo DNA methylation in Arabidopsis thaliana. Nat Struct Mol Biol 16:1325–1327

Axtell MJ (2009) The small RNAs of *Physcomitrella patens*: expression, function and evolution. Annu Plant Rev 36:113–142

Axtell MJ, Bartel DP (2005) Antiquity of microRNAs and their targets in land plants. Plant Cell 17:1658–1673

Axtell MJ, Jan C, Rajagopalan R et al (2006) A two-hit trigger for siRNA biogenesis in plants. Cell 127:565–577

Axtell MJ, Snyder JA, Bartel DP (2007) Common functions for diverse small RNAs of land plants. Plant Cell 19:1750–1769

Bartel DP (2004) MicroRNAs: genomics, biogenesis, mechanism, and function. Cell 116:281–297

Baulcombe D (2004) RNA silencing in plants. Nature 431:356–363

Baumberger N, Baulcombe DC (2005) Arabidopsis ARGONAUTE1 is an RNA Slicer that selectively recruits microRNAs and short interfering RNAs. Proc Natl Acad Sci USA 102:11928–11933

Bayne EH, White SA, Kagansky A et al (2010) Stc1: a critical link between RNAi and chromatin modification required for heterochromatin integrity. Cell 140:666–677

Bernstein E, Caudy AA, Hammond SM et al (2001) Role for a bidentate ribonuclease in the initiation step of RNA interference. Nature 409:363–366

Bezanilla M, Pan A, Quatrano RS (2003) RNA interference in the moss *Physcomitrella patens*. Plant Physiol 133:470–474

Bezanilla M, Perroud PF, Pan A et al (2005a) An RNAi system in *Physcomitrella patens* with an internal marker for silencing allows for rapid identification of loss of function phenotypes. Plant Biol 7:251–257

Bezanilla M, Perroud PF, Pan A et al (2005b) An RNAi system in *Physcomitrella patens* with an internal marker for silencing allows for rapid identification of loss of function phenotypes. Plant Biol (Stuttg) 7:251–257

Bohnsack MT, Czaplinski K, Gorlich D (2004) Exportin 5 is a RanGTP-dependent dsRNA-binding protein that mediates nuclear export of pre-miRNAs. RNA 10:185–191

Borsani O, Zhu J, Verslues PE et al (2005) Endogenous siRNAs derived from a pair of natural cis-antisense transcripts regulate salt tolerance in Arabidopsis. Cell 123:1279–1291

Bouche N, Lauressergues D, Gasciolli V et al (2006) An antagonistic function for Arabidopsis DCL2 in development and a new function for DCL4 in generating viral siRNAs. EMBO J 25:3347–3356

Brodersen P, Sakvarelidze-Achard L, Bruun-Rasmussen M et al (2008) Widespread translational inhibition by plant miRNAs and siRNAs. Science 320:1185–1190

Cai X, Hagedorn CH, Cullen BR (2004) Human microRNAs are processed from capped, polyadenylated transcripts that can also function as mRNAs. RNA 10:1957–1966

Cao X, Jacobsen SE (2002) Locus-specific control of asymmetric and CpNpG methylation by the DRM and CMT3 methyltransferase genes. Proc Natl Acad Sci USA 99(Suppl 4): 16491–16498

Cao X, Aufsatz W, Zilberman D et al (2003) Role of the DRM and CMT3 methyltransferases in RNA-directed DNA methylation. Curr Biol 13:2212–2217

Chan SW, Zilberman D, Xie Z et al (2004) RNA silencing genes control de novo DNA methylation. Science 303:1336

Chan SW, Henderson IR, Zhang X et al (2006) RNAi, DRD1, and histone methylation actively target developmentally important non-CG DNA methylation in Arabidopsis. PLoS Genet 2:e83

Chen X (2004) A microRNA as a translational repressor of APETALA2 in Arabidopsis flower development. Science 303:2022–2025

Chen HM, Chen LT, Patel K et al (2010) From the Cover: 22-nucleotide RNAs trigger secondary siRNA biogenesis in plants. Proc Natl Acad Sci USA 107:15269–15274

Cho SH, Addo-Quaye C, Coruh C et al (2008) *Physcomitrella patens* DCL3 is required for 22–24 nt siRNA accumulation, suppression of retrotransposon-derived transcripts, and normal development. PLoS Genet 4:e1000314

Dalmay T, Hamilton A, Rudd S et al (2000) An RNA-dependent RNA polymerase gene in Arabidopsis is required for posttranscriptional gene silencing mediated by a transgene but not by a virus. Cell 101:543–553

de la Luz Gutierrez-Nava M, Aukerman MJ, Sakai H et al (2008) Artificial trans-acting siRNAs confer consistent and effective gene silencing. Plant Physiol 147:543–551

Denli AM, Tops BB, Plasterk RH et al (2004) Processing of primary microRNAs by the Microprocessor complex. Nature 432:231–235

Dunoyer P, Himber C, Voinnet O (2005) DICER-LIKE 4 is required for RNA interference and produces the 21-nucleotide small interfering RNA component of the plant cell-to-cell silencing signal. Nat Genet 37:1356–1360

Dunoyer P, Brosnan CA, Schott G et al (2010) An endogenous, systemic RNAi pathway in plants. EMBO J 29:1699–1712

Elmayan T, Adenot X, Gissot L et al (2009) A neomorphic sgs3 allele stabilizing miRNA cleavage products reveals that SGS3 acts as a homodimer. FEBS J 276:835–844

Emery JF, Floyd SK, Alvarez J et al (2003) Radial patterning of Arabidopsis shoots by class III HD-ZIP and KANADI genes. Curr Biol 13:1768–1774

Fahlgren N, Montgomery TA, Howell MD et al (2006) Regulation of AUXIN RESPONSE FACTOR3 by TAS3 ta-siRNA affects developmental timing and patterning in Arabidopsis. Curr Biol 16:939–944

Fang Y, Spector DL (2007) Identification of nuclear dicing bodies containing proteins for microRNA biogenesis in living Arabidopsis plants. Curr Biol 17:818–823

Fattash I, Voss B, Reski R et al (2007) Evidence for the rapid expansion of microRNA-mediated regulation in early land plant evolution. BMC Plant Biol 7:13

Floyd SK, Bowman JL (2004) Gene regulation: ancient microRNA target sequences in plants. Nature 428:485–486

Gan J, Tropea JE, Austin BP et al (2006) Structural insight into the mechanism of double-stranded RNA processing by ribonuclease III. Cell 124:355–366

Gandikota M, Birkenbihl RP, Hohmann S et al (2007) The miRNA156/157 recognition element in the 3′ UTR of the Arabidopsis SBP box gene SPL3 prevents early flowering by translational inhibition in seedlings. Plant J 49:683–693

Gao Z, Liu HL, Daxinger L et al (2010) An RNA polymerase II- and AGO4-associated protein acts in RNA-directed DNA methylation. Nature 465:106–109

Garcia D, Collier SA, Byrne ME et al (2006) Specification of leaf polarity in Arabidopsis via the trans-acting siRNA pathway. Curr Biol 16:933–938

Gasciolli V, Mallory AC, Bartel DP et al (2005) Partially redundant functions of Arabidopsis DICER-like enzymes and a role for DCL4 in producing trans-acting siRNAs. Curr Biol 15:1494–1500

Golden TA, Schauer SE, Lang JD et al (2002) SHORT INTEGUMENTS1/SUSPENSOR1/CARPEL FACTORY, a Dicer homolog, is a maternal effect gene required for embryo development in Arabidopsis. Plant Physiol 130:808–822

Gregory RI, Yan KP, Amuthan G et al (2004) The Microprocessor complex mediates the genesis of microRNAs. Nature 432:235–240

Griffiths-Jones S, Grocock RJ, van Dongen S et al (2006) miRBase: microRNA sequences, targets and gene nomenclature. Nucleic Acids Res 34:D140–D144

Griffiths-Jones S, Saini HK, van Dongen S et al (2008) miRBase: tools for microRNA genomics. Nucleic Acids Res 36:D154–D158

Hall IM, Shankaranarayana GD, Noma K et al (2002) Establishment and maintenance of a heterochromatin domain. Science 297:2232–2237

Hamilton A, Voinnet O, Chappell L et al (2002) Two classes of short interfering RNA in RNA silencing. EMBO J 21:4671–4679

Hammond SM, Bernstein E, Beach D et al (2000) An RNA-directed nuclease mediates post-transcriptional gene silencing in Drosophila cells. Nature 404:293–296

Han J, Lee Y, Yeom KH et al (2004) The Drosha-DGCR8 complex in primary microRNA processing. Genes Dev 18:3016–3027

Havecker ER, Wallbridge LM, Hardcastle TJ et al (2010) The Arabidopsis RNA-directed DNA methylation argonautes functionally diverge based on their expression and interaction with target loci. Plant Cell 22:321–334

Howell MD, Fahlgren N, Chapman EJ et al (2007) Genome-wide analysis of the RNA-DEPENDENT RNA POLYMERASE6/DICER-LIKE4 pathway in Arabidopsis reveals dependency on miRNA- and tasiRNA-directed targeting. Plant Cell 19:926–942

Huettel B, Kanno T, Daxinger L et al (2006) Endogenous targets of RNA-directed DNA methylation and Pol IV in Arabidopsis. EMBO J 25:2828–2836

Huettel B, Kanno T, Daxinger L et al (2007) RNA-directed DNA methylation mediated by DRD1 and Pol IVb: a versatile pathway for transcriptional gene silencing in plants. Biochim Biophys Acta 1769:358–374

Hutvagner G, Simard MJ (2008) Argonaute proteins: key players in RNA silencing. Nat Rev Mol Cell Biol 9:22–32

Jeddeloh JA, Stokes TL, Richards EJ (1999) Maintenance of genomic methylation requires a SWI2/SNF2-like protein. Nat Genet 22:94–97

Juarez MT, Kui JS, Thomas J et al (2004) microRNA-mediated repression of rolled leaf1 specifies maize leaf polarity. Nature 428:84–88

Kanno T, Bucher E, Daxinger L et al (2010) RNA-directed DNA methylation and plant development require an IWR1-type transcription factor. EMBO Rep 11:65–71

Kasschau KD, Fahlgren N, Chapman EJ et al (2007) Genome-wide profiling and analysis of Arabidopsis siRNAs. PLoS Biol 5:e57

Kenrick P, Crane P (1997) The origin and early evolution of plants on land. Nature 389:33–39

Khraiwesh B, Ossowski S, Weigel D et al (2008) Specific gene silencing by artificial MicroRNAs in *Physcomitrella patens*: an alternative to targeted gene knockouts. Plant Physiol 148:684–693

Khraiwesh B, Arif MA, Seumel GI et al (2010) Transcriptional control of gene expression by microRNAs. Cell 140:111–122

Kim VN (2004) MicroRNA precursors in motion: exportin-5 mediates their nuclear export. Trends Cell Biol 14:156–159

Kim J, Jung JH, Reyes JL et al (2005) microRNA-directed cleavage of ATHB15 mRNA regulates vascular development in Arabidopsis inflorescence stems. Plant J 42:84–94

Kurihara Y, Watanabe Y (2004) Arabidopsis micro-RNA biogenesis through Dicer-like 1 protein functions. Proc Natl Acad Sci USA 101:12753–12758

Lagos-Quintana M, Rauhut R, Lendeckel W et al (2001) Identification of novel genes coding for small expressed RNAs. Science 294:853–858

Lanet E, Delannoy E, Sormani R et al (2009) Biochemical evidence for translational repression by Arabidopsis microRNAs. Plant Cell 21:1762–1768

Lau NC, Lim LP, Weinstein EG et al (2001) An abundant class of tiny RNAs with probable regulatory roles in *Caenorhabditis elegans*. Science 294:858–862

Law JA, Ausin I, Johnson LM et al (2010) A protein complex required for polymerase V transcripts and RNA-directed DNA methylation in Arabidopsis. Curr Biol 20:951–956

Lee RC, Ambros V (2001) An extensive class of small RNAs in *Caenorhabditis elegans*. Science 294:862–864

Lee RC, Feinbaum RL, Ambros V (1993) The *C. elegans* heterochronic gene lin-4 encodes small RNAs with antisense complementarity to lin-14. Cell 75:843–854

Lee Y, Jeon K, Lee JT et al (2002) MicroRNA maturation: stepwise processing and subcellular localization. EMBO J 21:4663–4670

Lee Y, Kim M, Han J et al (2004) MicroRNA genes are transcribed by RNA polymerase II. EMBO J 23:4051–4060

Liu Q, Rand TA, Kalidas S et al (2003) R2D2, a bridge between the initiation and effector steps of the Drosophila RNAi pathway. Science 301:1921–1925

Liu B, Chen Z, Song X et al (2007a) Oryza sativa dicer-like4 reveals a key role for small interfering RNA silencing in plant development. Plant Cell 19:2705–2718

Liu PP, Montgomery TA, Fahlgren N et al (2007b) Repression of AUXIN RESPONSE FACTOR10 by microRNA160 is critical for seed germination and post-germination stages. Plant J 52:133–146

Liu Q, Yao X, Pi L et al (2008) The ARGONAUTE10 gene modulates shoot apical meristem maintenance and leaf polarity establishment by repressing miR165/166 in Arabidopsis. Plant J 58:27–40

Lu C, Jeong DH, Kulkarni K et al (2008) Genome-wide analysis for discovery of rice microRNAs reveals natural antisense microRNAs (nat-miRNAs). Proc Natl Acad Sci USA 105:4951–4956

Luo QJ, Samanta MP, Koksal F et al (2009) Evidence for antisense transcription associated with microRNA target mRNAs in Arabidopsis. PLoS Genet 5:e1000457

MacRae IJ, Ma E, Zhou M et al (2008) In vitro reconstitution of the human RISC-loading complex. Proc Natl Acad Sci USA 105:512–517

Mallory AC, Bartel DP, Bartel B (2005) MicroRNA-directed regulation of Arabidopsis AUXIN RESPONSE FACTOR17 is essential for proper development and modulates expression of early auxin response genes. Plant Cell 17:1360–1375

Mallory AC, Hinze A, Tucker MR et al (2009) Redundant and specific roles of the ARGONAUTE proteins AGO1 and ZLL in development and small RNA-directed gene silencing. PLoS Genet 5:e1000646

Matzke MA, Birchler JA (2005) RNAi-mediated pathways in the nucleus. Nat Rev Genet 6:24–35

Matzke MA, Matzke AJ (2004) Planting the seeds of a new paradigm. PLoS Biol 2:E133

Matzke M, Kanno T, Huettel B et al (2006) RNA-directed DNA methylation and Pol IVb in Arabidopsis. Cold Spring Harb Symp Quant Biol 71:449–459

Melquist S, Bender J (2003) Transcription from an upstream promoter controls methylation signaling from an inverted repeat of endogenous genes in Arabidopsis. Genes Dev 17:2036–2047

Mlotshwa S, Pruss GJ, Peragine A et al (2008) DICER-LIKE2 plays a primary role in transitive silencing of transgenes in Arabidopsis. PLoS ONE 3:e1755

Moissiard G, Parizotto EA, Himber C et al (2007) Transitivity in Arabidopsis can be primed, requires the redundant action of the antiviral Dicer-like 4 and Dicer-like 2, and is compromised by viral-encoded suppressor proteins. RNA 13:1268–1278

Montgomery TA, Howell MD, Cuperus JT et al (2008) Specificity of ARGONAUTE7-miR390 interaction and dual functionality in TAS3 trans-acting siRNA formation. Cell 133:128–141

Morel JB, Mourrain P, Beclin C et al (2000) DNA methylation and chromatin structure affect transcriptional and post-transcriptional transgene silencing in Arabidopsis. Curr Biol 10:1591–1594

Morel JB, Godon C, Mourrain P et al (2002) Fertile hypomorphic ARGONAUTE (ago1) mutants impaired in post-transcriptional gene silencing and virus resistance. Plant Cell 14:629–639

Morin RD, Aksay G, Dolgosheina E et al (2008) Comparative analysis of the small RNA transcriptomes of Pinus contorta and *Oryza sativa*. Genome Res 18:571–584

Nagasaki H, Itoh J, Hayashi K et al (2007) The small interfering RNA production pathway is required for shoot meristem initiation in rice. Proc Natl Acad Sci USA 104:14867–14871

Nishikura K (2001) A short primer on RNAi: RNA-directed RNA polymerase acts as a key catalyst. Cell 107:415–418

Nishiyama T, Fujita T, Shin IT et al (2003) Comparative genomics of *Physcomitrella patens* gametophytic transcriptome and *Arabidopsis thaliana*: implication for land plant evolution. Proc Natl Acad Sci USA 100:8007–8012

Niu QW, Lin SS, Reyes JL et al (2006) Expression of artificial microRNAs in transgenic Arabidopsis thaliana confers virus resistance. Nat Biotechnol 24:1420–1428

Nobuta K, McCormick K, Nakano M et al (2010) Bioinformatics analysis of small RNAs in plants using next generation sequencing technologies. Methods Mol Biol 592:89–106

Noma K, Sugiyama T, Cam H et al (2004) RITS acts in cis to promote RNA interference-mediated transcriptional and post-transcriptional silencing. Nat Genet 36:1174–1180

Olmedo-Monfil V, Duran-Figueroa N, Arteaga-Vazquez M et al (2010) Control of female gamete formation by a small RNA pathway in Arabidopsis. Nature 464:628–632

Pak J, Fire A (2007) Distinct populations of primary and secondary effectors during RNAi in *C. elegans*. Science 315:241–244

Pandey SP, Gaquerel E, Gase K et al (2008) RNA-directed RNA polymerase3 from Nicotiana attenuata is required for competitive growth in natural environments. Plant Physiol 147:1212–1224

Parizotto EA, Dunoyer P, Rahm N et al (2004) In vivo investigation of the transcription, processing, endonucleolytic activity, and functional relevance of the spatial distribution of a plant miRNA. Genes Dev 18:2237–2242

Park W, Li J, Song R et al (2002) CARPEL FACTORY, a Dicer homolog, and HEN1, a novel protein, act in microRNA metabolism in *Arabidopsis thaliana*. Curr Biol 12:1484–1495

Park MY, Wu G, Gonzalez-Sulser A et al (2005) Nuclear processing and export of microRNAs in Arabidopsis. Proc Natl Acad Sci USA 102:3691–3696

Peragine A, Yoshikawa M, Wu G et al (2004) SGS3 and SGS2/SDE1/RDR6 are required for juvenile development and the production of trans-acting siRNAs in Arabidopsis. Genes Dev 18:2368–2379

Pontes O, Li CF, Nunes PC et al (2006) The Arabidopsis chromatin-modifying nuclear siRNA pathway involves a nucleolar RNA processing center. Cell 126:79–92

Pontes O, Costa-Nunes P, Vithayathil P et al (2009) RNA polymerase V functions in Arabidopsis interphase heterochromatin organization independently of the 24-nt siRNA-directed DNA methylation pathway. Mol Plant 2:700–710

Qi Y, Denli AM, Hannon GJ (2005) Biochemical specialization within Arabidopsis RNA silencing pathways. Mol Cell 19:421–428

Qin H, Chen F, Huan X et al (2010) Structure of the Arabidopsis thaliana DCL4 DUF283 domain reveals a noncanonical double-stranded RNA-binding fold for protein-protein interaction. RNA 16:474–481

Rajagopalan R, Vaucheret H, Trejo J et al (2006) A diverse and evolutionarily fluid set of microRNAs in *Arabidopsis thaliana*. Genes Dev 20:3407–3425

Reinhart BJ, Weinstein EG, Rhoades MW et al (2002) MicroRNAs in plants. Genes Dev 16:1616–1626

Rensing SA, Lang D, Zimmer AD et al (2008) The *Physcomitrella* genome reveals evolutionary insights into the conquest of land by plants. Science 319:64–69

Satoh N, Itoh J, Nagato Y (2003) The SHOOTLESS2 and SHOOTLESS1 genes are involved in both initiation and maintenance of the shoot apical meristem through regulating the number of indeterminate cells. Genetics 164:335–346

Schaefer DG, Zryd JP (1997) Efficient gene targeting in the moss *Physcomitrella patens*. Plant J 11:1195–1206

Schramke V, Allshire R (2004) Those interfering little RNAs! Silencing and eliminating chromatin. Curr Opin Genet Dev 14:174–180

Schwab R, Palatnik JF, Riester M et al (2005) Specific effects of microRNAs on the plant transcriptome. Dev Cell 8:517–527

Schwab R, Ossowski S, Riester M et al (2006) Highly specific gene silencing by artificial microRNAs in Arabidopsis. Plant Cell 18:1121–1133

Sempere LF, Cole CN, McPeek MA et al (2006) The phylogenetic distribution of metazoan microRNAs: insights into evolutionary complexity and constraint. J Exp Zool B Mol Dev Evol 306:575–588

Shaked H, Avivi-Ragolsky N, Levy AA (2006) Involvement of the Arabidopsis SWI2/SNF2 chromatin remodeling gene family in DNA damage response and recombination. Genetics 173:985–994

Sijen T, Steiner FA, Thijssen KL et al (2007) Secondary siRNAs result from unprimed RNA synthesis and form a distinct class. Science 315:244–247

Smith LM, Pontes O, Searle I et al (2007) An SNF2 protein associated with nuclear RNA silencing and the spread of a silencing signal between cells in Arabidopsis. Plant Cell 19:1507–1521

Song L, Axtell MJ, Fedoroff NV (2010) RNA secondary structural determinants of miRNA precursor processing in Arabidopsis. Curr Biol 20:37–41

Strepp R, Scholz S, Kruse S et al (1998) Plant nuclear gene knockout reveals a role in plastid division for the homolog of the bacterial cell division protein FtsZ, an ancestral tubulin. Proc Natl Acad Sci USA 95:4368–4373

Sugiyama T, Cam H, Verdel A et al (2005) RNA-dependent RNA polymerase is an essential component of a self-enforcing loop coupling heterochromatin assembly to siRNA production. Proc Natl Acad Sci USA 102:152–157

Tabara H, Yigit E, Siomi H et al (2002) The dsRNA binding protein RDE-4 interacts with RDE-1, DCR-1, and a DExH-box helicase to direct RNAi in *C. elegans*. Cell 109:861–871

Talmor-Neiman M, Stav R, Klipcan L et al (2006) Identification of trans-acting siRNAs in moss and an RNA-dependent RNA polymerase required for their biogenesis. Plant J 48:511–521

Timmermans MC, Juarez MT, Phelps-Durr TL (2004) A conserved microRNA signal specifies leaf polarity. Cold Spring Harb Symp Quant Biol 69:409–417

Tomari Y, Zamore PD (2005) Perspective: machines for RNAi. Genes Dev 19:517–529

Toriba T, Suzaki T, Yamaguchi T et al (2010) Distinct regulation of adaxial-abaxial polarity in anther patterning in rice. Plant Cell 22:1452–1462

Vaistij FE, Jones L, Baulcombe DC (2002) Spreading of RNA targeting and DNA methylation in RNA silencing requires transcription of the target gene and a putative RNA-dependent RNA polymerase. Plant Cell 14:857–867

Vaucheret H (2008) Plant ARGONAUTES. Trends Plant Sci 13:350–358

Vaucheret H, Vazquez F, Crete P et al (2004) The action of ARGONAUTE1 in the miRNA pathway and its regulation by the miRNA pathway are crucial for plant development. Genes Dev 18:1187–1197

Vaucheret H, Mallory AC, Bartel DP (2006) AGO1 homeostasis entails coexpression of MIR168 and AGO1 and preferential stabilization of miR168 by AGO1. Mol Cell 22:129–136

Vazquez F (2006) Arabidopsis endogenous small RNAs: highways and byways. Trends Plant Sci 11:460–468

Vazquez F, Vaucheret H, Rajagopalan R et al (2004) Endogenous trans-acting siRNAs regulate the accumulation of Arabidopsis mRNAs. Mol Cell 16:69–79

Verdel A, Jia S, Gerber S et al (2004) RNAi-mediated targeting of heterochromatin by the RITS complex. Science 303:672–676

Voinnet O (2009) Origin, biogenesis, and activity of plant microRNAs. Cell 136:669–687

Wang JW, Wang LJ, Mao YB et al (2005) Control of root cap formation by MicroRNA-targeted auxin response factors in Arabidopsis. Plant Cell 17:2204–2216

Wang JW, Schwab R, Czech B et al (2008) Dual effects of miR156-targeted SPL genes and CYP78A5/KLUH on plastochron length and organ size in Arabidopsis thaliana. Plant Cell 20:1231–1243

Wang JW, Czech B, Weigel D (2009) miR156-regulated SPL transcription factors define an endogenous flowering pathway in Arabidopsis thaliana. Cell 138:738–749

Warthmann N, Chen H, Ossowski S et al (2008) Highly specific gene silencing by artificial miRNAs in rice. PLoS ONE 3:e1829

Wierzbicki AT, Haag JR, Pikaard CS (2008) Noncoding transcription by RNA polymerase Pol IVb/Pol V mediates transcriptional silencing of overlapping and adjacent genes. Cell 135:635–648

Williams L, Carles CC, Osmont KS et al (2005) A database analysis method identifies an endogenous trans-acting short-interfering RNA that targets the Arabidopsis ARF2, ARF3, and ARF4 genes. Proc Natl Acad Sci USA 102:9703–9708

Wu G, Poethig RS (2006) Temporal regulation of shoot development in Arabidopsis thaliana by miR156 and its target SPL3. Development 133:3539–3547

Xie Z, Kasschau KD, Carrington JC (2003) Negative feedback regulation of Dicer-Like1 in Arabidopsis by microRNA-guided mRNA degradation. Curr Biol 13:784–789

Xie Z, Johansen LK, Gustafson AM et al (2004) Genetic and functional diversification of small RNA pathways in plants. PLoS Biol 2:E104

Xie Z, Allen E, Wilken A et al (2005) DICER-LIKE 4 functions in trans-acting small interfering RNA biogenesis and vegetative phase change in Arabidopsis thaliana. Proc Natl Acad Sci USA 102:12984–12989

Yamaguchi A, Wu MF, Yang L et al (2009) The microRNA-regulated SBP-Box transcription factor SPL3 is a direct upstream activator of LEAFY, FRUITFULL, and APETALA1. Dev Cell 17:268–278

Yang Z, Ebright YW, Yu B et al (2006) HEN1 recognizes 21–24 nt small RNA duplexes and deposits a methyl group onto the 2′ OH of the 3′ terminal nucleotide. Nucleic Acids Res 34:667–675

Yao X, Wang H, Li H et al (2009) Two types of cis-acting elements control the abaxial epidermis-specific transcription of the MIR165a and MIR166a genes. FEBS Lett 583:3711–3717

Yoshikawa M, Peragine A, Park MY et al (2005) A pathway for the biogenesis of trans-acting siRNAs in Arabidopsis. Genes Dev 19:2164–2175

Yu B, Yang Z, Li J et al (2005) Methylation as a crucial step in plant microRNA biogenesis. Science 307:932–935

Zeng Y, Wagner EJ, Cullen BR (2002) Both natural and designed micro RNAs can inhibit the expression of cognate mRNAs when expressed in human cells. Mol Cell 9:1327–1333

Zheng X, Zhu J, Kapoor A et al (2007) Role of Arabidopsis AGO6 in siRNA accumulation, DNA methylation and transcriptional gene silencing. EMBO J 26:1691–1701

Zheng Z, Xing Y, He XJ et al (2010) An SGS3-like protein functions in RNA-directed DNA methylation and transcriptional gene silencing in Arabidopsis. Plant J 62:92–99

Zilberman D, Cao X, Jacobsen SE (2003) ARGONAUTE4 control of locus-specific siRNA accumulation and DNA and histone methylation. Science 299:716–719

Zilberman D, Cao X, Johansen LK et al (2004) Role of Arabidopsis ARGONAUTE4 in RNA-directed DNA methylation triggered by inverted repeats. Curr Biol 14:1214–1220

Zong J, Yao X, Yin J et al (2009) Evolution of the RNA-dependent RNA polymerase (RdRP) genes: duplications and possible losses before and after the divergence of major eukaryotic groups. Gene 447:29–39

Small RNA Pathways and Their Function in the Male Gametophyte

Hugh Dickinson and Robert Grant-Downton

Contents

Abstract In animals, small RNAs are essential for reproductive development. Without the activity of members of the Piwi class of the Argonaute protein family, and their associated small RNAs, the formation and maintenance of the germline, and gametogenesis, cannot take place correctly. Reproductive development in plants is more complex than in animals; instead of the haploid products of meiosis leading directly to gametes, further mitotic divisions contribute to the formation of a separate haploid generation called the gametophyte. As might be expected, the formation of the gametophyte and gametogenesis also relies on small RNA systems. In plants, almost all of the information so far gathered on the gametophyte has come from studies of the male gametophyte (pollen). Here, recent studies have revealed all the families of small RNAs known from the somatic cells in the diploid sporophyte – microRNAs, *trans*-acting siRNAs, natural antisense siRNAs and siRNAs to be involved in RNA-dependent DNA methylation. Given the apparent simplicity of the development of the male gametophyte, microRNAs and tasiRNAs are unexpectedly diverse. As in animals, plants use small RNA systems to control transposable element activity in the germline.

H. Dickinson (✉) and R. Grant-Downton
Department of Plant Sciences, University of Oxford,
South Parks Road, OX1 3RB, Oxford, UK
e-mail: hugh.dickinson@plants.ox.ac.uk

V.A. Erdmann and J. Barciszewski (eds.), *Non Coding RNAs in Plants*,
RNA Technologies, DOI 10.1007/978-3-642-19454-2_11,

There is also recent evidence that a specific regulatory module of a natural antisense gene pair, that spawns nat-siRNAs in the sperm cells, has a key role in fertilisation.

Keywords Gametophyte • microRNA • Pollen • siRNA • small RNA • ta-siRNA

1 Introduction

Since their initial discovery, studies of epigenetic mechanisms based on small non coding RNAs eukaryotes have largely focused on their roles in the somatic cells and somatic development of eukaryotes. However, in the past 5 years there has been dramatic progress in understanding the importance of small non coding RNAs in the reproductive stage. Initially, the major discoveries of the unique nature and roles of small RNA systems were made in model animals such as *Drosophila melanogaster* and *Caenorhabditis elegans*. In this chapter, briefly we will review recent developments in animal reproductive development studies that have put small RNAs at the forefront of reproductive biology. We will then discuss new data emerging from studies of plant reproductive biology that has highlighted the significance of small RNA systems. By comparison with the plethora of data on small RNAs in animal reproduction, our knowledge of such molecules in plant reproduction remains limited with progress only being made possible through new generation sequencing technology and by advances in cell isolation techniques. Nevertheless, the more severe technical challenges of working with female reproductive cells in plants has resulted in virtually all data so far coming from studies of male reproductive development. Despite these limitations, interesting parallels and apparent differences between the use of small RNAs in animal and plant reproductive development are already emerging.

2 A Diversity of Small RNAs in Eukaryotic Reproductive Development

In eukaryotic reproduction, small RNAs have been recruited to serve different purposes. Ciliates have complex sexual cycles involving drastic nuclear changes (Durharcourt et al. 2009). For example, in the reproductive phase of ciliates such as *Tetrahymena* and *Paramecium*, a highly polyploid somatic macronucleus is produced from copies of the zygotic nucleus after fertilisation. During development of the new macronucleus, which is responsible for somatic gene expression, a reproducible, programmed DNA elimination occurs that eliminates repetitive DNA and other selected sequences. Interestingly, these patterns of sequence elimination are heritable, using the old, residual macronucleus in the zygote as a template. The residual macronucleus transcribes long, non coding RNAs whereas the germline micronucleus transcribes 28–29 nt small scan RNAs (scnRNAs) across its genome specifically in meiosis. Base-pairing interactions between the two types of non coding RNAs remove homologous scnRNAs via a subtractive mechanism, but those scnRNAs not "subtracted" by a genomic scan act on the new zygotic macronucleus. These scnRNAs guide chromatin modifications to homologous sites on the zygotic macronucleus, and these epigenetic marks appear to act as guide marks for excision of the corresponding DNA sequence. Through this RNA-guided mechanism, programmed DNA rearrangements of the macronucleus occur every sexual generation. Although these ciliates show constitutive expression of a smaller size class of small RNAs (23–24 nt), these can be distinguished from the scnRNAs as they lack 2′-*O*-methylation mediated by the HEN1 methyltransferase (Kurth and Mochizuki 2009).

Interestingly, all the Argonaute family proteins of *Tetrahymena* are of the Piwi subfamily (Kurth and Mochizuki 2009) which has been revealed to be of central importance to reproductive development in metazoans. The presence of a novel class of small RNAs in metazoan germline cells was first discovered by virtue of their larger size (~24–32 nt) as well as their unique 2′-*O*-methylation (Klattenhoff and Theurkauf 2008; Thomson and Lin 2009). These novel small RNAs were found to be associated with the Piwi subfamily of Argonaute proteins, hence they were named piRNAs. The genomic origin of piRNAs was found to be unusual, being derived from many loci spread throughout the genome. Some piRNAs (class I) are derived from clusters transcribing long primary transcripts that are processed into primary piRNAs by a mechanism that remains poorly understood (Thomson and Lin 2009; Lau 2010). Other piRNAs, class II piRNAs, are derived from more widespread loci (Lau 2010). A key to their role in the genome has been the recognition that many piRNAs are homologous to transposons, with binding of piRNA and the Piwi protein resulting in cleavage of the transposon mRNA into further 28 nt piRNAs. These act to amplify piRNA levels by inducing further cleavage of piRNA precursors into more piRNAs, in what is now called the "ping-pong mechanism". Whilst defending the germline against transposon activity by degradation of transposon RNA is a major function of piRNAs, piRNAs also act to guide repressive epigenetic marks in the genome that prevent further transcription. There is additional evidence that piRNAs have other regulatory roles in endogenous gene function, from setting up both activating and repressive chromatin states to post-transcriptional effects at the translational level and mRNA localisation (Thomson and Lin 2009; Lau 2010). A characteristic feature of germline cells in many different organisms, which has been long known but poorly understood, is the presence of electron-opaque nucleoprotein granules in the perinuclear cytoplasm (Arkov and Ramos 2010). Importantly, Piwi proteins localise to these granules, together with their bound piRNAs, along with other proteins such as Tudor-domain and DEAD-box helicases (Arkov and Ramos 2010). Mutations affecting these proteins, such as Piwi-class Argonautes, generate defects in meiosis, germline development and gametogenesis (Thomson and Lin 2009). Intriguingly, maternal deposition and inheritance of piRNAs in granules may have a protective role against transposon activity in early embryogenesis and even in later development in *Drosophila* (Brennecke et al. 2008).

In metazoan germlines, piRNAs are not the only class of non coding RNAs that have major roles in reproductive development. As might be expected from their pivotal importance in somatic development, microRNAs have been shown to have multiple roles in animal reproduction, at all the different stages of germline development and gametogenesis (Reynolds and Ruohola-Baker 2008). A consistent observation in different organisms has been that mutations in genes involved in the microRNA pathway have deleterious effects on the germline and gametogenesis (Lau 2010). However, the nature and degree of the effect of perturbing microRNA function varies from organism to organism depending on the overlap with other small RNA systems (Lau 2010). As with piRNAs, there is evidence that specific microRNAs make an important contribution to regulating chromatin states in the mouse germline (Benetti et al. 2008). Similar to piRNAs, it appears that microRNAs can be inherited from the previous generation via gametes, although their post-fertilisation effect remains little understood (e.g. Wagner et al. 2008). Although obscured by piRNAs, other more conventionally generated endogenous siRNAs of ~21 nt size have emerged as players in reproductive development different organisms. These siRNAs may have a function in targeting genes, for example those with roles in cytoskeletal organisation (Lau 2010; Tam et al. 2008; Watanabe et al. 2008).

Of all the animals studied so far, investigation of small RNA systems in *C. elegans* may have particular relevance to plants as, in common with them, its genome has an expanded number of Argonaute proteins. In reproductive development, this has engendered the evolution of even further small RNA classes, in addition to conventional microRNAs and siRNAs. *C. elegans* can boast three distinct classes in the gonads – 21U, 22G and 26G RNAs (Lau 2010). These are classified on both size and the preference for the 5′ residue. 21U small RNAs appear to be the closest to piRNAs, being different in size, genomic origin and mode of biogenesis (Lau 2010). However, 21U RNAs also localise to granular structures in the germline and are known to be essential for fertility, in particular temperature-dependent fertility (Batista et al. 2008). The 22G class are a complex population of small RNAs that are also required for fertility and appear to have an important role in genome surveillance, such as aberrant transcripts and transposons (Gu et al. 2009). Finally, the 26G small RNAs have been shown to have a role in regulating mRNAs; one class targets genes expressed in spermatogenesis whilst the other class are specific to the oocyte and are deposited into the zygote (Han et al. 2009). The role of both sets of 26G small RNAs appears to be repression of the transcripts; in the case of the maternal class of 26G, their role may be in the clearance of maternal transcripts during zygote development. These small RNAs would appear to have roles in fertility and gametogenesis as well as post-fertilisation development, as mutations that block their biogenesis have effects on gamete development and embryonic lethality (Gent et al. 2009).

Why does the animal germline require such an enhancement of the small RNA systems? Inevitably, generating and maintaining a specialised cell lineage will require significant input from small RNAs to co-ordinate the regulation of numerous transcripts during development. However, what makes the germline so special is the fact that the ultimate product, the gamete, initiates an entirely new generation after syngamy. From the perspective of survival and propagation of selfish genetic elements, such as transposons, from one generation to the next, this stage is critical, even if their resulting activity proves deleterious to the host. Interestingly, meiosis itself may indeed be the architect of this special need for regulation in the germline, as there is some evidence that there is a transient relaxation of transposon silencing at the onset of meiosis (van der Heijden and Bortvin 2009). Why such a derepression of transposon silencing occurs at such a critical time remains the subject of debate. It is possible that it is an unwanted consequence of altered chromatin states necessary for entry into and progression through meiosis. Equally, it could be that the repetitive component of the genome is utilised in guiding chromosome interactions in meiosis, and these regions are recognised by proteins involved in generating the architecture of meiosis, such as synaptonemal complexes (van der Heijden and Bortvin 2009; Renauld 1997). Recent work has shown that in the human genome germline transposition remains an event that is still not under complete control by epigenetic systems, as virtually all individual human genomes show evidence of germline transposition events (Iskow et al. 2010).

3 Ontogeny of the Angiosperm Male Gametophyte

The evolution of terrestrial plants is characterised by evolutionary changes to the “alternation of generations”, wherein the process of sexual reproduction the products of meiosis does not give rise directly to the gametes but instead the haploid cells undergo further mitotic divisions and development to generate a separate generation that gives rises to the gametes. In some land plants, for example the bryophytes (such as the moss *Physcomitrella patens*), the haploid gametophyte stage is the dominant phase of the plant life cycle (During 1979). In contrast to the leafy, photosynthetic gametophyte, the diploid sporophytic stage is far less prominent and dependent on the “host” gametophyte for anchoring and nutrients.

In seed plants, the diploid sporophyte is the dominant stage and through evolution there is evidence for progressive diminution of gametophyte development. In the male gametophyte, which gives rise to the sperm cells, there is a general evolutionary trend for a diminution in the number of cells, reduced developmental complexity and also an increasingly transitory time spent in this state (Rudall and Bateman 2007; Williams 2008). In angiosperms, the male gametophyte is characterised by two mitotic divisions that occur after mitosis. The haploid cells produced by mitosis, the microspores, undergo a highly asymmetric division at pollen mitosis 1 that produces a larger vegetative cell and a smaller generative cell (McCormick 2004; Borg et al. 2009). The generative cell then migrates from its position at the side of the larger vegetative cell, and into its cytoplasm. The generative cell can either remain in that state as the male gametophyte matures, or it can initiate a further mitotic division (pollen mitosis 2) to form two sperm cells before maturation: these differences result in pollen at the point of release from the anthers that is bicellular and tricellular respectively. Late pollen development is characterised by accumulation and storage of transcripts necessary for germination of the pollen grain on a receptive stigma and subsequent growth, and also by progressive storage of carbohydrate and lipid reserves to sustain post-germination growth. The final phase of pollen development before liberation from the anthers is dehydration, with progressive reduction in the water content of the pollen grain that facilitates long-distance dispersal and survival in hostile aerial environments.

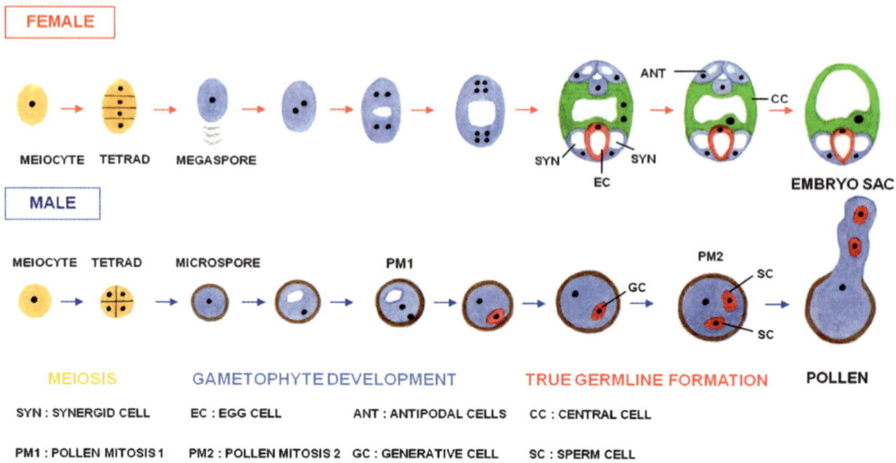

Fig. 1 Development of the male and female gametophytes of *Arabidopsis thaliana* (not to scale) showing the key developmental phases of meiosis, gametophyte development and germline formation. Reproduced with permission from Dickinson and Grant-Downton (2009)

Upon meeting a suitable stigmatic surface, pollen grains rapidly hydrate and initiate the formation of a highly polarised tip-growing structure, the pollen tube. The pollen tube penetrates the stigmatic tissue and rapidly grows through the style, carrying the generative cell (in bicellular pollen) or sperm cells (in tricellular pollen) within its cytoplasm. In the case of bicellular pollen, the generative cell undergoes pollen mitosis 2 as the pollen tube travels through the style. Studies in *Arabidopsis thaliana*, a species with tricellular pollen, have revealed that de novo transcription post-hydration and post-germination is quite limited (Wang et al. 2008; Qin et al. 2009). Pollen tube growth is guided by multiple environmental cues, for example within the style tissue as well as longer-distance signals emanating from the ovary sac (the female gametophyte) such as the small molecules GABA and nitric oxide, (e.g. Wu et al. 1995; Palanivelu et al. 2003; Yu and Sun 2007; Prado et al. 2008). Upon arrival at an unfertilised embryo sac, the pollen tube ruptures and liberates the two sperm cells which then undergo double fertilisation, fertilising the egg cell (to form a diploid zygote) and the two polar nuclei of the central cell (to form a triploid endosperm) (Fig. 1).

4 Angiosperm Pollen Development and the Epigenetics of Chromatin and DNA Methylation

The event of pollen mitosis 1 ushers in a sudden and significant change at both the level of the transcriptome and at the nuclear epigenetic level in the resulting cells. The two different cells diverge dramatically in terms of their transcriptome, with significant differences reflecting their different fates – the generative cell effectively becomes the plant "germline" which will undergo gametogenesis, whilst the vegetative cell develops into an "altruistic" cell that, after germination, forms the pollen tube which both sustains the sperm cells and delivers them to the site of fertilisation (Dickinson and Grant-Downton 2009). At the most fundamental level, dramatic differences manifest in the nuclei of the vegetative and generative cell that can be identified using basic microscopy. The vegetative nucleus has a relatively diffuse cloud-like nucleus with a relaxed chromatin structure, whereas the nucleus of the generative cell is compact with highly condensed chromatin. Several studies have examined the epigenetic differences between the two nuclei in more detail, with much of this work being performed in non-model species. In the vegetative nucleus, a set of epigenetic marks become established that may be important in conferring a state giving different transcriptional properties across the genome in this cell; such epigenetic differences may be a significant component in setting up the divergent transcriptomes of the generative and vegetative cells. The vegetative nucleus is generally associated with various histone marks that confer an "active" state, whilst the generative nucleus is generally associated with various histone marks that confer a more "silent" state (Janousek et al. 2000; Okada et al. 2006; Ribeiro et al. 2009) although differences between the species studied have been noted. Differences in DNA methylation between the two nuclei are also detectable, although it appears to depend on the species studied in which nucleus the DNA methylation marks are enriched (Janousek et al. 2000; Ribeiro et al. 2009; Oakeley et al. 1997).

A convincing demonstration that specific histone marks are essential for the function of the vegetative cell has come from work in *Arabidopsis*. Mutations in *SDG4*, a SET domain protein that acts as a histone methyltransferase, reduce the levels of H3K4 and H3K36 methylation in pollen vegetative cell nuclei, resulting in changes in gene expression and reduction in pollen tube growth but not pollen germination (Cartagena et al. 2008).

The uniqueness of the chromatin state in pollen has also been highlighted by recent work on the *Arabidopsis* vegetative nucleus (Schoft et al. 2009). Uniquely in plant cells, the decondensation of centromeric heterochromatin in this cell is accompanied by a great reduction in H3K9m2 levels. This loss of centromeric heterochromatin and alteration of chromatin marks is similar to *ddm1* mutants and, indeed, *DDM1* is not expressed in the vegetative nucleus (Slotkin et al. 2009). However, unlike *ddm1* mutants, the DNA methylation at centromeric repeats is not altered and this region with drastically altered chromatin retains its dense DNA methylation and transcriptional silencing, likely through the maintenance methylation activity of MET1, as the cell does not undergo a further mitotic division (Schoft et al. 2009).

Whilst histone marks and DNA methylation patterns undergo distinct changes in pollen development, another higher-level change in the organisation of the chromatin in pollen is also known to occur in different angiosperms. Pollen-specific histones are known to integrate into the chromatin and are likely to play a role in generating the novel chromatin structures seen in pollen cells, although this has yet to be experimentally demonstrated. Most of these histone variants are specific to the generative and sperm cells such as the sperm-specific H3 histone AtMGH3 in *Arabidopsis* (Okada et al. 2005) and gamete-specific H3 variants in lily (Okada et al. 2006) but also a vegetative cell histone H3.3 variant (MPH3) from lily has been identified (Sano and Tanaka 2005). In the vegetative nucleus of lily, this H3.3 variant is deposited into the chromatin by a replication-independent mechanism and, as in *Drosophila*, this H3.3 variant may maintain transcriptional activity in this cell. Studies of lily pollen have also revealed another global pattern in histone accumulation that occurs in the vegetative nucleus. In the vegetative nucleus, the levels of the linker histone H1 family declines in the vegetative nucleus after pollen mitosis 1, eventually leading to low levels of H1 in mature pollen (Tanaka et al. 1998). This active loss of H1 may be important in generating the diffuse chromatin state of the vegetative nucleus.

Although aspects of male gametophyte chromatin and DNA methylation have been explored, it is clear that much remains to be discovered about their regulatory effects on gene expression in pollen development. Even less is known about the mechanisms that set up the changes to DNA methylation and chromatin; as these establish so rapidly after pollen mitosis 1, it is clear that this is a very actively regulated mechanism rather than a passive change. In plants, small RNA systems are intrinsically linked to chromatin and DNA methylation changes through RNA-dependent DNA methylation (RdDM) pathways (Matzke and Birchler 2005). Remarkably, despite the epigenetic distinctiveness of pollen cells at the nuclear level and the importance of small RNA systems in sporophytic development, little is known about the diversity and role of non coding RNAs in the gametophyte. Only a handful of studies in recent years has illuminated this area of molecular epigenetics in plant reproduction. In view of the significance of such small molecules in the reproductive biology of other eukaryotes, the late emergence of this information seems all the more surprising.

5 Small RNA Pathways in the Male Gametophyte

After meiosis, the male gametophyte may be sufficiently isolated by an impermeable wall structure and the absence of cytoplasmic channels such as plasmodesmata to prevent effective transfer from soma of small RNAs that are known to be systemic (Dickinson and Grant-Downton 2009). There is also no evidence of inheritance of small RNAs through meiosis although this may be possible as some very abundant mRNAs can be inherited (Onodera et al. 2008). For small RNAs to be produced in the male gametophyte effectively, the components of the specific pathways leading to their biogenesis would need to be expressed in the cells. The first work to address this question examined the expression levels of genes involved in small RNA pathways using data from microarrays of *Arabidopsis* pollen (Pina et al. 2005). The general trend was for down-regulation of transcript levels during development and in mature pollen the transcripts were absent. Based on these findings, it was suggested that small RNA pathways became inactive in late pollen development and that, uniquely in the life cycle of flowering plants, small RNAs were not important in pollen.

However, more recent work has overturned this early hypothesis; for example a high level of complexity has now been demonstrated in the expression

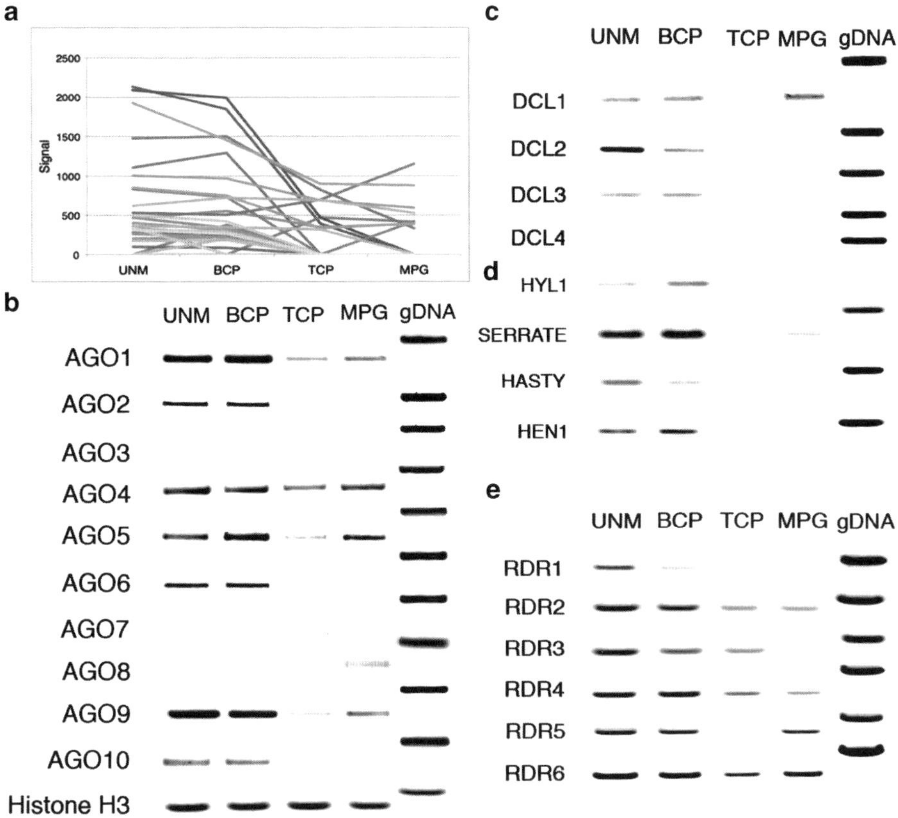

Fig. 2 Expression profiles of sRNA pathway-related genes at four stages of pollen development. (**a**) Transcriptomic profiles for 39 small RNA pathway related genes based on normalised ATH1 Affymetrix chip data. (**b–e**) RT-PCR analysis of small RNA pathway genes during pollen development. (**b**) Argonaute family members, (**c**) DCL1-4, (**d**) HASTY, SERRATE, HEN1, (**e**) six members of the RDR family; HISTONE H3 (AT4G40040) was used as a control. *UNM* unicellular microspores; *BCP* bicellular pollen; *TCP* tricellular pollen; *MPG* mature pollen grains (from Grant-Downton et al. 2009a, reprinted with permission)

patterns of many small RNA pathway genes throughout gametophytic development (Grant-Downton et al. 2009a; see Fig. 2). Importantly, mature pollen showed expression of a significant number of key genes such as *AGO1*, *AGO4*, *DCL1* and *RDR6*. Others were maintained until pollen mitosis II, raising the possibility that the corresponding proteins were retained in the sperms. The finding that some key transcripts were still detectable in mature pollen is supported by array-based transcriptomic work following the recent technical advance of sperm cell isolation from mature *Arabidopsis* pollen (Borges et al. 2008). This significant advance has allowed analysis of expression of small RNA pathway genes in the gametes alone. Using microarrays, number of small RNA pathway genes can be reliably called as present in sperm, including *DCL1*, *AGO6* and *RDR2*. Strikingly, two ARGONAUTE family members, *AGO5* and *AGO9*, and a dsRNA-binding protein, *DRB4*, are significantly enriched in sperm. *AGO9* has been implicated in RdDM and was recently shown to be important in female reproductive development as the loss of functional AGO9 resulted in effects on normal female gametophyte development (Olmedo-Monfil et al. 2010). Intriguingly, antibody-based detection in pollen localised the AGO9 protein to the vegetative cell rather than the sperm (Olmedo-Monfil et al. 2010). *AGO5* is also interesting as in *Arabidopsis* a recent study has implicated AGO5 with binding non-canonical size variants of microRNAs, including those with both +1 and +2 nucleotides at the 5′ end (Ebhardt et al. 2010). A rice homologue of *AGO5*, *MEL1*, plays an essential role in determining cell identity in premeiotic reproductive development, possibly by directing changes to the chromatin (Nonomura et al. 2007).

Any lingering notion that small RNA pathways have no role or importance in the male gametophyte is dispelled by these studies. Indeed, the unique profile of small RNA pathway gene expression in the male gametophyte, especially the sperm cells, suggests that they possess a novel small RNA transcriptome.

6 Small RNA Diversity in the Male Gametophyte

As predicted from the analysis of expression levels of small RNA pathway transcripts described above, next generation sequencing of the male gametophyte small RNA transcriptomes has provided important data. Primary indications that microRNAs were present in the male gametophyte, came from both in situ hybridisation with locked nucleic acid (LNA) probes (Válóczi et al. 2006; Sieber et al. 2007) and through RT-PCRs designed for small RNA amplification (Grant-Downton et al. 2009a). However, although these studies hinted at the diversity of microRNAs in the male gametophyte, a much fuller picture has been provided by 454 and Illumina sequencing of small RNAs from mature pollen. Considering mature *Arabidopsis* pollen consists of just three cells and two different cell types, profiling of the microRNAs has revealed an exceptional diversity of microRNAs that overlaps considerably with the sporophyte. 454 sequencing revealed 33 families of previously described microRNAs and the expression levels of a sub-set were quantified with qRT-PCR to validate this data (Grant-Downton et al. 2009b). Comparison of leaf and pollen material in qRT-PCR work revealed differences between sporophyte and gametophyte; whilst some of the tested microRNAs were expressed at a lower level in the gametophyte, and others appeared to be enriched. Another study combined detection of previously described microRNAs by miRCURY microarray and qRT-PCR analysis and validated these data by Illumina sequencing data from mature pollen (Chambers and Shuai 2009). Here, full agreement between the different detection methods was not achieved as the study found that only 22 microRNAs could be confirmed as present using both array-based and qRT-PCR detection, from a combined total of 45. However, the general trend in this data was for lower levels of microRNAs to be present in the gametophyte than the sporophyte.

A great benefit of small RNA sequencing by next generation methods is that it permits, with the correct bioinformatic analysis and validation steps, the identification of novel microRNAs. As might be expected for a unique stage of development, novel microRNAs have been identified even in a relatively limited dataset (Grant-Downton et al. 2009b). Several candidate microRNAs were identified that conformed to the strict criteria for annotation as microRNAs. One of them, miR2939, was not only shown to be highly enriched in the male gametophyte, but it also cleaved its predicted target – a sperm-expressed F-box superfamily transcript (At3g19890). This transcript is regulated by another previously described microRNA (miR774a) in mature pollen, as this microRNA was also found in pollen and a cleavage product corresponding to the activity of this miR was identified. The other novel pollen microRNAs identified were predicted to target coding transcripts but of those tested for cleavage in this study, none was found to be active. It is possible that these microRNAs act only through the translational suppression mechanism, or that the computational prediction of the target mRNAs was not accurate.

Certain microRNAs do not target coding mRNAs but instead act as the initiators for secondary siRNA production from long non coding *TAS* RNAs (Allen and Howell 2010). The initiation of secondary siRNA formation requires 22 nt size variants of microRNAs to recruit the activity of the RNA-dependent RNA polymerase RDR6 to generate dsRNA after miRNA targeting, and the RNA binding protein SGS3 for stabilisation, along with the activity of DCL proteins (principally DCL4) to cleave the dsRNA into ~21 nt siRNAs called *trans*-acting siRNAs (tasiRNAs) (Allen and Howell 2010; Cuperus et al. 2010; Chen et al. 2010). The DCLs act as molecular "ruler and scissors" from the specific miR-guided initiation point to give a specific "phased" pattern of siRNAs along the length of the transcript. These amplified tasiRNAs subsequently target coding transcripts, such as those from the PPR and MYB families (Allen and Howell 2010). In pollen, miR173 has been identified as present and, correspondingly, phased tasiRNAs derived from activity on its targets *TAS1a*, *TAS1b*, *TAS1c* and *TAS2* have been identified (Grant-Downton et al. 2009b).

The sequencing of small RNAs from *Arabidopsis* mature pollen and from isolated sperm cells has revealed details of how such regulatory systems affect transposable elements (TEs) in plant reproductive cells (Slotkin et al. 2009). In the vegetative cell, the loss of DDM1 activity and subsequent changes to chromatin structure and DNA methylation appears to result in the release of silencing of TEs. Reporter constructs have shown that the TEs are transcriptionally reactivated within this terminal, altruistic cell. However, in the sperm cells the silencing of TEs is maintained. Despite these differences between sperm and vegetative cells in transcription of TEs, small RNAs derived from processing of "aberrant" TE transcripts accumulate in both cells. Interestingly, instead of the 24 nt size normally associated with TEs, these cells show a predominance of ~21 nt siRNAs originating from TEs. A model has been proposed to explain the distribution of these siRNAs, with formation in the vegetative cell but subsequent export to the sperm cells. Although derived from a different mechanism of biogenesis, artificial microRNAs expressed in the vegetative cell appear to function in the sperm cells. How such small RNAs transit to the sperm cells remains a mystery. The mature sperm cell has no discernible cell wall and is bounded by two membranes without plasmodesmata or other channels to the vegetative cell, so there is no convincing ultrastructural evidence for cytoplasmic continuity; indeed, there is more evidence for directional loss of sperm cell cytoplasm *into* the vegetative cell (Mogensen 1996). Another intricacy is that the transposons alone lose DNA methylation and are derepressed, yet the centromeric repeats retain their DNA methylation and silencing in the same nucleus, so the change in the vegetative cells must be more complex than simply through the loss of DDM1 alone (Schoft et al. 2009; Slotkin et al. 2009). In apparent contrast to *Arabidopsis*, in maize sperm cells the silencing of TE sequences appears to be lifted, with TE transcripts accounting for a remarkably high proportion of the transcriptome (Engel et al. 2003).

7 Small RNA Function in the Male Gametophyte

The first unequivocal evidence of the necessity for small RNAs derived from the male gametophyte for reproductive development has come from the serendipitous identification of a unique *cis*-antisense gene pair in *Arabidopsis* (Ron et al. 2010). Although this natural antisense gene pair does appear to produce nat-siRNAs in some abundance, only a single read corresponding to siRNAs from this genomic location was identified in deep sequencing data. T-DNA insertion mutants in *KOKOPELLI*, one of the gene pair, produce pollen which develops normally, and has no defect until fertilisation. Pollen tubes from the mutants deliver two sperm cells but these do not undergo double fertilisation, and frequently result in single fertilisation events leading to defective early development and substantially reduced seed set. The expression pattern of the two genes gave an indication of their role: whilst *KOKOPELLI* appears specific to sperm cells, *ARIADNE14* is expressed in both the vegetative cell and the sperm. A role of the nat-siRNAs generated when both genes are transcribed in sperm would appear to be reduction or clearance of *ARIADNE14* transcripts in sperm cells. Transgenic plants expressing a siRNA-resistant *ARIADNE14* transcript in sperm impairs fertilisation, suggesting that siRNA-dependent down-regulation of ARIADNE14 in sperm is essential. The molecular mechanism by which over-production of ARIADNE14 generates defective sperm remains open to speculation. The ARIADNE family encodes putative E3 ubiquitin ligases, and *ARIADNE13*, *ARIADNE14* and *ARIADNE15* are all clustered on chromosome 5, likely the result of a recent gene duplication event, and all are expressed in sperm. However, *ARIADNE14* appears to be an inactive E3 ubiquitin ligase and the plausible explanation for the effects when over-expressed is from competition for substrates with the active E3 ubiquitin ligases, resulting in a higher level of target substrates remaining untagged and hence not degraded. The persistence of the as yet unidentified substrate(s) is proposed to cause the defects in sperm lacking proper regulation. There is certainly much evidence that proper control of the ubiquitination-dependent protein degradation pathway is necessary in male reproductive development; F-box transcripts are particularly enriched in sperm (Borges et al. 2008) whilst microRNAs targeting numerous F-box transcripts are found in mature pollen, including novel miRNAs regulating sperm-expressed F-box transcripts (Grant-Downton et al. 2009a, b; Chambers and Shuai 2009). However, the regulatory *KOKOPELLI/ARIADNE14* gene pair appears to have evolved recently; *KOKOPELLI* is a novel gene of unknown function and has no homologues outside of the Brassicaceae (Ron et al. 2010).

Surprisingly, given the dramatic phenotype of *kokopelli* mutants, *Arabidopsis* mutants impaired in siRNA biogenesis have not been shown to have this specific defect in sperm, although general reduced fertility is known for some mutants, for example *ago1* alleles (Morel et al. 2002). Nonetheless, a significant challenge with using plants homozygous for strong alleles of genes such as *ago1*, *dcl1* and *hen1* is their major effects on all aspects of sporophytic development, which are likely to confound any affect they may have on reproductive development. However, plants heterozygous for strong (but recessive) *ago1* alleles show a great reduction in male transmission of the mutant allele compared to the wild type allele (Kidner and Martienssen 2005). The precise nature and timing of this defect in transmission remains undetermined although it is possible that misregulation of *KOKOPELLI/ARIADNE14* is involved. As yet, there is no published evidence for similar male segregation distortion effects at other loci.

Although it remains without doubt that endogenous microRNAs function in the male gametophyte, with direct evidence for cleavage of target transcripts (Grant-Downton et al. 2009a, b), verification of their value in development has yet to be forthcoming. However, the presence in pollen of microRNAs that regulate AGO1 and DCL1 levels in an autoregulatory loop (Grant-Downton et al. 2009a, b; Chambers and Shuai 2009) suggests that an exquisitely sensitive control of these proteins at the heart of small RNA systems remains important in this stage of development. Cleavage of auxin response factor (ARF) transcripts *ARF16* and *ARF17* by miR160 also indicated that miRs in pollen may promote rapid clearance of selected transcripts (Grant-Downton et al. 2009a). Levels of both transcripts are very high prior to pollen mitosis II but afterwards suddenly drop below detectable levels. Some of the microRNAs in pollen, e.g. miR399, may have a role in cellular homeostasis. As pollen can still develop normally even when the plant is subjected to stresses, such as water and nutrient stress, and in post-germination development pollen tubes are exposed to a relatively hostile environment on the stigmatic surface and are exposed to the same stresses as the maternal parent when growing through the gynoecium, microRNAs are likely to be important in sensitive modulation of post-transcriptional gene expression under such conditions. Maternal guidance systems controlling pollen tube development might also demand such fine-tuning of gene expression. As AGO1 not only cleaves transcripts guided by microRNAs but also uses microRNAs to direct translational repression (Brodersen et al. 2008; Lanet et al. 2009), it is probable that both mechanisms operate in pollen. Indeed, in post-germination development where de novo expression of genes is highly reduced, it seems plausible that translational repression will be the more dominant form of control given its reversibility. Whether small RNAs and AGO1 (as well as other AGOs expressed in the male gametophyte) contribute through this mechanism to transcript storage in late pollen remains to be experimentally determined.

The biogenesis of small RNAs from TE transcripts and their location in sperm suggests that, at least in *Arabidopsis* with its small genome, there is a significant need to tightly regulate transposition in the plant "germline". As sperm cells have compact chromatin and relatively low levels of gene expression, these small RNAs may be a defence system to ensure that any "rogue" TE transcripts are immediately disabled at the post-transcriptional level. Whether these ~21 nt siRNAs reinforce RdDM remains to be determined; certainly, in the sporophyte, the 24 nt class is almost exclusively involved in this mechanism (Daxinger et al. 2009).

8 Conclusions

In this chapter, we have discussed the importance of non coding RNAs to the reproductive phase of development in flowering plants. As in animals, it has emerged from recent work that plants also depend on small RNA systems in this stage of the life cycle. Unlike animals, plants do not manage the development of the germline using Piwi class Argonautes and piRNAs. However, the gametophyte and plant germline is epigenetically distinctive and does utilise small RNA systems. So far, due to technical limitations, only thorough investigations of the male gametophyte and gametes (sperm) have been possible. MicroRNAs and tasiRNAs are abundant and diverse in the male gametophyte although their specific functional roles remain to be investigated in more detail – presenting a major challenge for future research. Another necessity for future work will be the investigation of small RNAs from the onset of meiosis through the different stages of pollen development. So far, only the small RNA populations of mature pollen (at the point of dehiscence from the anther) have been investigated. It remains to be discovered whether small RNAs are involved in directing changes to DNA methylation and chromatin during differentiation of the vegetative and generative cells, and whether these epigenetic marks themselves have reciprocal effects on small RNA production.

As in animals, the protection of the germline against the activity of TEs appears to be important in plants too. How small RNAs impact the function of the gametes at fertilisation and in post-fertilisation development has been revealed by the nat-siRNAs from the *KOKOPELLI/ARIADNE14* gene pair. Although this regulatory gene pair appears restricted to crucifers, the importance its small RNAs play in governing this critical junction in development presage the discovery of other small RNAs acting at the same point. It seems likely that, as in animals (Wagner et al. 2008; Rassoulzadegan et al. 2006), small RNAs from plant gametes are transgenerationally heritable. If so, they might have roles in regulating post-fertilisation development and parental conflict in both the fertilisation products.

References

Allen E, Howell MD (2010) miRNAs in the biogenesis of trans acting siRNAs in higher plants. Semin Cell Dev Biol. doi:10.1016/j.semcdb.2010.03.008

Arkov A, Ramos A (2010) Building RNA-protein granules: insight from the germline. Trends Cell Biol 20:482–490

Batista P, Ruby J, Claycomb J et al (2008) PRG-1 and 21U-RNAs interact to form the piRNA complex required for fertility in *C. elegans*. Mol Cell 31:67–78
Benetti R, Gonzalo S, Jaco I et al (2008) A mammalian microRNA cluster controls DNA methylation and telomere recombination via Rbl2-dependent regulation of DNA methyltransferases. Nat Struct Mol Biol 15:268–279
Borg M, Brownfield L, Twell D (2009) Male gametophyte development: a molecular perspective. J Exp Bot 60:1465–1478
Borges F, Gomes G, Gardner R et al (2008) Comparative transcriptomics of Arabidopsis sperm cells. Plant Physiol 148:1168–1181
Brennecke J, Malone C, Aravin A et al (2008) An epigenetic role for maternally inherited piRNAs in transposon silencing. Science 322:1387–1392
Brodersen P, Sakvarelidze-Achard L, Bruun-Rasmussen M et al (2008) Widespread translational inhibition by plant miRNAs and siRNAs. Science 320:1185–1190
Cartagena JA, Matsunaga S, Seki M et al (2008) The Arabidopsis SDG4 contributes to the regulation of pollen tube growth by methylation of histone H3 lysines 4 and 36 in mature pollen. Dev Biol 315:355–368
Chambers C, Shuai B (2009) Profiling microRNA expression in Arabidopsis pollen using microRNA array and real-time PCR. BMC Plant Biol 9:87
Chen HM, Chen LT, Patel K et al (2010) 22-nucleotide RNAs trigger secondary siRNA biogenesis in plants. Proc Natl Acad Sci USA 107:15269–15274
Cuperus JT, Carbonell A, Fahlgren N et al (2010) Unique functionality of 22-nt miRNAs in triggering RDR6-dependent siRNA biogenesis from target transcripts in *Arabidopsis*. Nat Struct Mol Biol 17:997–1003
Daxinger L, Kanno T, Bucher E et al (2009) A stepwise pathway for biogenesis of 24-nt secondary siRNAs and spreading of DNA methylation. EMBO J 28:48–57
Dickinson HG, Grant-Downton R (2009) Bridging the generation gap: flowering plant gametophytes and animal germlines reveal unexpected similarities. Biol Rev Camb Philos Soc 84:589–615
Durharcourt S, Lepère G, Meyer E (2009) Developmental genome rearrangements in ciliates: a natural genomic subtraction mediated by non coding transcripts. Trends Genet 25:344–350
During HJ (1979) Life strategies of bryophytes: a preliminary review. Lindbergia 5:2–18
Ebhardt HA, Fedynak A, Fahlman RP (2010) Naturally occurring variations in sequence length creates microRNA isoforms that differ in Argonaute effector complex specificity. Silence 1:12
Engel ML, Chaboud A, Dumas C et al (2003) Sperm cells of *Zea mays* have a complex complement of mRNAs. Plant J 34:697–707
Gent J, Schvarzstein M, Villeneuve A et al (2009) A *Caenorhabditis elegans* RNA-directed RNA polymerase in sperm development and endogenous RNA interference. Genetics 183:1297–1314
Grant-Downton RT, Hafidh S, Twell D et al (2009a) Small RNA pathways are present and functional in the angiosperm male gametophyte. Mol Plant 2:500–512
Grant-Downton RT, Le Trionnaire G, Schimd R et al (2009b) MicroRNA and ta-siRNA diversity in mature pollen of *Arabidopsis thaliana*. BMC Genomics 10:643
Gu W, Shirayama M, Conte D et al (2009) Distinct Argonaute-mediated 22G-RNA pathways direct genome surveillance in the *C. elegans* germline. Mol Cell 36:231–244
Han T, Manoharan A, Harkins T et al (2009) 26G endo-siRNAs regulate spermatogenic and zygotic gene expression in *Caenorhabditis elegans*. Proc Natl Acad Sci USA 106: 18674–18679
Iskow RC, McCabe MT, Mills RE et al (2010) Natural mutagenesis of human genomes by endogenous retrotransposons. Cell 141:1253–1261
Janousek B, Zluvova J, Vyscot B (2000) Histone H4 acetylation and DNA methylation dynamics during pollen development. Protoplasma 211:116–122
Kidner CA, Martienssen RA (2005) The role of ARGONAUTE1 (AGO1) in meristem formation and identity. Dev Biol 280:504–517

Klattenhoff C, Theurkauf W (2008) Biogenesis and germline functions of piRNAs. Development 135:3–9

Kurth H, Mochizuki K (2009) 2′-O-methylation stabilizes Piwi-associated small RNAs and ensures DNA elimination in *Tetrahymena*. RNA 15:675–685

Lanet E, Delannoy E, Sormani R et al (2009) Biochemical evidence for translational repression by Arabidopsis microRNAs. Plant Cell 21:1762–1768

Lau N (2010) Small RNAs in the animal gonad: guarding genomes and guiding development. Int J Biochem Cell Biol 42:1334–1347

Matzke M, Birchler J (2005) RNAi-mediated pathways in the nucleus. Nat Rev Genet 6:24–35

McCormick S (2004) Control of male gametophyte development. Plant Cell 16:S142–S153

Mogensen HL (1996) The hows and whys of cytoplasmic inheritance in seed plants. Am J Bot 83:383–404

Morel JB, Godon C, Mourrrain P et al (2002) Fertile hypomorphic ARGONAUTE (ago1) mutants impaired in post-transcriptional gene silencing and virus resistance. Plant Cell 14:629–639

Nonomura K, Morohoshi A, Nakano M et al (2007) A germ cell specific gene of the ARGONAUTE family is essential for the progression of premeiotic mitosis and meiosis during sporogenesis in rice. Plant Cell 19:2583–2594

Oakeley EJ, Podestà A, Jost J-P (1997) Developmental changes in DNA methylation of the two tobacco pollen nuclei during maturation. Proc Natl Acad Sci USA 94:11721–11725

Okada T, Endo M, Singh MB et al (2005) Analysis of histone H3 gene family in *Arabidopsis* and identification of male gamete-specific variant, AtMGH3. Plant J 44:557–568

Okada T, Singh MB, Bhalla PL (2006) Histone H3 variants in male gametic cells of lily and H3 methylation in mature pollen. Plant Mol Biol 62:503–512

Olmedo-Monfil V, Durán-Figueroa N, Arteaga-Vázquez M et al (2010) Control of female gamete formation by a small RNA pathway in *Arabidopsis*. Nature 464:628–632

Onodera Y, Nakagawa K, Haag J et al (2008) Sex-biased lethality or transmission of defective transcription machinery in *Arabidopsis*. Genetics 180:207–218

Palanivelu R, Brass L, Edlund AF et al (2003) Pollen tube growth and guidance is regulated by POP2, an Arabidopsis gene that controls GABA levels. Cell 114:47–59

Pina C, Pinto F, Feijó JA et al (2005) Gene family analysis of the Arabidopsis pollen transcriptome reveals biological implications for cell growth, division control, and gene expression regulation. Plant Physiol 138:744–756

Prado AM, Colaco R, Moreno N et al (2008) Targeting of pollen tubes to ovules is dependent on nitric oxide (NO) signaling. Mol Plant 1:703–714

Qin Y, Leydon AR, Manziello A et al (2009) Penetration of the stigma and style elicits a novel transcriptome in pollen tubes identifying genes critical for growth in the pistil. PLoS Genet 5:e1000621

Rassoulzadegan M, Grandjean V, Gounon P et al (2006) RNA-mediated non-mendelian inheritance of an epigenetic change in the mouse. Nature 441:469–474

Renauld H (1997) Heterochromatin: a meiotic matchmaker? Trends Cell Biol 7:201–205

Reynolds S, Ruohola-Baker H (2008) The role of microRNAs in germline differentiation. In: StemBook [Internet]. Harvard Stem Cell Institute, Cambridge

Ribeiro T, Viegas W, Morais-Cecílio L (2009) Epigenetic marks in mature pollen of *Quercus suber* L. (Fagaceae). Sex Plant Reprod 22:1–7

Ron M, Saez MA, Williams LE et al (2010) Proper regulation of a sperm-specific cis-nat-siRNA is essential for double fertilization in *Arabidopsis*. Genes Dev 24:1010–1021

Rudall PJ, Bateman RM (2007) Developmental bases for key innovations in the seed plant microgametophyte. Trends Plant Sci 12:317–326

Sano Y, Tanaka I (2005) A histone H3.3-like gene specifically expressed in the vegetative cell of developing lily pollen. Plant Cell Physiol 46:1299–1308

Schoft V, Chumak N, Mosiolek M et al (2009) Induction of RNA-directed DNA methylation upon decondensation of constitutive heterochromatin. EMBO Rep 10:1015–1021

Sieber P, Wellmer F, Gheyselinck J et al (2007) Redundancy and specialization among plant microRNAs: role of the MIR164 family in developmental robustness. Development 134:1051–1060

Slotkin RK, Vaughn M, Borges F et al (2009) Epigenetic reprogramming and small RNA silencing of transposable elements in pollen. Cell 136:461–472

Tam OH, Aravin AA, Stein P et al (2008) Nature 453:534–538

Tanaka I, Ono K, Fukuda T (1998) The developmental fate of angiosperm pollen is associated with a preferential decrease in the level of histone H1 in the vegetative nucleus. Planta 206:561–569

Thomson T, Lin H (2009) The biogenesis and function of PIWI proteins and piRNAs: progress and prospect. Annu Rev Cell Dev Biol 25:355–376

Válóczi A, Várallyay E, Kauppinen S et al (2006) Spatio-temporal accumulation of microRNAs is highly coordinated in developing plant tissues. Plant J 47:140–151

van der Heijden G, Bortvin A (2009) Transient relaxation of transposon silencing at the onset of mammalian meiosis. Epigenetics 4:76–79

Wagner KD, Wagner N, Ghanbarian H et al (2008) RNA induction and inheritance of epigenetic cardiac hypertrophy in the mouse. Dev Cell 14:962–969

Wang Y, Zhang WZ, Song LF et al (2008) Transcriptome analyses show changes in gene expression to accompany pollen germination and pollen tube growth in *Arabidopsis*. Plant Physiol 148:1201–1211

Watanabe T, Totoki Y, Kaneda M et al (2008) Endogenous siRNAs from naturally formed dsRNAs regulate transcripts in mouse oocytes. Nature 453:539–543

Williams JH (2008) Novelties of the flowering plant pollen tube underlie diversification of a key life history stage. Proc Natl Acad Sci USA 105:11259–11263

Wu HM, Wang H, Cheung AY (1995) A pollen tube growth stimulatory glycoprotein is deglycosylated by pollen tubes and displays a glycosylation gradient in the flower. Cell 82:395–403

Yu GH, Sun MX (2007) Deciphering a possible mechanism of GABA in tobacco pollen tube growth and guidance. Plant Signal Behav 2:393–395

Regulatory Roles of Novel Small RNAs from Pseudogenes

Xingyi Guo and Deyou Zheng

Contents

Abstract Arising from gene duplications or retrotranspositions, pseudogenes are genomic sequences with high sequence similarity to functional genes but unable to encode the same type of functional molecular products as what their parental sequences produce. For those that are copies of protein-coding genes, this means that they have lost the potential of encoding a functional protein due to disruption in their putative open reading frames. Several computational algorithms have been developed for detecting pseudogenes in recent years and their applications have annotated hundreds and thousands of pseudogenes in higher eukaryotic genomes, including the rice and *Arabidopsis* genomes. While conventional wisdom considers pseudogenes as dead and inactive sequences, emerging evidence indicates that a large number of

D. Zheng (✉)
Department of Neurology and
Departments of Genetics and Neuroscience, Albert Einstein College of Medicine,
1300 Morris Park Avenue, Bronx, NY 10461, USA
e-mail: deyou.zheng@einstein.yu.edu

X. Guo
Department of Neurology, Albert Einstein College of Medicine, 1300 Morris Park Avenue, Bronx, NY 10461, USA

V.A. Erdmann and J. Barciszewski (eds.), *Non Coding RNAs in Plants*,
RNA Technologies, DOI 10.1007/978-3-642-19454-2_12,

higher eukaryotic pseudogenes are transcriptionally alive and that furthermore many of the pseudogene transcripts may play a critical role in regulating gene expression. In particular, analyses of the RNAs from both plant and mammalian tissues or organs using deep-sequencing technology have uncovered scores of pseudogene-derived small RNAs. Their sequence features, together with carefully designed biochemical and genetic experiments, indicate that small RNAs from pseudogenes may function at different molecular levels, either as small interference RNAs directly regulating functional genes or modulating epigenomic silencing in the pseudogenic regions, or as decoy RNAs counteracting the inhibitory effectiveness of miRNAs supposedly targeting functional genes. These exciting discoveries suggest that pseudogenes may represent a hidden layer of regulatory elements in eukaryotic genomes, whose functional importance has just started to be unveiled and appreciated.

Keywords DCL • Dicer • Pseudogenes • RDR2 • siRNA • Small RNA

1 Introduction

The term "pseudogene" was first introduced in the 1970s by discovery that many tandem copies of DNA sequences showed high degree of similarity to the functional gene encoding 5S rRNA but contained genetic lesions such as truncations (Jacq et al. 1977). They called those sequences pseudogenes, and since then, the term has been used for a group of genomic sequences with the following two key characteristics: high sequence similarity to a functional gene and genetic defects that preclude it from generating a "functional" product (Vanin 1985; Mighell et al. 2000; Harrison et al. 2002a; Balakirev and Ayala 2003; Zheng and Gerstein 2007). Whereas sequence similarity is both conceptually and operationally easy to determine, the nonfunctional property of a pseudogene sequence is difficult to assert. As such, the latter feature of a pseudogene sequence has been suggested to be interpreted cautiously with respect to the final molecular type generated by the functional gene from which this pseudogene is derived (Zheng and Gerstein 2007). Accordingly, pseudogenes in this chapter refer to genomic sequences derived from protein coding genes but nonetheless unable to encode a functional protein peptide.

Pseudogenes are often divided into three types based on the molecular processes of their formations. A pseudogene resulting from gene death is called a unitary pseudogene, e.g., the primate *GULO* (Zhang et al. 2010); a pseudogene arising from a past DNA duplication is called a duplicated pseudogene, and a pseudogene derived from a retrotransposition event is called a processed pseudogene (also referred as retrotransposed pseudogene, as it goes through the process of transcription or reverse-transcription and integration) (Fig. 1). Conventionally, the term "parental gene" is used for the functional gene from which a duplicated pseudogene or processed pseudogene arises. While all pseudogenes would have accumulated various deleterious mutations over their evolutionary history, such as premature stop codons

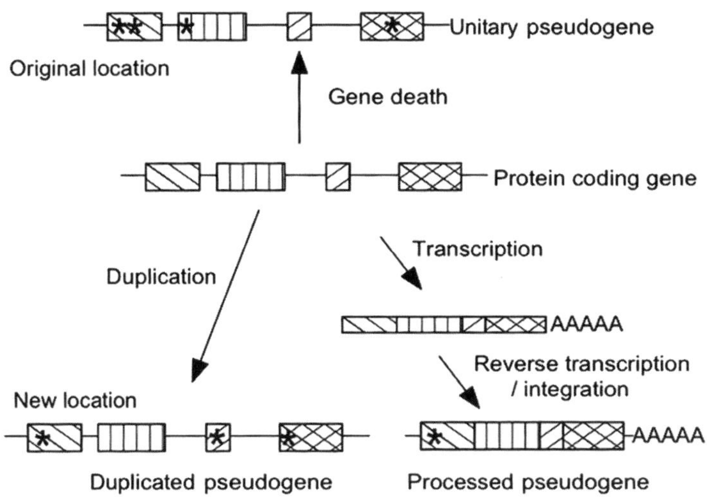

Fig. 1 Three distinct evolutionary mechanisms generate three classes of pseudogenes. *Boxes* represent exons while *lines* for introns. Disruptions (e.g., premature stop codons) in ORFs are shown as *asterisks*

or frameshift mutations in their hypothetical open reading frames (ORFs), unitary pseudogenes and duplicated pseudogenes typically contain exon–intron structure inherited from their ancestral genes, a feature that is usually absent in processed pseudogenes (Mighell et al. 2000; Balakirev and Ayala 2003; Zhang and Gerstein 2004; Zheng and Gerstein 2007). In terms of denotation, at least two symbols have been used, including the prefixed Greek symbol Ψ, for example ΨPGK-1, or by a capital "P" suffix, for example CYP21P (Mighell et al. 2000).

2 Prevalence of Pseudogenes in Eukaryotic Genomes

Pseudogenes are prevalent in eukaryotic genomes and present a major challenge to correct annotation of functional genes (Mighell et al. 2000; Balakirev and Ayala 2003; Zhang et al. 2003). Some pseudogenes might be misidentified as functional genes as they share many features with bona fide protein coding genes (Arabidopsis Genome Initiative 2000; International Rice Genome Sequencing Project 2005; Yu et al. 2005; van Baren and Brent 2006). Therefore, the value of pseudogene annotation goes beyond a simple collection of "genomic fossils" that can provide substantial information for inferring the evolutionary history of genes and genomes (Li et al. 1981; Gojobori et al. 1982; Balakirev and Ayala 2003; Zhang et al. 2003). The identification of pseudogenes is critical for comprehensive understanding of the structure, functional elements, and evolutionary processes of a genome. In the past decade, several computational algorithms have been developed to detect pseudogenes, including PseudoPipe, pseudoFinder, retroFinder, and many others (Harrison et al. 2002b; Ohshima et al. 2003; Torrents et al. 2003; Zhang et al. 2003, 2006;

Coin and Durbin 2004; Khelifi et al. 2005; Bischof et al. 2006; van Baren and Brent 2006), and the resulting data are often available in public databases, e.g., Yale pseudogene collection (http://www.pseudogene.org/) and Hoppsigen (http://pbil.univ-lyon1.fr/databases/hoppsigen.html). All these methods in principle contain steps for detecting sequences homologous to known genes (or proteins) and modules for evaluating nonfunctionality of a sequence, while they often differ in implementation especially for the latter. Also, some of them are specifically designed for processed pseudogenes. Applications of these methods have systematically identified and characterized large pseudogene populations in many sequenced genomes, including that of bacteria, plants, insects, and vertebrates.

2.1 Pseudogenes in Animals

In their pioneering work, Harrison et al. developed a prototype computational method and detected 2,168 pseudogenes, about one eighth of the total protein coding genes, in *Caenorhabditis elegans* (Table 1) (Harrison et al. 2001). Only a small fraction of them were processed pseudogenes, a manifestation of low retrotransposition activity in worms. The number of pseudogenes in *Drosophila* is even smaller;

Table 1 Annotated pseudogenes in some completely sequenced genomes

Organism	Genome size (Mb)	No. of genes	No. of pseudogenes	No. of processed pseudogenes	References
Rickettsia prowazekii	1.1	834	241	0	Ogata et al. (2001)
Mycobacterium leprae	3.3	1,604	1,116	0	Cole et al. (2001)
Yersinia pestis	4.6	4,061	160	0	Parkhill et al. (2001)
Esherichia coli, K-12	4.6	4,400	95	0	Homma et al. (2002)
E. coli, O157	5.5	6,000	101	0	Homma et al. (2002)
Saccharomyces cerevisiae	12.1	6,340	241	0	Harrison et al. (2002a)
Caenorhabditis elegans	102.9	20,009	2,168	208	Harrison et al. (2001)
Drosophila melanogaster	128.3	14,332	110	34	Harrison et al. (2003)
Arabidopsis thaliana	115.4	25,464	4,260	N/A	http://pseudogene.org/
			3,719		Zou et al. (2009)
Oryza sativa	389	37,544	11,956	3,392	Guo et al. (2009)
			7,902	675	Zou et al. (2009)
Homo sapiens	3,040	35,000	14,000	7,800	Zhang et al. (2003)
			3,600	3,600	Ohshima et al. (2003)
			19,000	13,300	Torrents et al. (2003)
Mus musculus	2,493	22,000	10,000	4,500	Zhang et al. (2004)
			13,000	N/A	Waterston et al. (2002)

initial survey of the fruit fly genome found only 100 pseudogenes (Harrison et al. 2003). Interestingly, it was shown that the parental genes of these pseudogenes were significantly longer than the average fly genes without a pseudogene relative and the majority of them coded for serine proteases, immunoglobulin-motif-containing proteins, or cytochromes P450 (Harrison et al. 2003), suggesting a clear bias in pseudogene generation.

Several family-based or genome-wide studies have been carried out for characterizing human pseudogenes. The family-based studies were focused on large gene families, including cytoplasmic and mitochondrial ribosomal protein genes (Zhang et al. 2002; Zhang and Gerstein 2003), nuclear mitochondrial genes (Tourmen et al. 2002; Woischnik and Moraes 2002), and genes encoding olfactory receptors (Glusman et al. 2001). These families constituted the groups of protein coding genes with the largest numbers of pseudogenes, ranging from several hundreds to a few thousands. Subsequently, three research groups have independently carried out systematical annotation of pseudogenes in the human genome. Using a homology-based approach and disruption of putative ORFs as evidence for nonfunctional, Zhang and his colleagues identified ~8,000 processed pseudogenes and ~3,000 duplicated pseudogenes (Zhang et al. 2003). These pseudogenes were derived from ~2,500 distinct functional genes, representing about 10% of the entire human coding genes. In the meantime, a study focusing on processed pseudogenes reported 3,664 pseudogenes by a relatively stricter criterion (Ohshima et al. 2003). Using an alternative model based on the lack of evolutionary constraint for evaluating nonfunctionality, ~20,000 potential human pseudogenes were identified. By analysis of sequence synteny with the mouse genome, they further estimated that 70% of them were processed pseudogenes (Torrents et al. 2003). Follow-up studies have largely corroborated these findings and consistently shown that there are about 20,000 pseudogenes in the human genome (Zhang et al. 2004).

The initial annotation of the mouse genome reported ~14,000 putative pseudogenes (Waterston et al. 2002). A subsequent analysis identified ~5,000 processed pseudogenes in mouse (Zhang et al. 2004). These results suggest that the number of processed pseudogenes is significantly smaller in mouse than in human, which may be explained by reduced retrotransposition activities or a faster sequence decay rate in mice (Graur et al. 1989; Waterston et al. 2002). The fact that the mouse genome has higher nucleotide substitution, insertion, and deletion rates than the human genome (Waterston et al. 2002) may be another confounding factor as these would reduce the computational power of detecting sequence similarity.

2.2 *Pseudogenes in Plants*

Studies of genome structure and dynamics have suggested that pseudogenes might be a significant component of plant genomes. First, it is reported that retrotransposons and their resulting remnants are abundant and can constitute 50–90% of higher plant genomes (Bennetzen et al. 2005; Sabot and Schulman 2006). Second, it is well documented that whole-genome, segmental, and tandem duplications all play key roles in the evolution of modern plant genomes (Guyot and Keller 2004; Wang et al. 2005; Guo et al. 2007). Third, many studies have described pseudogenes of several important gene families. The prevalence of plant pseudogenes, however, has only been recently appreciated after genome-wide pseudogene annotations were carried out for rice and *Arabidopsis* (Guo et al. 2009; Zou et al. 2009).

Most pioneering works were focused on the identification of retrotransposed pseudogenes in plants. For example, several plant processed pseudogenes have been identified for the actin gene family in potato (*Solanum tuberosum*) (Drouin and Dover 1987) and the alcohol dehydrogenase gene family in *Leavenworthia* (Charlesworth et al. 1998). In *Arabidopsis*, a recent search of the genome identified 69 retroposons, 22 of which were considered processed pseudogenes (Zhang et al. 2005). The first genome-wide analysis uncovered 411 retrotransposed genes in *Arabidopsis*, 376 of which contained frameshifts or premature stop codons and thus were likely bona fide pseudogenes (Benovoy and Drouin 2006). The more systematic annotation released by the *Arabidopsis* Information Resource (TAIR, version 8) identified 4,759 pseudogenes or transposable elements (Swarbreck et al. 2008).

Pseudogenes in rice have also been investigated. At the individual scale, a limited number of rice duplicated pseudogenes have been reported, including 99 pseudogenes in Cyt P450 family (Nelson et al. 2004) and many others arising from MADS-box genes, which encode a large family of transcription factors (Nam et al. 2004). Separately, in a detailed analysis of the Bric-a-Brac/Tramtrack/Broad domain family in rice, 43 out of its 192 annotated members were found to contain frameshifts and/or premature stops (Gingerich et al. 2007). After the genomes of two rice subspecies (*indica* and *japonica*) were sequenced in 2005 (International Rice Genome Sequencing Project 2005; Yu et al. 2005), an independent survey found that 1,439 of the annotated rice genes might indeed be pseudogenes based on the presence of frameshifts or premature stop codons (Thibaud-Nissen et al. 2009), highlighting the challenge in distinguishing functional genes from pseudogenes.

Two genome-scale pseudogene annotations are available for the rice genome (Guo et al. 2009; Zou et al. 2009). One study by Zou et al. reported a total of 5,608 pseudogenes, while the other by Guo et al. identified 11,956 nontransposon-related rice pseudogenes using the PseudoPipe initially developed for annotating mammalian pseudogenes (Table 1). Three thousand three hundred and ninety-two and 2,350 of the rice pseudogenes from Guo et al.'s study were further classified as processed pseudogenes and duplicated pseudogenes. The analysis of sequence identity and alignment coverage between rice pseudogenes and their parental genes indicated that most past retrotranspositions have left processed pseudogenes covering almost the full ORF of their parents. Interestingly, Guo and his colleague found that only 7.6% of singleton genes had a pseudogene copy, suggesting that coding genes of singleton families were less likely to have a detectable pseudogene. They also observed that the family size was negatively correlated with the number of pseudogenes in a family, suggesting that the large gene families did not necessarily have more "dead" (pseudogene) relatives, an interesting evolutional phenomenon for further investigation.

Comparison of animal pseudogenes and plant pseudogenes revealed some intriguingly distinct patterns. In mammals, a large fraction of their pseudogenes are derived from known gene families such as ribosomal protein genes and olfactory receptor genes (Zhang et al. 2002; Zheng et al. 2007). More specifically, the ribosomal protein genes have generated about 2,000 pseudogenes in humans, chimpanzees, dogs, and mice (Karro et al. 2007; Balasubramanian et al. 2009). This, however, is strikingly different from the observation in rice pseudogenes, which contained only 50 pseudogenes generated from ribosomal protein genes (Guo et al. 2009). Moreover, by investigating the top ten genes generating the most pseudogenes, Guo and his colleagues reported that eight of them have not been functionally annotated, while the other two appear to be housekeeping genes. As the enrichment of processed pseudogenes from housekeeping genes is considered to be relevant to high levels of expression of their parents in mammals, these differences between plant and mammalian pseudogenes suggest that distinct evolutionary events are perhaps responsible for the generation and subsequent retainment of pseudogene populations in plants and animals.

2.3 *Pseudogenes in Prokaryotes and Yeast*

In bacteria, many pseudogenes have also been reported, but most of them appear to be unitary pseudogenes. In an early study of *Escherichia coli* genome, 95 and 101 pseudogene candidates were identified in the strains K-12 and O157, respectively (Homma et al. 2002). Other studies have also indicated that the *E. coli* genome probably contained hundreds of pseudogenes (Harrison and Gerstein 2002; Lerat and Ochman 2005; Ochman and Davalos 2006). More comprehensive investigations of a variety of prokaryotic genomes have subsequently estimated that up to 5% of all gene-like sequences could be pseudogenes (Lerat and Ochman 2004, 2005; Liu et al. 2004). The *Saccharomyces cerevisiae* was also found to harbor about 200 pseudogenes or disabled ORFs (Harrison et al. 2002a).

3 Transcription and Functional Implication of Pseudogenes

As described above, pseudogenes are a critical component of genomes in all three kingdoms. Their preservation in modern genomes after millions of years of evolution has drawn many curiosities and speculations on their possible functions. The conventional definition of pseudogenes implies that pseudogenes would appear transcriptionally silent due to either the lack of functional promoters and auxiliary regulatory elements or the instability of the transcribed products. An accumulating body of evidence, however, shows that many pseudogenes actually can be transcribed to stable RNAs, a novel discovery that has been reported from both small-scale gene-centered studies (Balakirev and Ayala 2003) and genome-scale unbiased mapping of transcriptionally active regions in the human genome and the mouse genomes (Frith et al. 2006; Zheng et al. 2007). An analysis of microarray data showed that some pseudogenes of *S. cerevisiae* genome could be transcribed even though they carried multiple disablements (Harrison et al. 2002a). The pilot ENCODE (ENCyclopedia Of DNA Elements) project also reported that at least one fifth of human pseudogenes could be transcribed to various degrees based on a variety of empirical transcription evidence, such as those derived from 5′ RACE (Rapid Amplification of cDNA Ends), tiling microarray analysis, and high-throughput sequencing data (Zheng et al. 2007). Other studies have also corroborated this findings and collectively provided a conservative estimate that 5–20% of human pseudogenes could be transcriptionally active (Yano et al. 2004; Harrison et al. 2005; Zheng et al. 2005; Svensson et al. 2006). Additionally, in a report from the Functional Annotation of Mouse (FANTOM) project, 9,278 of the ~100,000 full-length mouse cDNA sequences were suggested to come from mouse pseudogenes, suggesting perhaps ~50% of mouse pseudogenes might be transcribed (Frith et al. 2006).

Several lines of evidence indicate that some plant pseudogenes can also be alive. In *Arabidopsis*, large-scale transcriptome projects and global expression studies using genome tiling arrays found that approximately 20% of annotated pseudogenes could be expressed (Yamada et al. 2003). More recently, evidence of pseudogene expression in *Arabidopsis* and rice has been provided using EST and massively parallel signature sequencing (MPSS) datasets, although relatively small percentages of pseudogenes, 2–5% and 2–3% in *Arabidopsis* and rice, respectively, were reported to produce RNA transcripts (Zou et al. 2009).

The biological and functional implications of these pseudogene transcripts are largely unexplored, but their importance have started to emerge (Ota and Nei 1995; Korneev et al. 1999; Mighell et al. 2000; Balakirev and Ayala 2003; Zheng and Gerstein 2007). Direct evidence has been established for a functional NOS (nitric oxide synthase) pseudogene that is transcribed specifically in selected neurons of *Limnaea stagnalis*, where its transcript forms a RNA duplex with the mRNA from its parental gene. This RNA duplex is able to curtain the production of NOS proteins in vivo (Korneev et al. 1999, 2005; Korneev and O'Shea 2002). Studies in mouse oocytes showed that some mouse pseudogenes produced small RNAs and those pseudogene-derived short interfering RNAs (siRNAs) were processed through Dicer, and more importantly the loss of Dicer significantly reduced the number of pseudogene siRNAs, which in turn led to upregulation of their targeted genes (often the parental genes) (Tam et al. 2008; Watanabe et al. 2008). More recently, a novel mechanism of regulation has been discovered for the PTENP1 and KRAS1P pseudogene (Poliseno et al. 2010). It was demonstrated that RNA transcribed from PTENP1, a pseudogene from the tumor suppressor gene PTEN, regulated the effects of microRNAs (miRNAs) targeting at PTEN by competing for miRNA binding. The authors found several binding sites for PTEN-targeting miRNAs in the 3′ UTR of PTENP1 RNA. Overexpression of PTEN-targeting miRNAs resulted in downregulation of both PTEN and PTENP1 RNAs, whereas overexpression of the PTENP1 3′ UTR increased the expression levels of PTEN transcript and protein. Thus, the PTENP1 (and likewise KRAS1P as well) may play a direct role in tumor development and disease pathogenesis.

4 Small RNAs Arising from Plant Pseudogenes

Although pseudogenes with newly evolved regulatory roles seems scarce, recent studies from deep sequencing of transcriptomes have provided strong evidence that this phenomenon may not be so anecdotal as widely assumed (Kasschau et al. 2007; Birchler and Kavi 2008; Sasidharan and Gerstein 2008; Tam et al. 2008; Watanabe et al. 2008; Guo et al. 2009). As mentioned above, the sequencing analysis of small RNAs in mouse oocytes has turned up many siRNAs that are derived from transcribed pseudogenes (Tam et al. 2008; Watanabe et al. 2008). The analyses of small RNA libraries from various rice and *Arabidopsis* tissues have also shed light on the existence of a significant number of plant pseudogenes that can produce novel regulatory RNAs (Kasschau et al. 2007; Guo et al. 2009). In their study of profiling small RNAs in the *Arabidopsis* RNA-dependent RNA polymerase 2 (RDR2) loss-of-function mutant, Lu and his colleagues found an enrichment of small RNAs derived from most miRNA loci but also noticed a significant reduction of 24-nt heterochromatic siRNAs associated with pseudogenes in the *rdr2* mutant relative to the wide type (Lu et al. 2006), suggesting that the biogenesis of small RNAs from pseudogenes is RDR2 dependent. Subsequently, genome-wide profiles and analyses of small RNAs have been carried out in *Arabidopsis* from wild-type (Col-0) and a variety of mutants with defects of three RDR and four Dicer-like (DCL) genes (Kasschau et al. 2007). The authors found that approximately 39% of annotated pseudogenes (excluding transposons/retroelements) could be associated with at least one uniquely mapped small RNA and that those pseudogene RNAs were particularly enriched for 24-nt siRNAs. By comparing Col-0 and each of their mutant plants, they showed that the numbers of small RNAs of all sizes from pseudogenes decreased specifically in the *rdr2* mutant. In *dcl3*, however, the 24-nt class of siRNAs was nearly lost, while the amounts of 21- and 22-nt classes from pseudogenes increased. Based on these results, the authors proposed that the biogenesis of small RNA from plant pseudogenes was dependent on RDR2 and DCL3, which were implicated in silencing of repetitive sequences such as retrotransposons, tandem repeats, and centromeric repeats.

The analysis of rice pseudogene siRNAs indicates that the actual mechanism is probably more complicated than what has been suggested previously. In their analysis of a library of small RNAs derived from developing rice grains, Guo and his colleagues found that 2,867 and 2,582 pseudogenes had at least one small RNA uniquely mapped to their sense and antisense strands, respectively (Guo et al. 2009). Consistent with the observation in *Arabidopsis* (Kasschau et al. 2007), the majority (53.4%) of these small RNAs were 24-nt long, a common signature feature of small RNAs derived from plant RDR2 pathway. However, a significant fraction of the rice pseudogene siRNAs was 21-nt long, which implied that a biogenetic process independent of RDR2/DCL3 might be involved. They further proposed and found evidence that small RNAs from rice pseudogenes might function as natural antisense siRNAs, either by interacting with the complementary sense mRNAs from functional parental genes or by forming double-strand RNAs with transcripts of adjacent paralogous pseudogenes (Fig. 2). The potential functional relevance of these rice pseudogene siRNAs was also supported by the observation that many pseudogene siRNAs exhibited tissue-specific expression. It would be interesting to investigate in the future if these stage-specific siRNAs play any special roles in rice development.

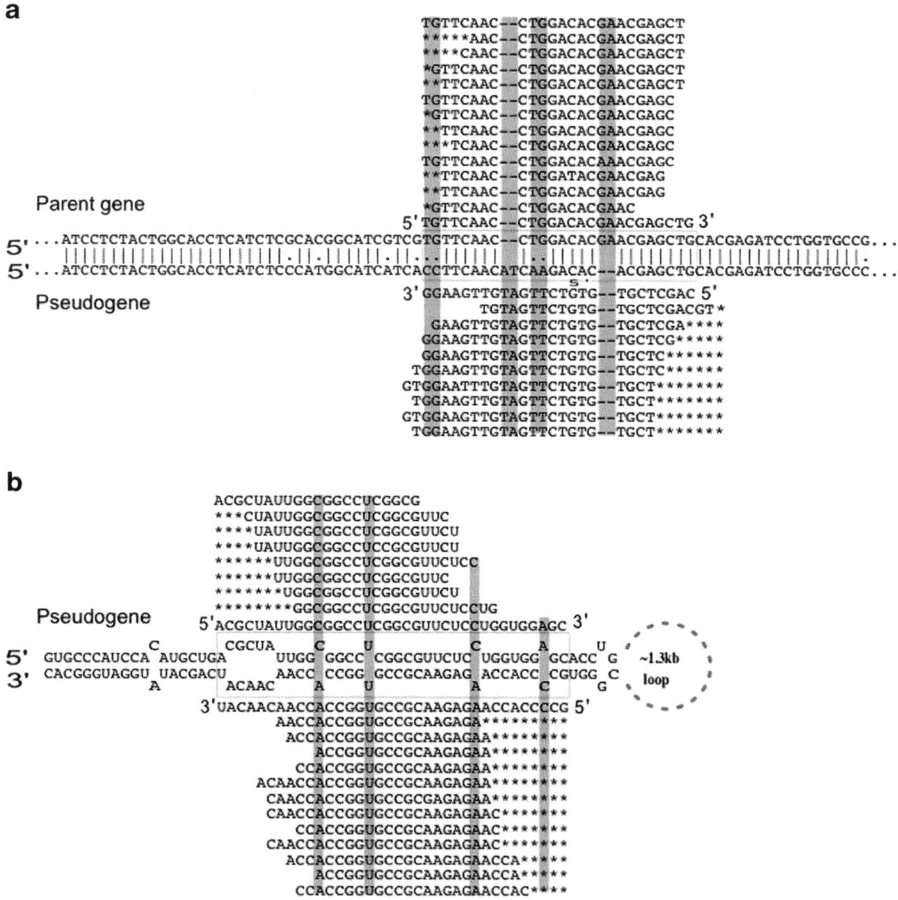

Fig. 2 Two examples of potential sources for pseudogene-derived siRNAs. (**a**) Many small RNAs in developing rice grains are mapped uniquely to the sense and antisense strands of a gene and its pseudogene, respectively. (**b**) A pseudogene transcript forms RNA hairpin structure, with small RNAs mapped to strands in the duplex region. In both cases, the double-strand RNAs can serve as substrates for producing siRNAs. *Gray bars* highlight unique sites or indels for placing small RNAs uniquely to one of the aligned strands (see details in a previous publication, Guo et al. 2009)

5 Potential Regulatory Roles of Pseudogene siRNAs

The discovery of a large number of small RNAs originated from pseudogenes in mammals and plants raises the speculations over what biochemical molecular roles these pseudogene siRNAs may have. Based on a limited number of functional studies, such as those focused on NOS pseudogene in snails, and the characteristics of pseudogene siRNAs, at least three mechanisms are conceivable for the potential regulatory functions of pseudogene siRNAs (Fig. 3). The first model is that pseudogene RNAs can interact with the mRNAs from their parental (or paralogous) genes and the complementary double-strand RNAs serve as substrates for the production of natural antisense siRNAs. Such antisense siRNAs can lead to downregulation of the expression of functional genes. This was initially proposed in the studies of pseudogene-derived siRNAs in mouse oocytes (Tam et al. 2008; Watanabe et al. 2008). It was shown that the occurrence of those siRNAs (called endo-siRNAs by the authors) was Dicer dependent, a protein essential for small RNA biogenesis, as the knockout of Dicer significantly reduced the number of pseudogene siRNAs, and furthermore the decrease of pseudogene siRNAs correlated with upregulation of their putative targets of protein coding genes. Although this potential mechanism has mainly been explored for mammalian pseudogene siRNAs, it is in line with the

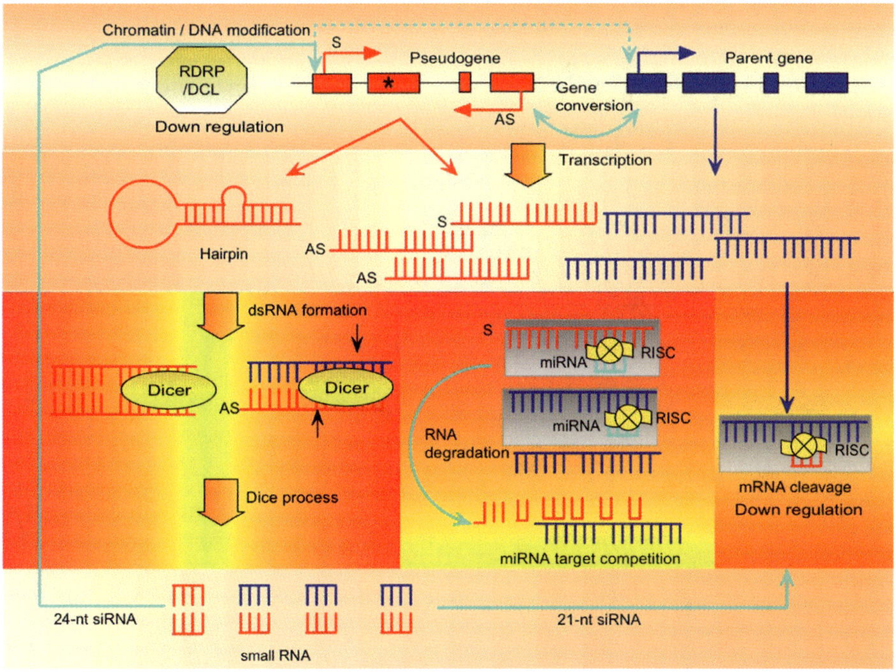

Fig. 3 Pseudogenes may generate RNAs with three distinct regulatory roles by three molecular mechanisms. RNAs from sense and antisense strands are marked with "S" and "AS," respectively

finding that many plant pseudogene RNAs are transcribed from the antisense strand and a significant portion of rice pseudogene-derived small RNAs are 21-nt long. It is known that plant *trans*-acting siRNAs (tasiRNAs) or nat-siRNAs exhibit a wide range of sizes; for example, rice nat-siRNAs can vary from 17 to 31 nt, with an enrichment around 21 nt (Borsani et al. 2005). It should be noted that such kind of functional siRNAs could be originated from paralogous pseudogenes as well if their RNA transcripts would form in vivo hairpin structure recognized by the cellular siRNA processing apparatus.

In plant, an alternative possibility has received much more attention and discussion in literature. Based on the features of small RNAs from *Arabidopsis* pseudogenes and their dependence on DCL3/RDR2, it was suggested that the biological process responsible for the production/function of repeat-associated siRNAs is also implicated in the biogenesis of pseudogene siRNAs. Under this scenario, pseudogene siRNAs may initiate and orchestrate the silencing of the regions where pseudogenes are located by RNA-directed DNA methylation (RdDM) process. There, pseudogene siRNAs were produced by RDR pathway and possibly guide the localization of epigenome modifying machinery to pseudogenic regions for modulating epigenetic silencing such as histone deacetylation, histone methylation, and DNA methylation. Such *cis*-acting siRNAs acting on several endogenous loci have been described, including retrotransposons, 5S rDNA, and centromeric repeats (Chan et al. 2005). The strong support for this regulatory mechanism comes from three key observations: the majority of rice and *Arabidopsis* pseudogene small RNAs are 24-nt long; a lot of them are produced from both DNA strands, and their production is dependent on the RDR2/DCL3 pathway. Moreover, plant 24-nt siRNAs and the Pol IV/RDR2/Pol V pathway are well known to be implicated in RdDM and heterochromatin formation, two processes important for silencing transposons and other retroelements in plants (Baulcombe 2004; Borsani et al. 2005; Chan et al. 2005; Brodersen and Voinnet 2006; Vaucheret 2006). Nevertheless, not all 24-nt RNAs are from RDR2. The first example of *trans*-acting nat-siRNAs derived from RNA duplex formed between SRO5 and P5CDH transcripts was 24-nt long but generated from the RDR6/DCL2 pathway (Borsani et al. 2005). The existence of RDR in mammals remains elusive to date, so no studies have addressed whether siRNAs from human and mouse pseudogenes are important for the silencing of pseudogene loci to prevent sporadic pseudogene transcription. Nevertheless, it is conceivable such a mechanism may exist in mammalian cells.

The third functional model involves a third party, e.g., miRNAs, in addition to pseudogenes and their parental genes. It is largely based on competitive inhibition but elegantly demonstrated for PTENP1 and KRAS1P pseudogene (Poliseno et al. 2010). When the transcripts from a pseudogene and its parent (or paralog) both contain target sites of a miRNA, their coexistence will interfere with each other as they both can interact with the same miRNA. The extent of this phenomenon has not been systematically examined at the genome-wise level, but it is conceivable that predominantly young pseudogenes may have this competitive role, as they are nearly identical to their parents at the sequence level. Certainly, this mechanism of pseudogene function is not restricted to miRNAs but likely extensible to any RNAs, proteins, or molecules that can functionally interact with RNAs, as long as the interacting sites are shared between a cellular mRNAs and a pseudogene RNA.

These are just some scenarios that have been documented for the functions of those pseudogenes that can produce RNA transcripts (which are not translated to proteins due to disrupted ORFs). Nevertheless, they clearly demonstrate that some pseudogenes have evolved novel functions by encoding non coding RNAs that can regulate their own and local expression (by *cis*-affect and RdDM) or the expression of their parental or paralogous genes (by *trans*-affect). It should be emphasized that these mechanisms are not necessarily exclusive and pseudogenes with multiple layers of functions may exist in the same biological system or even the same cell. Moreover, some pseudogenes may have experienced substantial sequence divergence and eventually evolve to functional sequences encoding miRNAs, which will regulate hundreds of potential targets (Sasidharan and Gerstein 2008). Indeed, the *mir*-161 and *mir*-163 in *Arabidopsis* were found to evolve from pseudogenes generated by inverted duplication (Allen et al. 2004). Similarly, the human XIST non coding gene, the key initiator of X chromosome inactivation, also emerged from the relic of an ancient pseudogene (Duret et al. 2006).

The prevalence of pseudogenes shows that pseudogenes are an important structural component of many eukaryotic and prokaryotic genomes. The pervasive transcription of pseudogenes adds a layer of complexity for decoding the function of many genomes. In this chapter, we focus our discussions on potential function of pseudogenes as non coding RNAs. There is plenty of evidence in literature that some pseudogenes may contribute to functional diversity of a genome at DNA level (see Zheng and Gerstein 2007 for more details). For instance, it was suggested that pseudogenes might serve as a sequence reservoir for increasing antibody diversity in humans, chickens, and other vertebrates, through gene conversion (Ota and Nei 1995; Balakirev and Ayala 2003). Conversely, pseudogenes may compromise the important function of their parental genes by introducing deleterious mutations through gene conversion. These indicate that the potential functional repository of pseudogenes is unlimited and the only limitation is perhaps our imagination.

References

Allen E, Xie Z, Gustafson AM et al (2004) Evolution of microRNA genes by inverted duplication of target gene sequences in *Arabidopsis thaliana*. Nat Genet 36:1282–1290

Arabidopsis Genome Initiative (2000) Analysis of the genome sequence of the flowering plant *Arabidopsis thaliana*. Nature 408:796–815

Balakirev ES, Ayala FJ (2003) Pseudogenes: are they "Junk" or functional DNA? Annu Rev Genet 37:123–151

Balasubramanian S, Zheng D, Liu YJ et al (2009) Comparative analysis of processed ribosomal protein pseudogenes in four mammalian genomes. Genome Biol 10:R2

Baulcombe D (2004) RNA silencing in plants. Nature 431:356–363

Bennetzen JL, Ma J, Devos KM (2005) Mechanisms of recent genome size variation in flowering plants. Ann Bot 95:127–132

Benovoy D, Drouin G (2006) Processed pseudogenes, processed genes, and spontaneous mutations in the Arabidopsis genome. J Mol Evol 62:511–522

Birchler JA, Kavi HH (2008) Molecular biology. Slicing and dicing for small RNAs. Science 320:1023–1024

Bischof JM, Chiang AP, Scheetz TE et al (2006) Genome-wide identification of pseudogenes capable of disease-causing gene conversion. Hum Mutat 27:545–552

Borsani O, Zhu J, Verslues PE et al (2005) Endogenous siRNAs derived from a pair of natural cis-antisense transcripts regulate salt tolerance in Arabidopsis. Cell 123:1279–1291

Brodersen P, Voinnet O (2006) The diversity of RNA silencing pathways in plants. Trends Genet 22:268–280

Chan SW, Henderson IR, Jacobsen SE (2005) Gardening the genome: DNA methylation in *Arabidopsis thaliana*. Nat Rev Genet 6:351–360

Charlesworth D, Liu FL, Zhang L (1998) The evolution of the alcohol dehydrogenase gene family by loss of introns in plants of the genus Leavenworthia (Brassicaceae). Mol Biol Evol 15:552–559

Coin L, Durbin R (2004) Improved techniques for the identification of pseudogenes. Bioinformatics 20:i94–i100

Cole ST, Eiglmeier K, Parkhill J et al (2001) Massive gene decay in the leprosy bacillus. Nature 409:1007–1011

Drouin G, Dover GA (1987) A plant processed pseudogene. Nature 328:557–558

Duret L, Chureau C, Samain S et al (2006) The Xist RNA gene evolved in eutherians by pseudogenization of a protein-coding gene. Science 312:1653–1655

Frith MC, Wilming LG, Forrest A et al (2006) Pseudo-messenger RNA: phantoms of the transcriptome. PLoS Genet 2:e23

Gingerich DJ, Hanada K, Shiu SH et al (2007) Large-scale, lineage-specific expansion of a bric-a-brac/tramtrack/broad complex ubiquitin-ligase gene family in rice. Plant Cell 19:2329–2348

Glusman G, Yanai I, Rubin I et al (2001) The complete human olfactory subgenome. Genome Res 11:685–702

Gojobori T, Ishii K, Nei M (1982) Estimation of average number of nucleotide substitutions when the rate of substitution varies with nucleotide. J Mol Evol 18:414–423

Graur D, Shuali Y, Li WH (1989) Deletions in processed pseudogenes accumulate faster in rodents than in humans. J Mol Evol 28:279–285

Guo X, Wang Y, Keightley PD et al (2007) Patterns of selective constraints in noncoding DNA of rice. BMC Evol Biol 7:208

Guo X, Zhang Z, Gerstein MB et al (2009) Small RNAs originated from pseudogenes: cis- or trans-acting? PLoS Comput Biol 5:e1000449

Guyot R, Keller B (2004) Ancestral genome duplication in rice. Genome 47:610–614

Harrison PM, Gerstein M (2002) Studying genomes through the aeons: protein families, pseudogenes and proteome evolution. J Mol Biol 318:1155–1174

Harrison PM, Echols N, Gerstein MB (2001) Digging for dead genes: an analysis of the characteristics of the pseudogene population in the *Caenorhabditis elegans* genome. Nucleic Acids Res 29:818–830

Harrison P, Kumar A, Lan N et al (2002a) A small reservoir of disabled ORFs in the yeast genome and its implications for the dynamics of proteome evolution. J Mol Biol 316:409–419

Harrison PM, Hegyi H, Balasubramanian S et al (2002b) Molecular fossils in the human genome: identification and analysis of the pseudogenes in chromosomes 21 and 22. Genome Res 12:272–280

Harrison PM, Milburn D, Zhang Z et al (2003) Identification of pseudogenes in the *Drosophila melanogaster* genome. Nucleic Acids Res 31:1033–1037

Harrison PM, Zheng D, Zhang Z et al (2005) Transcribed processed pseudogenes in the human genome: an intermediate form of expressed retrosequence lacking protein-coding ability. Nucleic Acids Res 33:2374–2383

Homma K, Fukuchi S, Kawabata T et al (2002) A systematic investigation identifies a significant number of probable pseudogenes in the *Escherichia coli* genome. Gene 294:25–33

International Rice Genome Sequencing Project (2005) The map-based sequence of the rice genome. Nature 436:793–800
Jacq C, Miller JR, Brownlee GG (1977) A pseudogene structure in 5S DNA of *Xenopus laevis*. Cell 12:109–120
Karro JE, Yan Y, Zheng D et al (2007) Pseudogene.org: a comprehensive database and comparison platform for pseudogene annotation. Nucleic Acids Res 35:D55–D60
Kasschau KD, Fahlgren N, Chapman EJ et al (2007) Genome-wide profiling and analysis of Arabidopsis siRNAs. PLoS Biol 5:e57
Khelifi A, Duret L, Mouchiroud D (2005) HOPPSIGEN: a database of human and mouse processed pseudogenes. Nucleic Acids Res 33:D59–D66
Korneev S, O'Shea M (2002) Evolution of nitric oxide synthase regulatory genes by DNA inversion. Mol Biol Evol 19:1228–1233
Korneev SA, Park JH, O'Shea M (1999) Neuronal expression of neural nitric oxide synthase (nNOS) protein is suppressed by an antisense RNA transcribed from an NOS pseudogene. J Neurosci 19:7711–7720
Korneev SA, Straub V, Kemenes I et al (2005) Timed and targeted differential regulation of nitric oxide synthase (NOS) and anti-NOS genes by reward conditioning leading to long-term memory formation. J Neurosci 25:1188–1192
Lerat E, Ochman H (2004) Psi-Phi: exploring the outer limits of bacterial pseudogenes. Genome Res 14:2273–2278
Lerat E, Ochman H (2005) Recognizing the pseudogenes in bacterial genomes. Nucleic Acids Res 33:3125–3132
Li WH, Gojobori T, Nei M (1981) Pseudogenes as a paradigm of neutral evolution. Nature 292:237–239
Liu Y, Harrison PM, Kunin V et al (2004) Comprehensive analysis of pseudogenes in prokaryotes: widespread gene decay and failure of putative horizontally transferred genes. Genome Biol 5:R64
Lu C, Kulkarni K, Souret FF et al (2006) MicroRNAs and other small RNAs enriched in the Arabidopsis RNA-dependent RNA polymerase-2 mutant. Genome Res 16:1276–1288
Mighell AJ, Smith NR, Robinson PA et al (2000) Vertebrate pseudogenes. FEBS Lett 468:109–114
Nam J, Kim J, Lee S et al (2004) Type I MADS-box genes have experienced faster birth-and-death evolution than type II MADS-box genes in angiosperms. Proc Natl Acad Sci USA 101:1910–1915
Nelson DR, Schuler MA, Paquette SM et al (2004) Comparative genomics of rice and Arabidopsis. Analysis of 727 cytochrome P450 genes and pseudogenes from a monocot and a dicot. Plant Physiol 135:756–772
Ochman H, Davalos LM (2006) The nature and dynamics of bacterial genomes. Science 311: 1730–1733
Ogata H, Audic S, Renesto-Audiffren P et al (2001) Mechanisms of evolution in *Rickettsia conorii* and *R. prowazekii*. Science 293:2093–2098
Ohshima K, Hattori M, Yada T et al (2003) Whole-genome screening indicates a possible burst of formation of processed pseudogenes and Alu repeats by particular L1 subfamilies in ancestral primates. Genome Biol 4:R74
Ota T, Nei M (1995) Evolution of immunoglobulin VH pseudogenes in chickens. Mol Biol Evol 12:94–102
Parkhill J, Wren BW, Thomson NR et al (2001) Genome sequence of *Yersinia pestis*, the causative agent of plague. Nature 413:523–527
Poliseno L, Salmena L, Zhang J et al (2010) A coding-independent function of gene and pseudogene mRNAs regulates tumour biology. Nature 465:1033–1038
Sabot F, Schulman AH (2006) Parasitism and the retrotransposon life cycle in plants: a hitchhiker's guide to the genome. Heredity 97:381–388
Sasidharan R, Gerstein M (2008) Genomics: protein fossils live on as RNA. Nature 453:729–731
Svensson O, Arvestad L, Lagergren J (2006) Genome-wide survey for biologically functional pseudogenes. PLoS Comput Biol 2:e46

Swarbreck D, Wilks C, Lamesch P et al (2008) The Arabidopsis Information Resource (TAIR): gene structure and function annotation. Nucleic Acids Res 36:D1009–D1014

Tam OH, Aravin AA, Stein P et al (2008) Pseudogene-derived small interfering RNAs regulate gene expression in mouse oocytes. Nature 453:534–538

Thibaud-Nissen F, Ouyang S, Buell CR (2009) Identification and characterization of pseudogenes in the rice gene complement. BMC Genomics 10:317

Torrents D, Suyama M, Zdobnov E et al (2003) A genome-wide survey of human pseudogenes. Genome Res 13:2559–2567

Tourmen Y, Baris O, Dessen P et al (2002) Structure and chromosomal distribution of human mitochondrial pseudogenes. Genomics 80:71–77

van Baren MJ, Brent MR (2006) Iterative gene prediction and pseudogene removal improves genome annotation. Genome Res 16:678–685

Vanin EF (1985) Processed pseudogenes – characteristics and evolution. Annu Rev Genet 19:253–272

Vaucheret H (2006) Post-transcriptional small RNA pathways in plants: mechanisms and regulations. Genes Dev 20:759–771

Wang X, Shi X, Hao B et al (2005) Duplication and DNA segmental loss in the rice genome: implications for diploidization. New Phytol 165:937–946

Watanabe T, Totoki Y, Toyoda A et al (2008) Endogenous siRNAs from naturally formed dsRNAs regulate transcripts in mouse oocytes. Nature 453:539–543

Waterston RH, Lindblad-Toh K, Birney E et al (2002) Initial sequencing and comparative analysis of the mouse genome. Nature 420:520–562

Woischnik M, Moraes CT (2002) Pattern of organization of human mitochondrial pseudogenes in the nuclear genome. Genome Res 12:885–893

Yamada K, Lim J, Dale JM et al (2003) Empirical analysis of transcriptional activity in the Arabidopsis genome. Science 302:842–846

Yano Y, Saito R, Yoshida N et al (2004) A new role for expressed pseudogenes as ncRNA: regulation of mRNA stability of its homologous coding gene. J Mol Med 82:414–422

Yu J, Wang J, Lin W et al (2005) The genomes of *Oryza sativa*: a history of duplications. PLoS Biol 3:e38

Zhang Z, Gerstein M (2003) Identification and characterization of over 100 mitochondrial ribosomal protein pseudogenes in the human genome. Genomics 81:468–480

Zhang ZL, Gerstein M (2004) Large-scale analysis of pseudogenes in the human genome. Curr Opin Genet Dev 14:328–335

Zhang Z, Harrison P, Gerstein M (2002) Identification and analysis of over 2000 ribosomal protein pseudogenes in the human genome. Genome Res 12:1466–1482

Zhang Z, Harrison PM, Liu Y et al (2003) Millions of years of evolution preserved: a comprehensive catalog of the processed pseudogenes in the human genome. Genome Res 13:2541–2558

Zhang Z, Carriero N, Gerstein M (2004) Comparative analysis of processed pseudogenes in the mouse and human genomes. Trends Genet 20:62–67

Zhang Y, Wu Y, Liu Y et al (2005) Computational identification of 69 retroposons in Arabidopsis. Plant Physiol 138:935–948

Zhang Z, Carriero N, Zheng D et al (2006) PseudoPipe: an automated pseudogene identification pipeline. Bioinformatics 22:1437–1439

Zhang ZD, Frankish A, Hunt T et al (2010) Identification and analysis of unitary pseudogenes: historic and contemporary gene losses in humans and other primates. Genome Biol 11:R26

Zheng DY, Gerstein MB (2007) The ambiguous boundary between genes and pseudogenes: the dead rise up, or do they? Trends Genet 23:219–224

Zheng D, Zhang Z, Harrison PM et al (2005) Integrated pseudogene annotation for human chromosome 22: evidence for transcription. J Mol Biol 349:27–45

Zheng D, Frankish A, Baertsch R et al (2007) Pseudogenes in the ENCODE regions: consensus annotation, analysis of transcription, and evolution. Genome Res 17:839–851

Zou C, Lehti-Shiu MD, Thibaud-Nissen F et al (2009) Evolutionary and expression signatures of pseudogenes in Arabidopsis and rice. Plant Physiol 151:3–15

Polyadenylation in RNA Degradation Processes in Plants

Heike Lange and Dominique Gagliardi

Contents

Abstract Although polyadenylation is best known for stabilizing eukaryotic mRNAs and promoting their translation, the primordial role of polyadenylation is to target RNAs for degradation by 3′→5′ exoribonucleases. This ancient mechanism is conserved among bacteria and eukaryotes, and in plants, polyadenylation-assisted RNA degradation operates in the nucleus, the chloroplast, and the mitochondrion. Polyadenylation-assisted RNA degradation contributes to maturation, turnover, and quality control of a variety of transcripts, the nature of which varies in the different genetic compartments of the plant cell. Moreover, polyadenylation-assisted RNA degradation rapidly removes a large variety of novel transcripts of unknown function that are produced by extensive transcription of extragenic regions, in particular from nuclear and mitochondrial genomes. In this chapter, we review the current knowledge of polyadenylation-assisted RNA degradation in plants, highlighting the different impact of this RNA degradation pathway on the expression of nuclear, plastidial, or mitochondrial genomes.

H. Lange (✉) and D. Gagliardi
Institut de Biologie Moléculaire des Plantes du CNRS, Université de Strasbourg, 12 rue du Général Zimmer, 67084 Strasbourg, France
e-mail: heike.lange@ibmp-cnrs.unistra.fr

V.A. Erdmann and J. Barciszewski (eds.), *Non Coding RNAs in Plants*, RNA Technologies, DOI 10.1007/978-3-642-19454-2_13,

1 Polyadenylation-Assisted RNA Degradation is a Conserved Mechanism

Ultimately, all mRNAs and the many types of non coding RNAs are degraded and recycled. But besides its role in RNA turnover, RNA degradation is crucial for the posttranscriptional regulation of genome expression because a cell's functional transcriptome is the result of both active transcription and RNA decay (Bickel and Morris 2006; Chekanova et al. 2007; Neil et al. 2009; Schmid and Jensen 2008; Wyers et al. 2005; Xu et al. 2009). In addition, RNA degradation mechanisms are also required for RNA processing, the generation of functional RNA from primary transcripts by removing for instance 3′ and 5′ extensions. Both partial and complete degradation of many RNAs can be triggered by the addition of poly-A tails, which serve as a landing pad for 3′→5′ exoribonucleases.

Poly(A)-assisted RNA degradation is a conserved mechanism that affects both coding and non coding RNAs in bacteria, in hyperthermophilic and some methanogenic Archaea, in chloroplasts and plant mitochondria (reviewed in Gagliardi et al. 2004; Portnoy and Schuster 2006; Régnier and Hajnsdorf 2009; Schuster and Stern 2009). It was, therefore, thought to be a mechanism restricted to bacteria or organelles derived from prokaryotic endosymbionts, from which canonical poly(A) polymerases (cPAPs) and stabilizing poly(A) binding proteins (PABPs) are absent. In fact, more recent research has demonstrated that polyadenylation-assisted RNA degradation occurs also in the nucleus of the yeast *Saccharomyces cerevisiae* (LaCava et al. 2005; Vanacova et al. 2005; Wyers et al. 2005). Since polyadenylated degradation intermediates of nuclear transcripts have been detected in trypanosomes, flies, mammals, and plants, this degradation pathway is likely conserved in all eukaryotes (Bühler et al. 2007; Chekanova et al. 2007; Cristodero and Clayton 2007; Etheridge et al. 2009; Nakamura et al. 2008; Slomovic et al. 2006; West et al. 2006; Win et al. 2006). The evolutionary conservation of this mechanism underlines that triggering degradation is probably the ancient and primordial role of polyadenylation, while the stabilizing function of poly(A) tails in conjunction with PABPs, as we know it from eukaryotic mRNAs, has evolved more recently.

2 Polyadenylating Enzymes in Plants

Three types of poly(A) polymerases (PAPs) coexist in plants (reviewed by Martin and Keller 2007). The *Arabidopsis* genome encodes four cPAPs, four bacterial-type poly(A) polymerases (btPAPs), and nine putative noncanonical poly(A) polymerases (ncPAPs) (Lange et al. 2009). The four canonical PAPs are highly processive enzymes that add long poly(A) tails to the 3′ end of RNA polymerase II transcripts (Addepalli et al. 2004). This is coupled to both transcription termination and binding of PABPs. The association of PABPs with the newly synthesized poly(A) tails is crucial for many subsequent steps of mRNA maturation such as release from the transcription site, export from the nucleus to the cytosol, and circularization via interaction with 5′ cap binding proteins (Moore and Proudfoot 2009). Furthermore, PABPs protect mRNAs against degradation by 3′→5′ exoribonucleases. By contrast, the addition of short poly(A) tails, likely catalyzed by noncanonical and bacterial-type PAPs, to non coding RNAs in the nucleus and both non coding and mRNAs in chloroplast and plant mitochondria, triggers RNA degradation (reviewed in Lange et al. 2009). In organelles, synthesis of such destabilizing poly(A) tails could also be catalyzed by polynucleotide phosphorylases (PNPases) because phosphorolysis is energetically close to equilibrium. Indeed, bacterial PNPase and the archaeal exosome (see below) can degrade RNAs and synthesize heteropolymeric A-rich tails (Slomovic et al. 2008). Such heteropolymeric tails have been observed in spinach chloroplast, indicating that in this organism, PNPase contributes to the polyadenylation of chloroplast RNA. In *Arabidopsis*, tails of both chloroplast and mitochondrial transcripts are homopolymeric and are, therefore, likely synthesized by a bona fide PAP activity (Schuster and Stern 2009; Stern et al. 2010). Moreover, downregulation of chloroplast or mitochondrial PNPase results in the accumulation of polyadenylated RNA (Holec et al. 2006; Perrin et al. 2004a; Walter et al. 2002). Hence, noncanonical or bacterial-type PAPs likely play a key role in polyadenylation-assisted RNA degradation in plant organelles. Actually, two of the four bacterial-type PAPs in *Arabidopsis*, encoded by At1g28090 and At5g23690, are targeted to mitochondria and possibly also to chloroplasts (Zimmer et al. 2009). The protein encoded by At1g28090 was shown to have PAP activity in vitro (Zimmer et al. 2009). Otherwise, none of the bacterial-type PAPs has been functionally characterized, and their contribution to polyadenylation-assisted RNA degradation in organelles remains largely unknown. Similarly, none of the noncanonical PAPs has been studied in *Arabidopsis* yet. The proteins encoded by At5g53770 and At4g00060 show significant sequence similarity to Trf4p and Trf5p, two ncPAPs involved in a nuclear polyadenylation-assisted RNA degradation pathway in *S. cerevisiae* and *Schizosaccharomyces pombe* (reviewed in Anderson and Wang 2009). Trf4p and Trf5p associate with the RNA helicase Mtr4p and the RNA binding proteins Air1p or Air2p to form the TRAMP (for Trf4/5-Air1/2-Mtr4-polyadenylation) complex (LaCava et al. 2005; Vanacova et al. 2005; Wyers et al. 2005). The TRAMP complex is responsible for the polyadenylation of nuclear transcripts destined for degradation by the nuclear exosome. Homologues of Trf4p, Trf5p, and Mtr4p are conserved in all eukaryotes, including plants. However, TRAMP-like complexes have only been characterized in *S. cerevisiae*, *S. pombe*, and *Trypanosoma brucei* (Bühler et al. 2007; Etheridge et al. 2009). Other ncPAPs may affect RNA stability by other means than polyadenylation: In fact, some ncPAPs can catalyze uridylation of 3′ ends, which appears to modulate the stability of nuclear-encoded RNAs, and possibly also of organellar RNAs (Ibrahim et al. 2010; Lehrbach et al. 2009; Li et al. 2005; Mullen and Marzluff 2008; Norbury 2010; Ramachandran and Chen 2008; Rissland and Norbury 2009; Shen and Goodman 2004; Song and Kiledjian 2007; Yu et al. 2005; Zimmer et al. 2009). The *Arabidopsis* protein encoded by At2g45620 has poly(U) polymerase activity and shares some sequence similarity with the *S. pombe* poly(U) polymerase Cid1p (Kwak and Wickens 2007). In *S. pombe*, Cid1p adds short poly(U) tails to the poly(A) tail of cytoplasmic mRNAs, which triggers decapping and subsequent degradation from the 5′-end (Rissland et al. 2007; Rissland and Norbury 2008, 2009). In both mice and plants, uridylation-triggered decapping may be involved in the degradation of 5′ mRNA fragments produced by RISC-induced cleavage (Shen and Goodman 2004). Interestingly, the addition of uridines to 3′ ends of nonmethylated plant miRNAs is thought to protect against degradation by SDN1 (for small RNA degrading nuclease1) (Ramachandran and Chen 2008). However, uridylation also stimulates the degradation of small RNAs, at least in plants (Ibrahim et al. 2010; Li et al. 2005; Yu et al. 2005, 2010). In *Chlamydomonas reinhardtii*, downregulation of the ncPAP MUT68 results in both decreased adenylation of RISC cleaved 5′ mRNA fragments and decreased uridylation of small RNAs, and is accompanied by an accumulation of both types of RNA (Ibrahim et al. 2006, 2010). Two *Arabidopsis* ncPAPs, encoded by At2g39740 and At3g45750, are related to MUT68, but their RNA substrates, catalytic activity, and biological roles are not known.

3 Polyadenylation-Stimulated Exoribonucleases

In contrast to the PAPs, the 3′→5′ exoribonucleases involved in polyadenylation-assisted RNA degradation processes in plants are relatively well characterized and have been identified by two criteria: accumulation of polyadenylated RNA substrates in exoribonuclease mutants, and/or stimulation of exoribonucleolytic activity by polyadenylation of RNA substrates. In chloroplast and mitochondria, polyadenylation-assisted RNA degradation is carried out by polynucleotide phosphorylases (PNPases) (reviewed in Gagliardi and Binder 2007; Gagliardi et al. 2004; Schuster and Stern 2009; Stern et al. 2010). PNPases are ring-shaped phosphorolytic enzymes composed of three PNPase monomers, each of which contains two RNase PH-domains alongside with a S1 and a KH RNA-binding domain. Chloroplast and mitochondrial PNPase are encoded by separate nuclear genes and imported into their respective target organelles (Perrin et al. 2004a; Walter et al. 2002). In addition, the *Arabidopsis* genome encodes a RNase II/RNase R protein, RNR1, which is dual-targeted to both chloroplast and mitochondria and has a hydrolytic 3′→5′ exoribonuclease activity (Perrin et al. 2004b). RNR1 is capable of degrading unstructured RNA, such as poly(A) tails, but is inhibited by structural 3′ determinants such as stem-loops (Perrin et al. 2004b). Dual-targeted RNRI is not required for the elimination of polyadenylated RNAs. By contrast, RNR1 contributes to rRNA maturation in chloroplast, and to mRNA maturation in mitochondria (Bollenbach et al. 2005; Perrin et al. 2004b). Although not demonstrated yet for an endogenous substrate, RNR1 can potentially counteract polyadenylation-assisted RNA degradation by removing the landing pad for PNPases, the poly(A) tail, as it has been observed in *Escherichia coli* (Marujo et al. 2000; Perrin et al. 2004b).

The main player of polyadenylation-assisted RNA degradation in plant nuclei is the exosome (Chekanova et al. 2007). The exosome is a protein complex of nine subunits found in both cytosol and nucleus of all eukaryotes, which is conserved in some Archaea, but absent from Bacteria (Houseley et al. 2006; Schmid and Jensen 2008, and references therein). Remarkably, the exosome is structurally and evolutionary related to bacterial and organellar PNPases: it consists of a ring-shaped core of six PH-domain proteins (RRP41, RRP42, RRP43, RRP45, RRP46, and MTR3) to which a cap of three RNA binding proteins with S1 or KH domains (RRP4, RRP40, and CSL4) is bound (Hartung and Hopfner 2009; Lin-Chao et al. 2007; Liu et al. 2006; Lorentzen et al. 2005). In yeast and humans, the core exosome is catalytically inactive (Dziembowski et al. 2007; Liu et al. 2006). Instead, both catalytic activity and substrate specificity are conferred by associated ribonucleases and other cofactors such as RNA binding proteins and RNA helicases (reviewed by Houseley and Tollervey 2009; Lebreton and Séraphin 2008; Vanacova and Stefl 2007). In *S. cerevisiae*, both cytosolic and nuclear exosomes are tightly associated with Rrp44p/Dis3p, an RNase II/RNase D-type protein that has both a hydrolytic exoribonucleolytic and an endonucleolytic activity conferred by a N-terminal PIN-domain (Lebreton et al. 2008; Mitchell et al. 1997; Schaeffer et al. 2009; Schneider et al. 2009). The PIN domain is also required for binding of Rrp44 to the exosome core complex. Additionally, the nuclear exosome binds Rrp6p, a hydrolytic 3′→5′ exoribonuclease belonging to the RNase D family (Allmang et al. 1999). In plants, the situation appears to be different, and more complex: First, in contrast to yeast and human exosomes, the plant core subunit AtRRP41 has a phosphorolytic activity that is stimulated by short poly(A) tails, at least in vitro (Chekanova et al. 2000). Second, downregulation of individual subunits of the plant core exosome results in differential plant phenotypes (Chekanova et al. 2007; Hooker et al. 2007; Xi et al. 2009). Moreover, downregulation of the core subunits AtRRP41, AtRRP4, or AtCSL4 resulted in accumulation of different polyadenylated RNAs, indicating that each of the proteins functions in polyadenylation-assisted RNA decay, albeit affects a distinct set of substrates (Chekanova et al. 2007). Third, the plant exosome may not tightly bind additional exoribonucleases because no further proteins were copurified with the nine subunits of the *Arabidopsis* core complex (at least under the purification conditions used) (Chekanova et al. 2007). However, two RRP44-like and three RRP6-like proteins are encoded in the *Arabidopsis* genome (Chekanova et al. 2002; Lange et al. 2008; Zhang et al. 2010). One of the plant RRP44-like proteins is located in the nucleus and possesses a PIN-domain (Zhang et al. 2010). Although its role in plant polyadenylation-assisted RNA degradation remains to be determined, this protein is a promising candidate for a functional orthologue of yeast Rrp44p. The second *Arabidopsis* RRP44-like protein, named SUPRESSOR OF VARICOSE (SOV), has recently been shown to function in cytosolic mRNA decay (Zhang et al. 2010). However, SOV lacks the PIN-domain and is therefore unlikely to interact with the cytoplasmic exosome. Whether its activity is stimulated by RNA modifications are not known yet. Interestingly, the three *Arabidopsis* RRP6-like proteins show also distinct localization patterns: RRP6L1 is predominantly in the nucleoplasm,

RRP6L2 is enriched in nucleoli, and RRP6L3, which belongs to a plant-specific subgroup, is located in the cytosol (Lange et al. 2008). While the roles of both RRP6L1 and RRP6L3 are unknown, the nucleolar protein RRP6L2 was shown to function in the polyadenylation-assisted degradation of the 5′ external transcribed spacer, an abundant by-product of rRNA maturation (Lange et al. 2008). Hence, both subunits of the core complex and putative cofactors of the plant exosome show an intriguing extent of functional specialization.

4 Polyadenylation of Plastid RNA Enhanced Degradation Efficiency

In contrast to mRNAs transcribed in the nucleus, mRNAs in chloroplast are not constitutively polyadenylated. But both mRNAs and non coding RNAs such as tRNA and rRNA can be polyadenylated, which accelerates their degradation by chloroplast polynucleotide phosphorylase (cpPNPase) (Komine et al. 2002; Schuster and Stern 2009; Stern et al. 2010; Walter et al. 2002). Polyadenylation occurs at mature and unprocessed 3′ ends, but the majority of polyadenylation sites map to endonucleolytic cleavage sites (Lisitsky et al. 1996; Kudla et al. 1996). This indicates that, as in bacteria, degradation is initiated by endonucleases such as CSP41 proteins or homologues of RNAseE and RNAseJ, and proceeds by polyadenylation-assisted exoribonucleolytic degradation of the cleavage products (Mudd et al. 2008; Régnier and Hajnsdorf 2009; Schein et al. 2008; Schuster and Stern 2009; Zimmer et al. 2008). The finding that degradation intermediates accumulate in chloroplast extracts upon inhibition of polyadenylation proved that polyadenylation is required for the efficient degradation of these fragments (Lisitsky et al. 1997). In vivo, downregulation of cpPNPase alone has only minor affects on mRNA steady-state levels (Walter et al. 2002; Marchive et al. 2009). By contrast, chloroplast mRNA abundance appears to be controlled by a 5′→3′ degradation pathway (Drager et al. 1999; Loiselay et al. 2008; Pfalz et al. 2009). Interestingly, plant genomes do not encode 5′→3′ exoribonucleases predicted to be imported into organelles. Instead, the main players of 5′→3′ degradation in chloroplasts are likely endonucleases, namely, homologues of RNase E and RNase J in *Arabidopsis thaliana*, and RNase J in *C. reinhardtii* (Mudd et al. 2008; Schein et al. 2008; Zimmer et al. 2008). As its bacterial counterpart, plant RNase E preferentially cleaves RNAs with 5′ monophosphates (Condon 2007; Mathy et al. 2007; Schein et al. 2008). In *E. coli*, conversion of the 5′ triphosphate to a 5′ monophosphate by the pyrophosphatase RppH facilitates initial cleavage by RNase E (Celesnik et al. 2007; Deana et al. 2008). As this first cleavage generates a new 5′ monophosphate end, degradation proceeds as a wave of endonucleolytic cleavages in 5′→3′ direction, a mechanism that appears to be conserved in chloroplasts (Stern et al. 2010). *Bacillus subtilis* RNase J1 has recently been shown to have both endonucleolytic and 5′-3′ exoribonucleolytic activity (Mathy et al. 2007, 2010; de la Sierra-Gallay et al. 2008). Whether plant RNase J homologues have also retained a 5′→3′ exoribonuclease activity is not known yet, but it is interesting to note that insertion alleles in the gene encoding chloroplast RNaseJ are embryo-lethal in *Arabidopsis* (Stern et al. 2010).

To some extent, polyadenylation-assisted RNA degradation participates also in RNA maturation: for instance, downregulation of cpPNAse results in incomplete 3′ end formation of *rblC* and *psbA* mRNAs (Walter et al. 2002). When generation of mature RNA termini involves 5′→3′ or 3′→5′ degradation pathways, complete destruction of the RNA may be prevented by specific stabilizing factors, many of which probably belong to the family of PPR (for pentatricopeptide repeat) proteins (Schmitz-Linneweber and Small 2008; Stern et al. 2010). For example, *Zea mays* PPR10 protects the 5′ end of *atpH* against endonucleolytic attacks, and the 3′ end of *psaJ* transcripts from 3′→5′ degradation (Pfalz et al. 2009). In other cases, rapid destruction of chloroplast transcripts is impeded by terminal stem-loops close to mature 3′ ends. However, multiple rounds of polyadenylation and degradation by PNPase can overcome the stabilizing effect of such structures (Schuster and Stern 2009).

In conclusion, RNA degradation in chloroplasts requires a combination of different mechanisms. Steady-state levels of chloroplast transcript are mainly controlled by specific stabilizing factors and endonucleases. Polyadenylation-assisted 3′→5′ degradation participates in 3′ end processing and contributes to the efficient elimination of degradation intermediates. This holds true for both plastid mRNAs and non coding RNAs, i.e., tRNAs and rRNAs. Whether polyadenylation contributes also to the stability of other non coding transcripts such as the 12 Ntc transcripts cloned from tobacco chloroplasts (Lung et al. 2006) has not been investigated yet.

5 RNA Degradation Shapes the Transcriptome of Plant Mitochondria

The only known role of polyadenylation in plant mitochondria is to promote RNA degradation (Lange et al. 2009; Schuster and Stern 2009). Intriguingly, especially in the light of the monophyletic origin of mitochondria, polyadenylation plays other roles in mitochondria of humans and trypanosomes and is absent in yeast mitochondria (Bobrowicz et al. 2008; Etheridge et al. 2008; Gagliardi et al. 2004; Kao and Read 2005; Nagaike et al. 2005; Schäfer 2005; Slomovic et al. 2005; Tomecki et al. 2004). However, in plant mitochondria, polyadenylation targets mRNAs, spliced introns, and a wide variety of non coding RNAs to rapid elimination by PNPase (Gagliardi and Binder 2007; Holec et al. 2006; Li-Pook-Than and Bonen 2006; Perrin et al. 2004a). Similar to chloroplast, specific *trans*-factors probably stabilize individual transcripts by protecting them from polyadenylation and degradation. A few factors involved in either processing or stability of mitochondrial RNAs have been identified and belong also to the family of PPR proteins (Schmitz-Linneweber and Small 2008). Another protein that probably contributes to 3′ processing of mitochondrial mRNAs is dual-targeted 3′→5′ exoribonuclease RNR1 (Perrin et al. 2004b). In contrast to bacteria and chloroplast, RNA degradation seems mostly initiated at 3′ ends, while the impact of endonucleases or 5′ determinants on RNA stability in plant mitochondria remains mostly speculative to date (Xiao et al. 2006; Gagliardi and Binder 2007). Although mtPNPase null mutants are inviable, the role of mtPNPase in vivo could be studied in individual *Arabidopsis* plants in which downregulation of mtPNPase was achieved via co-suppression (Holec et al. 2006, 2008a, b; Perrin et al. 2004a, b). These plants accumulate high levels of 3′ unprocessed mRNAs and rRNAs as polyadenylated species. Hence, mtPNPase could be involved in 3′ end formation of these RNAs. An alternative explanation is these transcripts are recognized as mis-processed RNAs and therefore targeted to rapid degradation. Similarly, PNPase probably degrades other type of misprocessed RNAs such as misfolded tRNA precursors (Placido et al. 2005). However, specific quality-control mechanism that monitors for example mRNA editing appear to be absent in plant mitochondria (Holec et al. 2008a) Another role of PNPase is the elimination of tRNA and rRNA maturation by-products (Holec et al. 2006). Finally, downregulation of mtPNPase in *Arabidopsis* revealed that a plethora of non coding transcripts is generated from intergenic regions or in antisense orientation to known genes. In contrast to the relative compact genomes of yeast or human mitochondria, plant mitochondrial genomes are often larger than 300 kb and are generally characterized by a substantial amount of nongenic DNA (Gagliardi and Binder 2007). In *Arabidopsis*, transcription can be initiated at numerous sites scattered across both strands of the 367 kb mitochondrial genome (Holec et al. 2006). The vast majority of the resulting transcripts have (1) no open reading frame, (2) no sequence similarity to other functional RNAs, and (3) are not conserved in other plant species. Moreover, these transcripts are frequently generated, but extremely short-lived, as they accumulate only upon downregulation of the main degradation pathway. Hence, although it cannot be excluded that individual cryptic transcripts have regulatory or other functions, the majority of them may well correspond to transcriptional noise caused by a relaxed transcription control. Moreover, some of them could have detrimental effects because they could compete with functional RNAs for limited available processing factors. The accumulation of these spurious transcripts in PNPase mutants indicates that an important role of polyadenylation and mtPNPase is actually to prevent their accumulation.

Owing to the absence of a 5′→3′ degradation pathway, degradation from the 3′ end is the main turnover and RNA surveillance pathway in plant mitochondria. Polyadenylation-assisted RNA degradation is essential for the elimination of misprocessed mRNAs, rRNAs, tRNAs, by-products of rRNA, and tRNA maturation, and transcripts generated from nongenic regions of the plant mitochondrial genome.

6 Polyadenylation-Assisted RNA Degradation by the Nuclear Exosome

In both cytosol and nuclei of eukaryotic cells, 3′→5′ exoribonucleolytic RNA degradation is mediated by the exosome complex and its cofactors (for recent reviews see Houseley and Tollervey 2009; Schmid and Jensen 2008; Tomecki et al. 2010a). Because the substrates of the *Arabidopsis* exosome have been identified by high-resolution tiling arrays, we have a rather complete picture of the roles of polyadenylation-assisted RNA degradation by the exosome in plants (Chekanova et al. 2007). This study identified about 1,100 RNAs that accumulated as polyadenylated species upon downregulation of the exosome subunits RRP41 or RRP4. Interestingly, about 300 transcripts accumulated either only in the RRP4-depleted or only in the RRP41-depleted sample. This confirmed that individual subunits of the core complex affect specific substrates and have different contributions to the in vivo activity of the exosome, as has already been suggested by the distinct phenotype of the mutants (Chekanova et al. 2007; Hooker et al. 2007; Xi et al. 2009). However, about 500 transcripts were upregulated in both samples and likely represent the common substrates of the exosome core complex (Chekanova et al. 2007). A major group of polyadenylated exosome substrates corresponded to rRNA precursor transcripts and maturation by-products removed during rRNA processing such as the external and internal transcribed spacer regions. This result confirmed that, as in plant mitochondria, one prominent function of polyadenylation-assisted RNA degradation is the removal of incompletely processed or misprocessed rRNA precursors and rRNA maturation by-products (Chekanova et al. 2007; Lange et al. 2008). In *Arabidopsis*, processing of mature rRNAs releases about one third of the common 18S-5.8S-23S precursor as maturation by-products. Since rRNAs are transcribed at high rates, the rapid removal of rRNA maturation by-products is a major task of RNA degradation in the nucleus. Other stable structural RNAs identified as exosome substrates include small nuclear RNAs (snRNA), small nucleolar RNAs (snoRNA), and MRP/7-2 RNA and 7SL RNA (Chekanova et al. 2007). In most cases, both correctly processed and 3′-extended precursor transcripts accumulated as polyadenylated transcripts, suggesting that polyadenylation-assisted RNA decay by the plant exosome mediates both turnover of stable structural RNAs, and removal of misprocessed species. Alternatively, these species could be intermediates of polyadenylation-assisted 3′ trimming by the exosome. Other substrates of the plant exosome include intermediates and by-products of miRNA processing (Chekanova et al. 2007). Interestingly, no tRNAs were observed among the polyadenylated exosome substrates, with the exception of tRNA-Tyr. This particular tRNA undergoes multiple base modification steps during its maturation. It was, therefore, suggested that the plant exosome may participate in quality control of highly modified tRNAs as it was also observed in yeast (Vanacova et al. 2005). Downregulation of plant exosome subunits also affected a number of mRNA loci, some of which were deregulated in either only in the RRP4-depleted or the RRP41-depleted sample. These could be both, true mRNA substrates of the exosome, or mRNAs upregulated or downregulated due to secondary effects caused by the depletion of functional exosomes. A significant fraction of upregulated mRNA-regions corresponded to sense and antisense transcripts derived from intronless pseudogenes, or appeared to be irregular read-through transcripts from protein-coding genes. Together, the genome-wide characterization of exosome mutants has shown that a major task of polyadenylation-assisted RNA degradation in plant nuclei is the degradation of all types of misprocessed and potentially nonfunctional transcripts including a large variety of non coding RNAs.

As mitochondrial PNPase mutants, plants depleted for RRP41 or RRP4 were also a valuable tool to explore spurious transcriptional activity across genic and nongenic regions of the nuclear genome. Indeed, a large variety of novel transcripts with no protein-coding potential and no predicted function, which have never been detected before, accumulated upon downregulation of the two exosome subunits (Chekanova et al. 2007). A considerable fraction of these new RNAs derived from centromeric and pericentromeric heterochromatic regions that give also rise to small RNAs. Another group of short-lived non coding transcripts of 100–600 nt corresponded to 5′ regions of known mRNAs. Similar non coding RNAs, termed CUTs (cryptic unstable transcripts) or PROMPTs (promoter upstream transcripts) have been observed in yeast and mammals, respectively (Belostotsky 2009; Preker et al. 2008; Wyers et al. 2005). Yeast CUTs have been proposed to be global regulators of gene expression because their synthesis might regulate transcription of neighboring genes (Neil et al. 2009; Xu et al. 2009). In plants, it remains to be explored to what extent the production of these novel transcripts has a regulatory function. Interestingly, at least one novel transcript first identified in exosome mutants was later found to be stress-induced in wild type, alongside with several other transcripts derived from unannotated regions (Matsui et al. 2010; Zeller et al. 2009). The accumulation of this "dark matter of the transcriptome" upon downregulation of the exosome demonstrates that, similar to the situation in plant mitochondria, polyadenylation-assisted RNA degradation efficiently eliminates transcripts from heterochromatic and intergenic regions of the nuclear genome. However, given the size and the complexity of the nuclear genome, the number of about 1,100 loci that were upregulated in plant exosome mutants is not as high as one could expect for the main 3′–5′ degradation machine of a eukaryotic cell. This is probably due to the presence of compensating RNA degradation pathways. In contrast to plant organelles, two bona fide 5′–3′ exoribonucleolases, XRN2 and XRN3, are present in plant nuclei (Kastenmayer et al. 2001). As the nuclear exosome, XRN2 and XRN3 are involved in both rRNA processing and degradation of rRNA maturation by-products (Zakrzewska-Placzek et al. 2010). Similar to the downregulation of the nuclear exosome, downregulation of XRN2 and XRN3 also results in the accumulation of intermediates and by-products of miRNA maturation (Gy et al. 2007). Interestingly, some of the aberrant rRNA processing intermediates detected in *xrn2* single and *xrn2/3* double mutants accumulated as polyadenylated species (Zakrzewska-Placzek et al. 2010). This is a strong indication that, although each of the main degradation pathways may have its preferred set of substrates, a large proportion of nuclear transcripts can be degraded from either their 5′- or their 3′ end. To what extent 5′→3′ and polyadenylation-assisted 3′→5′ exoribonucleolytic pathways are intertwined remains to be explored.

7 Concluding Remarks

Polyadenylation-assisted RNA degradation is a conserved process that operates in all three genetic compartments of a plant cell. In chloroplasts, the pathway contributes to mRNA processing and enhances degradation efficiency (for a recent review see Stern et al. 2010). In the nucleus, polyadenylation-assisted RNA degradation by the exosome contributes to turnover and quality control mainly of non coding RNAs and degrades numerous short-lived transcripts generated from heterochromatic regions or from 5′ regions to known genes (Chekanova et al. 2007). In plant mitochondria, polyadenylation and 3′→5′ degradation appears to be the main degradation pathway that controls the stability of both protein-coding and non coding RNAs and systematically prevents the accumulation of cryptic RNAs transcribed from nongenic regions (Holec et al. 2006). The different impact of polyadenylation-assisted RNA degradation on the transcriptomes of chloroplasts, mitochondria, and nuclei can be explained by (1) the complexity of the respective genome and its content of nongenic DNA and (2) the presence, or absence, of other degradation pathways. Particularly in plant nuclei, the relevance of RNA degradation for the functional transcriptome is probably much larger as what can be deduced from the characterization of exosome mutants.

Moreover, recent results obtained in human cells have indicated that polyadenylation may also assist degradation by the cytosolic exosome and hDIS3L (Slomovic et al. 2010; Staals et al. 2010; Tomecki et al. 2010b). Hence, it is probably worth to re-examine the substrates of the plant exosome using viable mutants of compartment-specific exosome cofactors instead of components of the core complex. Another aspect that remains to be studied is the link between polyadenylation-assisted RNA degradation and RNA silencing. 5′→3′ exoribonucleases in both nucleus and cytosol act as silencing suppressors, probably because they degrade aberrant RNAs that otherwise would become substrates of RNA-dependent RNA polymerases and amplified RNA silencing (Gregory et al. 2008; Gy et al. 2007; Souret et al. 2004). It is assumed that RNA quality control by 3′→5′ exoribonucleolytic pathways can similarly prevent the production of “false” siRNA (Voinnet 2008). In *S. pombe*, loss of the polymerase TRF4, which is responsible for the polyadenylation of nuclear exosome substrates, results in the generation of small RNAs predominantly from rRNAs and tRNAs (Bühler et al. 2008). These irregular small RNAs can compete with repeat-associated small RNAs for incorporation into Ago1 and have the potential to impair transcriptional silencing of heterochromatic and centromeric regions (Bühler et al. 2008). The overlap of plant exosome substrates with small-RNA generating loci suggests a similar relationship between silencing and 3′→5′ decay in plants (Chekanova et al. 2007; Voinnet 2008).

However, a role of the plant exosome as a silencing suppressor has not been directly demonstrated yet.

Probably, the least understood question is how functional coding or non coding RNAs are discriminated from misprocessed, defective, or otherwise unfunctional RNAs substrates of polyadenylation-assisted RNA degradation. Correctly processed and functional RNAs are always assembled into ribonucleoprotein (RNP) complexes. RNP assembly speed depends on binding energy and is further assisted by energy-consuming RNA helicases (for a detailed explanation of the kinetic proofreading concept, please see Houseley and Tollervey 2009). For misprocessed or otherwise defective transcripts, the formation of functional RNPs is slow, which renders the RNA more accessible to degradation. However, both dedicated RNA binding proteins and PAPs likely contribute to the specificity of RNA degradation pathways (Houseley and Tollervey 2009; Lebreton and Séraphin 2008; Vanacova and Stefl 2007). Characterizing these factors will be crucial to fully understand mechanisms of substrate recognition in polyadenylation-assisted RNA degradation in plants.

References

Addepalli B, Meeks LR, Forbes KP et al (2004) Novel alternative splicing of mRNAs encoding poly(A) polymerases in *Arabidopsis*. Biochim Biophys Acta 1679:117–128

Allmang C, Petfalski E, Podtelejnikov A et al (1999) The yeast exosome and human PM-Scl are related complexes of 3′ 5′ exonucleases. Genes Dev 13:2148–2158

Anderson JT, Wang X (2009) Nuclear RNA surveillance: no sign of substrates tailing off. Crit Rev Biochem Mol Biol 44:16–24

Belostotsky D (2009) Exosome complex and pervasive transcription in eukaryotic genomes. Curr Opin Cell Biol 21:352–358

Bickel KS, Morris DR (2006) Silencing the transcriptome's dark matter: mechanisms for suppressing translation of intergenic transcripts. Mol Cell 22:309–316

Bobrowicz AJ, Lightowlers RN, Chrzanowska-Lightowlers Z (2008) Polyadenylation and degradation of mRNA in mammalian mitochondria: a missing link? Biochem Soc Trans 36:517–519

Bollenbach TJ, Lange H, Gutierrez R et al (2005) RNR1, a 3′–5′ exoribonuclease belonging to the RNR superfamily, catalyzes 3′ maturation of chloroplast ribosomal RNAs in *Arabidopsis*. Nucleic Acids Res 33:2751–2763

Bühler M, Haas W, Gygi SP et al (2007) RNAi-dependent and -independent RNA turnover mechanisms contribute to heterochromatic gene silencing. Cell 129:707–721

Bühler M, Spies N, Bartel DP et al (2008) TRAMP-mediated RNA surveillance prevents spurious entry of RNAs into the *Schizosaccharomyces pombe* siRNA pathway. Nat Struct Mol Biol 15:1015–1023

Celesnik H, Deana A, Belasco JG (2007) Initiation of RNA decay in *Escherichia coli* by 5′ pyrophosphate removal. Mol Cell 27:79–90

Chekanova JA, Shaw RJ, Wills MA et al (2000) Poly(A) tail-dependent exonuclease AtRrp41p from *Arabidopsis thaliana* rescues 5.8 S rRNA processing and mRNA decay defects of the yeast ski6 mutant and is found in an exosome-sized complex in plant and yeast cells. J Biol Chem 275:33158–33166

Chekanova JA, Dutko JA, Mian IS et al (2002) *Arabidopsis thaliana* exosome subunit AtRrp4p is a hydrolytic 3′ 5′ exonuclease containing S1 and KH RNA-binding domains. Nucleic Acids Res 30:695–700

Chekanova JA, Gregory BD, Reverdatto SV et al (2007) Genome-wide high-resolution mapping of exosome substrates reveals hidden features in the *Arabidopsis* transcriptome. Cell 13:1340–1353
Condon C (2007) Maturation and degradation of RNA in bacteria. Curr Opin Microbiol 10:271–278
Cristodero M, Clayton CE (2007) Trypanosome MTR4 is involved in rRNA processing. Nucleic Acids Res 35:7023–7030
de la Sierra-Gallay IL, Zig L, Jamalli A et al (2008) Structural insights into the dual activity of RNase J. Nat Struct Mol Biol 15:206–212
Deana A, Celesnik H, Belasco JG (2008) The bacterial enzyme RppH triggers messenger RNA degradation by 5′ pyrophosphate removal. Nature 451:355–835
Drager RG, Higgs DC, Kindle KL et al (1999) 5′ to 3′ exoribonucleolytic activity is a normal component of chloroplast mRNA decay pathways. Plant J 19:521–531
Dziembowski A, Lorentzen E, Conti E et al (2007) A single subunit, Dis3, is essentially responsible for yeast exosome core activity. Nat Struct Mol Biol 14:15–22
Etheridge RD, Aphasizheva I, Gershon PD et al (2008) 3′ adenylation determines mRNA abundance and monitors completion of RNA editing in *T. brucei* mitochondria. EMBO J 27:1596–1608
Etheridge RD, Clemens DM, Gershon PD et al (2009) Identification and characterization of nuclear non-canonical poly(A) polymerases from *Trypanosoma brucei*. Mol Biochem Parasitol 164:66–73
Gagliardi D, Binder S (2007) Expression of the plant mitochondrial genome. In: Logan D (ed) Plant mitochondria. Annual plant reviews, vol 31. Blackwell Publishing, Oxford
Gagliardi D, Stepien PP, Temperley RJ et al (2004) Messenger RNA stability in mitochondria: different means to an end. Trends Genet 20:260–267
Gregory BD, O'Malley RC, Lister R et al (2008) A link between RNA metabolism and silencing affecting *Arabidopsis* development. Dev Cell 14:854–866
Gy I, Gasciolli V, Lauressergues D et al (2007) *Arabidopsis* FIERY1, XRN2, and XRN3 are endogenous RNA silencing suppressors. Plant Cell 19:3451–3461
Hartung S, Hopfner K (2009) Lessons from structural and biochemical studies on the archaeal exosome. Biochem Soc Trans 37:83–87
Holec S, Lange H, Kühn K et al (2006) Relaxed transcription in *Arabidopsis* mitochondria is counterbalanced by RNA stability control mediated by polyadenylation and polynucleotide phosphorylase. Mol Cell Biol 26:2869–2876
Holec S, Lange H, Canaday J et al (2008a) Coping with cryptic and defective transcripts in plant mitochondria. Biochim Biophys Acta 1779:566–573
Holec S, Lange H, Dietrich A et al (2008b) Polyadenylation-mediated RNA degradation in plant mitochondria. Methods Enzymol 447:439–461
Hooker TS, Lam P, Zheng H et al (2007) A core subunit of the RNA-processing/degrading exosome specifically influences cuticular wax biosynthesis in *Arabidopsis*. Plant Cell 19:904–913
Houseley J, Tollervey D (2009) The many pathways of RNA degradation. Cell 136:763–776
Houseley J, LaCava J, Tollervey D (2006) RNA-quality control by the exosome. Nat Rev Mol Cell Biol 7:529–539
Ibrahim F, Rohr J, Jeong W et al (2006) Untemplated oligoadenylation promotes degradation of RISC-cleaved transcripts. Science 314:1893
Ibrahim F, Rymarquis LA, Kim E et al (2010) Uridylation of mature miRNAs and siRNAs by the MUT68 nucleotidyltransferase promotes their degradation in *Chlamydomonas*. Proc Natl Acad Sci USA 107:3906–3911
Kao C, Read LK (2005) Opposing effects of polyadenylation on the stability of edited and unedited mitochondrial RNAs in *Trypanosoma brucei*. Mol Cell Biol 25:1634–1644
Kastenmayer JP, Johnson MA, Green PJ (2001) Analysis of XRN orthologs by complementation of yeast mutants and localization of XRN-GFP fusion proteins. Methods Enzymol 342:269–282
Komine Y, Kikis E, Schuster G et al (2002) Evidence for in vivo modulation of chloroplast RNA stability by 3′-UTR homopolymeric tails in *Chlamydomonas reinhardtii*. Proc Natl Acad Sci USA 99:4085–4090

Kudla J, Hayes R, Gruissem W (1996) Polyadenylation accelerates degradation of chloroplast mRNA. EMBO J 15:7137–7146

Kwak JE, Wickens M (2007) A family of poly(U) polymerases. RNA 13:860–867

LaCava J, Houseley J, Saveanu C et al (2005) RNA degradation by the exosome is promoted by a nuclear polyadenylation complex. Cell 121:713–724

Lange H, Holec S, Cognat V et al (2008) Degradation of a polyadenylated rRNA maturation by-product involves one of the three RRP6-like proteins in *Arabidopsis thaliana*. Mol Cell Biol 28:3038–3044

Lange H, Sement FM, Canaday J et al (2009) Polyadenylation-assisted RNA degradation processes in plants. Trends Plant Sci 14:497–504

Lebreton A, Séraphin B (2008) Exosome-mediated quality control: substrate recruitment and molecular activity. Biochim Biophys Acta 1779:558–565

Lebreton A, Tomecki R, Dziembowski A et al (2008) Endonucleolytic RNA cleavage by a eukaryotic exosome. Nature 456:993–996

Lehrbach NJ, Armisen J, Lightfoot HL et al (2009) LIN-28 and the poly(U) polymerase PUP-2 regulate let-7 microRNA processing in *Caenorhabditis elegans*. Nat Struct Mol Biol 16:1016–1020

Li J, Yang Z, Yu B et al (2005) Methylation protects miRNAs and siRNAs from a 3′-end uridylation activity in *Arabidopsis*. Curr Biol 15:1501–1507

Lin-Chao S, Chiou N, Schuster G (2007) The PNPase, exosome and RNA helicases as the building components of evolutionarily-conserved RNA degradation machines. J Biomed Sci 14:523–532

Li-Pook-Than J, Bonen L (2006) Multiple physical forms of excised group II intron RNAs in wheat mitochondria. Nucleic Acids Res 34:2782–2790

Lisitsky I, Klaff P, Schuster G (1996) Addition of destabilizing poly (A)-rich sequences to endonuclease cleavage sites during the degradation of chloroplast mRNA. Proc Natl Acad Sci USA 93:13398–13403

Lisitsky I, Klaff P, Schuster G (1997) Blocking polyadenylation of mRNA in the chloroplast inhibits its degradation. Plant J 12:1173–1178

Liu Q, Greimann JC, Lima CD (2006) Reconstitution, activities, and structure of the eukaryotic RNA exosome. Cell 127:1223–1237

Loiselay C, Gumpel NJ, Girard-Bascou J et al (2008) Molecular identification and function of cis- and trans-acting determinants for petA transcript stability in *Chlamydomonas reinhardtii* chloroplasts. Mol Cell Biol 28:5529–5542

Lorentzen E, Walter P, Fribourg S et al (2005) The archaeal exosome core is a hexameric ring structure with three catalytic subunits. Nat Struct Mol Biol 12:575–581

Lung B, Zemann A, Madej MJ et al (2006) Identification of small non-coding RNAs from mitochondria and chloroplasts. Nucleic Acids Res 34:3842–3852

Marchive C, Yehudai-Resheff S, Germain A et al (2009) Abnormal physiological and molecular mutant phenotypes link chloroplast polynucleotide phosphorylase to the phosphorus deprivation response in Arabidopsis. Plant Physiol 151:905–924

Martin G, Keller W (2007) RNA-specific ribonucleotidyl transferases. RNA 13:1834–1849

Marujo PE, Hajnsdorf E, Le Derout J et al (2000) RNase II removes the oligo(A) tails that destabilize the rpsO mRNA of *Escherichia coli*. RNA 6:1185–1193

Mathy N, Bénard L, Pellegrini O et al (2007) 5′-to-3′ exoribonuclease activity in bacteria: role of RNase J1 in rRNA maturation and 5′ stability of mRNA. Cell 129:681–692

Mathy N, Hébert A, Mervelet P et al (2010) *Bacillus subtilis* ribonucleases J1 and J2 form a complex with altered enzyme behaviour. Mol Microbiol 75:489–498

Matsui A, Ishida J, Morosawa T et al (2010) *Arabidopsis* tiling array analysis to identify the stress-responsive genes. Methods Mol Biol 639:141–155

Mitchell P, Petfalski E, Shevchenko A et al (1997) The exosome: a conserved eukaryotic RNA processing complex containing multiple 3′→5′ exoribonucleases. Cell 91:457–466

Moore MJ, Proudfoot NJ (2009) Pre-mRNA processing reaches back to transcription and ahead to translation. Cell 136:688–700

Mudd EA, Sullivan S, Gisby MF et al (2008) A 125 kDa RNase E/G-like protein is present in plastids and is essential for chloroplast development and autotrophic growth in *Arabidopsis*. J Exp Bot 59:2597–2610

Mullen TE, Marzluff WF (2008) Degradation of histone mRNA requires oligouridylation followed by decapping and simultaneous degradation of the mRNA both 5′ to 3′ and 3′ to 5′. Genes Dev 22:50–65

Nagaike T, Suzuki T, Katoh T et al (2005) Human mitochondrial mRNAs are stabilized with polyadenylation regulated by mitochondria-specific poly(A) polymerase and polynucleotide phosphorylase. J Biol Chem 280:19721–19727

Nakamura R, Takeuchi R, Takata K et al (2008) TRF4 is involved in polyadenylation of snRNAs in *Drosophila melanogaster*. Mol Cell Biol 28:6620–6631

Neil H, Malabat C, d'Aubenton-Carafa Y et al (2009) Widespread bidirectional promoters are the major source of cryptic transcripts in yeast. Nature 457:1038–1042

Norbury CJ (2010) 3′ Uridylation and the regulation of RNA function in the cytoplasm. Biochem Soc Trans 38:1150–1153

Perrin R, Lange H, Grienenberger J et al (2004a) AtmtPNPase is required for multiple aspects of the 18S rRNA metabolism in *Arabidopsis thaliana* mitochondria. Nucleic Acids Res 32:5174–5182

Perrin R, Meyer EH, Zaepfel M et al (2004b) Two exoribonucleases act sequentially to process mature 3′-ends of atp9 mRNAs in *Arabidopsis* mitochondria. J Biol Chem 279:25440–25446

Pfalz J, Bayraktar OA, Prikryl J et al (2009) Site-specific binding of a PPR protein defines and stabilizes 5′ and 3′ mRNA termini in chloroplasts. EMBO J 28:2024–2052

Placido A, Gagliardi D, Gallerani R et al (2005) Fate of a larch unedited tRNA precursor expressed in potato mitochondria. J Biol Chem 280:33573–33579

Portnoy V, Schuster G (2006) RNA polyadenylation and degradation in different Archaea; roles of the exosome and RNase R. Nucleic Acids Res 34:5923–5931

Preker P, Nielsen J, Kammler S et al (2008) RNA exosome depletion reveals transcription upstream of active human promoters. Science 322:1851–1854

Ramachandran V, Chen X (2008) Degradation of microRNAs by a family of exoribonucleases in *Arabidopsis*. Science 321:1490–1492

Régnier P, Hajnsdorf E (2009) Poly(A)-assisted RNA decay and modulators of RNA stability. Prog Mol Biol Transl Sci 85:137–185

Rissland OS, Norbury CJ (2008) The Cid1 poly(U) polymerase. Biochim Biophys Acta 1779:286–294

Rissland OS, Norbury CJ (2009) Decapping is preceded by 3′ uridylation in a novel pathway of bulk mRNA turnover. Nat Struct Mol Biol 16:616–623

Rissland OS, Mikulasova A, Norbury CJ (2007) Efficient RNA polyuridylation by noncanonical poly(A) polymerases. Mol Cell Biol 27:3612–3624

Schaeffer D, Tsanova B, Barbas A et al (2009) The exosome contains domains with specific endoribonuclease, exoribonuclease and cytoplasmic mRNA decay activities. Nat Struct Mol Biol 16:56–62

Schäfer B (2005) RNA maturation in mitochondria of *S. cerevisiae* and *S. pombe*. Gene 354:80–85

Schein A, Sheffy-Levin S, Glaser F et al (2008) The RNase E/G-type endoribonuclease of higher plants is located in the chloroplast and cleaves RNA similarly to the *E. coli* enzyme. RNA 14:1057–1068

Schmid M, Jensen TH (2008) The exosome: a multipurpose RNA-decay machine. Trends Biochem Sci 33:501–510

Schmitz-Linneweber C, Small I (2008) Pentatricopeptide repeat proteins: a socket set for organelle gene expression. Trends Plant Sci 13:663–670

Schneider C, Leung E, Brown J et al (2009) The N-terminal PIN domain of the exosome subunit Rrp44 harbors endonuclease activity and tethers Rrp44 to the yeast core exosome. Nucleic Acids Res 37:1127–1140

Schuster G, Stern D (2009) RNA polyadenylation and decay in mitochondria and chloroplasts. Prog Mol Biol Transl Sci 85:393–422

Shen B, Goodman HM (2004) Uridine addition after microRNA-directed cleavage. Science 306:997
Slomovic S, Laufer D, Geiger D et al (2005) Polyadenylation and degradation of human mitochondrial RNA: the prokaryotic past leaves its mark. Mol Cell Biol 25:6427–6435
Slomovic S, Laufer D, Geiger D et al (2006) Polyadenylation of ribosomal RNA in human cells. Nucleic Acids Res 34:2966–2975
Slomovic S, Portnoy V, Yehudai-Resheff S et al (2008) Polynucleotide phosphorylase and the archaeal exosome as poly(A)-polymerases. Biochim Biophys Acta 1779:247–255
Slomovic S, Fremder E, Staals RHG et al (2010) Addition of poly(A) and poly(A)-rich tails during RNA degradation in the cytoplasm of human cells. Proc Natl Acad Sci USA 107:7407–7412
Song M, Kiledjian M (2007) 3′ Terminal oligo U-tract-mediated stimulation of decapping. RNA 13:2356–2365
Souret FF, Kastenmayer JP, Green PJ (2004) AtXRN4 degrades mRNA in *Arabidopsis* and its substrates include selected miRNA targets. Mol Cell 15:173–183
Staals RHJ, Bronkhorst AW, Schilders G et al (2010) Dis3-like 1: a novel exoribonuclease associated with the human exosome. EMBO J 29:2358–2367
Stern DB, Goldschmidt-Clermont M, Hanson MR (2010) Chloroplast RNA metabolism. Annu Rev Plant Biol 61:125–155
Tomecki R, Dmochowska A, Gewartowski K et al (2004) Identification of a novel human nuclear-encoded mitochondrial poly(A) polymerase. Nucleic Acids Res 32:6001–6014
Tomecki R, Drazkowska K, Dziembowski A (2010a) Mechanisms of RNA degradation by the eukaryotic exosome. Chembiochem 11:938–945
Tomecki R, Kristiansen MS, Lykke-Andersen S et al (2010b) The human core exosome interacts with differentially localized processive RNases: hDIS3 and hDIS3L. EMBO J 29:2342–2357
Vanacova S, Stefl R (2007) The exosome and RNA quality control in the nucleus. EMBO Rep 8:651–657
Vanacova S, Wolf J, Martin G et al (2005) A new yeast poly(A) polymerase complex involved in RNA quality control. PLoS Biol 3:e189
Voinnet O (2008) Use, tolerance and avoidance of amplified RNA silencing by plants. Trends Plant Sci 13:317–328
Walter M, Kilian J, Kudla J (2002) PNPase activity determines the efficiency of mRNA 3′-end processing, the degradation of tRNA and the extent of polyadenylation in chloroplasts. EMBO J 21:6905–6914
West S, Gromak N, Norbury CJ et al (2006) Adenylation and exosome-mediated degradation of cotranscriptionally cleaved pre-messenger RNA in human cells. Mol Cell 21:437–443
Win TZ, Draper S, Read RL et al (2006) Requirement of fission yeast Cid14 in polyadenylation of rRNAs. Mol Cell Biol 26:1710–1721
Wyers F, Rougemaille M, Badis G et al (2005) Cryptic pol II transcripts are degraded by a nuclear quality control pathway involving a new poly(A) polymerase. Cell 121:725–737
Xi L, Moscou MJ, Meng Y et al (2009) Transcript-based cloning of RRP46, a regulator of rRNA processing and R gene-independent cell death in barley-powdery mildew interactions. Plant Cell 21:3280–3295
Xiao H, Zhang F, Zheng Y (2006) The 5′ stem-loop and its role in mRNA stability in maize S cytoplasmic male sterility. Plant J 47:864–872
Xu Z, Wei W, Gagneur J et al (2009) Bidirectional promoters generate pervasive transcription in yeast. Nature 457:1033–1037
Yu B, Yang Z, Li J et al (2005) Methylation as a crucial step in plant microRNA biogenesis. Science 307:932–935
Yu B, Bi L, Zhai J et al (2010) siRNAs compete with miRNAs for methylation by HEN1 in *Arabidopsis*. Nucleic Acids Res 38:5844–5850
Zakrzewska-Placzek M, Souret FF, Sobczyk GJ et al (2010) *Arabidopsis thaliana* XRN2 is required for primary cleavage in the pre-ribosomal RNA. Nucleic Acids Res 38:4487–4502
Zeller G, Henz SR, Widmer CK et al (2009) Stress-induced changes in the *Arabidopsis thaliana* transcriptome analyzed using whole-genome tiling arrays. Plant J 58:1068–1082

Zhang W, Murphy C, Sieburth LE (2010) Conserved RNaseII domain protein functions in cytoplasmic mRNA decay and suppresses *Arabidopsis* decapping mutant phenotypes. Proc Natl Acad Sci USA 107:15981–15985

Zimmer SL, Fei Z, Stern DB (2008) Genome-based analysis of *Chlamydomonas reinhardtii* exoribonucleases and poly(A) polymerases predicts unexpected organellar and exosomal features. Genetics 179:125–136

Zimmer SL, Schein A, Zipor G et al (2009) Polyadenylation in *Arabidopsis* and Chlamydomonas organelles: the input of nucleotidyltransferases, poly(A)-polymerases and polynucleotide phosphorylase. Plant J 59:88–99

Viroids and RNA Trafficking in Plants

Xiaorui Yang, Yuan Tian, and Biao Ding

Contents

Abstract Cell-to-cell trafficking of RNA is a newly discovered mechanism of gene regulation at the whole plant level. The RNAs that traffic within a plant range from mRNAs to non coding RNAs including microRNAs and small interfering RNAs. The mechanisms underlying such trafficking remain largely unknown. Recent studies on *Potato spindle tuber viroid* (PSTVd), a non coding, circular, and infectious RNA, have demonstrated that distinct three-dimensional (3D) structural motifs in an RNA mediate trafficking across specific cellular boundaries. This 3D motif-mediated cell-specific trafficking mechanism may well apply to the trafficking regulation of cellular RNAs. In this review, we summarize examples of RNA trafficking that functions in plant gene regulation and development and then discuss the utility of PSTVd to identify RNA motifs mediating trafficking. We finally highlight outstanding issues for future investigations.

X. Yang , Y. Tian, and B. Ding (✉)
Department of Molecular Genetics, The Ohio State University, 207 Rightmire Hall, 1060 Carmack Road, Columbus, OH 43210, USA
e-mail: ding.35@osu.edu

V.A. Erdmann and J. Barciszewski (eds.), *Non Coding RNAs in Plants*, RNA Technologies, DOI 10.1007/978-3-642-19454-2_14,

Keywords Phloem • Plasmodesmata • RNA trafficking • Viroid

1 Introduction

Cell-to-cell communication is a necessary function of all organisms, central to the coordinated gene expression and metabolism underlying morphogenesis and response to the environment. In plants, plasmodesmata and the vascular tissue phloem form continuous cytoplasmic channels for direct cell-to-cell communication of developmental and defense signals, as well as transport of photoassimilates (Lough and Lucas 2006; Turgeon and Wolf 2009).

Numerous mRNAs are detected in the phloem saps of several plant species, with some shown to traffic long distances within the phloem (Kehr and Buhtz 2008). Experimental studies showed that long-distance transport of some mRNAs can regulate distinct developmental processes (Hannapel 2010). Infectious RNAs such as viroids and viruses presumably utilize the endogenous trafficking system to spread within a plant to establish systemic infection (Taliansky et al. 2008; Ding 2009; Benitez-Alfonso et al. 2010; Harries et al. 2010). Gene silencing signals, including microRNAs (miRNAs) and small interfering RNAs (siRNAs), traffic intercellularly to trigger systemic silencing in plants and other organisms as a means of gene regulation and antiviral defense (Ding and Voinnet 2007; Jose and Hunter 2007; Voinnet 2009; Dunoyer and Voinnet 2009; Martienssen 2010).

These observations suggest that cell-to-cell trafficking of RNAs regulates many biological processes. Elucidating the underlying mechanisms is crucial to understand gene regulation as well as host–pathogen interactions at the organismal level. Outstanding mechanistic questions include the following: What RNA motifs direct trafficking between different cells? Do these motifs function individually or in combination to direct trafficking across a cellular boundary? What cellular factors are involved in the recognition and trafficking of an RNA? In this review, we summarize examples of RNA trafficking in plant gene regulation and development and then discuss the utility of *Potato spindle tuber viroid* (PSTVd) to identify RNA motifs mediating trafficking. We finally highlight outstanding issues to be investigated in future studies.

2 Cell-to-Cell Trafficking of RNA is a New Mechanism of Gene Regulation

2.1 Trafficking of mRNAs Regulates Plant Development

Numerous mRNAs are detected in the phloem saps of many plant species (Kehr and Buhtz 2008). Because mature sieve tube elements from which the saps were collected have no nuclei, these mRNAs are conceivably imported from surrounding nucleate cells. In situ hybridization localizes sucrose transporter *SUT1* mRNA in plasmodesmata at the sieve element–companion cell interface, indicating plasmodesmata as the pathway for mRNA transport (Kühn et al. 1997). A number of mRNAs have been shown to traffic long distances through graft unions. Some notable examples include *CmNACP* (Ruiz-Medrano et al. 1999), *CmPP16* (Xoconostle-Cázares et al. 1999), and *CmGAIP* (Haywood et al. 2005) from *Cucurbita maxima* (pumpkin).

Findings from a number of studies have shed light on the functional significance of mRNA trafficking. The tomato *PFP-LeT6* mRNA can be transported from rootstock to scion in a graft union to cause phenotypic changes in developing leaves (Kim et al. 2001). When expressed ectopically in tomato, the *Arabidopsis thaliana ΔDELLA-gai*, which carries a deletion of the putative gibberellic acid-regulatory DELLA domain in GIBBERELLIC ACID-INSENSITIVE (GAI) to have a gain-of-function phenotype, also traffics long-distances in graft unions to regulate leaf shapes (Haywood et al. 2005). The mRNAs of BEL-1-like family are involved in potato tuber formation (Hannapel 2010). A BEL1-like transcription factor in potato, StBEL5, is functional in regulating tuber formation by mediating hormone levels (Rosin et al. 2003; Chen et al. 2004). Under light regulation, the full-length *StBEL5* mRNA can be transported from leaves, where they are synthesized, to stolon tips to promote production of more and larger tubers (Banerjee et al. 2006).

These examples suggest that the role of mRNA trafficking in regulating plant development is likely extensive. Thus, identifying additional mRNAs that traffic to regulate distinct developmental processes, as well as elucidating the mechanisms of trafficking, will significantly expand our knowledge of whole plant signaling mechanisms that control how a plant grows and develops.

2.2 *Trafficking of Non Coding Small RNAs Regulates Systemic Gene Silencing*

Two classes of small non coding RNAs, siRNAs and miRNAs, play critical roles in gene regulation by mediating cleavage or translation inhibition of their target transcripts or by directing chromatin modifications in diverse organisms (Baulcombe 2005; Sontheimer and Carthew 2005; Wassenegger 2005; Siomi and Siomi 2009). In a very general sense, small RNAs are derived as duplexes (e.g., miRNA:miRNA* duplex) from precursor RNAs via cleavage by dicer (in animals) or dicer-like (DCL in plants). One of the strands from a duplex is eventually loaded into RISC (RNA-Induced Silencing Complex) to perform the gene silencing functions (Bartel 2004; Vaucheret 2006).

Early studies demonstrated that locally initiated gene silencing could spread throughout a plant to regulate the expression of a gene in a sequence-specific manner, implicating RNA as a component of the silencing signals (Voinnet and

Baulcombe 1997; Voinnet et al. 1998; Palauqui et al. 1997). Analyses of phloem saps collected from *C. maxima* (pumpkin), *Cucumis sativus* (cucumber), *Lupinus albus* (white lupin), *Ricinus communis* (castor bean), and *Yucca filamentosa* (yucca) showed the presence of endogenous small RNAs (18–25 nucleotides in length), including miRNAs and siRNAs, suggesting that a large group of miRNAs and siRNAs are mobile regulators (Yoo et al. 2004). miR399 has been identified as a long-distance signal, moving from shoot to root to regulate phosphate homeostasis in *A. thaliana* (Pant et al. 2008; Lin et al. 2008). siRNAs can be mobile signaling molecules that traffic from cell to cell and even from organ to organ to mediate gene silencing including RNA-dependent DNA methylation (Schwab et al. 2009; Chitwood et al. 2009; Molnar et al. 2010; Dunoyer et al. 2010a, b).

In an analysis of the pumpkin phloem sap, non coding RNAs distinct from miRNAs and siRNAs have been detected, which include tRNAs, rRNAs, and spliceosomal RNAs (Zhang et al. 2009). The specific cells where these RNAs are originated, how they traffic into the sieve tubes, whether they traffic long distances within the phloem, and whether they play any biological roles within the phloem remain to be investigated.

Altogether, these studies established that abundant non coding RNAs can traffic long distances within a plant to regulate gene expression. Further studies are expected to uncover the many potential biological functions of such trafficking.

3 Viroids are Simple Models for Studying Mechanisms of Cell-to-Cell RNA Trafficking

3.1 Basics of Viroids and Viroid Infection

Although the phenomenon of intercellular trafficking of mRNAs and non coding RNAs is firmly established and the functional significance is beginning to be understood, the mechanisms underlying this trafficking remain largely unknown. Here, we describe the utility of viroids as simple and tractable models to decipher the mechanisms, particularly the identification of RNA structural motifs that mediate trafficking across distinct cellular boundaries.

Viroids are the simplest form of RNA-based infectious agents. They are single-stranded and circular RNAs comprising 250–400 nucleotides that fold into distinct secondary structures with a series of loops and stems. In contrast to viruses, viroids are not packaged within any proteinaceous or membranous coat. Most significantly, viroids do not encode any proteins (Flores et al. 2005; Ding 2009). Therefore, a viroid RNA itself interacts with cellular factors to accomplish infection that includes systemic trafficking, making it a simple experimental system to investigate the RNA trafficking mechanisms (Ding and Wang 2009; Wang and Ding 2010).

Over 30 species of viroids have been discovered that are classified into two families: *Pospiviroidae* (type member PSTVd) and *Avsunviroidae* (type member *Avocado sunblotch viroid*, ASBVd) (Flores et al. 2004). When viroids are introduced into a cell mechanically, they are imported into a subcellular organelle to replicate (the nucleus for *Pospiviroidae* and the chloroplast for *Avsunviroidae*). The members of *Pospiviroidae* replicate via an asymmetric rolling circle mechanism (Fig. 1a). Briefly, the (+)-circular genomic RNA is first transcribed into concatemeric linear (−)-strand RNAs in the nucleoplasm. These (−)-strand RNAs then act as the templates for producing concatemeric linear (+)-RNAs. These RNAs presumably enter the nucleolus where they are cleaved into unit-length monomers and circularized by intramolecular ligation (Ding 2009; Flores et al. 2009). The members of *Avsunviroidae* replicate via a symmetric rolling circle (Fig. 1b). The (+)-circular genomic RNA is first transcribed into concatemeric linear (−)-strand RNAs. These are cleaved into unit-length (−)-RNAs and circularized. The circular (−)-RNAs then act as the templates to generate concatemeric linear (+)-RNAs, which are then cleaved into unit-length monomers and circularized (Ding 2009; Flores et al. 2009). After replication, viroids exit the organelles and traffic into neighboring cells and distant plant organs to spread infection.

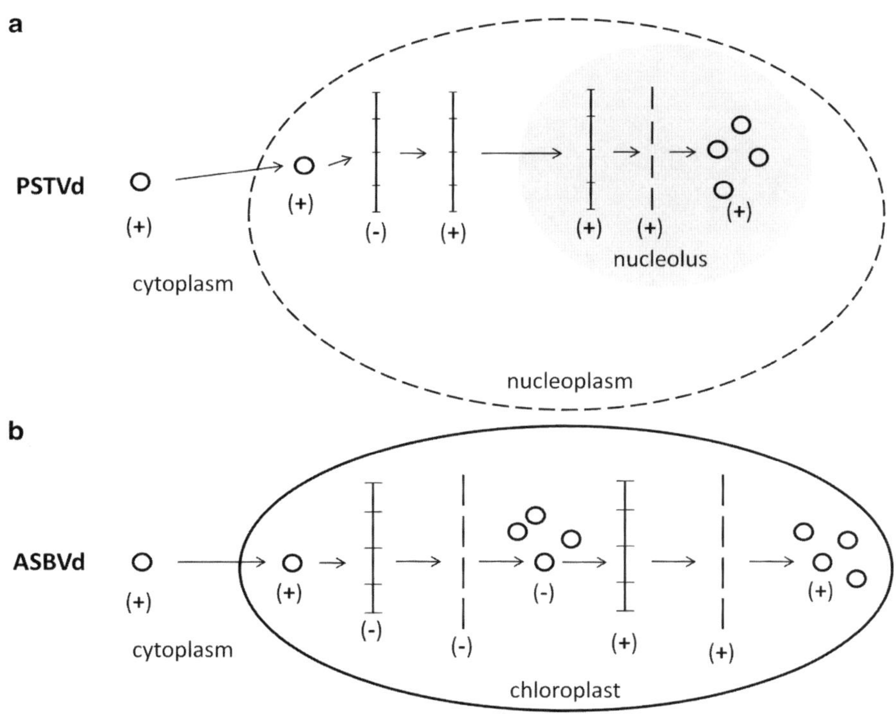

Fig. 1 Rolling circle replication models of (**a**) PSTVd (representative of family *Pospiviroidae*) and (**b**) ASBVd (representative of family *Avsunviroidae*). PSTVd replicates via an asymmetric rolling circle, whereas ASBVd replicates via a symmetric rolling circle

3.2 Distinct RNA Motifs Regulate Trafficking Across Different Cellular Boundaries

Studies on PSTVd indicate that viroids traffic from cell to cell through plasmodesmata (Ding et al. 1997) and from organ to organ through the phloem (Palukaitis 1987; Zhu et al. 2001). As an illustration of this process, from an initially infected upper epidermal cell of a leaf, viroid RNAs need to pass through the cellular boundaries between epidermis–palisade mesophyll, palisade mesophyll–spongy mesophyll, spongy mesophyll–bundle sheath, and bundle sheath–phloem for long-distance transport. In a systemic leaf, viroid RNAs will exit the phloem to traffic, in the reverse direction, into various nonvascular tissues (Fig. 2a). Viroid trafficking through plasmodesmata, based on studies on PSTVd, is regulated rather than occurring by diffusion (Ding et al. 1997). In the following paragraphs we discuss genetic identification of PSTVd structural motifs that mediate trafficking between specific cells.

As shown in Fig. 2b, the secondary structure of PSTVd comprises a series of short double helices formed by Watson–Crick base pairs interconnected by loops/bulges formed by non-Watson–Crick base pairs. Crystal structures of ribosomes and some RNA-ligand complexes provide overwhelming evidence that numerous loops/bulges

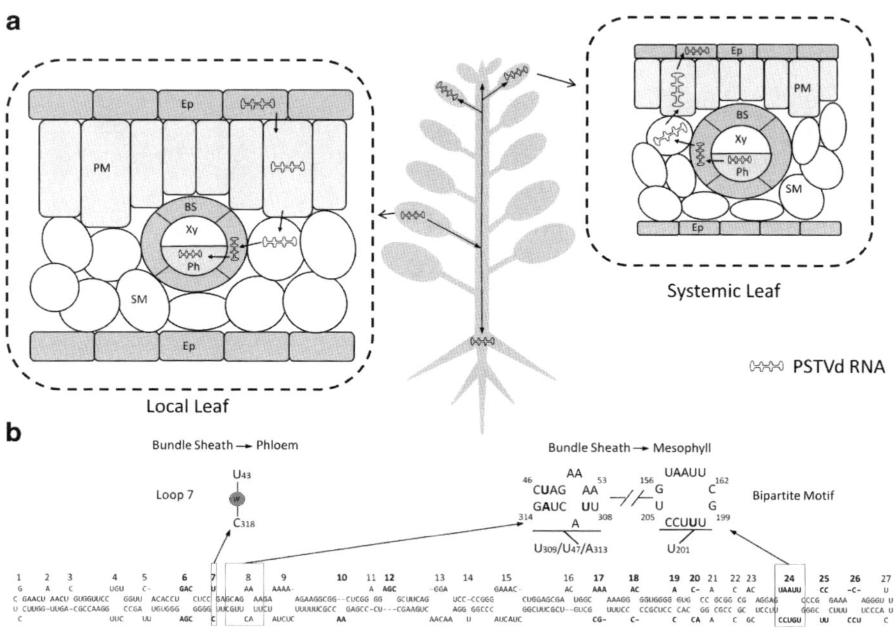

Fig. 2 Systemic trafficking of PSTVd in an infected plant. (**a**) Cellular boundaries that PSTVd passes during systemic trafficking, initiating in an epidermal cell in a local leaf and finishing in an epidermal cell in a systemic leaf. (**b**) PSTVd motifs identified to be responsible for trafficking between specific cells. See text for details

form three-dimensional (3D) motifs that serve as binding sites for RNA–RNA, RNA–protein, or RNA–small ligand interactions (Leontis et al. 2002, 2006). Importantly, 3D motifs often recur in different RNAs, as a result of convergent evolution. Thus, knowledge of the 3D structure of a motif in an RNA may be used to infer the 3D structure of a similar or identical motif in a different RNA, through comparative sequence analysis. Functionally, a motif that functions as a protein-binding site in an RNA may imply that a similar motif in a different RNA also functions as a protein-binding site. Therefore, structural studies of viroid motifs can be valuable for studying endogenous RNAs. The secondary structure of PSTVd, proposed by Gross et al. (1978), is well supported by a large body of biophysical and genetic studies (Ding 2009), making PSTVd an excellent model to investigate the structural motifs mediating cell-to-cell trafficking and other biological functions associated with infection.

Mutagenesis and functional assays of PSTVd identified a bipartite motif, which is composed of U201 and U309/U47/A313 (shown in Fig. 2b) that is necessary and sufficient to mediate unidirectional trafficking of PSTVd from the bundle sheath to mesophyll in a development-dependent manner in tobacco leaves (Qi et al. 2004). Whether this bipartite motif represents two individual 3D motifs that function together, or whether they form one single motif through viroid structural reorganization, to potentiate trafficking remains an outstanding question. In a subsequent study, loop U43/C318 (shown in Fig. 2b) was shown to be required for PSTVd to traffic from the bundle sheath to phloem in *Nicotiana benthamiana* leaves (Zhong et al. 2007). A combination of comparative sequence analysis, RNA crystal structure database analysis, mutagenesis, and covariation analysis suggest that this loop consists of *cis* Watson–Crick base pair with water insertion (Zhong et al. 2007). Because a similar structure in rRNAs functions as a protein-binding site, the PSTVd loop U43/C318 may well be recognized by a plant protein for trafficking (Zhong et al. 2007). Genome-wide mutagenesis altering each of the loops in conjunction with functional studies identified multiple loops critical for PSTVd systemic infection, but not required for replication (Fig. 2b) (Zhong et al. 2008). Further studies will determine the 3D structure each loop and how each loop functions in the cell-to-cell and long-distance trafficking of PSTVd.

4 Future Prospects

Many endogenous RNAs have the potential to traffic from cell to cell and even from organ to organ in a plant. The examples available to date suggest that trafficking of selective RNAs plays important roles in regulating gene expression and developmental processes. A future research focus is clearly the expanded investigation of the biological roles of the trafficking of many RNAs.

Knowledge of the molecular machinery controlling the trafficking of diverse RNAs is essential to understand the evolution and function of RNA trafficking. At the RNA level, evidence from viroid studies indicates that 3D structural motifs play a critical role in trafficking across specific cellular boundaries. Such motif-based trafficking can well be applied to cellular RNAs. Indeed, the untranslated regions (UTRs) of potato *StBEL5* mRNA appear to be important for long-distance trafficking (Banerjee et al. 2006, 2009). A *cis*-acting element in nucleotides 1–102 of the *A. thaliana FLOWERING LOCUS T* mRNA can mediate trafficking of a fused green fluorescent protein (GFP) mRNA or a modified viral RNA (Li et al. 2009). The coding sequences and 3′ UTR as well as the secondary structure of the *GAI* RNA can mediate trafficking of a fused GFP mRNA (Huang and Yu 2009). Further studies are needed to determine the precise structural features (secondary vs. tertiary) of the potential trafficking motifs. The conceptual and technical tools developed from viroid studies may prove useful in dissecting the cellular RNA trafficking motifs. With more of motifs identified, we should be able to also answer the question of whether one or multiple motifs are required for an RNA to traffic across a cellular boundary.

In a simple scenario, the RNA motifs likely function as recognition sites for cellular proteins to potentiate trafficking. CmPP16-1 isolated from pumpkin phloem exudates binds and traffics endogenous RNAs (Xoconostle-Cázares et al. 1999). Phloem SMALL RNA BINDING PROTEIN 1 (CmPSRP1) binds synthetic single-stranded siRNAs and potentiates trafficking of these RNAs from cell to cell in *N. benthamiana* mesophyll as shown by microinjection (Yoo et al. 2004). Ham et al. (2009) have recently identified an RNA–protein complex, from the pumpkin phloem sap, that contains a series of mRNAs and proteins. The key RNA-binding protein, RBP50, is a polypyrimidine tract binding protein. This complex is postulated to be the form for long-distance transport of RNAs (Ham et al. 2009). Further studies will determine the mechanisms as well as physiological functions of these proteins in RNA trafficking.

A number of proteins interacting with viroids in vitro or in vivo have been reported, but there is no experimental evidence yet to establish the role of these proteins in viroid trafficking (Ding 2009). Using viroids as models to identify cellular proteins in trafficking has clear advantages. First, because viroids do not encode any proteins, they can be simple models to fish out the cellular proteins. Second, assuming that viroids utilize the endogenous pathways for trafficking, a protein identified to facilitate viroid trafficking may well function in the trafficking of endogenous RNAs. Thus, elucidating the cellular proteins that participate in viroid trafficking should be yet another major effort in future studies in the field of RNA trafficking in plants.

Acknowledgments We thank Ying Wang and Ryuta Takeda for their comments on the manuscript. This work was supported by a grant from the US National Science Foundation (IOS-0840906).

References

Banerjee AK, Chatterjee M, Yu Y et al (2006) Dynamics of a mobile RNA of potato involved in a long-distance signaling pathway. Plant Cell 18:3443–3457

Banerjee AK, Lin T, Hannapel DJ (2009) Untranslated regions of a mobile transcript mediate RNA metabolism. Plant Physiol 151:1831–1843

Bartel DP (2004) MicroRNAs: genomics, biogenesis, mechanism and function. Cell 116:281–297
Baulcombe D (2005) RNA silencing. Trends Biochem Sci 30:290–293
Benitez-Alfonso Y, Faulkner C, Ritzenthaler C et al (2010) Plasmodesmata: gateways to local and systemic virus infection. Mol Plant Microbe Interact 23:1403–1412
Chen H, Banerjee AK, Hannapel DJ (2004) The tandem complex of BEL and KNOX partners is required for transcriptional repression of ga20ox1. Plant J 38:276–284
Chitwood DH, Nogueira FT, Howell MD et al (2009) Pattern formation via small RNA mobility. Genes Dev 23:549–554
Ding B (2009) The biology of viroid-host interactions. Annu Rev Phytopathol 47:105–131
Ding SW, Voinnet O (2007) Antiviral immunity directed by small RNAs. Cell 130:413–426
Ding B, Wang Y (2009) Viroids: uniquely simple and tractable models to elucidate regulation of cell-to-cell trafficking of RNA. DNA Cell Biol 28:51–56
Ding B, Kwon MO, Hammond R et al (1997) Cell-to-cell movement of potato spindle tuber viroid. Plant J 12:931–936
Dunoyer P, Voinnet O (2009) Movement of RNA silencing between plant cells: is the question now behind us? Trends Plant Sci 14:643–644
Dunoyer P, Brosnan CA, Schott G et al (2010a) An endogenous, systemic RNAi pathway in plants. EMBO J 29:1699–1712
Dunoyer P, Schott G, Himber C et al (2010b) Small RNA duplexes function as mobile silencing signals between plant cells. Science 328:912–916
Flores R, Delgado S, Gas ME et al (2004) Viroids: the minimal non coding RNAs with autonomous replication. FEBS Lett 567:42–48
Flores R, Hernández C, Martínez de Alba AE et al (2005) Viroids and viroid-host interactions. Annu Rev Phytopathol 43:117–139
Flores R, Gas M, Molina-Serrano D et al (2009) Viroid replication: rolling-circles, enzymes and ribozymes. Viruses 1:317–334
Gross HJ, Domdey H, Lossow C et al (1978) Nucleotide sequence and secondary structure of potato spindle tuber viroid. Nature 273:203–208
Ham BK, Brandom JL, Xoconostle-Cázares B et al (2009) A polypyrimidine tract binding protein, pumpkin RBP50, forms the basis of a phloem-mobile ribonucleoprotein complex. Plant Cell 21:197–215
Hannapel DJ (2010) A model system of development regulated by the long-distance transport of mRNA. J Integr Plant Biol 52:40–52
Harries PA, Schoelz JE, Nelson RS (2010) Intracellular transport of viruses and their components: utilizing the cytoskeleton and membrane highways. Mol Plant Microbe Interact 23:1381–1393
Haywood V, Yu TS, Huang NC et al (2005) Phloem long-distance trafficking of GIBBERELLIC ACID-INSENSITIVE RNA regulates leaf development. Plant J 42:49–68
Huang NC, Yu TS (2009) The sequences of Arabidopsis GA-INSENSITIVE RNA constitute the motifs that are necessary and sufficient for RNA long-distance trafficking. Plant J 59:921–929
Jose AM, Hunter CP (2007) Transport of sequence-specific RNA interference information between cells. Annu Rev Genet 41:305–330
Kehr J, Buhtz A (2008) Long distance transport and movement of RNA through the phloem. J Exp Bot 59:85–92
Kim M, Canio W, Kessler S et al (2001) Developmental changes due to long-distance movement of a homeobox fusion transcript in tomato. Science 293:287–289
Kühn C, Franceschi VR, Schulz A et al (1997) Macromolecular trafficking indicated by localization and turnover of sucrose transporters in enucleate sieve elements. Science 275:1298–1300
Leontis NB, Stombaugh J, Westhof E (2002) Motif prediction in ribosomal RNAs Lessons and prospects for automated motif prediction in homologous RNA molecules. Biochimie 84:961–973
Leontis NB, Lescoute A, Westhof E (2006) The building blocks and motifs of RNA architecture. Curr Opin Struct Biol 16:279–287
Li C, Zhang K, Zeng X et al (2009) A cis element within flowering locus T mRNA determines its mobility and facilitates trafficking of heterologous viral RNA. J Virol 83:3540–3548

Lin SI, Chiang SF, Lin WY et al (2008) Regulatory network of microRNA399 and PHO2 by systemic signaling. Plant Physiol 147:732–746

Lough TJ, Lucas WJ (2006) Integrative plant biology: role of phloem long-distance macromolecular trafficking. Annu Rev Plant Biol 57:203–232

Martienssen R (2010) Molecular biology. Small RNA makes its move. Science 328:834–835

Molnar A, Melnyk CW, Bassett A et al (2010) Small silencing RNAs in plants are mobile and direct epigenetic modification in recipient cells. Science 328:872–875

Palauqui JC, Elmayan T, Pollien JM et al (1997) Systemic acquired silencing: transgene-specific post-transcriptional silencing is transmitted by grafting from silenced stocks to non-silenced scions. EMBO J 16:4738–4745

Palukaitis P (1987) Potato spindle tuber viroid: investigation of the long-distance, intra-plant transport route. Virology 158:239–241

Pant BD, Buhtz A, Kehr J et al (2008) MicroRNA399 is a long-distance signal for the regulation of plant phosphate homeostasis. Plant J 53:731–738

Qi Y, Pélissier T, Itaya A et al (2004) Direct role of a viroid RNA motif in mediating directional RNA trafficking across a specific cellular boundary. Plant Cell 16:1741–1752

Rosin FM, Hart JK, Horner HT et al (2003) Overexpression of a knotted-like homeobox gene of potato alters vegetative development by decreasing gibberellin accumulation. Plant Physiol 132:106–117

Ruiz-Medrano R, Xoconostle-Cázares B, Lucas WJ (1999) Phloem long-distance transport of CmNACP mRNA: implications for supracellular regulation in plants. Development 126:4405–4419

Schwab R, Maizel A, Ruiz-Ferrer V et al (2009) Endogenous TasiRNAs mediate non-cell autonomous effects on gene regulation in *Arabidopsis thaliana*. PLoS One 4:e5980

Siomi H, Siomi MC (2009) On the road to reading the RNA-interference code. Nature 457:396–404

Sontheimer EJ, Carthew RW (2005) Silence from within: endogenous siRNAs and miRNAs. Cell 122:9–12

Taliansky M, Torrance L, Kalinina NO (2008) Role of plant virus movement proteins. Methods Mol Biol 451:33–54

Turgeon R, Wolf S (2009) Phloem transport: cellular pathways and molecular trafficking. Annu Rev Plant Biol 60:207–221

Vaucheret H (2006) Post-transcriptional small RNA pathways in plants: mechanisms and regulations. Genes Dev 20:759–771

Voinnet O (2009) Origin, biogenesis, and activity of plant microRNAs. Cell 136:669–687

Voinnet O, Baulcombe DC (1997) Systemic signalling in gene silencing. Nature 389:553

Voinnet O, Vain P, Angell S et al (1998) Systemic spread of sequence-specific transgene RNA degradation in plants is initiated by localized introduction of ectopic promoterless DNA. Cell 95:177–187

Wang Y, Ding B (2010) Viroids: small probes for exploring the vast universe of RNA trafficking in plants. J Integr Plant Biol 52:28–39

Wassenegger M (2005) The role of the RNAi machinery in heterochromatin formation. Cell 122:13–16

Xoconostle-Cázares B, Xiang Y, Ruiz-Medrano R et al (1999) Plant paralog to viral movement protein that potentiates transport of mRNA into the phloem. Science 283:94–98

Yoo BC, Kragler F, Varkonyi-Gasic E et al (2004) A systemic small RNA signaling system in plants. Plant Cell 16:1979–2000

Zhang S, Sun L, Kragler F (2009) The phloem-delivered RNA pool contains small non coding RNAs and interferes with translation. Plant Physiol 150:378–387

Zhong X, Tao X, Stombaugh J et al (2007) Tertiary structure and function of an RNA motif required for plant vascular entry to initiate systemic trafficking. EMBO J 26:3836–3846

Zhong X, Archual AJ, Amin AA et al (2008) A genomic map of viroid RNA motifs critical for replication and systemic trafficking. Plant Cell 20:35–47

Zhu Y, Green L, Woo YM et al (2001) Cellular basis of potato spindle tuber viroid systemic movement. Virology 279:69–77

Mechanisms Linking Cytosine Methylation to Histone Modification in *Arabidopsis thaliana*

Kyohei Arita, Tatsuo Kanno, Manabu Yoshikawa, and Yoshiki Habu

Contents

Abstract Histone modification is the key event in epigenetic regulation of gene expression in most eukaryotes. Plants, fungi, and vertebrates have machineries for DNA methylation as an additional layer of epigenetic modification, and the presence of CHH and CHG (H = A, T, or C) methylation sites are characteristics of plants. In *Arabidopsis thaliana*, both CHH and CHG methylation sites are enriched with sequences related to transposons in heterochromatin regions, and the RNA-directed DNA methylation pathway plays a major role in asymmetric cytosine (CHH) methylation. Heterochromatin regions are also rich in dimethylated histone

Y. Habu (✉) and M. Yoshikawa
Division of Plant Sciences, National Institute of Agrobiological Sciences, Tsukuba 305-8602, Japan
e-mail: habu@affrc.go.jp

K. Arita
Laboratory of Biomolecular Function Chemistry, Department of Molecular Engineering, Kyoto University, Kyoto 615-8510, Japan

T. Kanno
School of Natural Sciences, National University of Ireland, Galway, Ireland

V.A. Erdmann and J. Barciszewski (eds.), *Non Coding RNAs in Plants*, RNA Technologies, DOI 10.1007/978-3-642-19454-2_15,

H3 lysine 9 (H3K9me2) – a histone mark of inert chromatin. Although enzymes transferring methyl groups to DNA or H3K9 have been identified, mechanisms linking these two methylation steps specifically to inert chromatin are obscure. In this chapter, we review characteristics of the machineries governing DNA methylation and histone modification in *A. thaliana* and discuss the possible interplay between these two epigenetic marks in relation to suppression of non coding RNA production.

Keywords *Arabidopsis thaliana* • Cytosine methylation • Histone modification • MORPHEUS' MOLECULE1 • RNA-directed DNA methylation

Abbreviations

CMM2	Conserved MOM1 Motif 2
CMT3	CHROMOMETHYLASE3
DME	DEMETER
Dnmt1	DNA METHYLTRANSFERASE1
DRM2	DOMAINS REARRANGED METHYLTRANSFERASE2
H3K9me2	Di-methylated histone H3 lysine 9
HDAC	Histone deacetylase
IBM1	INCREASE IN BONSAI METHYLATION1
KYP	KRYPTONITE
MBD	Methyl-CpG-binding domain
MBT	Malignant brain tumor
MET1	METHYLTRANSFERASE1
MOM1	MORPHEUS' MOLECULE1
ORTH	ORTHRUS
PcG	*Polycomb* group
PHD	Plant homeo domain
Pol	RNA polymerase
PRC2	*Polycomb* repressive complex2
RdDM	RNA-directed DNA methylation
siRNA	Small interfering RNA
SRA	SET and RING associated
SUVH	SU(VAR)3-9 HOMOLOGUE
TGS	Transcriptional gene silencing
trxG	*Trithorax* group
UHRF1	UBIQUITIN-LIKE, CONTAINING PHD AND RING FINGER DOMAINS 1
VIM	VARIANT IN METHYLATION

1 Introduction

DNA and histone modifications are pivotal epigenetic marks in plants. In contrast to other model organisms, such as worms, yeast, and fruit flies, in which exclusively histone modification and small RNAs mediate epigenetic regulation of gene expression, plants and mammals adopt DNA methylation as another layer of regulation that is generally thought of as a stable memory of epigenetic states after DNA replication in mammals (Cedar and Bergman 2009; Cheng and Blumenthal 2010). However, the recent findings of epigenetic regulators in *Arabidopsis thaliana* suggest the presence of plant-specific epigenetic regulation that is clearly distinct from that of mammals (Law and Jacobsen 2010).

In this chapter, we review and discuss these phenomena and their possible function(s) in the interdependency of DNA and histone modifications that suppress production of non coding RNAs from endogenous templates in the genome. In the first part of the chapter, we present an overview of the components required for DNA and histone modification in plants. Thereafter, we review data obtained in structural analyses of protein modules recognizing DNA and histone modifications that are possible mediators between the two epigenetic markers. Because of the scarcity of such information in plant studies, we focus on data obtained from organisms other than plants but that are also conserved in plants. Next, we focus on a plant-specific epigenetic regulator that is required for transcriptional gene silencing (TGS) of endogenous loci and argue against the general idea that the role of DNA methylation is to serve as a stable memory of epigenetic state working downstream of histone modification. Finally, we discuss the emerging evidence indicating the involvement in chromatin silencing in plants of factors controlling the processing and stability of non coding RNAs.

2 Protein Factors Controlling DNA and Histone Modification in Plants

In *A. thaliana*, DNA methylation is introduced by three different methyltransferases at cytosines in three different sequence contexts (CG, CHG, and CHH). As in mammals, maintenance methyltransferase METHYLTRANSFERASE1 (MET1) – an ortholog of mouse DNA METHYLTRANSFERASE1 (Dnmt1) – maintains CG methylation patterns on newly synthesized DNA strands upon DNA replication (Finnegan and Dennis 1993; Kankel et al. 2003; Saze et al. 2003). Methylation in CHG sites depends mostly on the activity of plant-specific cytosine methyltransferase CHROMOMETHYLASE3 (CMT3), which contains a chromodomain module known to interact with chromatin (Bartee et al. 2001; Lindroth et al. 2001). CMT3 binds to tri-methylated histone H3 lysine 9 and lysine 27 (Jackson et al. 2002; Lindroth et al. 2004; Malagnac et al. 2002), indicating its role in maintenance of inert states of chromatin. The de novo DNA methyltransferase DOMAINS REARRANGED METHYLTRANSFERASE2 (DRM2), a counterpart of mammalian DNA METHYLTRANSFERASE3a/b, is responsible for introducing cytosine methylation in all sequence contexts with sequence complementarity to the 24-nt small interfering RNA (siRNA) during a process termed RNA-directed DNA methylation (RdDM) (Matzke et al. 2009). Template RNAs of siRNAs are thought to be transcribed by a plant-specific RNA polymerase (Pol) IV, and a recent study showed that non coding RNAs generated by either Pol II and/or Pol V are required for efficient siRNA production in a locus-specific manner (Wierzbicki et al. 2008; Zheng et al. 2009). Genome-wide analyses indicate that transposons and other repeats in both heterochromatic and euchromatic regions are the major endogenous RdDM targets (Cokus et al. 2008; Lister et al. 2008; Lu et al. 2005; Zhang et al. 2006; Zilberman et al. 2007). Thus, the function of RdDM is thought to be a mechanism of maintaining silencing of dispersed copies of repetitive elements in euchromatic regions that may be too small to be packed into heterochromatin. In addition, there is evidence to indicate that RdDM are involved in various biological processes including normal development and stress responses (Ahmad et al. 2010; Schoft et al. 2009; Vrbsky et al. 2010). In many, but not all, cases, RdDM directs repression/derepression of genes through nearby conventional targets such as transposon-related sequences or repetitive elements.

Demethylation of cytosine can be active or passive. Active demethylation is mediated by DNA glycosylases/lyases and involves base excision. To date, four proteins of *A. thaliana*, DEMETER (DME), REPRESSOR OF SILENCING (ROS1), DEMETER-LIKE2, and DEMETER LIKE3, with a demethylase activity have been characterized (Ikeda and Kinoshita 2009). It has been shown that the demethylase activity of DME is required for expression of imprinted genes, such as *MEDEA*, *FLOWERING WAGENINGEN*, and *FERTILIZATION INDEPENDENT SEED2* in central cells, while ROS1 demethylates a variety of transposons and repetitive sequences in the genome of *A. thaliana* (Ikeda and Kinoshita 2009).

An example of passive demethylation has been identified with the finding of a jumonjiC domain-containing protein, INCREASE IN BONSAI METHYLATION1 (IBM1), which is required to maintain a low DNA methylation state at the *BONSAI* locus (Saze et al. 2008). Further genome-wide analysis of DNA methylation showed that cytosine methylation at CHG sequences in genes, but not transposons, is increased in *ibm1*, indicating that IBM1 protects genes from CHG methylation (Miura et al. 2009). Since IBM1 belongs to the JHDM2/KDM3 family of H3K9 demethylases, recognition of H3K9me and subsequent DNA methylation by a combination of CMT3 and SU(VAR)3-9 HOMOLOGUE4 (SUVH4)/KRYPTONITE (KYP) (the latter being a H3K9 methyltransferase carrying the SET and RING associated [SRA] domain that binds to methylated cytosine (see next section for the details)), are expected to mediate the process. Other SUVH proteins related to SUVH4/KYP are also involved in regulation of DNA methylation with various target site specificities (Ebbs and Bender 2006; Johnson et al. 2008; Naumann et al. 2005). ORTHRUS (ORTH)/VARIANT IN METHYLATION (VIM) is another group of proteins containing the SRA domain. ORTH2/VIM1 is required for DNA methylation of centromeric repeats and binds to methylated cytosine and histones

(Woo et al. 2007). A recent study has shown that ORTH2/VIM1 possesses ubiquitin E3 ligase activity, implying that it regulates DNA methylation through ubiquitination of histones or other chromatin-bound proteins (Kraft et al. 2008).

Histone deacetylation is also linked to DNA methylation (Richards and Elgin 2002). Among a number of histone deacetylases (HDACs) in *A. thaliana* (Loidl 2004), HDA6 (a RPD3-type histone HDAC) and HDT1 (a plant-specific HDAC) have been characterized in detail in relation to DNA methylation: both are required for the maintenance of cytosine methylation on a transgene and/or endogenous repetitive sequences including the rRNA gene cluster (Aufsatz et al. 2002; Earley et al. 2006; Lawrence et al. 2004; Lippman et al. 2003). HDACs are often found in multiprotein complexes containing CHD3/Mi-2 family proteins and methyl-CpG-binding domain (MBD) proteins in mammalian cells (Bowen et al. 2004). It would be of great interest to find how HDA6 and HDT1 are recruited to their target regions through interaction with other proteins. As in mammals, a number of MBD proteins with various binding specificities to methylcytosine and other chromatin proteins including DECREASE IN DNA METHYLATION1 and ARGININE METHYLTRANSFERASE11 have been identified in plants (Grafi et al. 2007). MBD proteins, in addition to SRA domain proteins, may be key determinants directing DNA- or histone-modifying proteins to their target regions in the plant genome.

Another category of protein complexes required for the maintenance of negative states of chromatin through histone modification is the *Polycomb* group (PcG) proteins, which includes H3K27 methyltransferases (Pien and Grossniklaus 2007). The major PcG protein complex in *A. thaliana* is *Polycomb* repressive complex2 (PRC2), which is also conserved in other higher eukaryotes. The function of PRC2 resides in gene expression involved in developmental regulation and environmental responses. Although a related protein complex, PRC1, is responsible for ubiquitination of histone H2AK119 in fruit flies and mammals, PRC1 is thought to be lacking in *A. thaliana* (Pien and Grossniklaus 2007). The repressive effect of PcG proteins is counteracted by *Trithorax* group (trxG) proteins through maintaining active states of gene expression by their H3K4 methyltransferase activity (Avramova 2009). The function of trxG proteins is also conserved in *A. thaliana*, but recent studies have shown that plants have additional trxG proteins, ATRX5 and ATRX6, that monomethylate H3K27 (Jacob et al. 2009). The double mutant *atrx5 atrx6* reactivates transcription from endogenous transposons without changes in DNA and H3K9 methylation, suggesting that ATRX5/ATRX6 silences heterochromatic targets independently of the RdDM pathway.

3 Protein Modules for Recognition of DNA and Histone Modifications

DNA methylation and histone modifications are read out by protein modules recognizing each mark specifically or in a combinatorial manner. This section focuses on the structural features of protein modules that read and interpret DNA and histone modifications.

Fully methylated CG sites are recognized by MBD proteins and several zinc-finger proteins belonging to the Kaiso family (Kaiso/ZBTB33, ZBTB4, and ZBTB38) in mammals (Bogdanović and Veenstra 2009). The structure of the MBD domain binding to a fully methylated CG sequence has been reported (Ho et al. 2008; Ohki et al. 2001). DNA binding causes an induced fit at a β-hairpin region in the MBD domain, with a pair of arginine residues and an aspartic acid residue, which are fully conserved among MBD domain proteins, recognizing the methylated cytosines (Fig. 1a). Water-mediated hydrogen bonds also contribute to recognition of the methylated cytosine.

In mammals, UBIQUITIN-LIKE CONTAINING PHD AND RING FINGER DOMAINS 1 (UHRF1) recognizes the hemimethylated CG sequence produced upon DNA replication via the SRA domain and recruits Dnmt1 to the site (Bostick et al. 2007; Sharif et al. 2007). Crystal structures of mammalian SRA domains have revealed that the domain folds into a novel globular structure with a basic concave surface (Arita et al. 2008; Avvakumov et al. 2008; Hashimoto et al. 2008). DNA binding to the concave surface causes a loop and an N-terminal tail of the SRA domain to fold into the DNA interfaces at the major and minor grooves of the methylation site. Interestingly, in contrast to fully methylated CG sites recognized by MBDs, the methyl-cytosine base at the hemimethylated site in the SRA–DNA complex is flipped out of the DNA helix (Fig. 1b). The SRA domains in plants also bind to methylated cytosine at CHG and CHH sequences, but the structural basis for the recognition of these sequences remains to be elucidated.

Unlike DNA methylation, histone N-terminal tails protruding from the nucleosome core are subjected to many types of posttranslational modification, including acetylation, methylation, phosphorylation, and ubiquitination. Histone modifications are thought to act in combination as a "histone code." Here, we focus on the histone reading modules associated with lysine modification. Owing to space limitations, other histone reading modules are only briefly summarized in Table 1.

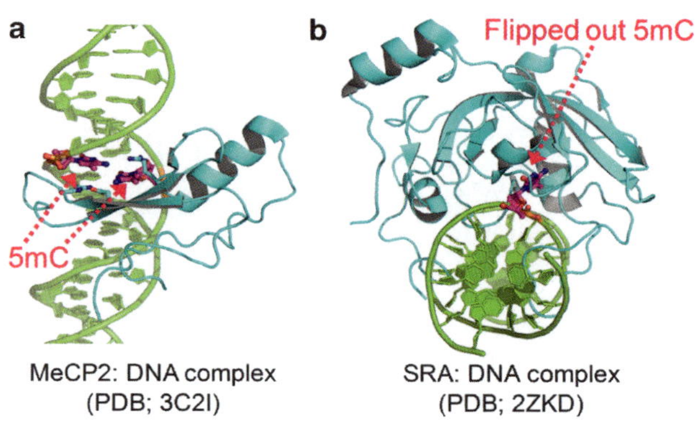

Fig. 1 Structures of methyl DNA binding protein. (**a**) MBD domain of MeCP2 in complex with full methylated CG (PDB; 3C2I). (**b**) UHRF1 SRA domain:hemimethylated DNA complex (PDB; 2ZKD). Methyl cytosines are shown as *magenta* sticks and indicated by *arrows*; other DNA bases are shown in *green*. The protein moiety is shown in *cyan*

Table 1 Reading modules of histone modification

Reader module	Histone modifications[a]
Bromodomain	Many acetylated lysine
Chromodomain	H3K9me2/3, H3K27me2/3
Double chromodomain	H3K4me1/2/3
Double/tandem Tudor	H3K4me3, H3K9me3, H4K20me3
MBT	H4K20me1/2
PHD finger	H3K4me3/0, R2me0
PWWP domain	H3K36me3
WD40 repeat	H3R2/K4me2
14-3-3	H3S10ph, H3S28ph
BRCT	H2AX S139ph

[a] *me* methyl; *ph* phosphoryl

Histone lysine acetylation is introduced by histone acetyltransferase domain proteins using the cofactor acetyl-CoA as an acetyl group donor, resulting in diminishing of the positive charge of the lysine side chain. The acetylated lysine residues are recognized and read by the bromodomain – a domain of about 110 amino acids that is conserved among many chromatin-associated proteins and transcription coactivator complexes.

Almost all histone lysine methyltransferases have a structurally conserved SET domain that catalyzes the methylation reaction using *S*-adenosyl L-methionine as a methyl group donor (Qian and Zhou 2006). Histone lysine methylation exists in three possible states – mono-, di-, and trimethylation – and all three methylation forms are thought to retain positive charge at physiological pH. Methylated lysine residues have been shown to be recognized by various histone reading modules, including the plant homeo domain (PHD) finger, the Royal Family, and WD40 repeats. The Royal Family includes the Tudor, chromodomain, and malignant brain tumor (MBT) domains (Table 1). A common feature of methyllysine recognition is the insertion of the methylammonium moiety into an aromatic cage consisting of two to four aromatic residues; methyl groups are then recognized via cation-π interactions (Taverna et al. 2007).

Histone modifications work as combinatorial codes and are multivalently and/or simultaneously recognized by various histone reader modules. However, the mechanism of combinatorial readout of histone modifications by multiple modules remains poorly understood. Recently, a mouse bromodomain protein, Brdt, has been shown to be responsible for selectively recognizing histone H4 tails harboring two or more acetylation marks. The crystal structure of the single bromodomain of Brdt has a wide keyhole-like pocket that recognizes acetylated H4K5 and H4K8 in one binding pocket simultaneously (Fig. 2a) (Morinière et al. 2009). In another case, an NMR study of the tandem PHD finger 3 (PHD1-PHD2) of human DPF3b showed that this tandem structure acts as one functional unit in the sequence-specific recognition of acetylated H3K14. The two PHD fingers are

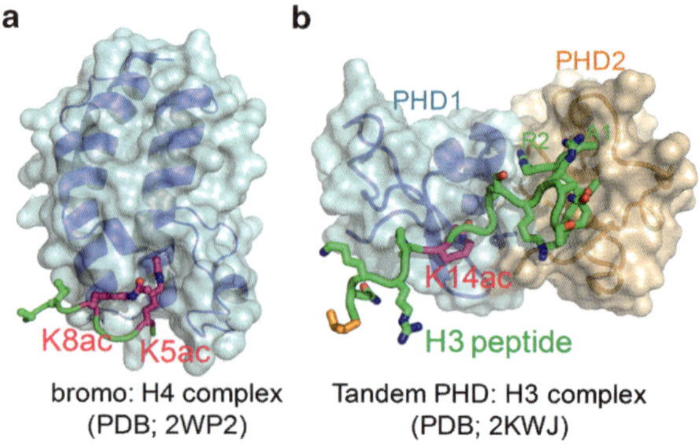

Fig. 2 Structures of histone reading modules. (**a**) Brdt bromodomain in complex with diacetylated H4 peptide (PDB; 2WP2). Acetylated lysines 5 and 8 are colored *magenta*. The protein moiety is shown in *blue* superimposed on a transparent surface model. (**b**) Tandem PHD of human DPF3b complex with acetylated H3K14 (PDB; 2KWJ). PHD1 and PHD2 are colored *blue* and *orange*, respectively, with transparent surface models. The H3 peptide is shown as a *green* stick model, and acetylated Lys 14 is colored *magenta*

positioned against one another in a face-to-back arrangement with the H3 peptide bound across the unified structure (Fig. 2b). The acetylated K14 and N-terminal residues of H3 are recognized simultaneously by PHD1 and PHD2, respectively (Zeng et al. 2010). These structural studies provide important clues as to how histone reading modules recognize the multivalent modifications on a single histone tail as a histone code.

Many chromatin-associated proteins have been shown to possess multiple histone reading modules, either in tandem or connected via a linker. Recently, it has been shown that *cis–trans* isomerization of a proline residue linking PHD3 and the bromodomain in human MLL1 that binds to histone H3K4me3/2 causes disruption of the interdomain contacts within the PHD3-bromodomain, resulting in alteration of the binding affinity for its associated protein, CyP33 (Wang et al. 2010). This result suggests that posttranslational modification of histone reading modules could alter specificities in interaction with other chromatin-associated proteins and, thus, play a regulatory role in the recognition of the histone code.

4 MORPHEUS' MOLECULE1 (MOM1) Links DNA Methylation to Intermediate Chromatin Modification

MORPHEUS' MOLECULE1 (MOM1) is a plant-specific regulator of TGS in *A. thaliana* that is distantly related to the CHD3/Mi-2 family of chromatin remodeling proteins (Amedeo et al. 2000; Čaikovski et al. 2008). In yeast and animals, the CHD3/Mi-2 proteins are components of multiprotein complexes containing HDACs and MBD proteins (Bowen et al. 2004), indicating that MOM1 may link DNA and histone modification. However, although CHD3/Mi-2 proteins are generally characterized by the presence of an ATPase/helicase domain, a chromodomain, and a PHD finger (Marfella and Imbalzano 2007), MOM1 protein of *A. thaliana* has only a region of partial similarity to the ATPase/helicase domain, and other domains characteristic of CHD3/Mi-2 proteins are missing (Amedeo et al. 2000; Čaikovski et al. 2008).

Most of the endogenous loci silenced by MOM1 are remnants of transposons that are clustered around centromeres, and therefore, RNAs transcribed from such loci accumulating in *mom1* are expected to be non coding RNAs (Numa et al. 2010; Steimer et al. 2000; Yokthongwattana et al. 2010). Indeed, cloning and sequence analyses of RNAs accumulating in *mom1* have proven this to be the case (Habu et al. 2006; Steimer et al. 2000). Accumulation of 24-nt siRNAs homologous to the MOM1-targets in wild-type plants and activation of MOM1-targets in mutants deficient in the RdDM pathway strongly indicate that at least a portion of MOM1-targets is also the target of RdDM (Numa et al. 2010; Yokthongwattana et al. 2010). An intriguing feature of MOM1-targets is their intermediate state of histone modification: the N-terminal tails of histone H3 at MOM1-targets show no strong bias toward the typical modification state of silent chromatin (Habu et al. 2006; Numa et al. 2010). Although *mom1* releases TGS without apparent changes in cytosine methylation at any sequence context (Amedeo et al. 2000; Habu et al. 2006), it induces a reduction in di-methylated histone H3 lysine 9 (H3K9me2) only in regions where CG-methylation depends on non-CG methylation (Numa et al. 2010). The mechanism of dependency of CG methylation on non-CG methylation is not known, but this observation possibly indicates an inability of MET1 to access hemimethylated CG sequences and that machineries for non-CG methylation (DRM2 and CMT3) maintain exclusively CG methylation in these regions. Together, the results suggest that MOM1 is required to maintain intermediate levels of histone modification downstream, or independently, of RdDM pathway at regions where MET1 is excluded from maintaining CG methylation. Release of TGS upon depletion of MOM1 by inducible RNAi in nondividing cells indicates that the action of MOM1 resides in maintaining the silent state of its targets that was established upon DNA replication (Tariq et al. 2002). These data, and those of other studies (Soppe et al. 2002; Tariq et al. 2003), imply the presence of mechanisms utilizing DNA methylation by RdDM to maintain histone modification in plants. MOM1 may specify pathways linking RdDM to intermediate states of histone modification.

As described above, MOM1 protein carries no structural features known to be characteristic of methylated cytosine- or histone-binding modules (Amedeo et al. 2000). Instead, a region consisting of approximately 200 amino acids named Conserved MOM1 Motif 2 (CMM2) is sufficient for MOM1 to silence endogenous repetitive sequences and a multicopy transgene (Čaikovski et al. 2008). CMM2 is conserved in MOM1 proteins in plant species other than *A. thaliana* and is located at the C-terminal region of the MOM1 protein, where domains for interaction with DNA or histones are found in other CHD family proteins (Marfella and Imbalzano 2007). These characteristics suggest that MOM1 might be the core protein of a multiprotein complex that mediates specific interaction with DNA and/or certain histone modifications to maintain the intermediate state of chromatin downstream or independently of RdDM.

5 Perspective

While plants have evolved to utilize non coding RNAs, and small RNA derivatives thereof, for maintaining normal development and genome integrity (Chen 2010), various strategies have also been adopted at the level of RNA processing for cleaning up aberrant RNAs and preventing subsequent accumulation of truncated and nonfunctional proteins (Belostotsky and Sieburth 2009); most of the machineries involved in these pathways are conserved in many eukaryotes (Garneau et al. 2007). By contrast, mutants of *A. thaliana* deficient in various epigenetic regulators have been shown to accumulate stable non coding RNAs that are produced from repetitive sequences including nonfunctional transposons (Gendrel et al. 2002; Lister et al. 2008; Numa et al. 2010). It is of great interest to consider why these non coding RNAs are not subject to the above-mentioned nuclear and cytoplasmic RNA surveillance machineries. There are several possible explanations: first, the accumulated RNAs, which are never produced in wild-type plants, are transported and accumulated in these mutants through pathways that are distinct from those used for normal mRNAs and non coding RNAs transcribed by Pol II and IV/V, respectively. Recently, components working in intracellular transport of RNA have been shown to be required for accumulation of small RNAs from endogenous repetitive sequences and transgenes (Yelina et al. 2010). This clearly indicates that non coding RNAs transcribed by Pol IV and/or V actively employ RNA processing machineries, which normally work for mRNAs, for the biogenesis of small RNAs thereof. The second, not mutually exclusive, explanation is that particular chromatin environments of the normally silent endogenous loci reduce the accessibility of RNA processing machineries functioning cotranscriptionally. It has been reported that chromatin states determine the mode of splicing (Schwartz and Ast 2010), and therefore, direct or indirect interaction between chromatin modification and RNA processing machineries should exist on loci that are transcribed normally, but may be lacking on normally silent loci.

As discussed above, cytosine and histone modifications are intimately linked via proteins carrying various modular domains specifically recognizing such modifications in the eukaryotic nucleus. Furthermore, there appear to be further links between chromatin modification and RNA processing. In addition to identifying individual components participating in these links by the genetic screening that is an advantage of plant biology and by recently emerging bioinformatic approaches, it will be undoubtedly important to identify whole molecular entities and their interactions by biochemical and biophysical approaches if we are to understand the detailed linkage between DNA methylation and histone modification in suppressing the production of non coding RNAs.

Note

Rajakumara et al (Genes Dev 2011, 25:137–152) recently reported that the SRA domain of SUVH5 of *Arabidopsis thaliana* recognizes fully or hemi-methylated cytosines in CG and CHH contexts by utilizing a dual flip-out mechanism in which both the methylated cytosine and a base in the partner strand simultaneously flipping out from the DNA duplex.

References

Ahmad A, Zhang Y, Cao XF (2010) Decoding the epigenetic language of plant development. Mol Plant 3:719–728
Amedeo P, Habu Y, Afsar K et al (2000) Disruption of the plant gene MOM releases transcriptional silencing of methylated genes. Nature 405:203–206
Arita K, Ariyoshi M, Tochio H et al (2008) Recognition of hemi-methylated DNA by the SRA protein UHRF1 by a base-flipping mechanism. Nature 455:818–821
Aufsatz W, Mette MF, van der Winden J et al (2002) HDA6, a putative histone deacetylase needed to enhance DNA methylation induced by double-stranded RNA. EMBO J 21:6832–6841
Avramova Z (2009) Evolution and pleiotropy of TRITHORAX function in Arabidopsis. Int J Dev Biol 53:371–381
Avvakumov GV, Walker JR, Xue S et al (2008) Structural basis for recognition of hemi-methylated DNA by the SRA domain of human UHRF1. Nature 455:822–825
Bartee L, Malagnac F, Bender J (2001) Arabidopsis cmt3 chromomethylase mutations block non-CG methylation and silencing of an endogenous gene. Genes Dev 15:1753–1758
Belostotsky DA, Sieburth LE (2009) Kill the messenger: mRNA decay and plant development. Curr Opin Plant Biol 12:96–102
Bogdanović O, Veenstra GJ (2009) DNA methylation and methyl-CpG binding proteins: developmental requirements and function. Chromosoma 118:549–565
Bostick M, Kim JK, Estève PO et al (2007) UHRF1 plays a role in maintaining DNA methylation in mammalian cells. Science 317:1760–1764
Bowen NJ, Fujita N, Kajita M et al (2004) Mi-2/NuRD: multiple complexes for many purposes. Biochim Biophys Acta 1677:52–57
Čaikovski M, Yokthongwattana C, Habu Y et al (2008) Divergent evolution of CHD3 proteins resulted in MOM1 refining epigenetic control in vascular plants. PLoS Genet 4:e1000165
Cedar H, Bergman Y (2009) Linking DNA methylation and histone modification: patterns and paradigms. Nat Rev Genet 10:295–304
Chen X (2010) Small RNAs – secrets and surprises of the genome. Plant J 61:941–958
Cheng X, Blumenthal RM (2010) Coordinated chromatin control: structural and functional linkage of DNA and histone methylation. Biochemistry 49:2999–3008
Cokus SJ, Feng S, Zhang X et al (2008) Shotgun bisulphite sequencing of the Arabidopsis genome reveals DNA methylation patterning. Nature 452:215–219
Earley KW, Pontvianne F, Wierzbicki AT et al (2006) Mechanisms of HDA6-mediated rRNA gene silencing: Suppression of intergenic Pol II transcription and differential effects on maintenance versus siRNA-directed cytosine methylation. Genes Dev 24:1119–1132
Ebbs ML, Bender J (2006) Locus-specific control of DNA methylation by the Arabidopsis SUVH5 histone methyltransferase. Plant Cell 18:1166–1176
Finnegan EJ, Dennis ES (1993) Isolation and identification by sequence homology of a putative cytosine methyltransferase from *Arabidopsis thaliana*. Nucleic Acids Res 21:2383–2388

Garneau NL, Wilusz J, Wilusz CJ (2007) The highways and byways of mRNA decay. Nat Rev Mol Cell Biol 8:113–126

Gendrel AV, Lippman Z, Yordan C et al (2002) Dependence of heterochromatic histone H3 methylation patterns on the Arabidopsis gene DDM1. Science 297:1871–1873

Grafi G, Zemach A, Pitto L (2007) Methyl-CpG-binding domain (MBD) proteins in plants. Biochim Biophys Acta 1769:287–294

Habu Y, Mathieu O, Tariq M et al (2006) Epigenetic regulation of transcription in intermediate heterochromatin. EMBO Rep 7:1279–1284

Hashimoto H, Horton JR, Zhang X et al (2008) The SRA domain of UHRF1 flips 5-methylcytosine out of the DNA helix. Nature 455:826–829

Ho KL, McNae IW, Schmiedeberg L et al (2008) MeCP2 binding to DNA depends upon hydration at methyl-CpG. Mol Cell 29:525–531

Ikeda Y, Kinoshita T (2009) DNA demethylation: a lesson from the garden. Chromosoma 118:37–41

Jackson JP, Lindroth AM, Cao X et al (2002) Control of CpNpG DNA methylation by the KRYPTONITE histone H3 methyltransferase. Nature 416:556–560

Jacob Y, Feng S, LeBlanc CA et al (2009) ATXR5 and ATXR6 are H3K27 monomethyltransferases required for chromatin structure and gene silencing. Nat Struct Mol Biol 16:763–768

Johnson LM, Law JA, Khattar A et al (2008) SRA-domain proteins required for DRM2-mediated de novo DNA methylation. PLoS Genet 4:e1000280

Kankel MW, Ramsey DE, Stokes TL et al (2003) Arabidopsis MET1 cytosine methyltransferase mutants. Genetics 163:1109–1122

Kraft E, Bostick M, Jacobsen SE et al (2008) ORTH/VIM proteins that regulate DNA methylation are functional ubiquitin E3 ligases. Plant J 56:704–715

Law JA, Jacobsen SE (2010) Establishing, maintaining and modifying DNA methylation patterns in plants and animals. Nat Rev Genet 11:204–220

Lawrence RJ, Earley K, Pontes O et al (2004) A concerted DNA methylation/histone methylation switch regulates rRNA gene dosage control and nucleolar dominance. Mol Cell 13:599–609

Lindroth AM, Cao X, Jackson JP et al (2001) Requirement of CHROMOMETHYLASE3 for maintenance of CpXpG methylation. Science 292:2077–2080

Lindroth AM, Shultis D, Jasencakova Z et al (2004) Dual histone H3 methylation marks at lysines 9 and 27 required for interaction with CHROMOMETHYLASE3. EMBO J 23:4286–4296

Lippman Z, May B, Yordan C et al (2003) Distinct mechanisms determine transposon inheritance and methylation via small interfering RNA and histone modification. PLoS Biol 1:E67

Lister R, O'Malley RC, Tonti-Filippini J et al (2008) Highly integrated single-base resolution maps of the epigenome in Arabidopsis. Cell 133:523–536

Loidl P (2004) A plant dialect of the histone language. Trends Plant Sci 9:84–90

Lu C, Tej SS, Luo S et al (2005) Elucidation of the small RNA component of the transcriptome. Science 309:1567–1569

Malagnac F, Bartee L, Bender J (2002) An Arabidopsis SET domain protein required for maintenance but not establishment of DNA methylation. EMBO J 21:6842–6852

Marfella CG, Imbalzano AN (2007) The Chd family of chromatin remodelers. Mutat Res 618:30–40

Matzke M, Kanno T, Daxinger L et al (2009) RNA-mediated chromatin-based silencing in plants. Curr Opin Cell Biol 21:367–376

Miura A, Nakamura M, Inagaki S et al (2009) An Arabidopsis jmjC domain protein protects transcribed genes from DNA methylation at CHG sites. EMBO J 28:1078–1086

Morinière J, Rousseaux S, Steuerwald U et al (2009) Cooperative binding of two acetylation marks on a histone tail by a single bromodomain. Nature 461:664–668

Naumann K, Fischer A, Hofmann I et al (2005) Pivotal role of AtSUVH2 in heterochromatic histone methylation and gene silencing in Arabidopsis. EMBO J 24:1418–1429

Numa H, Kim JM, Matsui A et al (2010) Transduction of RNA-directed DNA methylation signals to repressive histone marks in *Arabidopsis thaliana*. EMBO J 29:352–362

Ohki I, Shimotake N, Fujita N et al (2001) Solution structure of the methyl-CpG binding domain of human MBD1 in complex with methylated DNA. Cell 105:487–497

Pien S, Grossniklaus U (2007) Polycomb group and trithorax group proteins in Arabidopsis. Biochim Biophys Acta 1769:375–382

Qian C, Zhou MM (2006) Set domain protein lysine methyltransferases: structure, specificity and catalysis. Cell Mol Life Sci 63:2755–2763

Richards EJ, Elgin SC (2002) Epigenetic codes for heterochromatin formation and silencing: rounding up the usual suspects. Cell 108:489–500

Saze H, Mittelsten Scheid O, Paszkowski J (2003) Maintenance of CpG methylation is essential for epigenetic inheritance during plant gametogenesis. Nat Genet 34:65–69

Saze H, Shiraishi A, Miura A et al (2008) Control of genic DNA methylation by a jmjC domain-containing protein in *Arabidopsis thaliana*. Science 319:462–465

Schoft VK, Chumak N, Mosiolek M et al (2009) Induction of RNA-directed DNA methylation upon decondensation of constitutive heterochromatin. EMBO Rep 10:1015–1021

Schwartz S, Ast G (2010) Chromatin density and splicing destiny: on the cross-talk between chromatin structure and splicing. EMBO J 29:1629–1636

Sharif J, Muto M, Takebayashi S et al (2007) The SRA protein Np95 mediates epigenetic inheritance by recruiting Dnmt1 to methylated DNA. Nature 450:908–912

Soppe WJ, Jasencakova Z, Houben A et al (2002) DNA methylation controls histone H3 lysine 9 methylation and heterochromatin assembly in Arabidopsis. EMBO J 21:6549–6559

Steimer A, Amedeo P, Afsar K et al (2000) Endogenous targets of transcriptional gene silencing in Arabidopsis. Plant Cell 12:1165–1178

Tariq M, Habu Y, Paszkowski J (2002) Depletion of MOM1 in non-dividing cells of Arabidopsis plants releases transcriptional gene silencing. EMBO Rep 3:951–955

Tariq M, Saze H, Probst AV et al (2003) Erasure of CpG methylation in Arabidopsis alters patterns of histone H3 methylation in heterochromatin. Proc Natl Acad Sci USA 100:8823–8827

Taverna SD, Li H, Ruthenburg AJ et al (2007) How chromatin-binding modules interpret histone modifications: lessons from professional pocket pickers. Nat Struct Mol Biol 14:1025–1040

Vrbsky J, Akimcheva S, Watson JM et al (2010) siRNA-mediated methylation of Arabidopsis telomeres. PLoS Genet 6:e1000986

Wang Z, Song J, Milne TA et al (2010) Pro isomerization in MLL1 PHD3-bromo cassette connects H3K4me readout to CyP33 and HDAC-mediated repression. Cell 141:1183–1194

Wierzbicki AT, Haag JR, Pikaard CS (2008) Noncoding transcription by RNA polymerase Pol IVb/Pol V mediates transcriptional silencing of overlapping and adjacent genes. Cell 135:635–648

Woo HR, Pontes O, Pikaard CS et al (2007) VIM1, a methylcytosine-binding protein required for centromeric heterochromatinization. Genes Dev 21:267–277

Yelina NE, Smith LM, Jones AM et al (2010) Putative Arabidopsis THO/TREX mRNA export complex is involved in transgene and endogenous siRNA biosynthesis. Proc Natl Acad Sci USA 107:13948–13953

Yokthongwattana C, Bucher E, Čaikovski M et al (2010) MOM1 and Pol-IV/V interactions regulate the intensity and specificity of transcriptional gene silencing. EMBO J 29:340–351

Zeng L, Zhang Q, Li S et al (2010) Mechanism and regulation of acetylated histone binding by the tandem PHD finger of DPF3b. Nature 466:258–262

Zhang X, Yazaki J, Sundaresan A et al (2006) Genome-wide high-resolution mapping and functional analysis of DNA methylation in Arabidopsis. Cell 126:1189–1201

Zheng B, Wang Z, Li S et al (2009) Intergenic transcription by RNA polymerase II coordinates Pol IV and Pol V in siRNA-directed transcriptional gene silencing in Arabidopsis. Genes Dev 23:2850–2860

Zilberman D, Gehring M, Tran RK et al (2007) Genome-wide analysis of *Arabidopsis thaliana* DNA methylation uncovers an interdependence between methylation and transcription. Nat Genet 39:61–69

Biogenesis of Plant MicroRNAs

Julieta L. Mateos, Nicolás G. Bologna, and Javier F. Palatnik

Contents

Abstract MicroRNAs (miRNAs) are small RNAs of ~21 nt that regulate multiple biological pathways in complex organisms. They are defined by their specific biogenesis that involves the precise excision from an imperfect fold-back precursor. In plants, the ribonuclease III DICER-LIKE1 (DCL1) assisted by accessory proteins cleaves the precursor to release the miRNA. In general, the processing complex recognizes a 15-nt lower stem located below the miRNA in the precursors to produce the first cleavage, which is then followed by a second cut that releases the small RNA. Plant precursors are, however, very variable in size and shape, and not all of them are processed in the same way. The conserved miR319/159 precursors are cleaved in a

J.F. Palatnik (✉) and N.G. Bologna
Instituto de Biología Molecular y Celular de Rosario, Suipacha 531, 2000 Rosario, Argentina
e-mail: palatnik@ibr.gov.ar

J.L. Mateos
Instituto de Biología Molecular y Celular de Rosario, Suipacha 531, 2000 Rosario, Argentina

Max-Planck-Institute for Plant Breeding Research, Carl-von-Linné-Weg 10, 50829 Cologne, Germany

V.A. Erdmann and J. Barciszewski (eds.), *Non Coding RNAs in Plants*, RNA Technologies, DOI 10.1007/978-3-642-19454-2_16,

loop-to-base direction by several successive DCL1 cuts. The situation seems to be even more complex if newly evolved miRNAs are also taken into account. The emerging picture suggests a high plasticity of the miRNA processing machinery.

1 Introduction

Small RNAs are currently considered as major regulators of gene expression in eukaryotes. They are usually around 20–24 nt long and arise after the processing of longer RNAs. A complex spectrum of small RNAs exists in plants. They can be classified into several groups including microRNAs (miRNAs), small interfering RNAs (siRNAs), *trans*-acting siRNAs (tasiRNA), natural antisense siRNAs, and miRNAs (nat-siRNAs and nat-miRNAs) (reviewed in Chapman and Carrington 2007; Ramachandran and Chen 2008; Vaucheret 2006; Vazquez et al. 2010; Voinnet 2009). A common core step in the biogenesis of the small RNAs implicates the cleavage of a precursor that has a perfect or nearly perfect dsRNA structure by an RNAse III enzyme. The exact nature of the precursor as well as the specific RNAse III complex involved varies in each particular case and distinguishes the different small RNA pathways (Vaucheret 2006; Vazquez et al. 2010; Voinnet 2009). Known plant small RNAs act so far in the concert of an ARGONAUTE (AGO) protein, providing the complex with specificity to identify target RNAs through sequence complementarity.

MiRNAs are generated from endogenous *loci* in the cell. Although they exist in plants and animals, the current model indicates that they have appeared independently in the two lineages, deriving from more ancient small RNA pathways (Axtell 2008; Axtell and Bowman 2008). Still, they are essential regulators of gene expression in both systems and control key developmental processes and stress responses (Carthew and Sontheimer 2009; Voinnet 2009). They exert their function by controlling the abundance of their target genes, from fine-tuning protein levels to the clearance of RNA transcripts in the cell. The identity of the target genes is specified by the miRNA molecule, which recognizes its targets by base pairing. In plants, miRNAs have a good complementarity to their targets, and in many cases they regulate several genes of the same family (Chen 2009; Jones-Rhoades et al. 2006).

The biogenesis of miRNAs involves their excision from the stem of a fold-back precursor. During this process, an imperfect dsRNA structure is recognized by the miRNA processing machinery to release a 21-nt small RNA. The implicated process seems to be particularly complex in plants due to the wide range of RNA molecules that serve as precursors for miRNA biogenesis. In this chapter, we summarize some aspects of the current view of the biogenesis of plant miRNAs.

2 MiRNAs and Their Precursors in Plants

The development of the deep-sequencing technologies has caused a huge impact in the understanding of the genome structure and function, and the small RNA field has not been an exception. Several libraries of small RNAs has been prepared from *Arabidopsis thaliana*, using different tissues and mutants (Fahlgren et al. 2007; Lu et al. 2006; Rajagopalan et al. 2006). Libraries have recently been prepared from *Arabidopsis lyrata* (Fahlgren et al. 2010; Ma et al. 2010), which has allowed the direct comparison of the small RNA content in related species. Furthermore, studies have also been performed in plants with agronomic importance such as grape, maize, poplar, rice, soybean, and wheat (Johnson et al. 2009; Li et al. 2010; Mica et al. 2010; Wei et al. 2009; Xie et al. 2010; Zhang et al. 2009).

These studies have led to the identification of the small RNA content of many plant species. From these pools of sequences, consisting mostly of siRNAs, miRNAs represent around a third (Fahlgren et al. 2007; Rajagopalan et al. 2006). The identification and annotation of miRNAs follows an established criterion that is intimately related to its biogenesis process (Meyers et al. 2008). After the processing of the fold-back structure of a miRNA precursor, DCL1 produced two staggered cuts in the dsRNA, separated approximately 21 nucleotides of each other. The cleavages release the miRNA together with the opposing fragment of the precursor that is interacting with it, named miRNA*. The definition of a miRNA requires the sequencing of two small RNAs (miRNA and miRNA*) that can be located in the arms of a stem-loop structure (Meyers et al. 2008). The interaction between the miRNA and the miRNA* should also follow some rules (e.g., have less than 4 mismatches). However, not all the annotated miRNAs in the small RNA database (miRBase 16.0) necessarily follow this criterion.

Currently, there are 213 miRNAs from *A. thaliana* annotated in the database, and over 600 from plants. Most of the plant miRNAs present in a certain species are likely young miRNAs that have been recently appeared in evolution (Axtell 2008; Axtell and Bowman 2008). Some miRNAs are, however, deeply conserved. Twenty-one families are present throughout the angiosperms with some of them even conserved in the moss *Physcomitrella patens* (Axtell 2008; Axtell and Bowman 2008; Axtell et al. 2007). These ancient miRNAs have also been shown to regulate key biological aspects in plants (Chen 2009; Jones-Rhoades et al. 2006).

3 MiRNA Transcription and Processing

While many animal miRNAs are derived from introns or untranslated regions of coding messages (Carthew and Sontheimer 2009; Kim et al. 2009), plant miRNA genes resemble protein coding genes, as they are independent transcriptional units and may contain introns (Allen et al. 2004; Kim et al. 2009). One exception is miR838a, which is located in the intron of *DCL1* (Rajagopalan et al. 2006). Transcripts from miRNA genes are synthesized by RNA polymerase II, capped, spliced, and polyadenylated (Xie et al. 2005; Zhou et al. 2007). MiRNA primary transcripts are also known as pri-miRNAs.

In animals, pri-miRNAs are trimmed in the nucleus into stem-loop miRNA precursors (pre-miRNA) by a microprocessor complex formed by an RNase III-like enzyme termed Drosha, and the dsRNA binding protein DGCR8/Pasha (Denli et al. 2004; Gregory et al. 2004; Han et al. 2004a). Pre-miRNAs are then exported to the cytoplasm by Exportin 5 (Lund et al. 2004; Yi et al. 2003). Once there, they are cleaved again by another RNase III enzyme, Dicer, that releases the 20–22 nt RNA duplex (miRNA/miRNA* duplex).

In *A. thaliana*, there are four Dicer homologues, referred to as *DCL1*, *DCL2*, *DCL3*, and *DCL4* (Schauer et al. 2002), but no gene related to the animal Drosha. DCL1 mainly produces ~21-nt-long sRNA, while the products of DCL2, DCL3, and DCL4 are 22, 24, and 21 nt long, respectively (Hamilton et al. 2002; Qi et al. 2005). DCL1 is the one that orchestrate miRNA biogenesis (Park et al. 2002; Reinhart et al. 2002), while DCL2-4 are involved in other small RNA pathways (Bouche et al. 2006; Deleris et al. 2006; Gasciolli et al. 2005; Henderson et al. 2006; Tang et al. 2003). Null *dcl1* alleles of *Arabidopsis* are lethal, while partial loss-of-function alleles have reduced levels of miRNAs and severe developmental defects (Park et al. 2002; Reinhart et al. 2002; Schauer et al. 2002). Hypomorphic alleles of *dcl* do not necessarily affect all miRNAs; for example, miR165/166 and miR168 do not vary significantly in *dcl1-9* (Bouche et al. 2006).

Several aspects distinguish animal and plant miRNA biogenesis. The processing of plant miRNAs relies on one RNAse III enzyme, in contrast to animals, which have two complexes. In contrast to animals where miRNA biogenesis is compartmentalized between the nucleus and the cytoplasm, all the dicing events are thought to take place in the plant nucleus (Papp et al. 2003). These reactions take place in subnuclear bodies termed dicing bodies or D-bodies (Fang and Spector 2007; Fujioka et al. 2007; Song et al. 2007).

Recently, some miRNAs in rice have been shown to originate from the antisense strand of target genes and, therefore, were named natural antisense miRNA (nat-miRNA) (Lu et al. 2008). Their distinct feature is that the mature miRNA sequence derives from the antisense transcription of their targets that form a fold-back structure that is processed by DCL1 (Lu et al. 2008). Generally, their precursors are long transcripts with large introns, and it is only after removal of the introns that the hairpin precursors become evident (Lu et al. 2008).

4 The Processing Machinery

Once a precursor with an imperfect fold-back structure is generated, the plant processing machinery has the challenge to produce site-specific cleavages to release the 21-nt miRNA. Although the core component of this complex is DCL1, there are several assisting partners (Fig. 1). The double-strand RNA binding protein HYPONASTIC LEAVES1 (HYL1) (Han et al. 2004b; Vazquez et al. 2004), and the C2H2 zinc-finger protein SERRATE (SE) (Lobbes et al. 2006; Yang et al.

Fig. 1 (continued) DICER- LIKE1 (DCL1), assisted by HYPONASTIC LEAVES1 (HYL1) and SERRATE (SE) cleaves the precursor to release the miRNA/miRNA* duplex. The degradation of the small RNAs is prevented by the addition of a methyl group (*black dot*) by HEN1. The dsRNA small molecule is exported to the cytoplasm, and the miRNA strand is selected and incorporated into an ARGONAUTE (AGO) complex. The miRNA is indicated in *blue* and the miRNA* in *cyan*

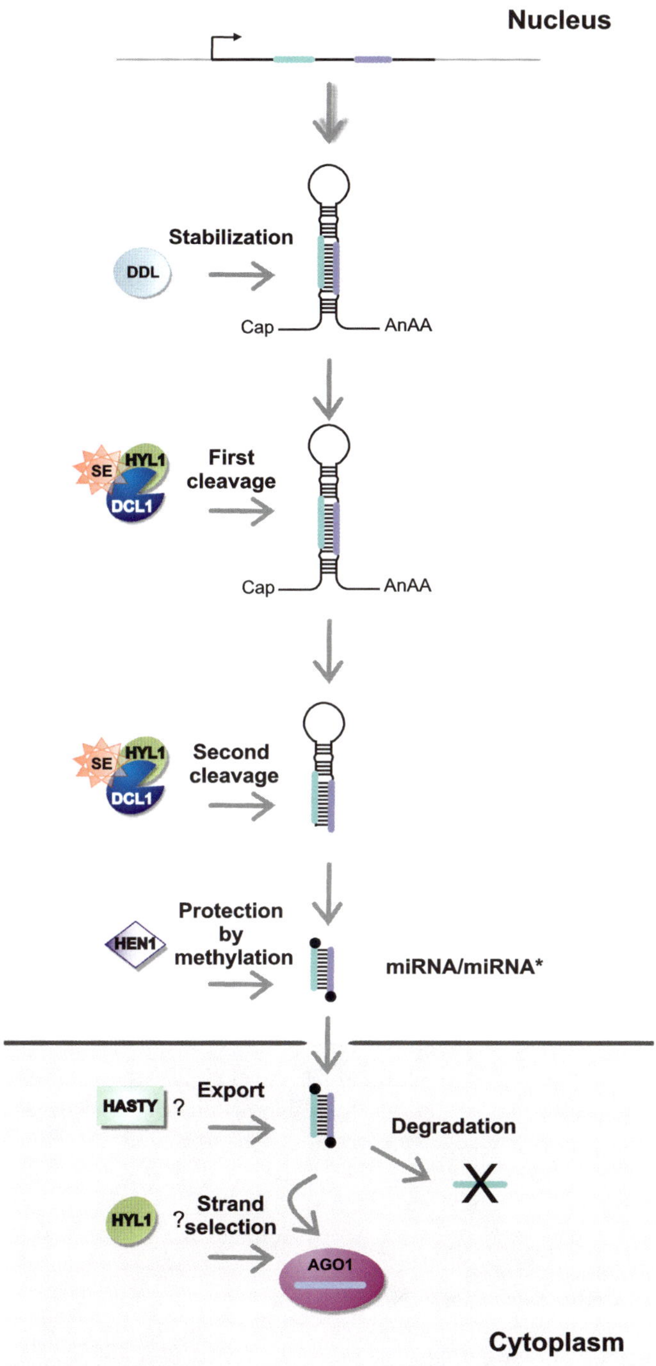

Fig. 1 MiRNA processing pathways in plants. MIRNA genes are transcribed by RNApol II and bear a 5′ cap and a poly(A) tail. Primary transcripts are stabilized by DAWDLE (DDL).

2006) facilitate miRNA processing by DCL1 (Kurihara et al. 2006; Kurihara and Watanabe 2004; Song et al. 2007). These three proteins physically interact and colocalize with miRNA primary in the nucleus, forming dicing bodies (Fang and Spector 2007; Fujioka et al. 2007; Song et al. 2007). Both, HYL1 and SE, improve the efficiency and precision of cleavage of DCL1 (Dong et al. 2008).

Although not lethal, *hyl1* null mutations severely impair miRNA maturation, which in turn causes developmental defects (Han et al. 2004b; Lu and Fedoroff 2000; Vazquez et al. 2004). HYL1 contains two tandem double-stranded RNA binding domains (dsRBD) separated by a short linker and a putative C-terminal nuclear localization domain. Both dsRBDs are sufficient to rescue *hyl1* phenotype and to generate 21-nt mature miRNA from precursors in vitro (Wu et al. 2007). Recently, structural analysis on HYL1 has revealed that both domains are indeed dsRBD that can bind miRNA precursors, though they have different affinities (Rasia et al. 2010; Yang et al. 2010). The first dsRBD of HYL1 is the main contributor to RNA binding, while the second domain, which seems to keep some primeval RNA binding capacity, could be involved in recognition of DCL1 (Qin et al. 2010; Rasia et al. 2010) or in the dimerization of HYL1 (Yang et al. 2010).

Mutations in *se* accumulate high levels of miRNA primary transcripts and less mature miRNAs (Lobbes et al. 2006; Yang et al. 2006). These mutants exhibit also general mRNA splicing defects (Laubinger et al. 2008). In this way, SE has dual roles in miRNA biogenesis and splicing, in contrast to specialized miRNA processing factors such as HYL1.

Other cellular components are also known to participate in plant miRNA biogenesis. DAWDLE associates with DCL1, and it is thought to stabilize miRNA primary transcripts (Morris et al. 2006: Yu et al. 2008). HASTY, an Exportin5 homologue, also contributes to the miRNA levels (Park et al. 2005). However, as all the dicing events are performed in the nucleus, the role of HASTY in the biogenesis of miRNAs is unclear in plants.

The miRNA/miRNA* duplex formed after DCL1 action contains 2-nt overhang on each 3′ end of the RNA. These ends are modified by the action of HUA ENHANCER 1 (HEN1), which methylates the 3′ end of the small RNA molecule (Huang et al. 2009). The action of HEN1 stabilizes the duplex and prevents its degradation (Boutet et al. 2003; Li et al. 2005a; Yu et al. 2005).

To complete miRNA maturation, the miRNA is loaded into the RISC complex and the miRNA* is degraded. Bioinformatic analysis in animals led to the proposal that the strand selected from the miRNA/miRNA* duplex is the one with less thermodynamic stability in the 5′ end (Matranga et al. 2005; Rand et al. 2005). More recently, HYL1 has been proposed to participate in miRNA strand selection in plants (Eamens et al. 2009).

Once the single-stranded miRNA molecule is loaded in the RISC complex, it is ready to scan for its targets. There are multiple AGO in plants, and the incorporation of the miRNA into the different complexes depends at least on the identity of the 5′ end nucleotide of the small RNA and its biogenesis pathway (Mi et al. 2008; Montgomery et al. 2008; Takeda et al. 2008; Wu et al. 2009).

5 RNA Recognition During MiRNA Precursor Processing

A key aspect during precursor processing is the recognition of fold-back structure by the processing machinery. The precision of this process will ultimately be responsible for the miRNA specificity, as a change in the cleavage positions would cause a concomitant change in the miRNA sequence.

Most of the specificity in the processing of animal precursors relies in the first cleavage performed by DROSHA, which cuts at approximately 11 nt, one helical turn, from the joint between the single-stranded RNA and the double-stranded stem. This mechanism is known as the "ssRNA-dsRNA junction anchoring" model for the processing of pri-miRNA in animals (Han et al. 2006). The dsRNA binding protein DGCR8/PASHA, which assists DROSHA, has a key role in this process.

The second cleavage is performed after the export from the nucleus, and it is mediated by Dicer with the assistance of the dsRNA binding domain protein Loquacious (Saito et al. 2005). Dicer cuts approximately 21 nt away from the end of the precursor, which was defined previously by DROSHA. This cleavage releases the miRNA/miRNA* duplex and defines the size of these two small RNAs.

Whereas animal precursors hold homogeneous 70–80-nt fold-back structures, plant miRNAs are harbored in a collection of hairpins with a wide range of size and shape (Bologna et al. 2009; Reinhart et al. 2002; Zhang et al. 2006). Owing to this reason, it has been harder to define the prototypical secondary structure of miRNA precursor in plants.

Beyond the variation in the stem-loop size and shape, detailed studies have demonstrated that several plants pri-miRNAs have structural similarities. One key feature is that many plant precursors have a lower stem of ~15 nt below the miRNA/miRNA* (Mateos et al. 2010; Song et al. 2010; Werner et al. 2010) (Fig. 2a). This section of the stem is not a perfect double-stranded region and unpaired bases or bulges are present. The processing of plant miRNA precursors appears to depend noticeably on this region (Cuperus et al. 2010; Mateos et al. 2010; Song et al. 2010; Werner et al. 2010). Experiments focused on miR167a, miR171a, and miR172a showed that mutations in the lower stem of these precursors dramatically affect their processing (Mateos et al. 2010; Song et al. 2010; Werner et al. 2010).

A random mutagenesis approach, performed on miR172a precursor revealed that point mutations located in the lower stem significantly affected its processing, while mutations located elsewhere such as the miRNA/miRNA* region or the terminal loop were largely neutral (Mateos et al. 2010). Interestingly, for many precursors a single change in the lower stem is sufficient to completely abolish its processing. In contrast to the relevance of the lower stem in plant miRNA processing, the size and structure of the loop is largely unimportant (Bologna et al. 2009; Mateos et al. 2010). However, a region that joins the miRNA/miRNA* and the terminal loop needs to be structured for a productive processing.

The emerging model indicates that the majority of the plant miRNA precursors consist of three regions: (1) a lower stem of 15 nt, which is partially structured, (2) the dsRNA region corresponding to the miRNA/miRNA*, and (3) a terminal loop

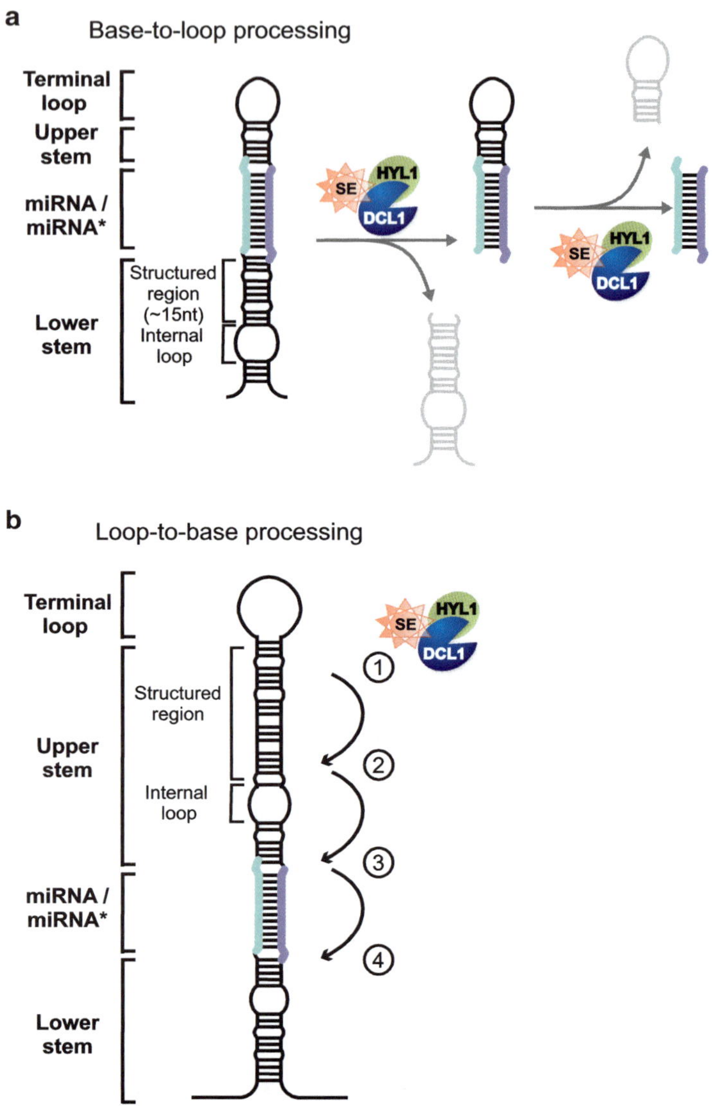

Fig. 2 Mechanisms of plant miRNA precursor processing. (**a**) Base-to-loop processing mechanism for plant miRNA biogenesis. The miRNA processing complex recognizes a 15-nt lower stem below the miRNA, and DCL1 performs the first cut. (**b**) Noncanonical loop-to-base processing of miR319 precursor. The first cleavage is performed below the terminal loop, and three additional cleavage events are required to release the miRNA. See legend of Fig. 1 for details

that has low structural restrictions, though it needs to be joined by a small structured segment to the miRNA region (Fig. 2a).

It is thought then that the miRNA processing complex can recognize these structural constraints during miRNA processing. For instance, the addition of extra nucleotides to the miR172a lower stem caused a concomitant shift in the cleavage position by DCL1 (Werner et al. 2010), further confirming that the plant processing machinery can "count" the length of the lower stem. These structural features have allowed the successful design of a completely artificial precursor in plants (Werner et al. 2010).

6 Noncanonical Processing of Plant MiRNAs

Interestingly, not all the miRNAs share the specific structural features described above. The family of miR319/159 is one of these examples. These miRNAs are widely distributed in plants, and copies of miR319 can even be found in mosses (Arazi et al. 2005; Axtell and Bartel 2005; Axtell et al. 2007), indicating their ancient origin. Their precursors have large fold-back sequences, with an upper stem over the miRNA/miRNA* duplex that is highly conserved (Axtell et al. 2007; Li et al. 2005b; Palatnik et al. 2003; Warthmann et al. 2008).

These unusually long precursors have been shown to harbor three regions from which potentially small RNAs can be generated, although the miRNA sequence accumulates much more than any other small RNA sequence (Fig. 2b) (Axtell et al. 2007; Rajagopalan et al. 2006; Talmor-Neiman et al. 2006).

Detailed mutagenesis studies have shown that the complete removal of the bases belonging to the lower stem below the miRNA/miRNA* did not affect the miR319 processing (Bologna et al. 2009). However, the biogenesis of miR319a was extremely susceptible to modifications at the top part of the precursor. Minor deletions or modifications in the bulges of the conserved upper stem segment were sufficient to reduce miR319a production. So, while precursor sequences below mi319/miR319* are not necessary for its processing, the upper stem segment is indispensable.

The analysis of the intermediates by a modified RACE PCR determined that miR319a biogenesis needs four dicing events for its processing, instead of the two described for animal and canonical plant miRNAs. In addition, the most striking finding was that the first cleavage takes place near the terminal loop, in an opposite manner to other known miRNAs. Then, three more cleavages are performed by DCL1 through the precursor until the miRNA is finally released (Addo-Quaye et al. 2009; Bologna et al. 2009) (Fig. 2b). This loop-to-base directionality of processing is consistent with the structural requirements found for this precursor. The same processing pattern was found for other precursors of the same family in *Arabidopsis* (miR319b and miR159a) and in other species such as *P. patens*, indicating that the origin of the miR319-processing mechanism is quite ancient (Addo-Quaye et al. 2009; Bologna et al. 2009).

7 Processing of Young MiRNA Precursors

The current model indicates that miRNAs are constantly being formed and dying in plants (Fahlgren et al. 2007; Rajagopalan et al. 2006). One of the most common birth mechanisms for plant miRNAs implicate inverted gene duplications (Allen et al. 2004). These genomic rearrangements would generate perfect dsRNA structures that could be processed by RNAse III enzymes to release small RNAs, which in turn can target related protein coding genes. These recently formed RNA structures do not necessarily have the requirements for miRNA biogenesis as described above. They are thought to be processed in a relaxed way by different DCLs to produce several small RNA sequences. The accumulation of mutations on the precursor sequences during evolution will allow the canalization into the canonical miRNA pathway (Axtell and Bowman 2008; Voinnet 2009).

This model of miRNA gene evolution is then intimately related to the processing pathway. First, the double-stranded RNA with perfect base pairing is initially generated and processed by DCL3 and DCL4 to release several siRNA (proto-miRNA). The model suggests that the proto-MIR suffered successive mutations during the evolution leading to important changes in the secondary structure of the precursors. These mutations could lead to the shortening of the hairpin stem, an increase on the number of unpaired bases in the stem, or an increase on the size of the distal loop. However, these young precursors might continue to be processes by DCL4 (Rajagopalan et al. 2006; Vazquez et al. 2008; Voinnet 2009). Continued selection would shape the young MIR gene into an ancient MIR gene, in which the size and various bulges in the stem loop produced a progressive shift in DCL usage (DCL4 to DCL1) (Rajagopalan et al. 2006). During this last step, the structural determinants recognized by the miRNA processing machinery are finally acquired. The processing of some *Arabidopsis* miRNAs by DCL4, such as miR822 and miR839, are in good agreement with this model (Rajagopalan et al. 2006).

DCL3 has also been shown to yield long-miRNAs (24 nt) from certain precursors in rice (Wu et al. 2010). For example, miR1850 has two miRNAs tandem in the precursor structure, one of 21 nt (miR1850.1) and the other of 24 nt (miR1850.2). This precursor is processed first by DCL1 to release the 21-nt species and subsequently by DCL3 to produce the 24-nt miRNA. Also, DCL1 and DCL3 can act in parallel on several precursors. This is found for miR168a, miR396e, and miR396f of rice (Wu et al. 2010). Finally, pri-miR820a/b/c give rise to two miRNA variants, miR820.1 (21 nt) and miR820.2 (24 nt), first diced by DCL3 and then either by DCL1 or DCL3 to produce the species of different size (Wu et al. 2010).

DCL3 has been also shown to process conserved miRNAs of *Arabidopsis* (Vazquez et al. 2008). The activity of DCL3 causes the accumulation of a 24-nt species for certain miRNAs in addition to the main 21-nt class. This is most obvious in inflorescences, which is also the tissue where DCL3 transcripts accumulate (Vazquez et al. 2008).

8 Strategies to Study MiRNA Biogenesis in Plants

8.1 Identification of Precursor Intermediates

Precise processing of miRNA/miRNA* duplexes from their precursors is a critical step in miRNA biogenesis, as it defines the sequence of the small RNA. Genetic analysis in *Arabidopsis* has identified multiple components that function in the miRNA biogenesis pathway, but mechanisms underlying miRNA biogenesis are largely unknown. Studies using site-directed mutagenesis of candidate precursor regions combined with in vitro processing systems have proven useful to uncover the structural determinants that are required for animal miRNA processing (Han et al. 2006; Zeng and Cullen 2005; Zeng et al. 2005). As miRNA processing in animals is physically separated between the nucleus and the cytoplasm, it is relatively easy to detect intermediates of precursor processing in RNA blots. By contrast, as plant processing seems to be a fast process occurring completely in the nucleus, intermediates are hardly detected in RNA blots. Therefore, more sensitive techniques are usually used to identify the processing intermediates in plants.

Kurihara et al. showed that miR163 biogenesis in *Arabidopsis* requires at least three cleavage steps by DCL1 at 21-nt-long intervals. They found that the exact position of the cleavage sites changed in a *dcl1-9* mutant and suggested that the dsRBDs of DCL1 determine the position of the cuts (Kurihara and Watanabe 2004). The authors identified the precise location of the cuts by RNA self-ligation of the precursors (Kurihara et al. 2006; Kurihara and Watanabe 2004). In this approach, RNA extracted from the plant tissue is self-ligated. Then, an RT-PCR with specific primers for the miRNA precursor is used to amplify the processing intermediates. Cloning and sequencing of the PCR products allows the precise identification of the cleavage sites (Fig. 3a).

RNA ligase-mediated 5′ rapid amplification of cDNA ends (RLM-5′-RACE) is a technique used to identify uncapped RNA fragment in cells. It has been applied to identify miRNA-mediated cleavage of target RNAs (Llave et al. 2002) and later to find intermediates of precursor processing (Addo-Quaye et al. 2009; Bologna et al. 2009; German et al. 2008). In this method, the RNA is ligated to a specific RNA adaptor. Only uncapped ends with a 5′ phosphate are susceptible to ligation. After cDNA synthesis, a PCR with generic and miRNA-specific oligos allows the amplification of the fragments. Cloning and sequencing of these fragments allows the identification of the precise ends of the RNA fragments (Fig. 3a).

These techniques can also be applied in mutants known to accumulate the desired fragments, such as *xrn4* and *fiery*, enriching the ligation in the by-products of the miRNA pathways (targets and/or precursors intermediates) (Bologna et al. 2009; German et al. 2008). Using these mapping strategies, the exact cleavage sites as well as direction of the dicing events have been inferred for several precursors, such as miR164, miR398, miR163, miR319, miR172, and miR171 (Bologna et al. 2009; Kurihara et al. 2006; Kurihara and Watanabe 2004; Mateos et al. 2010; Song et al. 2010; Werner et al. 2010).

Combining RLM-5′-RACE with high-throughput sequencing has improved the capacity to get information on miRNA targets and miRNA precursors metabolism (Addo-Quaye et al. 2008; German et al. 2008; Gregory et al. 2008). This method is also known as parallel analysis of RNA ends (PARE). By comparison of degradome tags to the genome, processing intermediates can be found for certain miRNA precursors (Addo-Quaye et al. 2008; German et al. 2008; Li et al. 2010). Deep-sequencing libraries of small RNAs provide the exact sequence of the different miRNAs, which then gives complementary information about the positions where the precursors have been cleaved.

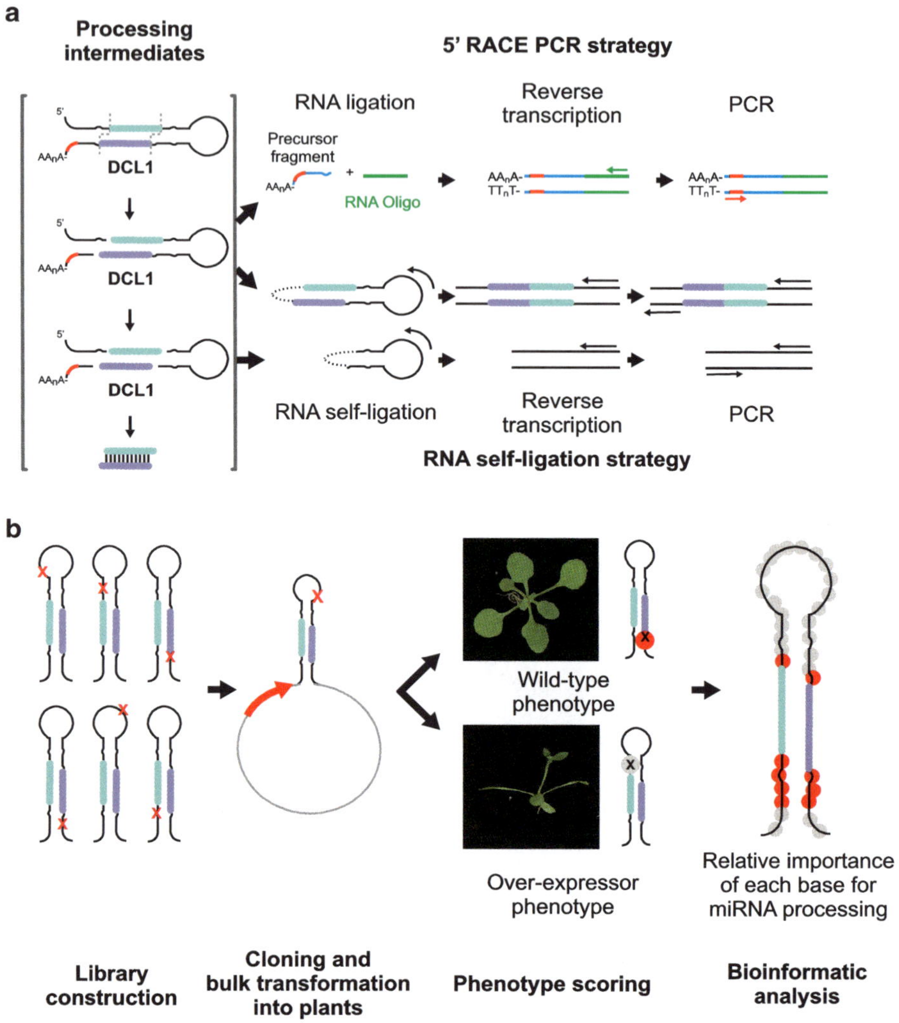

Fig. 3 Strategies to study miRNA precursor processing. (**a**) Mapping of precursor intermediates. *Left*: modified 5′ RACE-PCR. Using this strategy in a precursor that is processed in a canonical base-to-loop direction, the intermediate corresponding to the first DCL1 cleavage can be detected. *Right*: RNA self-ligation, which in a similar precursor allows detection of both intermediates. (**b**) Random mutagenesis on plant precursors. Variants of miRNA precursors constructed with nonequimolar oligos are cloned into an expression vector downstream of a selected promotor (*red arrow*). Plants are scored by their phenotype (wild-type looking plants correspond to transgenics harboring a precursor impaired in its processing). Sequencing of the transgenes and bioinformatic analysis allow the identification of bases that are important or neutral for precursor processing. See legend of Fig. 1 for details

In *Arabidopsis*, part of signals on pri-miR172b corresponded to the middle region of its miRNA*. These cuts matched to position 10 of miR172, suggesting that miR172 cleaves pri-miR172b (German et al. 2008).

Tiling arrays can also be used as a wide-genome approach to investigate miRNA processing, as miRNA precursors can be detected by this technique. Comparison of the expression profiles of wild-type and *se* mutants revealed a role of *SE* in RNA splicing (Laubinger et al. 2008).

8.2 Identification of Processing Determinants in Plant Precursors

Many plant miRNAs regulate transcription factors that have key roles in plant development. Overexpression of these miRNAs usually causes developmental defects, which are easily spotted by eye. This has allowed the exploration of miRNA processing efficiency in vivo by overexpressing wild-type and mutant precursors, as only a functional precursor can cause developmental defect. Selected plants with different phenotypes can then be specifically analyzed at a molecular level. This approach has been applied to miR171 (Song et al. 2010) and miR172 (Mateos et al. 2010; Werner et al. 2010), allowing the identification of key features in their precursors.

In these cases, structural variants of the stem-loop precursors created by site-directed mutagenesis were introduced into *Arabidopsis* plants. MiRNA biogenesis was then followed in vivo by scoring the developmental defects produced by the activity of the miRNA (Song et al. 2010; Werner et al. 2010). Mateos et al. used the same rationale, but applied a random mutagenesis approach. In this report, a randomized pre-miR172a library made with nonequimolar oligonucleotides was introduced into plants (Fig. 3b). The advantage of this random approach is that no previous assumption on the relevance of any region is required. With the data collected through this analysis, the contribution of each base of the precursor to its biogenesis could be inferred (Mateos et al. 2010). Anyway, these distinct strategies led to the identification of a common attribute of many plant precursors that represents a crucial element for miRNA biogenesis, a 15-nt lower stem.

8.3 In Vitro Studies

Currently, there is little information about the biochemical properties of the proteins involved in plant miRNA biogenesis. By contrast, in vitro systems have been widely used in animal systems (Han et al. 2004a, 2006; Zhang et al. 2002). In plants, wheat germ extracts have been shown to contain Dicer-like enzyme activity. When incubated with radiolabeled dsRNA, these extracts rendered small RNA molecules ranging from 21 to 25 nt length (Tang et al. 2003).

Antibodies against specific proteins or alternatively against their tagged versions can be used to isolate protein complexes from plant extracts and used as a source for biochemical analysis. First reports using this strategy showed that DCL1 was indeed responsible for generating the 21 nt RNA species (Qi et al. 2005). In a similar way, transformation of *hyl1* mutants with different constructs of a tagged HYL1 protein has allowed the isolation of complexes with precursor processing activity in vitro (Wu et al. 2007). Inmunoprecipitation assays were also useful to determine the physical interaction among components of the processing machinery (Kurihara et al. 2006).

An in vitro system to study miRNA processing in plants has been recently developed. By means of recombinant DCL1, SE, and HYL1, Dong et al. could achieve dicing activity when incubating these three proteins with a synthetic miRNA precursor. They demonstrated that HYL1 and SE are dispensable for cleavage but essential for the precision of the cuts performed by DCL1 (Dong et al. 2008). The use of in vitro systems such as this one might allow the biochemical characterization of RNA pathways in plants.

9 Future Perspectives

In the past few years, many components of the small RNA pathways have been identified in plants, mainly through genetic approaches. In general, their functions have been classified based on their ultimate effect on the different small RNAs. However, the molecular mechanisms underlying the functions of these genes are largely unknown. A combination of structural, biochemical, and genetic approaches might bring insights into their molecular role. This might be useful for the design of more efficient and specific tools for RNA silencing in plants.

Acknowledgments The work was supported by fellowships from CONICET to JM and NB; JP is a CONICET researcher and received grants from HFSP, ANPCyT, and HHMI.

References

Addo-Quaye C, Eshoo TW, Bartel DP et al (2008) Endogenous siRNA and miRNA targets identified by sequencing of the Arabidopsis degradome. Curr Biol 18:758–762

Addo-Quaye C, Snyder JA, Park YB et al (2009) Sliced microRNA targets and precise loop-first processing of MIR319 hairpins revealed by analysis of the *Physcomitrella patens* degradome. RNA 15:2112–2121

Allen E, Xie Z, Gustafson AM et al (2004) Evolution of microRNA genes by inverted duplication of target gene sequences in *Arabidopsis thaliana*. Nat Genet 36:1282–1290

Arazi T, Talmor-Neiman M, Stav R et al (2005) Cloning and characterization of micro-RNAs from moss. Plant J 43:837–848

Axtell MJ (2008) Evolution of microRNAs and their targets: are all microRNAs biologically relevant? Biochim Biophys Acta 1779:725–734

Axtell MJ, Bartel DP (2005) Antiquity of microRNAs and their targets in land plants. Plant Cell 17:1658–1673

Axtell MJ, Bowman JL (2008) Evolution of plant microRNAs and their targets. Trends Plant Sci 13:343–349

Axtell MJ, Snyder JA, Bartel DP (2007) Common functions for diverse small RNAs of land plants. Plant Cell 19:1750–1769

Bologna NG, Mateos JL, Bresso EG et al (2009) A loop-to-base processing mechanism underlies the biogenesis of plant microRNAs miR319 and miR159. EMBO J 28:3646–3656

Bouche N, Lauressergues D, Gasciolli V et al (2006) An antagonistic function for Arabidopsis DCL2 in development and a new function for DCL4 in generating viral siRNAs. EMBO J 25:3347–3356

Boutet S, Vazquez F, Liu J et al (2003) Arabidopsis HEN1: a genetic link between endogenous miRNA controlling development and siRNA controlling transgene silencing and virus resistance. Curr Biol 13:843–848

Carthew RW, Sontheimer EJ (2009) Origins and Mechanisms of miRNAs and siRNAs. Cell 136: 642–655

Chapman EJ, Carrington JC (2007) Specialization and evolution of endogenous small RNA pathways. Nat Rev Genet 8:884–896

Chen X (2009) Small RNAs and their roles in plant development. Annu Rev Cell Dev Biol 25:21–44

Cuperus JT, Montgomery TA, Fahlgren N et al (2010) Identification of MIR390a precursor processing-defective mutants in Arabidopsis by direct genome sequencing. Proc Natl Acad Sci USA 107:466–471

Deleris A, Gallego-Bartolome J, Bao J et al (2006) Hierarchical action and inhibition of plant Dicer-like proteins in antiviral defense. Science 313:68–71

Denli AM, Tops BB, Plasterk RH et al (2004) Processing of primary microRNAs by the microprocessor complex. Nature 432:231–235

Dong Z, Han MH, Fedoroff N (2008) The RNA-binding proteins HYL1 and SE promote accurate in vitro processing of pri-miRNA by DCL1. Proc Natl Acad Sci USA 105:9970–9975

Eamens AL, Smith NA, Curtin SJ et al (2009) The *Arabidopsis thaliana* double-stranded RNA binding protein DRB1 directs guide strand selection from microRNA duplexes. RNA 15(12):2219–2235

Fahlgren N, Howell MD, Kasschau KD et al (2007) High-throughput sequencing of Arabidopsis microRNAs: evidence for frequent birth and death of MIRNA genes. PLoS ONE 2:e219

Fahlgren N, Jogdeo S, Kasschau KD et al (2010) MicroRNA gene evolution in *Arabidopsis lyrata* and *Arabidopsis thaliana*. Plant Cell 22:1074–1089

Fang Y, Spector DL (2007) Identification of nuclear dicing bodies containing proteins for microRNA biogenesis in living Arabidopsis plants. Curr Biol 17:818–823

Fujioka Y, Utsumi M, Ohba Y et al (2007) Location of a possible miRNA processing site in SmD3/SmB nuclear bodies in Arabidopsis. Plant Cell Physiol 48:1243–1253

Gasciolli V, Mallory AC, Bartel DP et al (2005) Partially redundant functions of Arabidopsis DICER-like enzymes and a role for DCL4 in producing trans-acting siRNAs. Curr Biol 15:1494–1500

German MA, Pillay M, Jeong DH et al (2008) Global identification of microRNA-target RNA pairs by parallel analysis of RNA ends. Nat Biotechnol 26:941–946

Gregory RI, Yan KP, Amuthan G et al (2004) The microprocessor complex mediates the genesis of microRNAs. Nature 432:235–240

Gregory BD, O'Malley RC, Lister R et al (2008) A link between RNA metabolism and silencing affecting Arabidopsis development. Dev Cell 14:854–866

Hamilton A, Voinnet O, Chappell L et al (2002) Two classes of short interfering RNA in RNA silencing. EMBO J 21:4671–4679

Han J, Lee Y, Yeom KH et al (2004a) The Drosha-DGCR8 complex in primary microRNA processing. Genes Dev 18:3016–3027

Han MH, Goud S, Song L et al (2004b) The Arabidopsis double-stranded RNA-binding protein HYL1 plays a role in microRNA-mediated gene regulation. Proc Natl Acad Sci USA 101(4): 1093–1098

Han J, Lee Y, Yeom KH et al (2006) Molecular basis for the recognition of primary microRNAs by the Drosha-DGCR8 complex. Cell 125:887–901

Henderson IR, Zhang X, Lu C et al (2006) Dissecting *Arabidopsis thaliana* DICER function in small RNA processing, gene silencing and DNA methylation patterning. Nat Genet 38:721–725

Huang Y, Ji L, Huang Q et al (2009) Structural insights into mechanisms of the small RNA methyltransferase HEN1. Nature 461:823–827

Johnson C, Kasprzewska A, Tennessen K et al (2009) Clusters and superclusters of phased small RNAs in the developing inflorescence of rice. Genome Res 19:1429–1440

Jones-Rhoades MW, Bartel DP, Bartel B (2006) MicroRNAS and their regulatory roles in plants. Annu Rev Plant Biol 57:19–53

Kim VN, Han J, Siomi MC (2009) Biogenesis of small RNAs in animals. Nat Rev Mol Cell Biol 10:126–139

Kurihara Y, Watanabe Y (2004) Arabidopsis micro-RNA biogenesis through Dicer-like 1 protein functions. Proc Natl Acad Sci USA 101:12753–12758

Kurihara Y, Takashi Y, Watanabe Y (2006) The interaction between DCL1 and HYL1 is important for efficient and precise processing of pri-miRNA in plant microRNA biogenesis. RNA 12:206–212

Laubinger S, Sachsenberg T, Zeller G et al (2008) Dual roles of the nuclear cap-binding complex and SERRATE in pre-mRNA splicing and microRNA processing in *Arabidopsis thaliana*. Proc Natl Acad Sci USA 105:8795–8800

Li J, Yang Z, Yu B et al (2005a) Methylation protects miRNAs and siRNAs from a 3′-end uridylation activity in Arabidopsis. Curr Biol 15:1501–1507

Li Y, Li W, Jin YX (2005b) Computational identification of novel family members of microRNA genes in *Arabidopsis thaliana* and *Oryza sativa*. Acta Biochim Biophys Sin (Shanghai) 37:75–87

Li YF, Zheng Y, Addo-Quaye C et al (2010) Transcriptome-wide identification of microRNA targets in rice. Plant J 62:742–759

Llave C, Xie Z, Kasschau KD et al (2002) Cleavage of Scarecrow-like mRNA targets directed by a class of Arabidopsis miRNA. Science 297:2053–2056

Lobbes D, Rallapalli G, Schmidt DD et al (2006) SERRATE: a new player on the plant microRNA scene. EMBO Rep 7:1052–1058

Lu C, Fedoroff N (2000) A mutation in the Arabidopsis HYL1 gene encoding a dsRNA binding protein affects responses to abscisic acid, auxin, and cytokinin. Plant Cell 12:2351–2366

Lu C, Kulkarni K, Souret FF et al (2006) MicroRNAs and other small RNAs enriched in the Arabidopsis RNA-dependent RNA polymerase-2 mutant. Genome Res 16:1276–1288

Lu C, Jeong DH, Kulkarni K et al (2008) Genome-wide analysis for discovery of rice microRNAs reveals natural antisense microRNAs (nat-miRNAs). Proc Natl Acad Sci USA 105:4951–4956

Lund E, Guttinger S, Calado A et al (2004) Nuclear export of microRNA precursors. Science 303:95–98

Ma Z, Coruh C, Axtell MJ (2010) *Arabidopsis lyrata* small RNAs: transient MIRNA and small interfering RNA loci within the Arabidopsis genus. Plant Cell 22:1090–1103

Mateos JL, Bologna NG, Chorostecki U et al (2010) Identification of microRNA processing determinants by random mutagenesis of Arabidopsis MIR172a precursor. Curr Biol 20:49–54

Matranga C, Tomari Y, Shin C et al (2005) Passenger-strand cleavage facilitates assembly of siRNA into Ago2-containing RNAi enzyme complexes. Cell 123:607–620

Meyers BC, Axtell MJ, Bartel B et al (2008) Criteria for annotation of plant MicroRNAs. Plant Cell 20:3186–3190

Mi S, Cai T, Hu Y et al (2008) Sorting of small RNAs into Arabidopsis argonaute complexes is directed by the 5′ terminal nucleotide. Cell 133:116–127

Mica E, Piccolo V, Delledonne M et al (2010) Correction: high throughput approaches reveal splicing of primary microRNA transcripts and tissue specific expression of mature microRNAs in Vitis vinifera. BMC Genomics 11:109

Montgomery TA, Howell MD, Cuperus JT et al (2008) Specificity of ARGONAUTE7-miR390 interaction and dual functionality in TAS3 trans-acting siRNA formation. Cell 133:128–141

Morris ER, Chevalier D, Walker JC (2006) DAWDLE, a forkhead-associated domain gene, regulates multiple aspects of plant development. Plant Physiol 141:932–941

Palatnik JF, Allen E, Wu X et al (2003) Control of leaf morphogenesis by microRNAs. Nature 425:257–263

Papp I, Mette MF, Aufsatz W et al (2003) Evidence for nuclear processing of plant micro RNA and short interfering RNA precursors. Plant Physiol 132:1382–1390

Park W, Li J, Song R et al (2002) CARPEL FACTORY, a Dicer homolog, and HEN1, a novel protein, act in microRNA metabolism in *Arabidopsis thaliana*. Curr Biol 12:1484–1495

Park MY, Wu G, Gonzalez-Sulser A et al (2005) Nuclear processing and export of microRNAs in Arabidopsis. Proc Natl Acad Sci USA 102:3691–3696

Qi Y, Denli AM, Hannon GJ (2005) Biochemical specialization within Arabidopsis RNA silencing pathways. Mol Cell 19:421–428

Qin H, Chen F, Huan X et al (2010) Structure of the *Arabidopsis thaliana* DCL4 DUF283 domain reveals a noncanonical double-stranded RNA-binding fold for protein-protein interaction. RNA 16:474–481

Rajagopalan R, Vaucheret H, Trejo J et al (2006) A diverse and evolutionarily fluid set of microRNAs in *Arabidopsis thaliana*. Genes Dev 20:3407–3425

Ramachandran V, Chen X (2008) Small RNA metabolism in Arabidopsis. Trends Plant Sci 13:368–374

Rand TA, Petersen S, Du F et al (2005) Argonaute2 cleaves the anti-guide strand of siRNA during RISC activation. Cell 123:621–629

Rasia RM, Mateos J, Bologna NG et al (2010) Structure and RNA interactions of the plant microRNA processing-associated protein HYL1. Biochemistry 49:8237–8239

Reinhart BJ, Weinstein EG, Rhoades MW et al (2002) MicroRNAs in plants. Genes Dev 16: 1616–1626

Saito K, Ishizuka A, Siomi H et al (2005) Processing of pre-microRNAs by the Dicer-1-Loquacious complex in Drosophila cells. PLoS Biol 3:e235

Schauer SE, Jacobsen SE, Meinke DW et al (2002) DICER-LIKE1: blind men and elephants in Arabidopsis development. Trends Plant Sci 7:487–491

Song L, Han MH, Lesicka J et al (2007) Arabidopsis primary microRNA processing proteins HYL1 and DCL1 define a nuclear body distinct from the Cajal body. Proc Natl Acad Sci USA 104:5437–5442

Song L, Axtell MJ, Fedoroff NV (2010) RNA secondary structural determinants of miRNA precursor processing in Arabidopsis. Curr Biol 20:37–41

Takeda A, Iwasaki S, Watanabe T et al (2008) The mechanism selecting the guide strand from small RNA duplexes is different among argonaute proteins. Plant Cell Physiol 49:493–500

Talmor-Neiman M, Stav R, Frank W et al (2006) Novel micro-RNAs and intermediates of microRNA biogenesis from moss. Plant J 47:25–37

Tang G, Reinhart BJ, Bartel DP et al (2003) A biochemical framework for RNA silencing in plants. Genes Dev 17:49–63

Vaucheret H (2006) Post-transcriptional small RNA pathways in plants: mechanisms and regulations. Genes Dev 20:759–771

Vazquez F, Gascioli V, Crete P et al (2004) The nuclear dsRNA binding protein HYL1 is required for microRNA accumulation and plant development, but not posttranscriptional transgene silencing. Curr Biol 14:346–351

Vazquez F, Blevins T, Ailhas J et al (2008) Evolution of Arabidopsis MIR genes generates novel microRNA classes. Nucleic Acids Res 36:6429–6438

Vazquez F, Legrand S, Windels D (2010) The biosynthetic pathways and biological scopes of plant small RNAs. Trends Plant Sci 15:337–345

Voinnet O (2009) Origin, biogenesis, and activity of plant microRNAs. Cell 136:669–687

Warthmann N, Das S, Lanz C et al (2008) Comparative analysis of the MIR319a microRNA locus in Arabidopsis and related Brassicaceae. Mol Biol Evol 25:892–902

Wei B, Cai T, Zhang R et al (2009) Novel microRNAs uncovered by deep sequencing of small RNA transcriptomes in bread wheat (*Triticum aestivum* L.) and *Brachypodium distachyon* (L.) Beauv. Funct Integr Genomics 9:499–511

Werner S, Wollmann H, Schneeberger K et al (2010) Structure determinants for accurate processing of miR172a in *Arabidopsis thaliana*. Curr Biol 20:42–48

Wu F, Yu L, Cao W et al (2007) The N-terminal double-stranded RNA binding domains of Arabidopsis HYPONASTIC LEAVES1 are sufficient for pre-microRNA processing. Plant Cell 19:914–925

Wu L, Zhang Q, Zhou H et al (2009) Rice MicroRNA effector complexes and targets. Plant Cell 21:3421–3435

Wu L, Zhou H, Zhang Q et al (2010) DNA methylation mediated by a microRNA pathway. Mol Cell 38:465–475

Xie Z, Allen E, Fahlgren N et al (2005) Expression of Arabidopsis MIRNA genes. Plant Physiol 138:2145–2154

Xie F, Frazier TP, Zhang B (2010) Identification and characterization of microRNAs and their targets in the bioenergy plant switchgrass (*Panicum virgatum*). Planta 232:417–434

Yang L, Liu Z, Lu F et al (2006) SERRATE is a novel nuclear regulator in primary microRNA processing in Arabidopsis. Plant J 47:841–850

Yang SW, Chen HY, Yang J et al (2010) Structure of Arabidopsis HYPONASTIC LEAVES1 and its molecular implications for miRNA processing. Structure 18:594–605

Yi R, Qin Y, Macara IG et al (2003) Exportin-5 mediates the nuclear export of pre-microRNAs and short hairpin RNAs. Genes Dev 17:3011–3016

Yu B, Yang Z, Li J et al (2005) Methylation as a crucial step in plant microRNA biogenesis. Science 307:932–935

Yu B, Bi L, Zheng B et al (2008) The FHA domain proteins DAWDLE in Arabidopsis and SNIP1 in humans act in small RNA biogenesis. Proc Natl Acad Sci USA 105:10073–10078

Zeng Y, Cullen BR (2005) Efficient processing of primary microRNA hairpins by Drosha requires flanking nonstructured RNA sequences. J Biol Chem 280:27595–27603

Zeng Y, Yi R, Cullen BR (2005) Recognition and cleavage of primary microRNA precursors by the nuclear processing enzyme Drosha. EMBO J 24(1):138–148

Zhang H, Kolb FA, Brondani V et al (2002) Human Dicer preferentially cleaves dsRNAs at their termini without a requirement for ATP. EMBO J 21:5875–5885

Zhang B, Pan X, Cannon CH et al (2006) Conservation and divergence of plant microRNA genes. Plant J 46:243–259

Zhang L, Chia JM, Kumari S et al (2009) A genome-wide characterization of microRNA genes in maize. PLoS Genet 5:e1000716

Zhou X, Ruan J, Wang G et al (2007) Characterization and identification of microRNA core promoters in four model species. PLoS Comput Biol 3(3):e37

MicroRNAs in Plant Roots: Current Understanding and Future Perspectives

Yijun Meng, Ping Wu, and Ming Chen

Contents

Abstract Since the tip of the iceberg of plant small RNA (sRNA) world emerged at the end of the last century, diverse sRNA species with various biological roles at either transcriptional or posttranscriptional level have been widely recognized. However, much more research efforts are required to thoroughly understand these sRNA molecules, even for the most sophisticatedly characterized species, microRNAs (miRNAs). The miRNAs, ~21 nucleotides (nt) in length, are generated from stem-loop structured precursors, most of which are transcribed from RNA polymerase (Pol) II-dependent miRNA genes. Elaborate transgenic experiments and other wet-lab results showed that miRNAs play a myriad of essential biochemical or physiological roles in both plants and animals, although they only occupy a very small portion of the enormous sRNA population. In this chapter, we focus on the implication of miRNAs in plant root development. First, we review the inspiring progress that has been made in this research area recently, providing an integrated map of miRNA-mediated and multisignal-involved modulation of plant root development. Then, we present a complicated scene where numerous signals including nutrients, hormones, stress, and biotic stimuli are likely to be involved in the

M. Chen (✉), Y. Meng, and P. Wu
College of Life Sciences, Zhejiang University, Hangzhou 310058, China
e-mail: mchen@zju.edu.cn

V.A. Erdmann and J. Barciszewski (eds.), *Non Coding RNAs in Plants*, RNA Technologies, DOI 10.1007/978-3-642-19454-2_17,

miRNA-centered regulatory networks based on the current reports. Next, we introduce some useful experimental and bioinformatics approaches that can be employed for the functional studies on miRNAs. Owing to the recently developed next-generation sequencing (NGS) technology, the plant degradome libraries can be sequenced by the PARE (*p*arallel *a*nalysis of *R*NA *e*nds) method to perform a transcriptome-wide identification of miRNA–target pairs, the result of which could be used to build a comprehensive miRNA-mediated gene regulatory network. Finally, some concerns on the plant miRNA research are discussed.

Keywords Degradome • Hormone • MicroRNA • MicroRNA–target pair • Network • Next-generation sequencing • Nutrition • PARE (parallel analysis of RNA ends) • Root development • Small RNA • Stress

1 Introduction

As an underground organ, the root plays an indispensable role in plant growth and development, such as seedling fixation, and water and nutrient absorption. More importantly, the roots are essential for plant survival under various unforeseeable environmental changes and stimuli, considering the relatively high phenotypic plasticity of the root system architecture (RSA). After decades of fundamental research efforts, the molecular mechanisms underlying root development and RSA transformation in response to environmental or endogenous cues have been partially uncovered (see reviews Benfey et al. 2010; Benfey and Scheres 2000; Casimiro et al. 2003; Coudert et al. 2010; de Dorlodot et al. 2007; Hardtke 2006; Hochholdinger and Tuberosa 2009; Ishida et al. 2008; Iyer-Pascuzzi et al. 2009; Monshausen and Gilroy 2009; Osmont et al. 2007; Peret et al. 2009; Scheres 1997; Schiefelbein 2000). Several novel genes have been cloned and demonstrated to be key players in root patterning, such as *CRL1* (*crown rootless 1*)/*ARL1* (*adventitious rootless 1*) (Inukai et al. 2005) and OsGNOM1 (a guanine nucleotide exchange factor for ADP-ribosylation factor) (Liu et al. 2009) involved in rice (*Oryza sativa*) adventitious root (AR) development. Besides, numerous signals, such as plant hormones, nutrients, and biotic/abiotic stress, can affect RSA based on either local response or long-distance signaling mechanisms (see reviews Desnos 2008; Gojon et al. 2009; Liu et al. 2009; Lopez-Bucio et al. 2003; Monshausen and Gilroy 2009; Oka-Kira and Kawaguchi 2006; Stougaard 2001). Although the tap root systems of dicots are phenotypically distinct from the fibrous root systems of monocots, these two major root types share partially conserved molecular basis (Hochholdinger and Zimmermann 2008). Together, these signal-induced, protein-coding gene-mediated pathways constitute a quite complex picture of regulatory networks involved in root growth modulation in plants (Birnbaum and Benfey 2004; Ishida et al. 2008; Montiel et al. 2004).

At the end of the last century, the ~21-nt small non coding RNA molecules, known as microRNAs, caught scientists' eyes, since more and more emerging evidences pointed to their indispensable roles in gene expression control at either posttranscriptional or translational level (Carthew and Sontheimer 2009; Jones-Rhoades et al. 2006;

Voinnet 2009). In plants, a number of miRNA families, either highly conserved or species-specific, have been computationally predicted and/or experimentally cloned either by traditional genetic approaches or by the currently developed next-generation sequencing (NGS) strategy (Hsieh et al. 2009; Jones-Rhoades and Bartel 2004; Joshi et al. 2010; Morozova and Marra 2008; Nakano et al. 2006; Pantaleo et al. 2010; Simon et al. 2009a, b; Sunkar et al. 2005; Zhang et al. 2006, 2007, 2009). Meanwhile, numerous factors were identified to be implicated in the plant miRNA biogenesis pathway, which showed conservation with that of the animals to some extent (Carthew and Sontheimer 2009; Voinnet 2009). To date, a number of miRNA targets have been validated (Jones-Rhoades et al. 2006). Functional studies revealed that miRNAs play essential roles in various biochemical and physiological processes along the plant life cycle (Chen 2009; Jones-Rhoades et al. 2006; Voinnet 2009). For example, *miR160*, *miR164*, *miR165*/*miR166*, *miR167*, *miR390*, and *miR393* were confirmed to be master regulators involved in growth modulation of different root components (Carlsbecker et al. 2010; Guo et al. 2005; Gutierrez et al. 2009; Vidal et al. 2010; Wang et al. 2005; Yoon et al. 2009). Additionally, a dozen of internal and external signals have been integrated into miRNA-guided regulatory pathways (Fujii et al. 2005; Kawashima et al. 2009; Scheres 2010; Sunkar et al. 2006; Vidal et al. 2010; Yamasaki et al. 2007; Yang et al. 2006; Yoon et al. 2009; Zhao et al. 2007), suggesting an inconceivably complicated networks in plant root systems.

Here, we focus on the miRNAs and related signal transduction pathways that have either verified control roles or deduced regulatory potential in plant root development, based on the current reports (Bari et al. 2006; Carlsbecker et al. 2010; Franco-Zorrilla et al. 2007; Fujii et al. 2005; Gifford et al. 2008; Guo et al. 2005; Gutierrez et al. 2009; Hsieh et al. 2009; Jones-Rhoades and Bartel 2004; Kawashima et al. 2009; Meng et al. 2009; Sunkar et al. 2006; Vidal et al. 2010; Wang et al. 2005; Yamasaki et al. 2007; Yang et al. 2006; Yoon et al. 2009). Besides, both the bioinformatics and experimental strategies for plant miRNA studies are partially mentioned, especially for the NGS, although a few related reviews have already been available (Chen et al. 2010; Fahlgren et al. 2009). During the past decade, significant progress has been made in the plant miRNA research area. However, studies by using NGS technology uncovered that millions of small non coding RNAs, largely represented by endogenous small interfering RNAs (siRNAs), existed in various plant species (Hsieh et al. 2009; Joshi et al. 2010; Morozova and Marra 2008; Nakano et al. 2006; Pantaleo et al. 2010; Simon et al. 2009a, b). From this point of view, only a tip of the iceberg of plant sRNA world has emerged at the current stage, and much more efforts are needed.

2 MicroRNA, a Master Regulator in the Roots

So far, several miRNA families have been confirmed to play key roles in plant root development. Intriguingly, nearly all parts of the root systems, such as root caps (Wang et al. 2005), lateral roots (Guo et al. 2005; Yoon et al. 2009), ARs (Gutierrez et al. 2009), and root cells for radial tissue organization (Carlsbecker et al. 2010), have shown to be influenced by miRNA-mediated regulatory pathways.

In 2000, *NAC1* (*NAM/ATAF/CUC 1*), a member of the *NAC* family, was identified to promote lateral root development by Chua's group (Xie et al. 2000). It acts as a transcription activator downstream of *TIR1* (*transport inhibitor response 1*) (Ruegger et al. 1998). Five years later, *NAC1* was demonstrated to be the target of *miR164* in *Arabidopsis* (*Arabidopsis thaliana*) (Guo et al. 2005). Hence, all the evidences point to the notion that the *NAC1*-mediated auxin signaling pathway involved in lateral root formation in *Arabidopsis* is targeted by *miR164*. In the same year, Wang and colleagues reported that *ARF10* (*auxin response factor 10*) and *ARF16*, both of which were targeted by *miR160*, acted as key controllers of root-cap cell formation in *Arabidopsis* (Wang et al. 2005). Accordingly, our previous transgenic result has showed that overexpression of *miR160* could result in severe root-tip defect in rice, indicating that *miR160* is also functional in the rice root-cap cells (unpublished data). Thus, it is clear that *miR160*-mediated regulation of root-cap formation is quite conserved between dicots and monocots.

In the past 2 years, the understanding of the regulatory roles of miRNAs in plant root development has been greatly advanced. In addition to its key role in root-cap formation, *miR160*, targeting *ARF17*, along with *miR167* that targets *ARF6* and *ARF8* is reported to be involved in AR initiation, forming a quite complex regulatory network including a feedback regulation of miRNA homeostasis (Gutierrez et al. 2009). *TAS3*, *trans*-acting small interfering RNA (tasiRNA) gene, is targeted by *miR390* to produce tasiRNAs targeting *ARF3* and *ARF4*, through a RDR6 (*R*NA-*d*ependent *R*NA polymerase 6) and DCL4 (*D*icer-*l*ike 4)-dependent pathway (Allen et al. 2005; Chen 2009; Williams et al. 2005). More recently, the *miR390–TAS3–ARF4* pathway has been confirmed to modulate the lateral root developmental processes (Yoon et al. 2009). Besides these auxin signal-involved pathways, it has been observed that the regulatory module *miR393–AFB3* (*auxin F-box 3*) could control RSA, including primary and lateral roots, in response to external and internal nitrate availability in *Arabidopsis* (Vidal et al. 2010). Recently, it has been identified that the regulatory pathway, *SHR* (*SHORT ROOT*)/*SCR* (*SCARECROW*)–*miR165*/*miR166–HD-ZIP III* (*class III homeodomain-leucine zipper*) transcription factors (TFs), is essential for cell patterning and radial tissue organization in the roots of *Arabidopsis* (Carlsbecker et al. 2010). The functionalities of the *miR166–HD-ZIP III* pathway, involved in lateral rooting, vascular bundle development, and symbiotic nodule formation, had been observed in *Medicago truncatula* earlier (Boualem et al. 2008). Our own study on an auxin-resistant rice mutant, *osaxr*, with plethoric root defects showed that the expression levels of *miR164*, *miR167*, *miR171*, and *miR390* were significantly changed compared to the wild type rice (Meng et al. 2009). However, the exact roles of these miRNA genes in establishing RSA remain elusive.

We have noticed that nearly all the miRNA families implicated in plant root development tend to be highly conserved between dicots and monocots (Table 1), except for the *miR165* family, which was specific in *Arabidopsis*. Together, we could conclude that several evolutionarily conserved miRNA-mediated regulatory pathways should exist in the angiosperms to modulate the growth and patterning of the key components of the root systems.

Table 1 Summary of microRNA-mediated pathways involved in plant root development

Species	MicroRNA family	Regulatory module(s)	Function(s) in root development	Involved signal(s)	References
Arabidopsis thaliana	*miR160*	*miR160–ARF10* (*auxin response factor 10*)/*ARF16*	Root cap formation	Auxin	Wang et al. (2005)
		miR160–ARF17	Adventitious rooting	Auxin	Gutierrez et al. (2009)
	miR164	*miR164–NAC1* (*NAM/ATAF/CUC 1*)	Lateral rooting	Auxin	Guo et al. (2005), Ruegger et al. (1998), and Xie et al. (2000)
	miR165/miR166	*SHR* (*SHORT ROOT*)/*SCR* (*SCARECROW*) –*miR165/miR166–HD-ZIP III* (*class III homeodomain-leucine zipper*)	Radial tissue organization	Signal for cell patterning	Carlsbecker et al. (2010)
	miR167	*miR167–ARF6/ARF8*	Adventitious rooting	Auxin	Gutierrez et al. (2009)
	miR390	*miR390–TAS3* (one *trans*-acting small interfering RNA gene)–*ARF4*	Lateral root development	Auxin	Allen et al. (2005), Chen (2009), Williams et al. (2005), and Yoon et al. (2009)
	miR393	*miR393–AFB3* (*auxin F-box 3*)	Primary and lateral root development	N (nitrogen) and auxin (P, potentially)	Vidal et al. (2010)
Medicago truncatula	*miR166*	*miR166–HD-ZIP III*	Lateral rooting, vascular bundle development, and symbiotic nodule formation	N (P) and signal for cell patterning (P)	Boualem et al. (2008)
Oryza sativa	*miR160*	Unidentified	Root cap formation	Auxin (P)	Meng et al. (2009) and unpublished data
	miR164		Unidentified	N (P) and auxin (P)	
	miR167			Auxin (P)	
	miR171			Unidentified	
	miR390			Auxin (P)	

Potentially (P) means that the signals likely involved in root development have not been confirmed yet according to current reports

3 Signal Interactions and MicroRNA-Mediated Regulatory Networks of Plant Root Development

Table 1 provides summary of the biological roles of miRNAs in root development. However, we have to be aware that all the single miRNA-guided pathways identified by respective studies should not be isolated ones in the plant root systems. For the single pathways themselves, the TFs targeted by certain miRNAs, may in turn modulate the expression of the miRNA genes directly or indirectly, thus forming feedback circuits. The *miR160/miR167*-mediated and *ARF6/ARF8/ARF17*-involved regulatory network that controls adventitious rooting is a good example and partially reflects their feedback complexity (Gutierrez et al. 2009). Additionally, many distinct pathways may interact with each other based on the shared nodes, such as the same miRNA genes or the same miRNA targets. Hence, one miRNA could affect different parts of the root systems, and different miRNA families may cooperate with each other to maintain the normality of a specific root component. For example, *miR160* has been shown to play important roles in both root-cap formation and adventitious rooting in *Arabidopsis* (Gutierrez et al. 2009; Wang et al. 2005). Besides, adventitious rooting was regulated by *miR160* and *miR167* (Gutierrez et al. 2009), and *miR164*, *miR390*, and *miR393* were all indicated to modulate lateral rooting in *Arabidopsis* (Guo et al. 2005; Vidal et al. 2010; Yoon et al. 2009). Imaginably, more such interlaced cases will be verified in the near future, since a suit of targets could be recognized by one miRNA, and one TF may regulate several miRNA genes from different families.

It is quite amazing that the phenotypic characteristics of certain plant organs can be prominently changed in response to the environmental or endogenous stimuli. This is a high-efficient strategy for plant survival, since it can compensate for the limitation of the plant mobility. Recent studies on miRNAs have integrated a number of signals, such as nutrition, hormone, and biotic or abiotic stress, into the miRNA-mediated root developmental processes, and some other signals showed great potential to affect RSA through miRNA actions (Meng et al. 2010b).

The miRNAs are implicated in various nutrition signaling pathways. The most elaborately elucidated one is the *miR399*-involved phosphate (Pi) signaling. *miR399* was first found to be highly induced by low-Pi stress, and it has functional roles in *Arabidopsis* to cope with fluctuations in Pi availability through its target *PHO2* encoding a ubiquitin-conjugating E2 enzyme 24 (Fujii et al. 2005). Subsequently, *PHR1* (*phosphate starvation response 1*) (Rubio et al. 2001) was confirmed to be necessary for *miR399* expression, placing *miR399–PHO2* in a branch of the Pi-signaling network downstream of *PHR1* (Bari et al. 2006). Accordingly, the *PHR1–miR399* regulatory module was also discovered in the roots of common bean (*Phaseolus vulgaris*) (Valdes-Lopez et al. 2008). Besides, the regulatory mechanism of *PHR1–miR399–PHO2* via long-distance signaling in response to Pi deficiency has also been discussed recently (Lin et al. 2008). Again, this scenario has been confirmed in studies on other plant species, e.g., rapeseed (*Brassica napus*) and pumpkin (*Cucurbita maxima*) (Pant et al. 2008). In addition to the major part of *miR399*-mediated Pi signaling pathway, a novel regulatory mechanism, balancing the level of active *miR399* gene products in response to the exogenous Pi availability, was proposed by Paz-Ares's group (Franco-Zorrilla et al. 2007). They discovered that different from the classical targets, the non coding transcript of *IPS1* (*induced by phosphate starvation 1*) (Martin et al. 2000), had a *miR399* recognition motif with a mismatched loop at the expected miRNA cleavage site. As a result, the *IPS1* transcript can be targeted by *miR399* but is not cleavable. In other words, *miR399* can be sequestered by the *IPS1* transcript (Franco-Zorrilla et al. 2007). Considering the high sensitivity of IPS1 to Pi availability (Martin et al. 2000), the novel mechanism, called "target mimicry," provides another layer for regulating the level of mature *miR399* in planta correlating with Pi homeostasis (Chitwood and Timmermans 2007). Recent advent of NGS technology has enabled us to gain deeper insights into the huge sRNA world in plants (Morozova and Marra 2008; Simon et al. 2009b). Small RNA (sRNA) deep sequencing carried out by Chiou's group uncovered that a number of miRNAs and other sRNAs were differentially expressed in response to Pi deprivation in *Arabidopsis*. Besides Pi starvation-inducible *miR399*, the expression of *miR156*, *miR778*, *miR827*, and *miR2111* was highly induced upon Pi stress, whereas *miR169*, *miR395*, and *miR398* were significantly repressed (Hsieh et al. 2009). Notably, this study also demonstrated that cross talk could be coordinated by the miRNAs under different nutrient deficiencies. They found that *miR156* can be upregulated by Pi, nitrogen (N), or potassium (K) starvation in the roots of *Arabidopsis*, but unaffected in the shoots. *miR169* was downregulated under Pi, N, or sulfur (S) deficiency, which was more prominent in the roots. The expression of *miR395* was downregulated under –Pi or –N condition, but induced by –S treatment. Moreover, *miR398* was repressed by Pi, N, K, or iron (Fe) deprivation, but was upregulated by copper (Cu) starvation. Some of the observations above are supported by previous reports (Jones-Rhoades and Bartel 2004; Kawashima et al. 2009; Sunkar et al. 2006; Yamasaki et al. 2007). In addition to that, the authors also confirmed that *miR399*, *miR778*, and *miR827* were specifically induced by Pi starvation in the roots and/or the shoots of *Arabidopsis* (Hsieh et al. 2009). Based on those results related to Pi signaling, we could make a conclusion that certain miRNAs, such as *miR399*, *miR778*, and *miR827*, are specifically involved in Pi signaling, whereas some other miRNAs, *miR156*, *miR169*, *miR395*, and *miR398*, for example, may play a role as a hub for the interaction among multiple nutrition signals.

Another miRNA-mediated nutrition signaling pathway that involved in plant root development is N signaling. It was proposed that the regulatory pathway, *miR167–ARF8*, is essential for lateral root initiation and emergence in *Arabidopsis*. At the same time, the expression of *miR167* was suppressed by exogenous N signal. They also indicated that the plant hormones, auxin and cytokinin, are candidate signaling cues for cell-specific nitrogen responses (Gifford et al. 2008). Another study demonstrated that the regulatory module *miR393–AFB3* that is implicated in RSA control is also N-responsive in *Arabidopsis* (Vidal et al. 2010). Moreover, in our studies on the rice mutant *osaxr* (Meng et al. 2009), we found that the expression of all the *miR164* family members could be significantly repressed by –N treatment in *osaxr*, although whether it is the same case for wild-type rice remains unclear (unpublished data). All these data point to the fact that the miRNA-mediated N signaling could be as intricate as miRNA-mediated Pi signaling network mentioned above. This complexity may also extend to the other nutrition signals, such as S, K, Fe, and Cu.

In addition to the nutrition, several miRNA families were reported to participate in other biotic or abiotic stimulus-initiated signaling pathways. Gene expression profiling uncovered that *miR169* in rice, and *miR156*, *miR166*, *miR171*, and *miR408* in barley (*Hordeum vulgare*) were drought-inducible (Kantar et al. 2010; Zhao et al. 2007). In rice, *miR169g* was the only member of *miR169* family that can be highly induced by drought treatment, and its induction was found to be more prominent in the roots compared to the shoots, indicating its potential role in the root systems for water availability sensing (Zhao et al. 2007). It has been showed that a number of miRNAs, such as *miR159*, *miR166*, *miR167*, *miR171*, *miR395*, *miR396*, *miR474*, and *miR528*, were affected by submergence treatment in the maize (*Zea mays*) roots (Zhang et al. 2008). Also in the maize roots, the expression of 98 miRNAs from 27 families was shown to be significantly altered after salt treatment (Ding et al. 2009). As a result, potential regulatory networks of submergence-responsive and salt-responsive miRNAs in the maize roots were proposed respectively (Ding et al. 2009; Zhang et al. 2008). Interestingly, in these networks, *miR164* and *miR167* that are critical for auxin signal transduction, *miR395* that is involved in S homeostasis, and *miR399*, a key performer in Pi signaling, were included (Ding et al. 2009; Zhang et al. 2008). Another hot research topic that has been extensively investigated is miRNA-involved nodulation in the root systems of legumes, such as soybean (*Glycine max*) and *M. truncatula* (Simon et al. 2009a). Several studies indicate that a portion of miRNAs participate in the rhizobia legume symbiosis, such as *miR159*, *miR160*, *miR164*, *miR166*, *miR167*, *miR172*, *miR319*, *miR393*, *miR482*, *miR1512*, and *miR1515* (Boualem et al. 2008; Li et al. 2010; Subramanian et al. 2008; Wang et al. 2009). These miRNAs were indicated to cover several signals, such as auxin, cytokinin, ABA (abscisic acid), GA (gibberellin), JA (jasmonic acid), ethylene, and N, into one regulatory network of nodulation (Simon et al. 2009a). Moreover, the *miR166–HD-ZIP III* module was involved in both nodulation and lateral rooting (Boualem et al. 2008).

4 Toward miRNA-Mediated Networks: New Technologies and Bioinformatics Tools

The above described results present an extremely complicated multisignal-involved, miRNA-guided gene regulatory network (Fig. 1). However, more research efforts are required to obtain a more comprehensive view, which should be restricted not only to new miRNA identification but also to functional characterization of the existing miRNA genes. One of the most straightforward approaches is to predict and verify more miRNA targets, through which, miRNA-guided pathways could be integrated into the currently established regulatory networks. So far, several bioinformatics tools

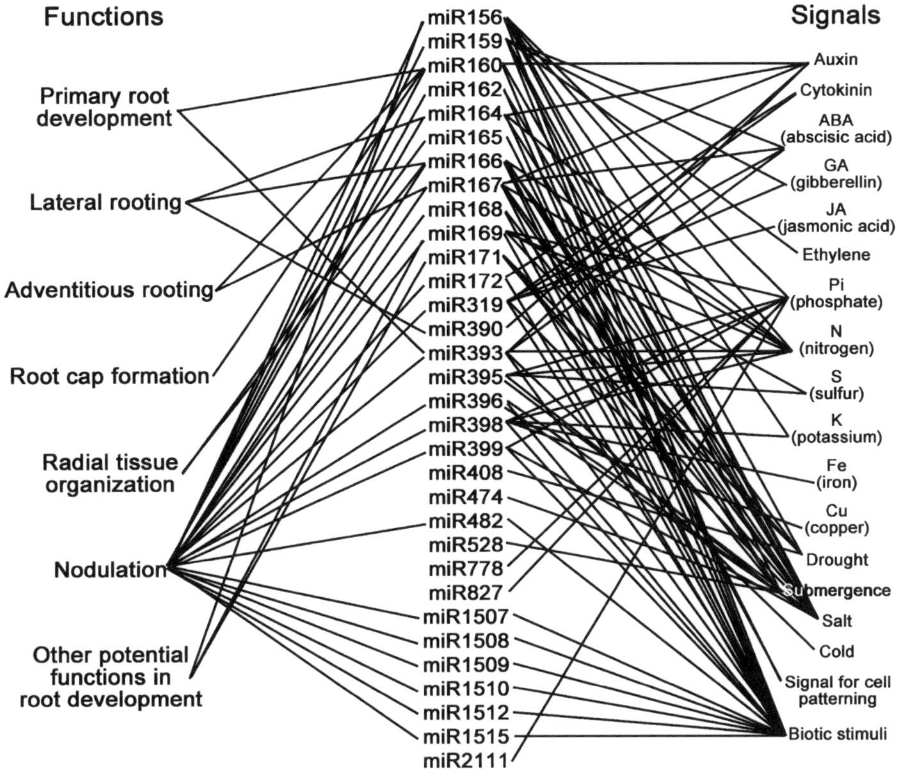

Fig. 1 Graphic presentation of the connections between microRNA-mediated regulation of root development and biotic/abiotic signals. All the root-related functions and microRNA-involved signaling shown in this figure are based on the current reports and our unpublished experimental results in various plant species (Allen et al. 2005; Bari et al. 2006; Boualem et al. 2008; Carlsbecker et al. 2010; Ding et al. 2009; Fujii et al. 2005; Gifford et al. 2008; Guo et al. 2005; Gutierrez et al. 2009; Hsieh et al. 2009; Jones-Rhoades and Bartel 2004; Kantar et al. 2010; Kawashima et al. 2009; Li et al. 2010; Lin et al. 2008; Meng et al. 2009; Pant et al. 2008; Phillips et al. 2007; Rubio et al. 2001; Simon et al. 2009a, b; Subramanian et al. 2008; Sunkar et al. 2006; Sunkar and Zhu 2004; Valdes-Lopez et al. 2008; Vidal et al. 2010; Wang et al. 2005, 2009; Williams et al. 2005; Xie et al. 2000; Yamasaki et al. 2007; Yoon et al. 2009; Zhang et al. 2008; Zhao et al. 2007)

have been developed and made available for miRNA target prediction in plants, such as Target Finder (Allen et al. 2005), miRU (Zhang 2005), and the tool provided by CSRDB (*c*ereal *s*mall *R*NAs *d*ata*b*ase) (Johnson et al. 2007). However, none of these tools could guarantee that all the prediction results are correct. It requires experimental validation. In plants, most identified miRNAs exert their regulatory effects on their targets through cleavage actions, resulting in 3′ cleavage products with relatively high stability in vivo. Taking advantage of these 3′ remnants, a method called modified 5′ RACE (*r*apid *a*mplification of *c*DNA *e*nds) was widely adopted for miRNA target validation (Jones-Rhoades et al. 2006). However, it istedious, time-consuming, and costly if a large amount of miRNA–target candidate pairs are to be verified. Fortunately, a high-throughput method, PARE (*p*arallel *a*nalysis of *R*NA *e*nds), combining the modified 5′ RACE with NGS technology has recently been developed, allowing researchers to search for miRNA targets on a whole-transcriptome scale (German et al. 2008). Because of the high-efficient strategy, more and more degradome sequencing data have been generated to identify miRNA–target pairs in plants (Grigg et al. 2009; Joshi et al. 2010; Pantaleo et al. 2010). However, such huge NGS datasets are indeed a great challenge lying ahead for the bioinformaticians, although we have noticed that these valuable data remain to be fully exploited. To deal with this issue, we have developed a database, named PmiRKB (*p*lant *mi*cro*R*NA *k*nowledge *b*ase). By utilizing publicly available PARE data, the functional module "MiR–Tar" provides users with graphic presentation of the likelihood of certain miRNA–target pairs (Meng et al. 2010a). Recently, we have extracted all the miRNA–target pairs with relatively high reliability based on the PARE signals, hoping to generate comprehensive and valid miRNA-mediated regulatory networks in *Arabidopsis* and rice. Besides, we also started a large-scale analysis of miRNA promoters. Since most miRNA genes were demonstrated to be transcribed by RNA Pol II (Lee et al. 2004; Meng et al. 2009), the analytical tools designed for Pol II-dependent promoters, such as PLACE (a database of plant *cis*-acting regulatory DNA elements) (Higo et al. 1998), PlantCARE (also a database of plant *cis*-acting regulatory elements) (Lescot et al. 2002), PlantTFDB (a comprehensive plant TF database) (Guo et al. 2008), PlnTFDB (also an integrative plant TF database) (Riano-Pachon et al. 2007), PlantProm (a database of plant promoter sequences) (Shahmuradov et al. 2003), and TSSP (a tool provided by PlantProm) (Shahmuradov et al. 2003), are applicable to miRNA promoter analysis (Table 2; also see Chen et al. 2010). However, there is still a long way to go

Table 2 Bioinformatics tools available for plant miRNA promoter analysis

Tool	Web site	Description	References
PlantProm	http://mendel.cs.rhul.ac.uk/mendel.php?topic=plantprom	Plant promoter database	Shahmuradov et al. (2003)
TSSP	http://www.softberry.ru/berry.phtml?group=programs&subgroup=promoter&topic=tssp	Search for TSS combined with TATA-box in an RNA polymerase II-dependent promoter	Shahmuradov et al. (2003)
PlnTFDB	http://plntfdb.bio.uni-potsdam.de/v3.0/	Plant TF database	Riano-Pachon et al. (2007)
PlantTFDB	http://planttfdb.cbi.pku.edu.cn	Plant TF database	Guo et al. (2008)
WebLogo	http://weblogo.berkeley.edu/logo.cgi	Sequence conservation analysis	Crooks et al. (2004)
PLACE	http://www.dna.affrc.go.jp/PLACE/	For *cis*-element analysis	Higo et al. (1998) and Higo et al. (1999)
PlantCARE	http://bioinformatics.psb.ugent.be/webtools/plantcare/html/	For *cis*-element analysis	Lescot et al. (2002)

TSS transcription start site; *TF* transcription factor

to draw a comprehensive miRNA-centered regulatory network in a plant species because numerous signals including nutrients, hormones, stress, and biotic stimuli are involved in this network, making it extraordinarily complicated.

5 Perspectives

In this chapter, we summarize the current status of the research area on miRNA-mediated regulation of plant root systems (Table 1 and Fig. 1). Although we should appreciate the achievements resulting from the previous research efforts, a great challenge still lies ahead for us to uncover the half-opened black box. Here, we raise several issues that need to be addressed.

To our knowledge, no miRNA family has been suggested to play a role in root-hair development, although a number of genes were demonstrated to be important for root-hair cell formation (Gilroy and Jones 2000; Samaj et al. 2004; Schiefelbein 2000). Whether these root-hair-related genes or other unidentified genes are targeted by certain miRNAs need to be further explored. Considering the in-depth detection feature of NGS, it is convincible that much more miRNAs or other sRNA species that are implicated in plant root development could be cloned. However, as mentioned above, these huge biological datasets generated by NGS still need powerful bioinformatics tools to deal with.

For the functional studies on miRNAs, several elegant methods, such as target mimicry (Franco-Zorrilla et al. 2007) and site-directed silent mutations of miRNA target genes (Mallory et al. 2005), have been developed to inhibit the activities of certain miRNAs. However, these methods cannot completely distinguish one miRNA from another within the same family. The same problem will also occur when hybridization-based methods, including northern blotting, in situ hybridization, and microarray, are employed for miRNA expression profiling. Sequence-based approaches, such as NGS and quantitative real-time PCR, can partially solve this problem for miRNA expression detection. But, for functional characterization, there is still a requirement for efficient fine-scale methods to avoid the interference of functional redundancy introduced by the different members belonging to the same family.

Moreover, the question whether the miRNAs of the same family have distinct biological functions remains to be elucidated. Therefore, the transcription activities and the tissue-specific expression patterns need to be thoroughly investigated for each family member. This may result in a reannotation of miRNA genes according to the current information provided by miRBase (Griffiths-Jones et al. 2006), since some miRNA family members may be pseudo-miRNA genes. Furthermore, we noticed that one miRNA gene may possess multiple distinct biological functions. This notion is reasonable, since a number of genes could be targeted by a single miRNA. For example, in addition to the essential role in lateral rooting (Guo et al. 2005), *miR164* was also indicated to be important for maintaining normal embryonic, vegetative, and floral development (Mallory et al. 2004). It reminds us that the miRNA genes that currently seem to be root-unrelated may be potential regulators for root development.

Last but not the least, a lot of miRNA genes discussed in this chapter tend to be highly conserved, but the species-specific miRNA genes and the conserved ones with unidentified species-specific regulatory roles need to be further investigated.

Together, we present an overview of the miRNAs involved in plant root development, and we hope that this brief review could be timely and useful for the scientists with related research interests. However, we would like to emphasize that the studies on sRNA-mediated modulation of plant root development should not restrict to the miRNAs, considering the enormous endogenous sRNA population reflected by the high-throughput sequencing data. And it may also extend to the other non coding RNAs such as long non coding RNAs, based on the current results (Ben Amor et al. 2009). It is foreseeable that a quite comprehensive map of non coding RNA-involved gene regulatory network will be drawn in the plant root systems in the near future.

Acknowledgments This work was supported by the MoST of China (Grant No. 2005CB120901, 2008AA10Z125, 2009DFA32030), the NSFC (Grant No. 30771326, 30971743), and the MoE of China (Grant No. NCET-07-0740).

References

Allen E, Xie Z, Gustafson AM et al (2005) MicroRNA-directed phasing during *trans*-acting siRNA biogenesis in plants. Cell 121:207–221

Bari R, Datt Pant B, Stitt M et al (2006) PHO2, microRNA399, and PHR1 define a phosphate-signaling pathway in plants. Plant Physiol 141:988–999

Ben Amor B, Wirth S, Merchan F et al (2009) Novel long non-protein coding RNAs involved in *Arabidopsis* differentiation and stress responses. Genome Res 19:57–69

Benfey PN, Scheres B (2000) Root development. Curr Biol 10:R813–R815

Benfey PN, Bennett M, Schiefelbein J (2010) Getting to the root of plant biology: impact of the *Arabidopsis* genome sequence on root research. Plant J 61:992–1000

Birnbaum K, Benfey PN (2004) Network building: transcriptional circuits in the root. Curr Opin Plant Biol 7:582–588

Boualem A, Laporte P, Jovanovic M et al (2008) MicroRNA166 controls root and nodule development in *Medicago truncatula*. Plant J 54:876–887

Carlsbecker A, Lee JY, Roberts CJ et al (2010) Cell signalling by microRNA165/6 directs gene dose-dependent root cell fate. Nature 465:316–321

Carthew RW, Sontheimer EJ (2009) Origins and mechanisms of miRNAs and siRNAs. Cell 136:642–655

Casimiro I, Beeckman T, Graham N et al (2003) Dissecting *Arabidopsis* lateral root development. Trends Plant Sci 8:165–171

Chen X (2009) Small RNAs and their roles in plant development. Annu Rev Cell Dev Biol 25:21–44

Chen M, Meng Y, Mao C et al (2010) Methodological framework for functional characterization of plant microRNAs. J Exp Bot 61:2271–2280

Chitwood DH, Timmermans MC (2007) Target mimics modulate miRNAs. Nat Genet 39:935–936

Coudert Y, Perin C, Courtois B et al (2010) Genetic control of root development in rice, the model cereal. Trends Plant Sci 15:219–226

Crooks GE, Hon G, Chandonia JM et al (2004) WebLogo: a sequence logo generator. Genome Res 14:1188–1190

de Dorlodot S, Forster B, Pages L et al (2007) Root system architecture: opportunities and constraints for genetic improvement of crops. Trends Plant Sci 12:474–481

Desnos T (2008) Root branching responses to phosphate and nitrate. Curr Opin Plant Biol 11:82–87

Ding D, Zhang L, Wang H et al (2009) Differential expression of miRNAs in response to salt stress in maize roots. Ann Bot 103:29–38

Fahlgren N, Sullivan CM, Kasschau KD et al (2009) Computational and analytical framework for small RNA profiling by high-throughput sequencing. RNA 15:992–1002

Franco-Zorrilla JM, Valli A, Todesco M et al (2007) Target mimicry provides a new mechanism for regulation of microRNA activity. Nat Genet 39:1033–1037

Fujii H, Chiou TJ, Lin SI et al (2005) A miRNA involved in phosphate-starvation response in *Arabidopsis*. Curr Biol 15:2038–2043

German MA, Pillay M, Jeong DH et al (2008) Global identification of microRNA-target RNA pairs by parallel analysis of RNA ends. Nat Biotechnol 26:941–946

Gifford ML, Dean A, Gutierrez RA et al (2008) Cell-specific nitrogen responses mediate developmental plasticity. Proc Natl Acad Sci USA 105:803–808

Gilroy S, Jones DL (2000) Through form to function: root hair development and nutrient uptake. Trends Plant Sci 5:56–60

Gojon A, Nacry P, Davidian JC (2009) Root uptake regulation: a central process for NPS homeostasis in plants. Curr Opin Plant Biol 12:328–338

Griffiths-Jones S, Grocock RJ, van Dongen S et al (2006) miRBase: microRNA sequences, targets and gene nomenclature. Nucleic Acids Res 34:D140–D144

Grigg SP, Galinha C, Kornet N et al (2009) Repression of apical homeobox genes is required for embryonic root development in *Arabidopsis*. Curr Biol 19:1485–1490

Guo HS, Xie Q, Fei JF et al (2005) MicroRNA directs mRNA cleavage of the transcription factor NAC1 to downregulate auxin signals for *Arabidopsis* lateral root development. Plant Cell 17:1376–1386

Guo AY, Chen X, Gao G et al (2008) PlantTFDB: a comprehensive plant transcription factor database. Nucleic Acids Res 36:D966–D969

Gutierrez L, Bussell JD, Pacurar DI et al (2009) Phenotypic plasticity of adventitious rooting in *Arabidopsis* is controlled by complex regulation of AUXIN RESPONSE FACTOR transcripts and MicroRNA abundance. Plant Cell 21:3119–3132

Hardtke CS (2006) Root development – branching into novel spheres. Curr Opin Plant Biol 9:66–71

Higo K, Ugawa Y, Iwamoto M et al (1998) PLACE: a database of plant *cis*-acting regulatory DNA elements. Nucleic Acids Res 26:358–359

Higo K, Ugawa Y, Iwamoto M et al (1999) Plant *cis*-acting regulatory DNA elements (PLACE) database: 1999. Nucleic Acids Res 27:297–300

Hochholdinger F, Tuberosa R (2009) Genetic and genomic dissection of maize root development and architecture. Curr Opin Plant Biol 12:172–177

Hochholdinger F, Zimmermann R (2008) Conserved and diverse mechanisms in root development. Curr Opin Plant Biol 11:70–74

Hsieh LC, Lin SI, Shih AC et al (2009) Uncovering small RNA-mediated responses to phosphate-deficiency in *Arabidopsis* by deep sequencing. Plant Physiol 151:2120–2132

Inukai Y, Sakamoto T, Ueguchi-Tanaka M et al (2005) Crown rootless1, which is essential for crown root formation in rice, is a target of an AUXIN RESPONSE FACTOR in auxin signaling. Plant Cell 17:1387–1396

Ishida T, Kurata T, Okada K et al (2008) A genetic regulatory network in the development of trichomes and root hairs. Annu Rev Plant Biol 59:365–386

Iyer-Pascuzzi A, Simpson J, Herrera-Estrella L et al (2009) Functional genomics of root growth and development in *Arabidopsis*. Curr Opin Plant Biol 12:165–171

Johnson C, Bowman L, Adai AT et al (2007) CSRDB: a small RNA integrated database and browser resource for cereals. Nucleic Acids Res 35:D829–D833

Jones-Rhoades MW, Bartel DP (2004) Computational identification of plant microRNAs and their targets, including a stress-induced miRNA. Mol Cell 14:787–799

Jones-Rhoades MW, Bartel DP, Bartel B (2006) MicroRNAs and their regulatory roles in plants. Annu Rev Plant Biol 57:19–53

Joshi T, Yan Z, Libault M (2010) Prediction of novel miRNAs and associated target genes in *Glycine max*. BMC Bioinformatics 11(Suppl 1):S14

Kantar M, Unver T, Budak H (2010) Regulation of barley miRNAs upon dehydration stress correlated with target gene expression. Funct Integr Genomics 10:493–507

Kawashima CG, Yoshimoto N, Maruyama-Nakashita A et al (2009) Sulphur starvation induces the expression of microRNA-395 and one of its target genes but in different cell types. Plant J 57:313–321

Lee Y, Kim M, Han J et al (2004) MicroRNA genes are transcribed by RNA polymerase II. EMBO J 23:4051–4060

Lescot M, Dehais P, Thijs G et al (2002) PlantCARE, a database of plant *cis*-acting regulatory elements and a portal to tools for in silico analysis of promoter sequences. Nucleic Acids Res 30:325–327

Li H, Deng Y, Wu T et al (2010) Misexpression of miR482, miR1512, and miR1515 increases soybean nodulation. Plant Physiol 153:1759–1770

Lin SI, Chiang SF, Lin WY et al (2008) Regulatory network of microRNA399 and PHO2 by systemic signaling. Plant Physiol 147:732–746

Liu TY, Chang CY, Chiou TJ (2009) The long-distance signaling of mineral macronutrients. Curr Opin Plant Biol 12:312–319

Lopez-Bucio J, Cruz-Ramirez A, Herrera-Estrella L (2003) The role of nutrient availability in regulating root architecture. Curr Opin Plant Biol 6:280–287

Mallory AC, Dugas DV, Bartel DP et al (2004) MicroRNA regulation of NAC-domain targets is required for proper formation and separation of adjacent embryonic, vegetative, and floral organs. Curr Biol 14:1035–1046

Mallory AC, Bartel DP, Bartel B (2005) MicroRNA-directed regulation of *Arabidopsis* AUXIN RESPONSE FACTOR17 is essential for proper development and modulates expression of early auxin response genes. Plant Cell 17:1360–1375

Martin AC, del Pozo JC, Iglesias J et al (2000) Influence of cytokinins on the expression of phosphate starvation responsive genes in *Arabidopsis*. Plant J 24:559–567

Meng Y, Huang F, Shi Q et al (2009) Genome-wide survey of rice microRNAs and microRNA-target pairs in the root of a novel auxin-resistant mutant. Planta 230:883–898

Meng Y, Gou L, Chen D et al (2010a) PmiRKB: a plant microRNA knowledge base. Nucleic Acids Res. doi:10.1093/nar/gkq721

Meng Y, Ma X, Chen D et al (2010b) MicroRNA-mediated signaling involved in plant root development. Biochem Biophys Res Commun 393:345–349

Monshausen GB, Gilroy S (2009) The exploring root – root growth responses to local environmental conditions. Curr Opin Plant Biol 12:766–772

Montiel G, Gantet P, Jay-Allemand C et al (2004) Transcription factor networks. Pathways to the knowledge of root development. Plant Physiol 136:3478–3485

Morozova O, Marra MA (2008) Applications of next-generation sequencing technologies in functional genomics. Genomics 92:255–264

Nakano M, Nobuta K, Vemaraju K et al (2006) Plant MPSS databases: signature-based transcriptional resources for analyses of mRNA and small RNA. Nucleic Acids Res 34:D731–D735

Oka-Kira E, Kawaguchi M (2006) Long-distance signaling to control root nodule number. Curr Opin Plant Biol 9:496–502

Osmont KS, Sibout R, Hardtke CS (2007) Hidden branches: developments in root system architecture. Annu Rev Plant Biol 58:93–113

Pant BD, Buhtz A, Kehr J et al (2008) MicroRNA399 is a long-distance signal for the regulation of plant phosphate homeostasis. Plant J 53:731–738

Pantaleo V, Szittya G, Moxon S et al (2010) Identification of grapevine microRNAs and their targets using high-throughput sequencing and degradome analysis. Plant J 62:960–976

Peret B, De Rybel B, Casimiro I et al (2009) *Arabidopsis* lateral root development: an emerging story. Trends Plant Sci 14:399–408

Phillips JR, Dalmay T, Bartels D (2007) The role of small RNAs in abiotic stress. FEBS lett 581:3592–3597

Riano-Pachon DM, Ruzicic S, Dreyer I et al (2007) PlnTFDB: an integrative plant transcription factor database. BMC Bioinformatics 8:42

Rubio V, Linhares F, Solano R et al (2001) A conserved MYB transcription factor involved in phosphate starvation signaling both in vascular plants and in unicellular algae. Genes Dev 15:2122–2133

Ruegger M, Dewey E, Gray WM et al (1998) The TIR1 protein of *Arabidopsis* functions in auxin response and is related to human SKP2 and yeast grr1p. Genes Dev 12:198–207

Samaj J, Baluska F, Menzel D (2004) New signalling molecules regulating root hair tip growth. Trends Plant Sci 9:217–220

Scheres B (1997) Cell signaling in root development. Curr Opin Genet Dev 7:501–506

Scheres B (2010) Developmental biology: roots respond to an inner calling. Nature 465:299–300

Schiefelbein JW (2000) Constructing a plant cell. The genetic control of root hair development. Plant Physiol 124:1525–1531

Shahmuradov IA, Gammerman AJ, Hancock JM et al (2003) PlantProm: a database of plant promoter sequences. Nucleic Acids Res 31:114–117

Simon SA, Meyers BC, Sherrier DJ (2009a) MicroRNAs in the rhizobia legume symbiosis. Plant Physiol 151:1002–1008

Simon SA, Zhai J, Nandety RS et al (2009b) Short-read sequencing technologies for transcriptional analyses. Annu Rev Plant Biol 60:305–333

Stougaard J (2001) Genetics and genomics of root symbiosis. Curr Opin Plant Biol 4:328–335

Subramanian S, Fu Y, Sunkar R et al (2008) Novel and nodulation-regulated microRNAs in soybean roots. BMC Genomics 9:160

Sunkar R, Zhu JK (2004) Novel and stress-regulated microRNAs and other small RNAs from Arabidopsis. Plant Cell 16:2001–2019

Sunkar R, Girke T, Jain PK et al (2005) Cloning and characterization of microRNAs from rice. Plant Cell 17:1397–1411

Sunkar R, Kapoor A, Zhu JK (2006) Posttranscriptional induction of two Cu/Zn superoxide dismutase genes in *Arabidopsis* is mediated by downregulation of miR398 and important for oxidative stress tolerance. Plant Cell 18:2051–2065

Valdes-Lopez O, Arenas-Huertero C, Ramirez M et al (2008) Essential role of MYB transcription factor: PvPHR1 and microRNA: PvmiR399 in phosphorus-deficiency signalling in common bean roots. Plant Cell Environ 31:1834–1843

Vidal EA, Araus V, Lu C et al (2010) Nitrate-responsive miR393/AFB3 regulatory module controls root system architecture in *Arabidopsis thaliana*. Proc Natl Acad Sci USA 107: 4477–4482

Voinnet O (2009) Origin, biogenesis, and activity of plant microRNAs. Cell 136:669–687

Wang JW, Wang LJ, Mao YB et al (2005) Control of root cap formation by microRNA-targeted auxin response factors in *Arabidopsis*. Plant Cell 17:2204–2216

Wang Y, Li P, Cao X et al (2009) Identification and expression analysis of miRNAs from nitrogen-fixing soybean nodules. Biochem Biophys Res Commun 378:799–803

Williams L, Carles CC, Osmont KS et al (2005) A database analysis method identifies an endogenous *trans*-acting short-interfering RNA that targets the *Arabidopsis* ARF2, ARF3, and ARF4 genes. Proc Natl Acad Sci USA 102:9703–9708

Xie Q, Frugis G, Colgan D et al (2000) *Arabidopsis* NAC1 transduces auxin signal downstream of TIR1 to promote lateral root development. Genes Dev 14:3024–3036

Yamasaki H, Abdel-Ghany SE, Cohu CM et al (2007) Regulation of copper homeostasis by microRNA in *Arabidopsis*. J Biol Chem 282:16369–16378

Yang JH, Han SJ, Yoon EK et al (2006) Evidence of an auxin signal pathway, microRNA167-ARF8-GH3, and its response to exogenous auxin in cultured rice cells. Nucleic Acids Res 34:1892–1899

Yoon EK, Yang JH, Lim J et al (2009) Auxin regulation of the microRNA390-dependent transacting small interfering RNA pathway in Arabidopsis lateral root development. Nucleic Acids Res. doi:10.1093/nar/gkp1128

Zhang Y (2005) miRU: an automated plant miRNA target prediction server. Nucleic Acids Res 33:W701–W704

Zhang B, Pan X, Anderson TA (2006) Identification of 188 conserved maize microRNAs and their targets. FEBS Lett 580:3753–3762

Zhang B, Wang Q, Wang K et al (2007) Identification of cotton microRNAs and their targets. Gene 397:26–37

Zhang Z, Wei L, Zou X et al (2008) Submergence-responsive microRNAs are potentially involved in the regulation of morphological and metabolic adaptations in maize root cells. Ann Bot 102:509–519

Zhang W, Luo Y, Gong X et al (2009) Computational identification of 48 potato microRNAs and their targets. Comput Biol Chem 33:84–93

Zhao B, Liang R, Ge L et al (2007) Identification of drought-induced microRNAs in rice. Biochem Biophys Res Commun 354:585–590

Computational Identification of MicroRNAs and Their Targets in Wheat (*Triticum aestivum* L.)

Zhu Benzhong

Contents

Abstract MicroRNAs (miRNAs) are non coding small RNAs that regulate gene expression by translational repression or transcript degradation. A large number of miRNAs have been identified from model plant species; however, the character of conserved miRNAs is poorly understood. Conserved miRNAs in wheat are identified using ESTs (Expressed Sequence Tags) and GSS analysis. All previously known miRNAs in other plant species were blasted against wheat EST and GSS sequences to select novel miRNAs in wheat by a series of filtering criteria. From total of 37 conserved miRNAs belonging to 18 miRNA families, 10 conserved miRNAs comprising 4 families were reported in wheat. MiR395 is found to be a special family, as three members belonging to the same miR395 family are clustered together. MiRNA targets are transcription factors involved in wheat growth and development, metabolism, and stress responses.

Keywords EST • GSS • MicroRNAs • Target genes • Wheat

Z. Benzhong (✉)
Laboratory of Fruit Biotechnology, College of Food Science and Nutritional Engineering, China Agricultural University, Beijing 100083, China
e-mail: cauzbz@yahoo.com.cn

V.A. Erdmann and J. Barciszewski (eds.), *Non Coding RNAs in Plants*, RNA Technologies, DOI 10.1007/978-3-642-19454-2_18,

1 Introduction

MicroRNAs (miRNAs) are a class of small and non coding RNAs that regulate gene expression at the posttranscriptional stage by degrading target mRNAs or by repressing target gene translation in a wide range of eukaryotes, such as *Arabidopsis thaliana*, *Caenorhabditis elegans*, mice, and human beings (Bartel and Bartel 2003). They are found in intronic, exonic, and intergenic regions and initially transcribed as long, capped, and polyadenylated primary-miRNA (pri-miRNA), mostly by RNA polymerase II (Ambros and Lee 2004), but some by RNA polymerase III (Borchert et al. 2006). Pri-miRNAs are first processed into *70-nt pre-miRNAs with hairpin structures by DCL1 and subsequently cleaved into miRNA (Bartel 2004). The active miRNA strand of miRNA:miRNA* duplex are incorporated into the RNA induced silencing complex (RISC) to exert their function through perfect complementarity to mRNAs of target gene (Sempere et al. 2004). The lengths of mature miRNAs reported so far vary from 17 to 29 nucleotides (nt), and the majority of miRNAs are about 21–25 nts in length. Most miRNAs have the typical hairpin structure. In general, miRNAs are highly conserved not only among closely related species but also among different species (Arteaga-Vazquez et al. 2006).

Computational prediction is a powerful method for rapid and large-scale identification of miRNAs from different plants. However, because of the distribution of miRNAs predominantly in the intergenic regions or introns of coding genes, prediction of miRNAs by computational methods requires detailed genome sequence information. They are only used in model plants such as rice, *A. thaliana*, and tobacco. For many other important plants such as common wheat (*Triticum aestivum* L.), because sequencing of wheat genome is not finished now, it is difficult for us to predict their miRNAs by computational methods. The new approach has been developed to identify conserved miRNAs by analyzing the expressed sequence tag (EST), which is based on the conservation of botanical miRNAs. In this method, miRNAs from the model plant species as *Arabidopsis* and rice were used to search against EST sequences of other new plant species through a series of criteria, including the numbers of mismatches, and hairpin structure. By this method, Zhang et al. (2005) found a lot of conserved miRNAs in various plant species, such as cotton and maize (Zhang et al. 2006a–c; Qiu et al. 2007), but not in wheat. Zhang et al. (2005) paper indicates that the limited genome sequences as EST can be used to identify conserved miRNAs in plants.

Wheat (*T. aestivum* L.) is one of the most extensively grown crops throughout the world, occupying 17% of all cultivated land and providing approximately 55% of the world's carbohydrates (Yin and Shen 2010). There are four papers to report miRNAs in wheat by different methods such as EST analysis (Zhang et al. 2005; Jin et al. 2008), direct cloning from a small RNA library (Yao et al. 2007), and data mining (Dryanova et al. 2008). Few miRNAs in wheat were identified, and most of them were conserved (Zhang et al. 2005; Jin et al. 2008) because there are fewer conserved miRNAs in 2005 and restriction of query miRNAs in *Arabidopsis* and rice. Dryanova et al. (2008) found a lot of miRNAs in wheat by data mining. Yao et al. (2007) constructed a small wheat RNA library to clone miRNAs and identified

58 miRNAs belonging to 43 families, of which 23 are wheat-specific. However, these methods have not sufficiently identified miRNAs in wheat. Here, we used all miRNAs registed in the miRNA Registry Database publicly available at http://miRNA.sanger.ac.ck (Release 10.0, November 2007) (Griffiths-Jones et al. 2006), including newly identified and species-specific miRNAs to search against wheat EST and GSS sequences. Ten conserved miRNAs and 27 previously reported miRNAs were found in wheat. We used potential miRNAs to predict their targets genes in wheat and found 361 target genes encoding transcription factors, enzymes implicated in metabolic processes and in stress responses.

2 Results and Discussion

2.1 Computational Approaches to miRNA Gene Finding in Plants

Filter-based approaches. One of the first methods for identifying miRNAs in plants is described in Wang et al. (2004). The authors proposed a workflow that began by identifying all potential hairpins in the intergenic regions of *A. thaliana*. The hairpins were found by looking for imperfect inverted repeats of 21 nt, representing the putative mature miRNA and corresponding star sequence, which were separated by a distance within a given window. The candidate hairpins were then filtered according to criteria concerning GC content and loop length. The putative miRNA sequences were checked against the rice genome and only those showing high conservation were retained. Finally, the remaining precursor candidates and their orthologues were folded to validate the characteristic stem-loop secondary structure.

Target-centered approaches. A single-genome approach called FINDMIRNA (Adai et al. 2005) replaced the sieve of cross-species conservation of candidate stem-loops with the detection of potential targets within transcripts of the same species. The algorithm starts by indexing all the 7-mers of the intergenic regions, excluding repeats and low GC-content sequences. For each transcript, its overlapping 7-mers are tentatively matched against the index previously computed. For each match, an uncapped alignment of the surrounding areas is produced. The best length-normalized alignment score of size 18–25 is marked as a potential miRNA. If the score is above a given threshold, a dynamic programming algorithm is used to search for a complementary sequence in the vicinity. A secondary structure prediction algorithm is used to verify the presence of a stem-loop structure and whether the length-normalized MFE is below a given threshold. An additional filter is then used for higher specificity, which exploits the expected typical divergence pattern of miRNA precursors of the same family, whose members have presumably arisen by duplication events. Precursor candidates are put in the same family cluster if they target the same transcript region. Clusters are then scored according to the degree of conservation of the miRNA, miRNA_ and intervening sequence, using a scoring function that privileges conservation of the miRNA sequence and penalizes conservation in the intervening region.

A similar approach described in Lindow and Krogh (2005), unlike the previous method, does not require that miRNAs be clustered into families. This method takes each mRNA and a genome-wide search is performed to identify regions of 20–27 nt that match a portion of the mRNA with at most two mismatches. These matches, termed micromatches, are then used to identify miRNA candidates. The candidates are passed through six filters: (1) high sequence complexity, (2) no overlap with annotated exons, (3) no overlap with repeat sequences, (4) stable miRNA:mRNA duplex, (5) no more than ten identical copies in the genome, and (6) the putative miRNA is contained in a stable precursor stem-loop structure exhibiting some typical features. An additional sieve is then added that includes only miRNA candidates with more than one target, which is thought to be typical of most plant miRNAs.

Homology-based searches. Upon the identification of an ever-increasing number of plant miRNA genes in several species, homology-based search methods begun to be developed, seeking the complete enumeration of miRNAs in model organisms (Li et al. 2005; Dezulian et al. 2006). In general terms, these methods first identify genome hits matching known miRNA mature sequences and then extract the genomic context of such hits and align the candidates with their putative miRNA families followed by the application of some criteria to determine a final list of candidate homologues. More recently, these protocols have been adapted to search for new miRNAs by analyzing EST data (Zhang et al. 2006a–c).

Other approaches. Other methods for plant miRNA gene identification have been developed using a combination of high-throughput sequencing, filtering, and machine learning approaches in similar ways to those discussed for animal miRNA prediction (Sunkar et al. 2008).

2.2 Previous Identified Potential miRNAs

In order to identify conserved miRNAs in wheat, according to their botanical conservation botanical conservation, Han et al. performed the EST and GSS analysis to predict wheat miRNAs and to compare them with other plant species (Han et al. 2009). They found 37 in their study, including 27 miRNAs previously reported (Zhang et al. 2005; Jin et al. 2008; Yao et al. 2007; Dryanova et al. 2008) and ten novel conserved miRNAs in wheat and then 27 conserved miRNAs are grouped into 15 miRNA families (Table 1), with the numbers of each member in individual families being different. There have not got any information about MiR165, miR168, miR390, miR393, miR397, and miR479 by EST and GSS analysis, though they were cloned from a small RNA library by Yao et al. (2007), because of the limited numbers of wheat EST and GSS sequences. Among the 37 miRNAs, only three members were obtained from the GSS database, namely, miR166, miR169*, and miR319a. Others are all from EST database.

Jin et al. (2008) used the following method to search the miRNA in wheat. The first, 613,015 wheat EST sequences were downloaded from the GenBank database. Second, the conserved miRNAs from other plants such rice, maize, and *A. thaliana*

Table 1 Previously reported conserved miRNAs in wheat identified in our study (Han et al. 2009)

miRNAs	Sequence	EST or GSS	LP	Location	(A+U)%	MFEs	MEFIs
miR156a	UGACAGAAGAGAGAGAGCAC	CD454302	306	3′	46.7	115.6	0.91
miR156b	UGACAGAAGAGAGAGAGCAU	BJ319553	52	5′	51.92	42.30	0.85
miR156c	CGACAGAGGAAAGAGAGCAC	CK205751	42	3′	59.52	31.20	0.87
miR159a	UUUGGAUUGAAGGGAGCUCUG	CJ898977	173	3′	46.82	80.70	0.99
miR159b	UUUGGAUUGAAGGGAGCUCUG	CA484819	252	3′	53.57	103.2	0.87
miR160	UGCCUGGCUCCCUGUAUGCCA	CJ641547	50	5′	66	51.80	1.35
miR164a	UGGAGAAGCAGGGCACGUGCA	CA704421	142	5′	33.09	92.50	0.97
miR164b	UGGAGAAGCAGGUCACGUGCG	CD899685	60	5′	35.89	25.0	0.84
miR166	CCGGACCAGGCUUCAAUUUGC	EI674180	58	3′	44.83	31.30	0.86
miR167a	UGAAGCUGCCAGCAUGAUCUA	CK209908	60	5′	48.33	28.70	0.92
miR167b	UGAAGCUGCCAGCAUGAUCUA	CJ833771	90	5′	46.67	35.90	0.91
miR167c	UGAAGCUGACAGCAUGAUCUA	CK209889	60	5′	50	33.30	1.11
miR169*	UUCAGCAAGUUGACCUUGGCA	EI673529	46	3′	43.47	41.20	1.29
miR171	UGAUUGAGCCGUGCCAAUAUC	CD910903	94	3′	37.24	44.98	0.87
mIR172a	AGAAUCCUGAUGAUGCUGCAA	CA694228	46	3′	43.47	25.80	0.96
miR172b	AGAAUCCUGAUGAUGCUGCCU	CJ942609	90	3′	38.89	32.20	0.85
miR319a	UUGGACUGAAGGGUGCUCCCG	CV774688	47	3′	39.78	22.80	0.79
miR319b	UUGGACUGAAGGGAGCUCCCU	CZ889526	198	3′	48.99	97.40	0.94
miR395	GUGAAGUGUUUGGGGGAACUC	CV763592	66	3′	54.55	34.90	1.16
miR396	UCCACAGGCUUUCUUGAACUG	CJ776495	137	5′	45.26	59.03	0.78
miR398	UGUGUUCUCAGGUCGCCCCCG	CJ711035	86	3′	44.19	44.30	0.93
miR399a	UGCCAAAGGAGAGUUGCCCUG	CJ667854	79	3′	41.77	44.90	1.13
miR399b	UGCCAAAGGAGAAUUGCCCUG	CJ666653	114	3′	40.35	58.18	0.85
miR408a	CUGCACUGCCUCUUCCCUGGC	BE419354	113	3′	47.79	40.23	0.92
miR408b	CUGCACUGCCUCUGCCCUGGC	CD907090	115	3′	44.35	40.56	0.88
miR444a	UUGCUGCCUCAAGCUUGCUGC	CK200584	151	3′	50.33	72.69	0.96
miR444b	UUGCUGUCUCAAGCUUGCUGA	DR738640	113	3′	54.87	56.10	1.10

Underlined letters display the mutation sites with their query sequences

LP length of precursors

were used to analyse the EST sequences. There have two major procedures to be included in this method: searching premiRNA-like sequences and identifying pre-miRNAs. The first step was to search for potential hairpin structures in the wheat EST sequences, which yielded 551,129 qualified sequences. The secondary structures of the wheat ESTs were predicted using the RNA fold program (Hofacker 2003). According to their nucleotide composition and free energy of the secondary structure, the second step was to further search miRNA precursor-like sequences and yielded 129,957 miRNA precursor-like sequences. The third step was to remove repeat elements and protein-coding sequences with the BLASTN and BLASTX programs, which yielded 5,834 sequences. The fourth step was to apply Genomics to identify pre-miRNAs in the precursor-like sequences. Consequently, 79 pre-miRNA candidates were obtained. As a part of the paper, they randomly selected 22 candidates from 79 predicted miRNAs for experimental verification. The result show that there are nine novel miRNA genes confirmed. Therefore, the prediction accuracy is about 40% (Fig. 1, Table 1).

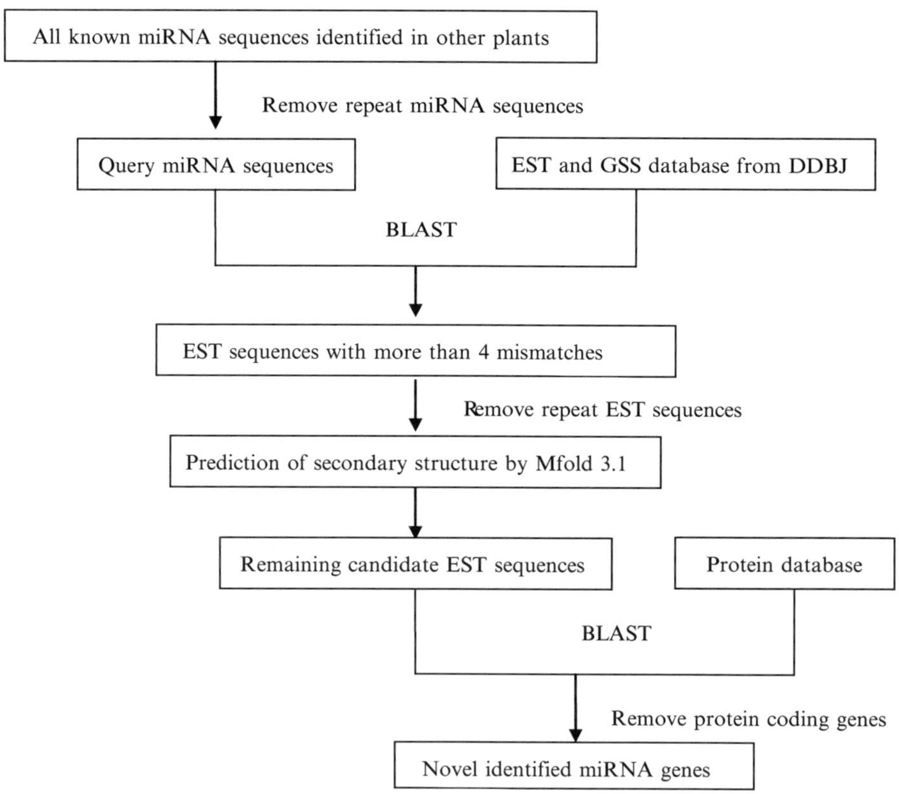

Fig. 1 Procedure of miRNA identification in wheat by EST and GSS analysis (Han et al. 2009)

2.3 Novel Conserved miRNAs Identified by EST and GSS Analysis in Plant

For Han et al. (2009) paper, apart from the 27 previously reported conserved miRNAs (Jin et al. 2008; Yao et al. 2007; Zhang et al. 2005), all miRNAs deposited in the miRNA Registry Database from other plant species including *Arabidopsis*, rice, and maize were used to search for other conserved miRNAs in wheat. As a result, ten conserved miRNAs, including four miRNA families (Table 2), were found. To confirm that the newly identified small RNAs were miRNAs from other small RNAs such as siRNAs, their salient characteristics, e.g., the A+U content, the miRNA precursor length, the minimal folding free energy index (MFEI) for each miRNA precursor, all of which are the primary main traits for differentiating miRNAs from other small RNAs.

All of the precursors for those mature miRNAs fold near ideal hairpin structures (Figs. 2 and 4). The lengths of the precursors vary in a larger range from 49 to 252 nt compared to their counterparts in animals (Fig. 3). This phenomenon is similar to other plant species, such as *Arabidopsis* and rice. A previous report has demonstrated that miRNA precursors contain more A+U than other small RNAs (Zhang et al. 2006a–c). The results showed that the contents of A+U of miRNA precursors in wheat are also higher, with an average of 58.33%. Although mature miRNAs are equally distributed at each arm of hairpin structures, only miR395a and miR395c are located at the 5′ end of their hairpin structures, with the remaining eight located at the 3′ ends. In view of the 37 identified miRNAs in wheat, only ten mature miRNAs are located at the 5′ end, but its remains unknown whether or not this phenomenon is wheat-specific. Another significant trait for distinguishing miRNAs from other RNAs is the MFE index (MFEI). The average MFEI of those novel conserved miRNAs is 1.11, which is considerably greater than that of other RNAs such as rRNAs (0.59) and tRNAs (0.64) (Meyers et al. 2008). All of these characteristics indicate that these ten novel small RNAs are probably miRNAs. The distribution of newly identified wheat miRNAs is similar to their counterparts in other plant species. The largest group, miR818s, has five members, miR395s have four members, and miR156 and miR167 each has three members. Seven miRNA families contain two members, i.e., miR159, miR164, miR172, miR319, miR399, miR408, and miR444, but only one member is contained in each of the other families.

2.4 The Cluster of Three Members of Wheat miR395 in Wheat

Although it is a universal characteristic for animal miRNAs belonging to the same family to cluster together in one polycistronic mRNA (Yu et al. 2006), only some examples of clustered miRNAs have been reported in plants (Talmor-Neiman et al. 2006; Lu et al. 2007). MiR395 is the most frequently observed clustered family in plants such as *Arabidopsis* and rice, which is most likely to be animal miRNA.

Table 2 Novel conserved miRNAs identified in wheat (Han et al. 2009)

miRNAs	Sequence	EST	LP (nt)	Location	(A+U)%	MFEs	MEFIs
miR395a	CUGAAGUGUUUGGGGGAACUC	CK194045	162	5′	56.25	57.31	0.81
miR395b	AUGAAGUGUUUGGGGGAACUC	CK194045	252	3′	57.94	96.61	0.91
miR395c	GUGAAGUGUUUGGGGGAACUC	CK194045	123	5′	50.41	49.60	0.83
miR414	UCAUCAUCAUCAUCAUCGUCA	CJ967189	49	3′	61.22	13.60	0.72
miR818a	AACGUCUUAUAUUAUGGGACGG	EB513609	69	3′	68.12	27.1	1.23
miR818b	ACAUUCUUAUAUUAUGGGACGG	DR739383	72	3′	63.89	28.50	1.10
miR818c	AUGUUCUUAUAUUAUGGGACGG	AL816538	69	3′	63.77	45.80	1.83
miR818d	ACGCUCUUAUAUUAUGGGACGG	CN010552	69	3′	62.33	41.40	1.59
miR818e	AACGUCUUAUAUUAUGGGACGG	CK210390	68	3′	64.71	33.70	1.40
miR835	UUCUUGCAUAUGUUCUUUAAC	CK171746	124	3′	34.68	58.40	0.72

Underlined letters display the mutation sites with their query sequences
LP length of precursors; *MFE* minimal folding free energy; *MFEI* minimal folding free energy indexes

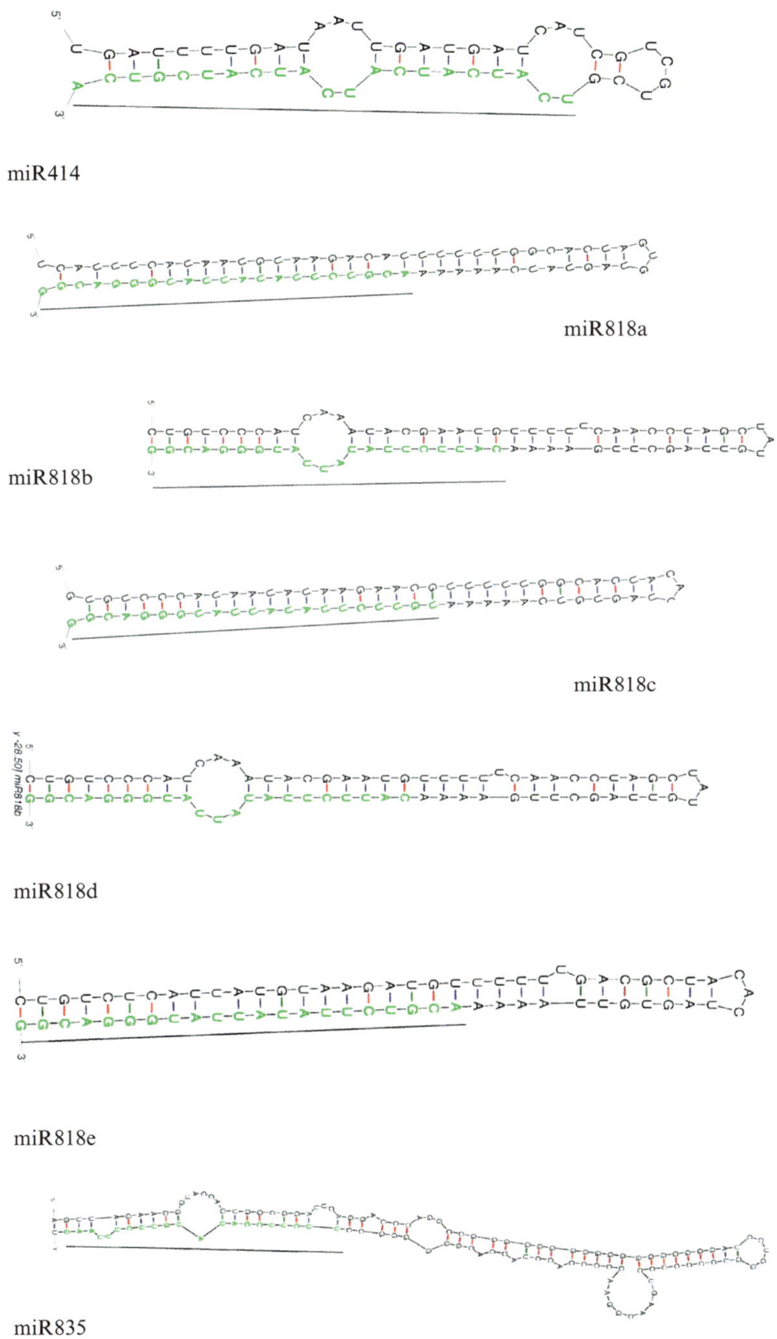

Fig. 2 The predicted secondary stem-loop structures of newly identified wheat miRNAs. The mature miRNA sequences are *underlined* (Han et al. 2009)

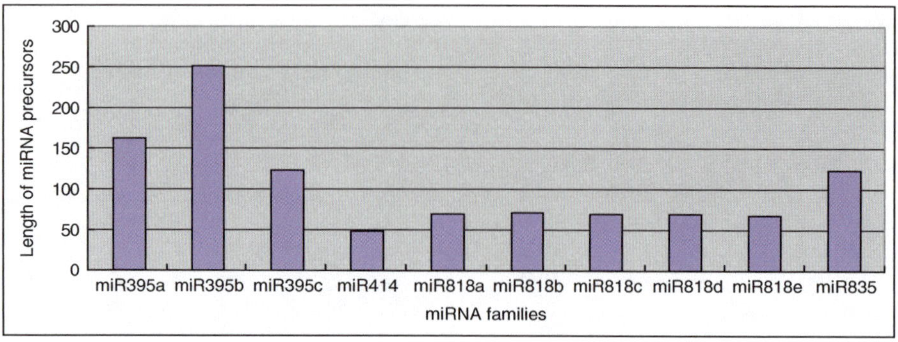

Fig. 3 Diversity of lengths of newly identified wheat miRNA precursors (Han et al. 2009)

Three members of miR395s were observed to be clustered together in *Arabidopsis*. Another example is miR1219 (miR1219a and miR1219b, separated by a distance of approximately 200 bp), also clustered in moss (Lu et al. 2007). Two members of miR950s (miR950a and miR950b) were also found clustered in a single contig (Talmor-Neiman et al. 2006). In Han et al. (2009) research, miR395s (miR395a, b, and c) were also found to be clustered in one wheat EST sequence (CK194045, Fig. 4). MiR395a and miR395b share the same precursor, but at opposite directions, and miR395c has its own precursor. The complex organization of miR395 in wheat may reflect the complicated mechanism of botanical miRNA maturation.

Han et al. (2009) used these newly identified wheat miRNAs to compare their conservation with their counterparts in other plant species. The results suffused that mature miR395 is highly conserved among plant species, though it is more likely to mutate at specific sites and the site at 5′ end is the point that mutated most frequently (Fig. 5). The highly conserved sites may be essential for recognizing their target sites known as key sites.

2.5 *Predicted Targets of miRNAs in Wheat*

Previous studies have predicted the targets of several miRNA families and confirmed their roles in plant model species, especially *Arabidopsis*. The functions have been shown to be involved in organ development processes such as floral organ identity, leaf morphogenesis, root development, various stress responses, and signal transduction (Allen et al. 2005; Guo et al. 2005; Mallory and Bartel 2005; Palatnik et al. 2003).

Han et al. (2009) used identified miRNAs in wheat to search against wheat mRNAs by miRU for potential targets of wheat miRNAs and identified 361 potential targets for 37 wheat miRNAs (data not shown). Similar to previous reports, the major potential target genes refer to various types of wheat development and are grouped

Fig. 4 Cluster of miR395 in wheat. Three members of the mi395 family, mir395a, miR395b, and miR295c, are clustered together in a single EST sequence CK194045 comparable like animal miRNAs (Han et al. 2009)

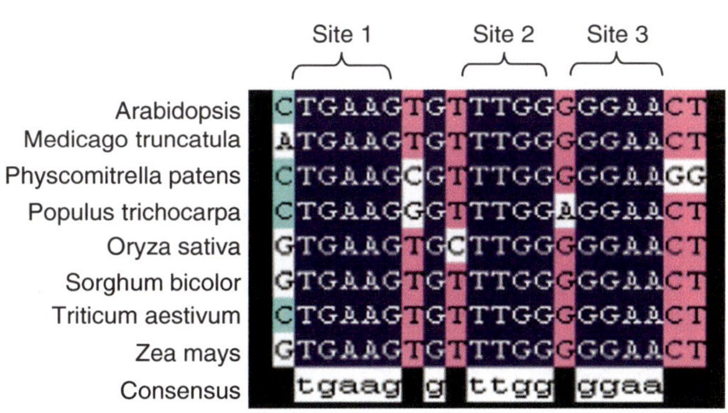

Fig. 5 Comparison of miR395a in wheat with its counterparts in other plant species. Three domains are highly conserved among these plant species, indicating the conserved functions of botanical miR395 (Han et al. 2009)

into several gene families according to their functions (Table 3). Wheat miRNAs preferred to target the transcription factors involved in wheat development, consistent with the miRNA tendency in other plant species, which comprise a large part of those target genes. Another target gene family is involved in wheat stress responses, which may greatly influence the quantity and quality of wheat production. A huge family containing 120 target genes for miR414 was first reported in wheat in our study (data not shown), with many lacking functional descriptions.

Jin et al. (2008) used nine miRNAs to search for wheat ESTs to determine potential regulatory targets. In the paper, the screening criterion was that there were no

Table 3 Major potential target genes for newly identified miRNAs in wheat (Han et al. 2009)

miRNAs	Targeted proteins	Functions	Target genes
miR159	MYB33	Transcription factor	TC194473
	Cell cycle-associated protein Mobl-like protein	Development	TC224747
	Nucleotide-binding leucine-rich-repeat protein 1	Metabolism	TC229221
	Type I topoisomerase	Metabolism	TC226217
miR160	ARF10	Transcription factor	TC220268, TC223417
	UDP-glucose:sinapate glucosyltransferase	Metabolism	TC215559
miR164	NAC1	Transcription factor	TC224059, TC224062
	NAC2	Transcription factor	TC224408,
	MAP kinase 2	Metabolism	CA681504, TC203027
	Homeodomain leucine zipper protein	Transcription factor	TC213254, TC214742
miR167	ARF8	Transcription factor	TC195542
miR169	Calcium-proton antiporter	Metabolism	TC215924
miR172	ORF2	Transcription factor	CA424651
	APETALA2-like protein	Transcription factor	TC221786, TC209305, TC209306
miR319	PCF8	Transcription factor	TC199472
	PCF6	Transcription factor	TC230286
miR395	SAC domain protein 2	Transcription factor	TC220251
	Poly(A) polymerase	Metabolism	TC189870
	Plastidic ATP sulfurylase	Metabolism	TC191731
miR396	Growth-regulating factor	Transcription factor	CK209519
miR414	Nucleosome assembly protein I	Metabolism	TC191077
	IIA large subunit (TFIIA-L1)	Transcription factor	TC194829
	DNA-binding protein-like protein	Transcription factor	CD891478
	A differentially expressed in relation to the extent of cell elongation	Development	CK158764
	Nuclear polyadenylated RNA-binding protein NAB3	Development	TC189588
miR818	ATPase subunit 6	Metabolism	TC219270
	Pathogenesis-related protein 4	Stress response	CD936868, TC213259

more than four mismatches in the complementary region between miRNAs and their miRNA targets. Any gap and G:U as well as other noncanonical pairs were not allowed in the complementary region and were considered as mismatches according to the description in the methods section. Finally, Jin et al. found 59 target ESTs for the nine miRNAs (data not shown). There were 34 target ESTs encoding functional proteins and another 25 target ESTs coding unknown proteins.

In Xin et al. (2010) study, by using Solexa high-throughput sequencing, they identified a diverse set of wheat small RNAs that are responsive to powdery mildew infection and heat stress. A total of 51 known conserved miRNAs and 81 new identified miRNAs were obtained (Xin et al. 2010).

2.6 Functions of Wheat miRNAs

The potential target genes of wheat miRNAs are involved in a diverse range of roles in wheat development, and responses to various environmental stresses. A large part of miRNAs target genes consist of transcription factors in model plant species, e.g., *Arabidopsis* (Lee et al. 2004). Wheat miR159, also known as miR-JAW, was predicted to primarily target MYB transcription factors (*MYB33*), which control leaf development, though it may also target genes in other organ development processes. By overexpression of the miR159-resistant version of *MYB33*, transgenic plants exhibit curled leaves (Palatnik et al. 2003), indicating that controlling of *MYB33* by miR159 is necessary for healthy leaf development. Previous research has indicated that several auxin response factors (ARFs) were involved in plant development and growth, containing potential miRNAs sites, such as ARF10, ARF16, and ARF17 for miR160 (Rhoades et al. 2002), and ARF6 and ARF8 for miR167 (Bartel and Bartel 2003). In wheat, only ARF10 is predicted to be targeted by miR160, and ARF8 by miR167. Mallory et al. (2005) showed the process of this miR160-directed ARF17 degradation by expressing a miR160-resistant version of ARF17 in *Arabidopsis*, which displayed increased ARF17 mRNA accumulation and root growth defects. MiR172 was predicted to target two *APETALA2* (*APE2*)-like proteins in wheat. *APE2* families belong to a class A gene in the ABC model of floral organ development. Transgenic plants with overexpression of miR172-resistant version of APE2 display early flowering (Chen 2004). In addition to APE2, miR172 also targets several APE2-like genes, such as *TARGET OF EAT1* (*TOE1*), *TOE2*, *TOE3* in *A. thaliana*, and *INDETERMINATE SPIKE-LET1* and GLOY15 in *Zea mays* (Aukerman and Sakai 2003; Park et al. 2002). We believed that these two APE2-like genes were potential genuine target genes for miR172 in wheat. Two members of *NAC*-domain transcription factors (*NAC1* and *NAC2*) were predicted to be targeted by wheat miR164, and overexpression of miR164 driven by 35S promoter led *A. thaliana* to exhibit floral organ fusion and cotyledon fusion (Mallory et al. 2004), suggesting that miR164 is quite essential for wheat development.

In addition to targeting transcription factors, wheat miRNAs also prefer to bind target genes coding proteins involved in metabolism and stress responses.

For example, a protein transport protein SEC61 beta subunit is predicted to be targeted by miR818, which still may target several pathogenesis-related proteins.

In Xin et al. (2010) study, they found that many of these wheat miRNAs showed differential expression in response to powdery mildew infection and heat stress. In addition, 149 genes were predicted as potential targets for novel wheat miRNAs, which included transcription factors implicated in development as well as genes involved in other physiological processes, such as stress responses (Xin et al. 2010).

References

Adai A, Johnson C, Mlotshwa S et al (2005) Computational prediction of miRNAs in *Arabidopsis thaliana*. Genome Res 15:78–91

Allen E, Xie Z, Gustafson AM (2005) MicroRNA-directed phrasing during trans-acting siRNA biogenesis in plants. Cell 121:207–221

Ambros V, Lee RC (2004) Identification of microRNAs and other tiny noncoding RNAs by cDNA cloning. Meth Mol Biol 265:131–158

Arteaga-Vazquez M, Caballero-Perez J, Vielle-Calzada JP (2006) A family of microRNAs present in plants and animals. Plant Cell 18:3355–3369

Aukerman MJ, Sakai H (2003) Regulation of flowering time and floral organ identity by a microRNA and its APETALA2-like target genes. Plant Cell 15:2730–2741

Bartel DP (2004) MicroRNAs: genomics, biogenesis, mechanism, and function. Cell 116: 281–297

Bartel B, Bartel DP (2003) MicroRNAs: at the root of plant development. Plant Physiol 132: 709–717

Borchert GM, Lanier W, Davidson BL (2006) RNA polymerase III transcribes human microRNAs. Nat Struct Mol Biol 13:1097–1101

Chen XA (2004) MicroRNA as a translational repressor of APETALA2 in *Arabidopsis* flower development. Science 303:2022–2025

Dezulian T, Remmert M, Palatnik JF et al (2006) Identification of plant microRNA homologs. Bioinformatics 22:359–360

Dryanova A, Zakharov A, Gulick PJ (2008) Data mining for miRNAs and their targets in the Triticeae. Genome 51:433–443

Griffiths-Jones S, Grocock RJ, van Dongen S (2006) miRBase: microRNA sequences, targets and gene nomenclature. Nucleic Acids Res 34:D140–D144

Guo HS, Xie Q, Fei JF et al (2005) MicroRNA directs mRNA cleavage of the transcription factor NAC1 to downregulate auxin signals for *Arabidopsis* lateral root development. Plant Cell 17:1376–1386

Han YS, Luan FL, Zhu HL et al (2009) Computational identification of microRNAs and their targets in wheat (*Triticum aestivum* L.). Sci China C Life Sci 52:1091–1100

Hofacker IL (2003) Vienna RNA secondary structure server. Nucleic Acids Res 31:3429–3431

Jin W, Li N, Zhang B et al (2008) Identification and verification of microRNA in wheat (*Triticum aestivum*). J Plant Res 121:351–355

Lee Y, Kim M, Han J et al (2004) MicroRNA genes are transcribed by RNA polymerase II. EMBO J 23:4051–4060

Li Y, Li W, Jin Y-X (2005) Computational identification of novel family members of microRNA genes in *Arabidopsis thaliana* and *Oryza sativa*. Acta Biochim Biophys Sin 37:75–87

Lindow M, Krogh A (2005) Computational evidence for hundreds of non-conserved plant microRNAs. BMC Genomics 6:119

Lu S, Sun YH, Amerson H et al (2007) MicroRNAs in loblolly pine (*Pinus taeda* L.) and their association with fusiform rust gall development. Plant J 51:1077–1098
Mallory AC, Bartel DP (2005) microRNA-directed regulation of *Arabidopsis* AUXIN RESPONSE FACTOR17 is essential for proper development and modulate expression of early auxin response genes. Plant Cell 17:1360–1375
Mallory AC, Dugas DV, Bartel DP et al (2004) MicroRNA regulation of NAC-domain targets is required for proper formation and separation of adjacent embryonic, vegetative, and floral organs. Curr Biol 14:1035–1046
Meyers BC, Axtell MJ, Bartel B et al (2008) Criteria for annotation of plant microRNAs. Plant Cell 20:3186–3190
Palatnik JF, Allen E, Wu X et al (2003) Control of leaf morphogenesis by microRNAs. Nature 425:257–263
Park W, Li J, Song R et al (2002) CARPEL FACTORY, a Dicer homolog, and HEN1, a novel protein, act in microRNA metabolism in *Arabidopsis thaliana*. Curr Biol 12:1484–1495
Qiu CX, Xie FL, Zhu YY et al (2007) Computational identification of microRNAs and their targets in *Gossypium hirsutum* expressed sequence tags. Gene 395:49–61
Rhoades MW, Reinhart BJ, Lim LP et al (2002) Prediction of plant microRNA targets. Cell 110:513–520
Sempere LF, Freemantle S, Pitha-Rowe I et al (2004) Expression profiling of mammalian microRNAs uncovers a subset of brain-expressed microRNAs with possible roles in murine and human neuronal differentiation. Genome Biol 5:R13
Sunkar R, Zhou X, Zheng Y et al (2008) Identification of novel and candidate miRNAs in rice by high throughput sequencing. BMC Plant Biol 8:25
Talmor-Neiman M, Stav R, Frank W et al (2006) Novel microRNAs and intermediates of microRNA biogenesis from moss. Plant J 47:25–37
Wang X-J, Reyes JL, Chuan H et al (2004) Prediction and identification of *Arabidopsis thaliana* microRNAs and their mRNA targets. Genome Biol 5:R65
Xin M, Wang Y, Yao Y et al (2010) Diverse set of microRNAs are responsive to powdery mildew infection and heat stress in wheat (*Triticum aestivum* L.). BMC Plant Biol 10:123
Yao Y, Guo G, Ni Z et al (2007) Cloning and characterization of microRNAs from wheat (*Tricitum aestivum* L.). Genome Biol 8:R96
Yin ZJ, Shen FF (2010) Identification and characterization of conserved microRNAs and their target genes in wheat (*Triticum aestivum*). Genet Mol Res 9:1186–1196
Yu J, Wang F, Yang GH et al (2006) Human microRNA clusters: genomic organization and expression profile in leukemia cell lines. Biochem Biophys Res Commun 349:59–68
Zhang BH, Pan XP, Wang QL (2005) Identification and characterization of now plant microRNAs using EST analysis. Cell Res 15:336–360
Zhang BH, Pan XP, Anderson TA (2006a) Identification of 188 conserved maize microRNAs and their targets. FEBS Lett 580:3753–3762
Zhang BH, Pan XP, Cox SB et al (2006b) Evidence that miRNAs are different from other RNAs. Cell Mol Life Sci 63:246–254
Zhang B, Pan X, Cannon CH et al (2006c) Conservation and divergence of plant microRNA genes. Plant J 46:243–259

MicroRNAs in Cotton

Guiling Sun, Qinglian Wang, and Baohong Zhang

Contents

Abstract MicroRNAs (miRNAs) are an extensive class of newly discovered small regulatory RNAs with 20–24 nucleotides in length. Since it was first discovered in plants in 2002, it has been attracting the attention of many plant scientists because of its importance in plant growth and development. However, most of our knowledge about plant miRNAs comes from the investigation of model plant species, including *Arabidopsis* and rice. Only very recently, several reports have been published on the identification and expression analysis of miRNAs in cotton. Based on the conservation and unique characteristics of miRNAs, several tens of miRNAs have been predicted by comparative genome-based EST/GSS analysis. The next-generation deep-sequencing technologies, such as Solexa, provide a new platform to identify cotton miRNAs,

B. Zhang (✉)
Department of Biology, East Carolina University, Greenville, NC 27858, USA

Henan Institute of Science and Technology, Xinxiang, Henan 453003, China
e-mail: zhangb@ecu.edu

G. Sun
Department of Biology, East Carolina University, Greenville, NC 27858, USA

Q. Wang
Henan Institute of Science and Technology, Xinxiang, Henan 453003, China

V.A. Erdmann and J. Barciszewski (eds.), *Non Coding RNAs in Plants*,
RNA Technologies, DOI 10.1007/978-3-642-19454-2_19,

including cotton-specific miRNAs. However, due to the lack of cotton genome sequence, the identification of cotton miRNAs is significantly limited. Currently, only a few cotton-specific miRNAs have been indentified, although several millions of small RNA sequences have been obtained from different cotton tissues. In silico analysis shows that miRNAs target transcription factor as well as other protein-coding genes in cotton, which control cotton development, phage change, and response to environmental stress. Several target genes were also validated for both conserved and cotton-specific miRNAs by 5′RACE. Many miRNAs are differentially expressed in cotton tissues, and several miRNAs are differentially expressed during cotton fiber differentiation and development, suggesting that miRNAs may play a role in cotton fiber development.

Keywords Comparative genome • Cotton • Deep sequencing • EST • MicroRNA

1 Introduction

Cotton is one of the most important fiber crops as well economic and oil crops around the world, which has been widely cultivated in many developed and developing countries, including USA, China, and India. The major product of cotton is fiber, which differentiates from a single epidermal cell (Basra and Malik 1984; Kim and Triplett 2001; Lee et al. 2007; Tiwari and Wilkins 1995). Cotton fiber differentiation and development is a complicated biological process, and it consists of four overlapped stages: fiber initiation, elongation, secondary wall deposition, and fiber maturation (Basra and Malik 1984; Haigler et al. 2005; Kim and Triplett 2001; Lee et al. 2007; Tiwari and Wilkins 1995). Fiber cells are usually initiated around the day of anthesis; it may start from 2 days prior to the day of anthesis (–2 DPA) and continue up to 3 days post anthesis (DPA). After it is initiated, cotton fiber cell immediately elongates, and within a very short time period, usually 10 days, the length of cotton fiber can reach 30 mm. However, many factors influence the cotton fiber elongation, and it seems that the time point of differentiation of an epidermal cell into a fiber cell is most critical for this process; the early initiated epidermal cells usually become the commercially important lint fibers; however, the later differentiated epidermal cells will stop elongation very quickly and only develop into very short fibers, called fuzz. At the later stage of elongation, the secondary cell wall starts to form by quickly biosynthesizing cellulose and its deposition into the cell wall. Cellulose biosynthesis and the secondary wall deposition constitute an amazing biological process; many investigations have demonstrated that cotton fiber is almost the purest cellulose in the world, and about 95% of the mature cotton fiber is cellulose. Thus, cotton fiber is always a great model system to investigate the cellulose biosynthesis and biofuel production. After about 25 days of quick cellulose deposition and secondary wall biosynthesis, cotton fiber enters the last stage of development, i.e., fiber maturation, which will take about up to 60 days before cotton fiber is harvested.

During cotton fiber development, particularly at the early stage of fiber differentiation, elongation, and secondary cell wall biosynthesis, a dramatic change is observed at the cellular, molecular, morphological, and physiological level. Many studies have shown the change in gene expression during cotton fiber initiation and development, and some genes have been studied for their functions during cotton fiber development (Asif et al. 2008; Chaudhary et al. 2008, 2009; FeiFei et al. 2009; Gao et al. 2007; Gou et al. 2007; He et al. 2008; Hovav et al. 2008; Huang et al. 2008a, b; Iqbal et al. 2008; Li et al. 2003, 2005a, b; Liu et al. 2009; Michailidis et al. 2009; Schwartz and Smith 2008; Taliercio and Boykin 2007; Tu et al. 2007; Wu et al. 2008; Xu et al. 2008). However, the mechanism of controlling cotton fiber differentiation and development is still unknown. Besides the protein-coding genes, recently identified microRNAs (miRNAs) may also play some roles during cotton fiber development.

2 Brief History, Biogenesis, and Functions of miRNAs

miRNAs are an extensive class of newly identified small regulatory RNAs with 20–22 nt in length (Bartel 2004). miRNAs negatively regulate gene expression at the posttranscriptional levels (Bartel 2004). The first miRNA *lin-4* was discovered about 20 years ago by two independent laboratories, which regulates *Caenorhabditis elegans* development by targeting protein-coding gene Lin-14 (Lee et al. 1993; Wightman et al. 1993). However, at that time, scientists were not aware of its importance. Ten years later, hundreds of similar small RNAs in humans, flies, and worms, and their important functions of miRNAs were first recognized (Lagos-Quintana et al. 2001; Lau et al. 2001; Lee and Ambros 2001). Since then, the investigations on miRNA identification and functional analysis become one of the hottest research topics in biological and biomedical fields.

Similar to other RNAs, miRNAs are also coded by miRNA genes. miRNA genes can be located anywhere in the genome; however, a majority of miRNA genes are located in intergenic regions in plants (Bartel 2004). A majority of miRNA genes are transcribed by RNA polymerase II. Although miRNAs are very short, the miRNA genes are pretty long. A miRNA gene is first transcribed into a long transcript, termed as primary miRNA (pri-miRNA). Then, a pri-miRNA is formed into a stem-looped hairpin structure and is sequentially cleaved into a miRNA precursor (pre-miRNA) and miRNA:miRNA* duplex by an enzyme called dicer-like 1 (*dcl 1*). Then, the miRNA:miRNA* duplex is translocated from the nucleus into the cytoplasm by HASTY, an homolog of animal Exportin 5. In the cytoplasm, the miRNA:miRNA* duplex is separated by enzyme helicase and the miRNA is introduced into the RNA-induced silence complex (RISC) for targeting the expression of a protein-coding gene and the miRNA* sequence is degraded by an unknown mechanism. During miRNA biogenesis, several other proteins, such as HEN, are also involved (Chen 2005).

There are two major mechanisms for miRNAs regulating gene expression (Bartel 2004). One is mRNA cleavage and another one is translation repression. At the early stage, many people think that mRNA cleavage is the primary mechanism for miRNA-mediated gene regulation in plants; however, there are several current investigations show that miRNA-mediated translation repression may also widely exist in plant kingdom (Brodersen et al. 2008; Yu and Wang 2010).

Enormous evidences show that miRNAs are almost involved in any biological and metabolic process in plant. The major functions of miRNA include, but not limited to controlling plant growth and development, organ initiation and morphology, developmental stage and phase change (Zhang et al. 2006c). miRNAs also respond to abiotic and biotic environmental stress, such as salinity, drought stress and pathogen infection (Zhang et al. 2006c).

Plant miRNA-related research is far behind the animal miRNA-related research. The first plant miRNAs were identified in 2002 (Rhoades et al. 2002); almost all studies were focused on model species, such as *Arabidopsis*, rice and maize at the early stage. As the application of comparative genomics and the development of deep sequencing technology, the identification and functional analysis of miRNAs has been becoming extremely exciting in the past couple of years. According to the public available miRNA database, miRBase, there are currently 2,570 plant miRNAs, from 37 plant species, identified and deposited in the database (Release 15 April 2010) (Griffiths-Jones et al. 2008). Of them, 1,363 are from dicots, 832 from monocots and 375 from lower plant species. There are currently eight plant species with more than 100 miRNAs (Table 1), which are rice (*Oryza sativa*, 447), barrel (*Medicago truncatula*, 375), black cottonwood (*Populus trichocarpa*, 234), mosses (*Physcomitrella patens*, 230), *Arabidopsis thaliana* (199), maize (*Zea mays*, 170), sorghum (*Sorghum bicolor*, 148), and wine grape (*Vitis vinifera*, 137). A total of 1,940 (75.5%) plant miRNAs were obtained from these eight plant species, suggesting that plant miRNA research is still limited to a few plant species with a majority of them being model species.

There are a total of 40 miRNAs from cotton deposited in the miRBase database, which include 34 from upland cotton (*Gossypium hirsutum*), one from the Asian cotton (*Gossypium arboreum*), one from the Arabian cotton (*Gossypium herbaceum*) and the rest four from a wild cotton species, *Gossypium raimondii*. However, much more miRNAs have been identified from cotton but have not deposited in the public available database, including the miRBase.

Table 1 Plant species with more than 100 miRNAs currently identified[a]

Plant species	Latin name		Number of miRNAs
Rice	*Oryza sativa*	Monocot	447
Barrel	*Medicago truncatula*	Monocot	375
Black cottonwood	*Populus trichocarpa*	Eudicot	234
Moss	*Physcomitrella patens*	Mosses	230
Thale cress	*Arabidopsis thaliana*	Eudicot	199
Maize	*Zea mays*	Monocot	170
Sorghum	*Sorghum bicolor*	Monocot	148
Wine grape	*Vitis vinifera*	Eudicot	137

[a]According to the microRNA database miRBase (release 15 April 2010)

3 Identification of Cotton miRNAs

Both computational and experimental approaches have been employed to identify cotton miRNAs.

3.1 Comparative Genome-Based Approach

Evidence shows that many miRNAs are highly evolutionarily conserved across all major lineages of plant species, from mosses to gymnosperms to higher flowering plants. This provides a powerful strategy to identify miRNAs from a new species using the miRNAs already known in another plant species (Zhang et al. 2006b). Zhang et al. developed an expressed sequence tag (EST)-based analysis to identify conserved miRNAs (Zhang et al. 2005). In that analysis, they first search for sequence homolog of a known miRNA in an EST database, and then they identify miRNAs based on the major characteristics of miRNAs and pre-miRNAs, which include miRNA conservation, the stem-looped hairpin structure, high negative minimal free enzyme (MFE), high minimal free enzyme index (MFEI), and the nucleotide content. Later, they expanded this approach to other nucleotide sequences, including genome survey sequence (GSS) (Pan et al. 2007). Currently, this approach has been widely adopted to identify plant miRNAs in many plant species, including maize (Zhang et al. 2006a), soybean (Zhang et al. 2008), wheat (Yin and Shen 2010), oilseed (Xie et al. 2007), switchgrass (Xie et al. 2010), tomato (Yin et al. 2008), and apple (Gleave et al. 2008).

Comparative genome-based analysis is also employed to identify miRNAs in cotton (Barozai et al. 2008; Qiu et al. 2007; Zhang et al. 2007). Although Zhang and colleagues obtained the first miRNAs in cotton, they did not systematically investigate cotton miRNAs (Zhang et al. 2005, 2006b). Later on, two independent laboratories identified 37 (belong to 20 families) and 30 miRNAs (belong to 22 families) from cotton EST and GSS databases, respectively (Qiu et al. 2007; Zhang et al. 2007). Zhang et al. not only identified miRNAs from upland cotton but also from the Asian cotton as well a wild cotton species *G. raimondii* (Zhang et al. 2007). Following these two studies, Barozai et al. also found 22 cotton miRNAs belonging to 13 families (Barozai et al. 2008). Although a similar approach was used for all these three studies, a majority of their results are difference. Of the identified miRNAs, only three miRNA families (miR156/157, miR171, and miR390) were identified in all three studies. One possible reason is that they used different nucleotide database. Barozai et al. (2008) used EST datasets and Zhang et al. (2007) used GSS datasets but Qiu et al. (2007) used both datasets. Another reason is that this difference may be caused by Blastn search. Traditional Blastn search was designed for homolog search or sequence comparison for long nucleotide sequence. One big issue for searching miRNA homology in public available databases is that current BLASTn search only give the sequences with a continued 7 nt match. If there is two or three nt variation, this sequence will be missed. But in most cases, these sequences may be potential miRNAs. To avoid this from happening, when using Blastn search, we need to manually compare the potential sequences with the query miRNA sequences and find how many nucleotides change in the subjected sequences.

3.2 Experimental Approaches

Although comparative genome-based EST/GSS analysis has identified dozens of cotton miRNAs, this approach only can be used to identify conserved miRNAs. Another big problem is that limited cotton nucleotide sequence dataset limits the identification of cotton miRNAs using this strategy. As sequencing technology, particularly the next generation deep sequencing technology become available, scientists start to use sequencing technology to identify cotton miRNAs.

Abdurakhmonov et al. employed direct cloning and sequencing technology for the first time to identify miRNAs in cotton (Abdurakhmonov et al. 2008). They first cloned small RNA sequences from cotton ovules at stage of 0–10 days of post anthesis (DPA). After sequencing 6,691 individual colonies, 2,482 small RNAs were obtained with a total of 583 unique sequence signatures. However, they only identified three miRBase-confirmed plant miRNAs (miR172, miR390 and miR853) with nine different sequences. Devor et al. used a similar strategy to conduct an initial survey of small RNAs in the 3–5 days old root tissues of TM-1 and they found eight conserved miRNA families existing in cotton; these eight families were miR156, miR166, miR167, miR168, miR169, miR171, miR396 and miR457 (Devor et al. 2009). In these direct cloning, the authors observed that only 1.5% of small RNA sequences are miRNAs (Abdurakhmonov et al. 2008), suggesting that only a small part of cloned small RNAs are identified to be miRNAs. One potential reason is that they may miss some miRNAs because of the limited resources. Among different plant species, the mature miRNA sequences may have 1–2 nt change. If the authors re-do the Blastn search and allow 1–2 nt difference, they may find more miRNAs from their cloning sequences. In the reported study, however, they did not try to find novel miRNAs (Abdurakhmonov et al. 2008). It is also potential for them to find some novel, potentially cotton-species specific miRNAs.

Compared with the direct cloning and sequencing technique, the next generation high through-put deep sequencing technology is much more powerful for identifying plant small RNAs and miRNAs. Deep sequencing technology can generate million sequence reads at a very short time. Thus, it not only identifies which gene is expressed but also the differentially expressed genes through their read number. Kwak et al. employed Solexa sequencing technology to identify miRNAs and their expression in the developing ovules of two cotton genotypes (wildtype and fuzz/lintless mutant). Of the total of over 12 million sequence reads, they identified 22 conserved miRNAs as well two novel cotton-specific miRNAs (miRNVL1 and miRNVL2) in cotton (Kwak et al. 2009). Pang et al. also employed the Solexa sequencing technology to sequence and analyze miRNAs from fiber and nonfiber tissues in cotton cultivar TM-1. After analyzing over four million small RNAs, they identified a total of 31 miRNA families in cotton. Among the 31 identified miRNA families, 27 are conserved in other plant species and four cotton-specific novel miRNAs (miR2947, miR2949, miR2950 and miRcand 1) were identified from at least one of the examined tissues (Pang et al. 2009). At the same time, Ruan et al. obtained 3,129,095 small RNA sequences from 6-day seedlings of cotton cultivar Coker 312 grown on the MS medium. By homology search, they found 34 conserved miRNA families in cotton; however, they only identified eight miRNAs (miR156, miR157a, miR157b, miR162, miR164, miR393, miR399, and miR827) with potential precursor sequences from the NCBC EST databases (Ruan et al. 2009).

4 Expression and Function of miRNAs in Cotton

Gene function requires the expression of a gene at a specific time with a certain level. A first report on miRNA expression in cotton was published in 2009. In that study, Zhang and Pan employed quantitative real-time PCR (qRT-PCR) to detect and compare the expression profiles of four miRNAs in eight different cotton organs at different developmental stages (Zhang and Pan 2009). Their results demonstrated that miRNAs were differentially expressed with certain classes expressed preferentially in an organ-specific manner. Based on their result, miR172 was highly expressed in several organs, including young leaves at fruit branch, young flower buds, 0 DPA ovules, and 0 DPA petals; however, miR-156 was highly expressed in cotyledon. It has been known that miR172 play an important function at flower development and miR172 was highly expression in flower during flowering time; interestingly, Zhang and Pan (2009) found that miR172 was not highly expressed in all parts of flowers. By contrast, miR172 was highly expressed in petal but not in stamen and carpel. Surprisingly, they found another miRNA, miR162 beside miR172, was highly expressed in 2 DPA ovules, immature fiber, and mixtures of 0 DPA stamen and carpel. This expression pattern suggests that both miR172 and miR162 may play a role in cotton fiber differentiation and development.

As mentioned in the previous section, the next generation deep sequencing not only identifies the potential miRNAs but also allow us to investigate their expression levels based on their sequence reads. Ruan et al. (2009) observed that miR156/157 was most highly expressed in cotton seedlings with 22,560 reads per million, following by miR168 (11,627 reads per million) and miR167 (6,030 reads per million). However, Pang et al. (2009) only observed that miR165/166 was highly expressed in cotton leaves with the highest number of sequencing reads (16,234 reads per million); by contrast, other miRNAs were only sequenced for less than couple hundreds of time per million read. They also observed that miR165/166 was most highly expressed in ovule with 7,712, 11,342, and 30,027 reads per million at 0, +3, and −3 DPA, respectively. Other highly expressed miRNAs in ovules include miR167 and miR168. Kwak et al. (2009) observed that miR167, miR156/157, and miR172 were highly expressed in cotton ovules. From these three reported deep-sequencing data, we can clearly see that the miRNA expression profile is varied from one study to another. It is unclear that what caused this difference.

Deep sequencing, miRNA microarrays, qRT-PCR, and miRNA blot analysis show that many miRNAs are differentially expressed during cotton fiber differentiation and development. Pang et al. (2009) identified four sets of cotton miRNAs (leaf, −3 DPA, 0 DPA and +3 DPA) for four different small RNA libraries. Based on the sequence read, several miRNAs (miR156/157, miR170/171, miR172, miR472/282, miR535 and miR2947) are highly expressed in leaf than in ovules. By contrast, miR167, miR168, and miR393 were highly expressed in certain stage of ovules. By comparing the expression of miRNAs between normal cotton ovule and fiberless mutant ovule, Kwak et al. (2009) found that 24 of 34 identified miRNAs are significantly differentially expressed in cotton developing ovules. miR160 and miR165/166 were significantly induced in the fiber mutant; by contrast, the expression of miR399, miR397 and miR395 were significantly reduced in the mutant. Pang et al. (2009) observed that many miRNAs accumulate at lower levels in fiber-bearing ovules (3 DPA) and fibers (10 DPA) than immature ovules (3 DPA). miR156, miR159, miR165, miR166, miR167, miR168, miR171, and miR172 were highly expressed in the immature ovules (−3 DPA) but not in fiber-bearing ovules and fibers.

5 Function of Specific miRNAs in Cotton

The biological function of miRNAs in cotton is largely unknown. miRNAs negatively regulate gene expression by targeting protein-coding genes for mRNA cleavage or translation inhibition. Thus, it is always the first step to investigate the function of a specific miRNA by identifying their targets. In plants, a majority of miRNAs bind to their targeted mRNAs through perfect or near-perfect complementary sequences, which provide a powerful approach to identify plant miRNA targets by simply Blastn search. Several cotton miRNA targets have been predicted using computational approach (Qiu et al. 2007; Zhang et al. 2007). Similar with other plant species, many cotton miRNAs target transcription factors, which further control plant growth and development. It is well studied that miR156 and miR172 target squamosa-promoter binding protein-like protein (SBP) and apetal 2 (AP2) transcription factor and further control leaf and flower development in *Arabidopsis*, respectively (Rhoades et al. 2002). Currently, almost all studies on cotton miRNAs have shown that miR156 and miR172 also target SBP and AP2 gene in cotton (Pang et al. 2009; Qiu et al. 2007; Zhang et al. 2007). In silico studies also show that cotton miRNAs may be involved in other cellular and metabolic processes, such as signal transduction and stress response; some miRNA function may be unique to cotton, such as involving in secondary wall synthesis and deposition. (Pang et al. 2009; Qiu et al. 2007; Zhang et al. 2007). It has been predicted that at least three miRNA families (miR396, 414, and 782) target cotton fiber-related genes, such as callous synthase, fiber protein Fb23, and fiber quinoneoxidoreductase, which suggests that

Table 2 Confirmed miRNA targets in cotton by experimental RACE approach (Pang et al. 2009)

miRNA	Target	Cleavage site
miR159	TC128888 (unknown protein)	10
miR160	TC118163 (–) (putative ARF10 or 16)	10
miR160	TC82706 (+) (ARF3-like)	10
miR164	TC116985 (+) (no apical meristem, NAM-like)	10
miR165/166	TC128553 (–) (class III HD-Zip protein 8)	10
miR165/166	ES810681 (–) (class III HD-Zip protein 5)	10
miR165/166	TC107071 (–) (class III HD-Zip protein 5)	10
miR165/166	TC116644 (–) (class III HD-Zip protein 4)	9, 10
miR167	TC119855 (–) (ARF4-like)	10
miR172	TC116912 (–) (APETALA2 protein homolog)	10
miR390	DW502659 (–) (TAS3-like)	10
Gh-miRcand1	TC121013 (+) (NAC-like transcription factor)	10

miRNAs may play a role in cotton fiber differentiation and development (Zhang et al. 2007).

Although hundreds of targets have been predicted for miRNAs in cotton, only one study reported the evidence for miRNA targets in cotton (Pang et al. 2009). A good strategy to test the target is to determine whether a specific target is cleaved by a corresponding miRNA. Pang et al. (2009) employed RACE-PCR to determine the miRNA-guided target cleavage site for 12 targets of eight miRNAs (miR159, miR160, miR164, miR165/166, miR167, miR172, miR390 and cotton specific miRcand1). Their results have confirmed all the predicted cleavage sites. The expression levels of some targets and its corresponding miRNAs were inversely correlated. Table 2 lists the miRNA and mRNA target pairs, which have been experimentally confirmed.

6 Conclusions

Although cotton is one of the most important crops, its miRNA-related research is far beyond other plant species such as rice and maize. Currently, there are nine papers published; although all of these work are considerate and provide useful information, the results were mainly descriptive. Thus, the complexity of miRNAs in cotton has just been realized.

The next-generation deep-sequencing technology provides a powerful approach to investigate small RNAs, including miRNAs in cotton as well in other plant species. Although it is a very short time since Solexa sequencing technology was used to identify miRNAs, there have been three reports on identification of miRNAs in cotton using this technology, and we believe more papers are on the way. It is undoubtful that deep sequencing produces lots of datasets and genome information; however, the analysis of the data was significantly limited because of the lack of cotton genome sequence data. Currently, no finished genome sequence is available for any cotton species. This limits the identification of cotton miRNAs, particularly for the cotton-specific miRNAs. The full miRNA content of cotton will be achieved when the full genome of cotton is available. Several institutes are currently working together to sequence the cotton genome. We believe that great progress will be made when this full sequence is elucidated.

It is clear that miRNAs play an important role in many biological and metabolic processes, including plant development, organ differentiation and morphology, phage change from vegetative growth to reproductive growth, and response to abiotic and biotic stress. However, no study has been performed on the function of miRNAs in cotton fiber initiation and development although several studies have shown that certain miRNAs are differentially expressed during cotton fiber development and in silico analysis shows that miRNAs target fiber-related genes. Understanding the biological function of miRNAs involved in cotton fiber development requires a lot more investigation in this field. However, studying the miRNA-mediated mechanism in cotton fiber may provide a novel insight into the mechanism of controlling cotton fiber development and may also provide a potential new biotechnology for improving cotton fiber quality and yield through a new miRNA-based biotechnology.

References

Abdurakhmonov IY, Devor EJ, Buriev ZT et al (2008) Small RNA regulation of ovule development in the cotton plant, *G. hirsutum* L. BMC Plant Biol 8:93

Asif M, Mirza JI, Zafar Y (2008) Genetic analysis for fiber quality traits of some cotton genotypes. Pakistan J Bot 40:1209–1215

Barozai MYK, Irfan M, Yousaf R et al (2008) Identification of micro-RNAs in cotton. Plant Physiol Biochem 46:739–751

Bartel DP (2004) MicroRNAs: genomics, biogenesis, mechanism, and function. Cell 116:281–297

Basra AS, Malik CP (1984) Development of the cotton fiber. Int Rev Cytol 89:65–113

Brodersen P, Sakvarelidze-Achard L, Bruun-Rasmussen M et al (2008) Widespread translational inhibition by plant miRNAs and siRNAs. Science 320:1185–1190

Chaudhary B, Hovav R, Rapp R et al (2008) Global analysis of gene expression in cotton fibers from wild and domesticated Gossypium barbadense. Evol Dev 10:567–582

Chaudhary B, Hovav R, Flagel L et al (2009) Parallel expression evolution of oxidative stress-related genes in fiber from wild and domesticated diploid and polyploid cotton (Gossypium). BMC Genomics 10:378

Chen XM (2005) microRNA biogenesis and function in plants. FEBS Lett 579:5923–5931

Devor EJ, Huang L, Abdukarimov A et al (2009) Methodologies for in vitro cloning of small RNAs and application for plant genome(s). Int J Plant Genomics 2009:915061

FeiFei L, ShenJie W, Fenni L et al (2009) Modified fiber qualities of the transgenic cotton expressing a silkworm fibroin gene. Chinese Sci Bull 54:1210–1216

Gao P, Zhao PM, Wang J et al (2007) Identification of genes preferentially expressed in cotton fibers: a possible role of calcium signaling in cotton fiber elongation. Plant Sci 173:61–69

Gleave AP, Ampomah-Dwamena C, Berthold S et al (2008) Identification and characterisation of primary microRNAs from apple (Malus domestica cv. Royal Gala) expressed sequence tags. Tree Genet Genomics 4:343–358

Gou JY, Wang LJ, Chen SP et al (2007) Gene expression and metabolite profiles of cotton fiber during cell elongation and secondary cell wall synthesis. Cell Res 17:422–434

Griffiths-Jones S, Saini HK, van Dongen S et al (2008) miRBase: tools for microRNA genomics. Nucleic Acids Res 36:D154–D158

Haigler CH, Zhang D, Wilkerson CG (2005) Biotechnological improvement of cotton fibre maturity. Physiol Plant 124:285–294

He YJ, Guo WZ, Shen XL et al (2008) Molecular cloning and characterization of a cytosolic glutamine synthetase gene, a fiber strength-associated gene in cotton. Planta 228:473–483

Hovav R, Udall JA, Hovav E et al (2008) A majority of cotton genes are expressed in single-celled fiber. Planta 227:319–329

Huang GQ, Xu WL, Gong SY et al (2008a) Characterization of 19 novel cotton FLA genes and their expression profiling in fiber development and in response to phytohormones and salt stress. Physiol Plant 134:348–359

Huang QS, Wang HY, Gao P et al (2008b) Cloning and characterization of a calcium dependent protein kinase gene associated with cotton fiber development. Plant Cell Rep 27:1869–1875

Iqbal S, Bashir A, Naseer HM et al (2008) Identification of differentially expressed genes in developing cotton fibers (*Gossypium hirsutum L.*) through differential display. Electron J Biotechnol 11:1–10

Kim HJ, Triplett BA (2001) Cotton fiber growth in planta and in vitro. Models for plant cell elongation and cell wall biogenesis. Plant Physiol 127:1361–1366

Kwak PB, Wang QQ, Chen XS et al (2009) Enrichment of a set of microRNAs during the cotton fiber development. BMC Genomics 10:457

Lagos-Quintana M, Rauhut R, Lendeckel W et al (2001) Identification of novel genes coding for small expressed RNAs. Science 294:853–858

Lau NC, Lim LP, Weinstein EG et al (2001) An abundant class of tiny RNAs with probable regulatory roles in *Caenorhabditis elegans*. Science 294:858–862

Lee RC, Ambros V (2001) An extensive class of small RNAs in *Caenorhabditis elegans*. Science 294:862–864

Lee RC, Feinbaum RL, Ambros V (1993) The *C. elegans* heterochronic gene lin-4 encodes small RNAs with antisense complementarity to lin-14. Cell 75:843–854

Lee JJ, Woodward AW, Chen ZJ (2007) Gene expression changes and early events in cotton fibre development. Ann Bot 100:1391–1401

Li YL, Sun J, Li CH et al (2003) Specific expression of a beta-tubulin gene (GhTub1) in developing cotton fibers. Sci China C Life Sci 46:235–242

Li XB, Fan XP, Wang XL et al (2005a) The cotton *ACTIN1* gene is functionally expressed in fibers and participates in fiber elongation. Plant Cell 17:859–875

Li YL, Sun J, Xia GX (2005b) Cloning and characterization of a gene for an LRR receptor-like protein kinase associated with cotton fiber development. Mol Genet Genomics 273:217–224

Liu HW, Wang XF, Pan YX et al (2009) Mining cotton fiber strength candidate genes based on transcriptome mapping. Chinese Sci Bull 54:4651–4657

Michailidis G, Argiriou A, Darzentas N et al (2009) Analysis of xyloglucan endotransglycosylase/hydrolase (XTH) genes from allotetraploid (*Gossypium hirsutum*) cotton and its diploid progenitors expressed during fiber elongation. J Plant Physiol 166:403–416

Pan XP, Zhang BH, Francisco MS et al (2007) Characterizing viral MicroRNAs and its application on identifying new MicroRNAs in viruses. J Cell Physiol 211:10–18

Pang MX, Woodward AW, Agarwal V et al (2009) Genome-wide analysis reveals rapid and dynamic changes in miRNA and siRNA sequence and expression during ovule and fiber development in allotetraploid cotton (*Gossypium hirsutum* L.). Genome Biol 10:R122

Qiu CX, Xie FL, Zhu YY et al (2007) Computational identification of microRNAs and their targets in *Gossypium hirsutum* expressed sequence tags. Gene 395:49–61

Rhoades MW, Reinhart BJ, Lim LP et al (2002) Prediction of plant microRNA targets. Cell 110:513–520

Ruan MB, Zhao YT, Meng ZH et al (2009) Conserved miRNA analysis in *Gossypium hirsutum* through small RNA sequencing. Genomics 94:263–268

Schwartz BM, Smith CW (2008) Genetic gain in fiber properties of upland cotton under varying plant densities. Crop Sci 48:1321–1327

Taliercio EW, Boykin D (2007) Analysis of gene expression in cotton fiber initials. BMC Plant Biol 7:22

Tiwari SC, Wilkins TA (1995) Cotton (*Gossypium hirsutum*) seed trichomes expand via diffuse growing mechanism. Canad J Bot 73:746–757

Tu LL, Zhang XL, Liang SG et al (2007) Genes expression analyses of sea-island cotton (*Gossypium barbadense* L.) during fiber development. Plant Cell Rep 26:1309–1320

Wightman B, Ha I, Ruvkun G (1993) Posttranscriptional regulation of the heterochronic gene lin-14 by lin-4 mediates temporal pattern-formation in *C. elegans*. Cell 75:855–862

Wu ZD, Soliman KM, Bolton JJ et al (2008) Identification of differentially expressed genes associated with cotton fiber development in a chromosomal substitution line (CS-B22sh). Funct Integr Genomics 8:165–174

Xie FL, Huang SQ, Guo K et al (2007) Computational identification of novel microRNAs and targets in *Brassica napus*. FEBS Lett 581:1464–1474

Xie FL, Frazier TP, Zhang BH (2010) Identification and characterization of microRNAs and their targets in the bioenergy plant switchgrass (*Panicum virgatum*). Planta 232:417–434

Xu ZY, Kohel RJ, Song GL et al (2008) Gene-rich islands for fiber development in the cotton genome. Genomics 92:173–183

Yin ZJ, Shen FF (2010) Identification and characterization of conserved microRNAs and their target genes in wheat (*Triticum aestivum*). Genet Mol Res 9:1186–1196

Yin ZJ, Li CH, Han ML et al (2008) Identification of conserved microRNAs and their target genes in tomato (*Lycopersicon esculentum*). Gene 414:60–66

Yu B, Wang H (2010) Translational inhibition by microRNAs in plants. Prog Mol Subcell Biol 50:41–57

Zhang BH, Pan XP (2009) Expression of microRNAs in cotton. Mol Biotechnol 42:269–274

Zhang BH, Pan XP, Wang QL et al (2005) Identification and characterization of new plant microRNAs using EST analysis. Cell Res 15:336–360

Zhang BH, Pan XP, Anderson TA (2006a) Identification of 188 conserved maize microRNAs and their targets. FEBS Lett 580:3753–3762

Zhang BH, Pan XP, Cannon CH et al (2006b) Conservation and divergence of plant microRNA genes. Plant J 46:243–259

Zhang BH, Pan XP, Cobb GP et al (2006c) Plant microRNA: a small regulatory molecule with big impact. Dev Biol 289:3–16

Zhang BH, Wang QL, Wang KB et al (2007) Identification of cotton microRNAs and their targets. Gene 397:26–37

Zhang BH, Pan XP, Stellwag EJ (2008) Identification of soybean microRNAs and their targets. Planta 229:161–182

Nutrient-Responsive Plant microRNAs

Wolf-Rüdiger Scheible, Bikram Datt Pant, Magdalena Musialak-Lange, and Przemyslaw Nuc

Contents

Abstract Being sessile, plants have to cope with many adverse environmental changes, including changing nutrient availability. Adequate availability of mineral macronutrients (e.g., N, P, K, S) and micronutrients (e.g., Cu, Fe, Zn) in the soil and their acquisition are vitally important for plant growth, development, and reproduction. Too little or too much of the nutrients negatively affects these traits and hence plant fitness and

W.-R. Scheible (✉), B.D. Pant, and M. Musialak-Lange
Max-Planck Institute for Molecular Plant Physiology, Am Mühlenberg 1, 14476 Potsdam, Germany
e-mail: scheible@mpimp-golm.mpg.de

P. Nuc
Max-Planck Institute for Molecular Plant Physiology, Am Mühlenberg 1, 14476 Potsdam, Germany

Institute of Molecular Biology and Biotechnology, Adam Mickiewicz University, Poznan, Poland

V.A. Erdmann and J. Barciszewski (eds.), *Non Coding RNAs in Plants*,
RNA Technologies, DOI 10.1007/978-3-642-19454-2_20,

survival. Therefore, cellular concentrations of these nutrients have to be maintained at physiological levels, and adaptive physiological or developmental responses need to be initiated if necessary. Over the last few years, some nutrient-responsive microRNAs (miRs) have been discovered primarily in the model plant species *Arabidopsis thaliana* (*A. thaliana*), and several quickly evolved as critical components in scenarios such as regulation of plant P, S, or Cu homeostasis, or plant developmental adaptations to N availability. Several of these nutrient-responsive miRs also were found in phloem, i.e., the conductive vessels of plants, and a few were shown to act as systemic, long-distance signals. The presence of these miRs, their conserved nutrient response, and target genes in many higher plant species suggest deep conservation of the regulatory mechanisms. This chapter highlights and summarizes these discoveries.

Keywords Copper • Homeostasis • microRNA • Nitrogen • Phloem transport • Phosphorus • Sulfur • Systemic signal

Abbreviations

AFB	Auxin F-box receptor
APS	ATP sulfurylase
ARF	Auxin-responsive transcription factor
A. thaliana	*Arabidopsis thaliana*
BAH	Benzoic acid hypersensitive
CCS1	Copper chaperone CCS1
COX5b	Cytochrome c oxidase subunit V
C. reinhardtii	*Chlamydomonas reinhardtii*
CSD	Copper/zinc (Cu/Zn) superoxide dismutases
GFP	Green fluorescent protein
GUS	Beta glucuronidase
IPS1	INDUCED BY PHOSPHATE STARVATION 1
LNA	Locked nucleic acid
LR	Lateral root
miR	microRNA
miR*	miR star strand
NF	Nuclear factor
NLA	Nitrogen limitation adaptation
PC	Plastocyanin
PHO2	E2 ubiquitin ligase PHO2
Pi	(Inorganic) phosphate
PSI	Phosphate starvation induced
qRT-PCR	Quantitative real-time polymerase chain reaction
RACE	Rapid amplification of cDNA ends
RNA	Ribonucleic acid
ROS	Reactive oxygen species

SPL	SQUAMOSA promoter binding transcription factor
SPX	Eucaryotic protein domain named after Syg1, Pho81 and XPR1
UBC	Ubiquitin conjugase
UTR	Untranslated region

1 Phosphorus-Responsive miRs

Phosphorus (P) is an essential plant macronutrient and is involved in many biological processes such as photosynthesis, sugar and energy metabolism, or signal transduction. P is also a constituent of major cellular components such as nucleic acids or membrane lipids. P-limitation is a common stress for plants in most parts of the world, thus adaptive mechanisms evolved, such as (1) mechanisms to improve the acquisition of external phosphate (Pi) by inducing the expression of Pi transporter genes, secretion of organic acids, ribonucleases, or phosphatases into the soil, and altering root architecture, (2) engaging into symbiotic interactions with e.g., mycorrhizal fungi leading to a better Pi provision in exchange for sugars, and (3) mechanisms to make more efficient use of internal Pi pools, such as exchange of phospholipids in cell membranes against glycolipids, accelerated senescence and Pi remobilization from old plant parts, and/or reduction of cellular ribonucleic acid (RNA) content (Poirier and Bucher 2002). Moreover, (4) reallocation and translocation of the spare resource to growing and reproductive organs are crucial to ensure reproduction and survival. MiR399 and presumably also other recently identified P-starvation inducible miRs play key roles in plant adaptive responses to P-deprivation and maintenance of P homeostasis.

1.1 MiR399 is a Key Component for Regulation of Plant Pi Homeostasis

MiR399 was first described in *Arabidopsis thaliana* and rice (Jones-Rhoades and Bartel 2004; Sunkar and Zhu 2004) and is encoded by gene-families (e.g., 6 and 11 genes in *A. thaliana* and rice, respectively). All six primary *MIRNA399* gene (pri-miR399) transcripts in *A. thaliana* are strongly and specifically induced by P-limitation, whereas no or only very low expression of the genes is detectable in P-replete conditions (Bari et al. 2006; Pant et al. 2009). The strong induction leads to a high concentration of mature miR399 in *A. thaliana* during P-limitation (Bari et al. 2006; Chiou et al. 2006; Fujii et al. 2005; Pant et al. 2009), whereas miR399 and the corresponding primary transcripts are not or only barely detectable in other nutrient stress conditions (Bari et al. 2006; Fujii et al. 2005; Hsieh et al. 2009) and abiotic stress conditions such as cold, salt, or drought (Jones-Rhoades and Bartel 2004; Sunkar and Zhu 2004). MiR399, thus, resembles a molecular switch that is turned on during P limitation only.

MiR399 has five near-perfect binding sites in the 5′-untranslated region (UTR) of the *PHO2* gene, which encodes an ubiquitin-conjugase (UBC24) in *A. thaliana* (Aung et al. 2006; Bari et al. 2006; Chiou et al. 2006; Fujii et al. 2005; Sunkar and Zhu 2004), making the *PHO2* gene the best and a special miR399 target. A prerequisite for miRNA action is colocalization of microRNA (miR) and its target transcript. Presence of *PHO2* transcript and miR399 in the vascular cylinder of the root was indeed observed (Aung et al. 2006). Direct confirmation of this miR-target gene connection has been obtained first by 5′-rapid amplification of cDNA ends (RACE) analysis, where miR399-guided transcript cleavage at the second and third binding site was found (Allen et al. 2005). Furthermore, *PHO2/UBC*24 transcript level was decreased down to 10–20% by Pi limitation, i.e., conditions in which miR399 expression is high, and in miR399 overexpressing *A. thaliana* plants (Bari et al. 2006; Chiou et al. 2006; Fujii et al. 2005), whereas the transcript level of a 5′-UTR-less *PHO2/UBC*24 transgene was not affected by low-Pi condition (Fujii et al. 2005). In addition, transgenic *A. thaliana* plants with constitutive expression of miR399 show all physiological and molecular phenotypes of *pho2* null mutants (see Bari et al. 2006; Chiou et al. 2006; Pant et al. 2009).

Regulation of *PHO2/UBC24* by miR399 not only is restricted to posttranscriptional transcript cleavage but includes translational repression too (Bari et al. 2006; Pant et al. 2009): first, very high expression of miR399 primary transcripts and stoichiometric levels of mature miR399 are necessary to fully suppress the activity of the target *PHO2* transcript (Bari et al. 2006). Second, miR399 overexpressing seedlings still have 10–20% of wild-type *PHO2* transcript, but their physiological and molecular phenotypes are indistinguishable from those of *pho2* null-mutant. Third, in transgenic *A. thaliana* seedlings expressing beta glucuronidase (*GUS*) under control of the *PHO2* promoter-5′-UTR, GUS activity and staining rises quickly after Pi readdition to previously Pi-starved transgenic seedlings, and this happens although the endogenous *PHO2* gene or the transgene transcripts remain unchanged (Bari et al. 2006).

As compared to wild type, *pho2* mutant plants or miR399 overexpressing plants growing in the presence of Pi in their rooting medium have unaltered Pi levels in roots but accumulate 4–5-fold higher Pi levels in leaves, leading to symptoms of Pi toxicity, i.e., leaf necrotic lesions (Aung et al. 2006; Bari et al. 2006; Chiou et al. 2006; Delhaize and Randall 1995). These phenotypes are due to (1) increased Pi uptake, (2) increased translocation of Pi from roots to shoots, (3) retention of Pi in the shoots, and (4) impaired remobilization of Pi from old to young leaves (Aung et al. 2006; Chiou et al. 2006). In *pho2* mutants or miR399 overexpressers, Pi does not repress a set of Pi starvation-induced (PSI) genes, including induced by phosphate starvation 1 (*IPS1*) (*At3g09922*), *AT4* (*At5g03545*), or the Pi transporter genes *Pht1;4*, *Pht1;8*, and *Pht1;9* (Aung et al. 2006; Bari et al. 2006; Pant et al. 2009) (cf. Fig. 1). The continuous upregulation of these *PSI* genes in P-replete *pho2* mutants or miR399 overexpressing plants suggests that loss of *PHO2* mimics P-starvation responses which are causal for the observed physiological phenotypes. How PHO2, in its function as an ubiquitin conjugase, affects *PSI* gene expression is not yet known. It is clear, however, that PHO2

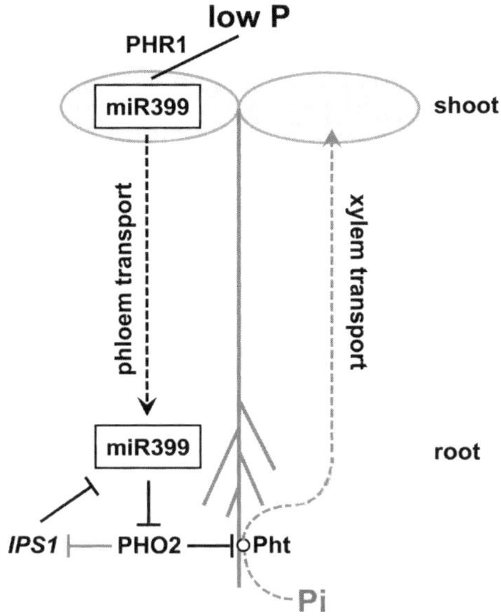

Fig. 1 Regulation of plant phosphate homeostasis by systemic miR399 and PHO2. Low P concentration in the shoot leads to a strong, PHR1-dependent induction of miR399. Via the phloem, miR399 reaches the root, where it inhibits PHO2 via transcript cleavage and translational repression. This results in derepression of Pi transporter genes (*Pht*), increased Pi uptake capacity and translocation via the xylem to the shoot, thereby closing the homeostatic circle. The activity of miR399 is controlled by *IPS1* via target mimicry (cf. text), and *IPS1* expression itself is repressed by PHO2, revealing autoregulation of miR399 and PHO2 activity. The model is derived from Aung et al. (2006), Bari et al. (2006), Chiou et al. (2006), Franco-Zorrilla et al. (2007), Lin et al. (2008), and Pant et al. (2008)

activity in P-sufficient conditions is required to maintain proper Pi uptake, translocation, concentration, and homeostasis.

Factors that control *MIRNA* gene expression and miR activity are often unknown. MiR399 is an exemption in this regard (Fig. 1). On the one hand, PHOSPHATE RESPONSE1 (PHR1), a MYB-type transcription factor central for Pi-starvation signaling in *A. thaliana* (Rubio et al. 2001), was shown to be required for expression of *MIRNA399* genes in *A. thaliana*, as Pi-deprived *phr1* mutants show strongly reduced *MIRNA399* gene transcript levels (Bari et al. 2006). Furthermore, *MIRNA399* gene promoters also contain several copies of the PHR1 binding motif GNATATNC. This places miR399 and PHO2 in a branch of the Pi-signaling network downstream of PHR1 (Bari et al. 2006). On the other hand, miR399 activity was shown to be controlled by the expression of *IPS1*, a long non coding RNA that harbors a near-perfect binding site for miR399 (Franco-Zorrilla et al. 2007). *IPS1* is bound but not cleaved by miR399, and like a sponge, thus, sequesters miR399 in the argonaute complex. This reduces the activity of miR399 in cell types and tissues where *IPS1* is expressed, leading to protection of *PHO2* transcript (Franco-Zorrilla et al. 2007). The term "target mimicry" was coined for this unique mechanism of non coding RNA action. Artificial target mimicry presents a promising and straightforward approach to knock down entire *MIRNA* gene families (Todesco et al. 2010).

Genes encoding miR399 exist in all (examined) mono- and dicotyledonous plant species (http://www.mirbase.org), but not in lower unicellular photosynthetic organism such as the moss *Physcomitrella patens* or the green alga *Chlamydomonas reinhardtii*. MiR399 is frequently encoded by surprisingly large gene families (e.g., six genes in *A. thaliana*, 10 in corn, 11 in rice, 12 in poplar, and 17 known genes in *Medicago*). The presence of six *MIRNA399* genes in *A. thaliana* and further expansion of the gene family, for example in barrel medic, suggests functional specialization and new functions, for example in fungal symbiosis (Branscheid et al. 2010). A satisfactory understanding as to why, for example, *A. thaliana* has six *MIRNA399* genes, encoding five miR399 isoforms that differ by 1–2 nucleotides, is not available. It was speculated that the sequence variation at nucleotide 13 could affect cleavage of *PHO2* and binding to *IPS1* (Lin and Chiou 2008). Another possibility is that some isoforms cleave or repress translation of yet unknown or additional predicted target gene transcripts, e.g., *At4g09730* or *At3g54700* (encoding DNA-dependent RNA helicase and a phosphate transporter, respectively) (Adai et al. 2005; Jones-Rhoades and Bartel 2004; Pant et al. 2009).

Similar to the omnipresence of *MIRNA399* genes in higher plants, the induction of the genes and mature miR399 by Pi-limitation is conserved, as was demonstrated in rice (Bari et al. 2006), tobacco (B. Pant and W. Scheible, unpublished), tomato (Chiou et al. 2006), bean (Valdés-López et al. 2008), or barrel medic (Jagadeeswaran et al. 2009). *PHO2* orthologs with intron/exon structures identical to *AtPHO2* and harboring five miR399 binding sites in their 5′-UTR were described for poplar, barrel medic, rice (Bari et al. 2006), or bean (Valdés-López et al. 2008). Furthermore, constitutive overexpression of miR399 in tobacco results in a decrease of *NtPHO2* transcript and Pi accumulation in leaves (Branscheid et al. 2010). These results show that the miR399/PHO2 regulatory mechanism is probably conserved across angiosperms and indicate that it may have emerged during the evolution of higher plants.

1.2 Additional Pi-Responsive microRNAs

MiR399 isoforms are not the only P-responsive miRs known. Pant et al. used quantitative real-time polymerase chain reaction (qRT-PCR) to investigate the expression levels of 177 annotated *A. thaliana MIRNA* gene primary (pri-miR) transcripts during P-starvation (Pant et al. 2009). Although the pri-miR transcripts are not known to be biologically active molecules, the approach was shown to be a useful discovery tool, as many of the changes observed for the pri-miR transcripts were confirmed for the encoded mature miRs. In this study, miR399s, miR827, miR778, and miR398 were found to be highly responsive, and miR169 and miR408 were still considerably responsive to P-limitation (Table 1). Pant et al. (2009) also used high-throughput sequencing of small RNA libraries produced from P-replete, P-starved, and P-starved/resupplied *A. thaliana* seedlings. The results from this approach confirmed the strong induction of miR399s, miR827, and miR778, as well as the repression of miR398 and miR169 during P-limitation. It further revealed that some miR star strands (miR*s; the strand of a miR/miR* duplex that is supposedly not incorporated in the argonaute complex and degraded) accumulate to high levels during P-limitation. Also, target genes for these miR* species can be predicted (Table 1), raising the question whether they have functions in gene regulation, too. Regulatory activity of miR*s and their presence in argonaute complexes were demonstrated (Mi et al. 2008; Okamura et al. 2008). These results (Pant et al. 2009) were confirmed in a second high-throughput sequencing study of Pi-sufficient or Pi-deficient root and shoot samples (Hsieh et al. 2009). The latter study also reported miR156 to be induced and miR395 to be repressed during P-starvation in roots (cf. Sect. 2). Using miRCURY LNA™ microarrays, the results for miR827 were further confirmed, and a few more miRs were suggested to be moderately induced by P limitation (Lundmark et al. 2010).

Small RNA sequencing data together with a plant-adapted version of the miRDeep algorithm (Friedländer et al. 2008) provided the possibility to predict novel potential miR stem loop structures, which were then filtered for novel P status-dependent miR candidates (Pant et al. 2009). This yielded an additional strongly P-responsive miR (miR2111) expressed from two genes in *A. thaliana*. *MIRNA2111* gene homologs were also reported from *Lotus japonicus* and *Vitis vinifera*, but were not detected in monocotyledonous plants (Hsieh et al. 2009). Moreover, miR2111 is P-limitation induced and encoded by larger gene families in *Brassica napus* (8 loci) (Pant et al. 2009, P. May and W. Scheible, unpublished) or *Medicago truncatula* (19 loci) (Lelandais-Brière et al. 2009), indicating that miR2111 is conserved in dicots and plays important role(s) during the adaptation to P limitation.

Specificity of the response was tested for the additional P-starvation-responsive miRs (Hsieh et al. 2009; Pant et al. 2009). These experiments revealed that, similar to miR399, miR827, miR778, and miR2111 are all highly specific for P limitation, as neither sugar, N, K, S Cu, or Fe limitations were able to induce (or repress) these miRs. By contrast, miR169 was repressed not only by P- but also by N-limitation, and miR398 was induced by Cu- and repressed by P-, sugar-, N-, K-, and Fe-limitation.

1.3 Target Genes and Potential Biological Functions of the Additional P-Responsive miRs

Various algorithms can be used to predict target genes of plant miRs (cf. Pant et al. 2009) based on near perfect miR-target gene complementarity (Jones-Rhoades and Bartel 2004; Rhoades et al. 2002). Such predictions require experimental confirmation by testing miR-mediated transcript cleavage, for instance. They are also not

Table 1 MicroRNA/target mRNA modules in nutrient sensing and homeostasis

miR/miR* family and conservation	Stimulus and response up (↑) down (↓)	TF required for miR expression	Target gene	Gene product	Target verification
156 (1–4)	–P↑ (root)	–	AT5G43270	Transcription factor SPL2	R; D
			AT2G33810	Transcription factor SPL3	R; D
			AT1G53160	Transcription factor SPL4	R; D
			AT3G15270	Transcription factor SPL5	R; D
			AT1G69170	Transcription factor SPL6	R; D
			AT2G42200	Transcription factor SPL9	R
			AT1G27370	Transcription factor SPL10	R; D
			AT1G27360	Transcription factor SPL11	D
167 (1, 4)	N↓ (root)	–	AT5G37020	Auxin response factor ARF8	R; D; oex; RT
			AT1G30330	Auxin response factor ARF6	R; D; oex; RT
			AT5G60360	Senescence associated gene SAG2	D
			AT1G51760	IAA-amino acid conjugate hydrolase IAR3	D
169 (1)	–N↓ –P↓	–	AT3G05690	Subunit of CCAAT-binding complex AtHAP2B	R; D
			AT1G17590	Subunit of CCAAT-binding complex NFY-A8	R; D
			AT1G54160	Subunit of CCAAT-binding complex NFY-A5	R; D; oex
			AT1G72830	Subunit of CCAAT-binding complex AtHAP2C	D
			AT5G06510	Subunit of CCAAT-binding complex NFY-A10	D
			AT5G12840	Subunit of CCAAT-binding complex AtHAP2A	R; D
			AT3G20910	Subunit of CCAAT-binding complex NFY-A9	PAREdb
2111 (d)	–P↑	–	AT3G27150	Kelch repeat containing F-box protein	D; oex
2111a* (b)	–P↑	–	AT1G11990	Unknown protein	–
2111b* (b)	–P↑	–	AT2G26650	AKT1 (potassium transporter)	–
393 (1)	N↑	–	AT4G03190	Auxin signaling F-box1 protein AFB1	R; D
			AT3G26810	Auxin signaling F-box1 protein AFB2	R; D
			AT1G12820	Auxin signaling F-box1 protein AFB3	R; D; oex; RT
			AT3G23690	bHLH family transcription factor	R; D
			AT3G62980	Auxin receptor TIR1	R; D

395 (1, 4)	–S↑ –P↓	SLIM1	AT3G22890	ATP sulfurylase APS1	R; D; oex
			AT5G10180	Low affinity S transporter SULTR2;1	R; D; oex
			AT5G43780	ATP sulfurylase APS4	R; D; oex
			AT4G14680	ATP sulfurylase APS3	R; D; oex
397 (1)	–Cu↑	SPL7	AT2G29130	Laccase LAC2	R; D
			AT2G38080	Laccase LAC4	R; D
			AT5G60020	Laccase LAC17	R; D
398 (1, 2)	–Cu↑	SPL7	AT1G08830	Cu/Zn superoxide dismutase CSD1	R; D; cs
			AT2G28190	Cu/Zn superoxide dismutase CSD2	R; D; cs; RT
			AT3G15640	Cytochrome c oxidase subunit V COX5b	R; D
			AT1G12520	Cu/Zn SOD copper chaperone CCS1	R; D
399 (1)	–P↑	PHR1	AT2G33770	E2 ubiquitin ligase PHO2	R; D; oex; RT
399c* (b)	–P↑	PHR1	AT5G64470	Unknown protein DUF 231	–
399d* (b)	–P↑	PHR1	AT5G49430	WD-40 transducin family protein	–
399f* (b)	–P↑	PHR1	AT3G25905	Clavata3 homolog CLE27	–
408 (1–4)	–Cu↑	SPL7	AT2G30210	Laccase LAC3	R
			AT5G05390	Laccase LAC12	R
			AT5G07130	Laccase LAC13	R
			AT2G02850	Plantacyanin	R; D
827 (1)	–P↑	–	AT1G02860	SPX-domain containing E3 ligase NLA	R; D; oex
778 (a)	–P↑	–	AT2G22740	Histone methyltransferase SUVH6	D; oex
			AT2G35160	Histone methyltransferase SUVH5	PAREdb
857 (a)	–Cu↑	SPL7	AT3G09220	Laccase LAC7	R; D

MiRNA conservation: (a) *A. thaliana*, (b) *Brassicaceae*, (1) angiosperms, (2) gymnosperms, (3) lycopods, (4) mosses. Experimental evidence for miRNA-directed target mRNA cleavage: (R) gene-specific RACE, (D) degradome tag, (oex) miRNA overexpressor line, (RT) plant transformed with a miR-resistant target mRNA, (cs) cosuppression, (PAREdb) listed in PARE database (http://mpss.udel.edu/at_pare/). If cleavage is not experimentally confirmed, the best target as predicted by miRU2 (http://bioinfo3.noble.org/psRNATarget/) is given necessarily comprehensive as (1) transcript cleavage can occur at poor complementarity (Bouché 2010; Li et al. 2010a), and (2) translational repression by miRs may also require lower complementarity to the target transcript (Brodersen et al. 2008; Brodersen and Voinnet 2009).

The best predicted target of miR827, *At1g02860* (Fahlgren et al. 2007; Pant et al. 2009) was confirmed by 5′-RACE analysis (Fahlgren et al. 2007; Hsieh et al. 2009). *At1g02860* transcript also dropped 2–3-fold during P-limitation (Morcuende et al. 2007) or miR827 overexpressing plants (M. Musialak-Lange and W. Scheible, unpublished), and transcript fragments characteristic for miR827-mediated cleavage were found to be abundant in a degradome library from P-limited *A. thaliana* seedlings, but absent in a degradome library from P-replete seedlings (P. Nuc and W. Scheible, unpublished). *At1g02860* encodes a SPX (eucaryotic protein domain named after Syg1, Pho81 and XPR1)-domain E3 ligase protein. In yeast, SPX domain proteins were shown to participate in Pi transport or sensing (Lenburg and O'Shea 2006), and similar roles were now also established in plants (Rouached et al. 2010). *At1g02860* however was first described as nitrogen limitation adaptation (*NLA*) gene with a role in the N-limitation adaptation response (Peng et al. 2007). In contrast to wild-type plants, N-limited *nla* mutant plants had strongly reduced levels of anthocyanin in leaves and displayed early leaf senescence, suggesting that *NLA* is important for anthocyanin production in these conditions. During P-limitation or simultaneous N- and P-limitation, however, *nla* mutants showed wild type-like, high anthocyanin production and no early leaf senescence (Peng et al. 2008), indicating that a signal derived from P limitation is sufficient to induce anthocyanin production in the *nla* mutant. The link between P limitation and *NLA* provided by miR827 suggests that NLA activity is actively downregulated during P limitation. This could indicate that plants select one or the other input signal depending on nutrient conditions and, therefore, the existence of hierarchies in the interplay of macronutrient regulatory networks (cf. Pant et al. 2009). Yaeno and Iba (2008) isolated another mutant allele of *At1g02860/NLA*, i.e., *benzoic acid hypersensitive1-dominant* (*bah1-D*). These authors found that the E3 ligase encoded by *BAH1/NLA* is also involved in the regulation of salicylic acid accumulation and immune responses to the pathogen *Pseudomonas syringae* DC3000, possibly linking P-limitation, via miR827-BAH1/NLA, with plant pathogen resistance.

Although encoded by only one gene in *A. thaliana*, miR827 appears to be conserved. *MIRNA827* genes were found in rice, cotton, or rapeseed (http://www.mirbase.org, Sunkar and Jagadeeswaran 2008). Similarly, orthologs of *BAH1/NLA* were reported for rice and barrel medic (Yaeno and Iba 2008), and miR827-mediated cleavage of the rice ortholog (*Os04g0573000*) was experimentally verified (Lacombe et al. 2008).

The target gene of miR2111 (*At3g27150*, Hsieh et al. 2009; Pant et al. 2009), encodes a Kelch repeat-containing F-box protein for which a biological function is not known yet. The gene shows root-specific expression in large-scale transcriptome databases with a preferential localization in roots pericycle cells (http://bar.utoronto.ca/efp/cgi-bin/efpWeb.cgi), suggesting that it functions in the root. A potential ortholog of this gene containing a miR2111 binding site was identified in rapeseed and barrel medic (P. Nuc and W. Scheible, unpublished). It is noteworthy that the confirmed target genes of three strongly and specifically P-starvation inducible miRs (miR399, miR827 and miR2111; Table 1) encode proteins involved in ubiquitin-dependent protein degradation, suggesting a major role of this posttranslational mechanism in controlling P-limitation responses (Hsieh et al. 2009; Pant et al. 2009).

Target genes of P-limitation induced miR778 are *At2g22740/SUVH6* and *At2g35160/SUVH5*, encoding SET domain containing histone methyltransferases (Table 1), involved in regulation of histone methylation (Ebbs and Bender 2006) and thus playing an important role in heterochromatin formation and reprogramming of gene expression. The biological role for upregulation of miR778 in P-limitation still needs to be established, and this is also the case for upregulation of miR156 in P-, N-, or K-limitation. MiR156 regulates transcripts from the *SQUAMOSA-BINDING PROTEIN-LIKE* (*SPL*) gene family of transcription factors, which are known to be involved in juvenile-to-adult vegetative phase transition and other developmental processes (Wang et al. 2008, 2009; Wu and Poethig 2006). Potential biological roles of P-responsive miR*s, miR398, miR169, and their likely targets are discussed later in this chapter.

2 The Role of microRNA395 in the Regulation of Sulfur Assimilation and Allocation

Sulfur is another macronutrient with key importance for plant growth and development. Sulfur is found in amino acids (methionine and cysteine) and proteins, in sulfated polysaccharides, or in coenzymes and vitamins essential for metabolism (Leustek 2002; Saito 2004). Sulfur is also found in the molecules glutathione and thioredoxin and is, thus, important for redox processes. Plants acquire most of the sulfur as inorganic sulfate, which is taken up by plant roots from soil and transported into various tissues for assimilation. In *A. thaliana*, two high-affinity sulfate transporters, SULTR1;1 and SULTR1;2, are responsible for root uptake from soil, whereas SULTR2;1 and SULTR2;2 mediate translocation of sulfate within the plant. Intracellular sulfate reduction and assimilation into cysteine then occurs in a multistep reaction, in which plastidic ATP sulfurylase (APS) catalyzes the initial activation step by converting sulfate into 5′-adenylylsulfate (Leustek 2002; Saito 2004).

In *A. thaliana*, genes encoding plastidic APS (*APS1*, *APS3*, *APS4*) and the *SULTR2;1* gene are predicted and experimentally confirmed targets of miR395 (Allen et al. 2005; Jones-Rhoades and Bartel 2004; Kawashima et al. 2009; Liang et al. 2010), indicating that miR395 is involved in sulfate assimilation and translocation (Fig. 2). MiR395 is encoded by six genes, *MIRNA395a-f*, in *A. thaliana*. The six primary *MIRNA395* gene transcripts and the two processed mature miR395 species, that differ by one nucleotide, are strongly induced in –S conditions (Bari et al. 2006; Jones-Rhoades and Bartel 2004; Kawashima et al. 2009; Pant et al. 2009), although to different levels and in different patterns. For example, promoter-green fluorescent protein (GFP) experiments

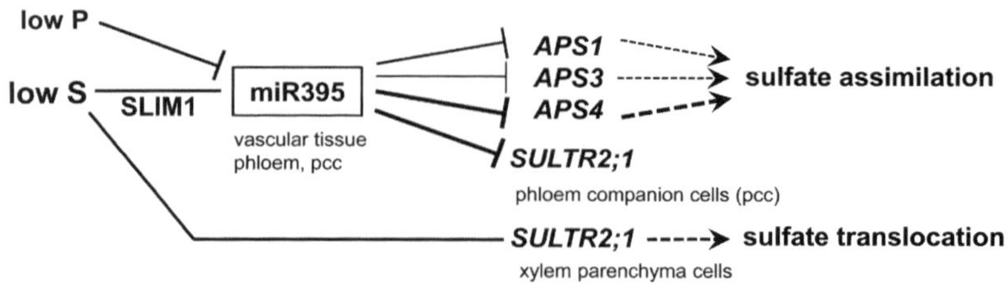

Fig. 2 Regulation of sulfate assimilation and translocation by miR395. Low S concentration leads to a strong, SLIM1-dependent induction of miR395, whereas low P represses miR395. MiR395 targets and hence inhibits expression of *APS* genes to various extents (*arrow* weight indicates the degree of inhibition) and expression of the sulfate transporter gene *SULTR2;1* in root-phloem companion cells (pcc). Expression of *SULTR2;1*, but not miR395, in root-xylem parenchyma cells probably ensures sulfate translocation from roots to shoots during S limitation. Weak inhibition of APS3 by miR395 allows sulfate assimilation to continue at low rates during S limitation. The model is derived from Hsieh et al. (2009), Kawashima et al. (2009), Liang et al. (2010), and Takahashi et al. (1997, 2000)

revealed strong expression of *MIRNA395c* and *e* in the same tissue- and cell-type specific pattern in roots and leaves, more moderate expression and similar patterns for *MIRNA395a* and *b*, and only low expression for *MIRNA395d* in roots but strong expression in root tips (Kawashima et al. 2009). Expression of *MIRNA395f* was not detected by this approach or RNA blot analysis, but a strong induction of *MIRNA395f* is easily revealed by more sensitive qRT-PCR (Pant et al. 2009).

Induction of *MIRNA395* genes and miR395 in –S conditions and in both roots and shoots is dependent on SULFUR LIMITATION1 (SLIM1) (Kawashima et al. 2009), a transcription factor from the ETHYLENE-INSENSITIVE-LIKE (EIL) family (Maruyama-Nakashita et al. 2006). *Slim1* mutants were initially found in a screen of *SULTR1;2* promoter-*GFP* reporter gene lines by their inability to induce *SULTR1;2* transcripts under low-sulfur (–S) conditions.

Given the upregulation of miR395, one would expect downregulation of the target gene transcripts during sulfur limitation, as was shown for *APS1* (Jones-Rhoades and Bartel 2004), and more recently also for *APS3* and *APS4* using miR395 overexpressing lines (Liang et al. 2010). The *SULTR2;1* transcript, however, was only slightly downregulated in S-limited wild-type leaves, and strongly increased in S-limited roots (Kawashima et al. 2009; Liang et al. 2010; Takahashi et al. 2000). Still, *SULTR2;1* is a miR395 target gene as was shown by 5′ RACE analysis (Kawashima et al. 2009) and by the use of miR395 overexpressers (Liang et al. 2010). Vice versa, a S-limited *slim1* mutant with strongly reduced miR395 levels displayed strong induction of *SULTR2;1* in leaves. High *SULTR2;1* expression in S-limited wild-type roots is the result of a rather cell-type specific localization of *SULTR2;1* transcript and miR395. Whereas miR395 was mainly expressed in root-phloem companion cells (Kawashima et al. 2009), the target *SULTR2;1* transcript was expressed in root-xylem parenchyma cells and less in phloem companion cells (Takahashi et al. 1997, 2000; Voinnet 2009) (Fig. 2). In accordance, the *slim1* mutation did not affect *SULTR2;1* expression in roots. The function of miR395 in this scenario, therefore, would be to restrict *SULTR2;1* expression to the root xylem, thus channeling sulfate flux to the xylem and favoring organ translocation (Kawashima et al. 2009). In general, it is also important to note that a negative temporal correlation between the expression level of a miR and its target gene in a complex tissue is not a requirement for target gene validation, as mutual exclusive spatial expression patterns might prevent transcript cleavage.

More details about the function of miR395 in the regulation of sulfate assimilation and translocation have been obtained recently. Overexpression of miR395 not just decreased target gene transcripts but also resulted in sulfur deficiency symptoms and overaccumulation of sulfate in the shoot (but not the root), as well as impaired relocation of sulfate between leaves (Liang et al. 2010). The combination of these phenotypes can also be achieved by simultaneous inhibition of the target genes in *aps1-1 sultr2;1 APS4*-RNAi plants (Liang et al. 2010). These results clearly showed that miR395 regulates the accumulation and distribution of sulfate between leaves by repressing *APS* genes and *SULTR2;1*, respectively.

MiR395 was shown to be responsive to P status too (Hsieh et al. 2009), being repressed during P limitation (Fig. 2), whereas the target *APS4* and *SULTR2;1* transcripts increased in these conditions. It was suggested that suppression of miR395 during P-limitation may contribute to this upregulation and increase sulfate assimilation and translocation for sulfolipid biosynthesis, thus compensating for the reduced phospholipids during P limitation (Hsieh et al. 2009).

Similar to miR399/*PHO2*, the miR395/*APS-SULTR2;1* regulatory module appears to be conserved in plants (Huang et al. 2010; Liang et al. 2010). Gene families encoding miR395 were reported from many higher plant species (http://www.mirbase.org), ranging between two (tomato) and 24 (rice) known loci. One gene was also found in the moss *P. patens*. Induction of miR395 was established in rapeseed (Buhtz et al. 2008; Liang et al. 2010) and barrel medic (Jones-Rhoades and Bartel 2004). Furthermore, miR395 was predicted to target both *APS* and sulfate transporter genes (*SULTR*) in rice, barrel medic, tomato, and sorghum, whereas, may be due to limited sequence data, only *APS* but not *SULTR* is a predicted target in poplar, rapeseed, corn, and grapevine (Liang et al. 2010). Downregulation of *APS* genes during S-limitation was demonstrated in rapeseed (Huang et al. 2010).

3 N-Responsive miRs Affect Root Growth and Architecture *in A. thaliana*

On a quantity basis, nitrogen (N) is the most important mineral macronutrient for plants. N is a major component of amino acids, proteins, nucleic acids, and a myriad of secondary metabolites. Plant N-status, thus, has deep impact on plant metabolism, growth, and development. N is taken up by roots predominantly as nitrate and ammonium. Nitrate per se and N-metabolites (e.g., glutamine) are thought to act as signals and to trigger plant responses, including nitrate uptake and assimilation (Crawford 1995), changes in primary and secondary metabolism, and developmental changes (Fritz et al. 2006; Stitt 1999). The signaling pathways underlying such nitrate or N-status dependent changes are not well understood, and only recently several molecular players have emerged (Krouk et al. 2010). MiR167 and miR393 have also been recently discovered and found to be involved in regulation of root growth in response to N. Another N-responsive miR species, miR169, was revealed by qRT-PCR analysis and small RNA sequencing (Pant et al. 2009) and RNA blot analysis (Hsieh et al. 2009).

3.1 MiR167 Mediates Lateral Root Outgrowth in Response to N

MiR167 is conserved across higher plant species (http://www.mirbase.org) and encoded by four and ten loci in *A. thaliana* and rice, respectively. In *A. thaliana*, transcripts of two auxin response transcription factor genes, *ARF6* and *ARF8*, are miR167-targets (Allen et al. 2005; Jones-Rhoades and Bartel 2004; Ru et al. 2006). MiR167 and its targets were shown to have developmental roles in the shoot, specifically in the development of male and female floral organs (Wu et al. 2006), as well as in lateral root (LR) formation (Tian et al. 2004).

By microarray analysis of cell-type specific responses after treatment of N-limited *A. thaliana* roots with nitrate, *ARF8* transcript was found to be induced in pericycle and LR cap cells (Gifford et al. 2008). Subsequent experiments using qRT-PCR and GUS-fusion constructs confirmed induction of *ARF8* and revealed repression of miR167 in response to N in the same cell types, consistent with an antagonist effect on *ARF8*. Consistently, an *ARF8-GUS* fusion with a mutated miR167-binding site showed loss of N regulation. N-repression of miR167 thus permits *ARF8* transcript to accumulate in the pericycle upon N treatment. In *A. thaliana* wild-type plants, N stimulates LR initiation in the pericycle, whereas N limitation leads to LR emergence and outgrowth (Gifford et al. 2008). The ratio of initiating to emerging LRs was, thus, higher in N-treated wild-type roots (~1) than in untreated roots (~0.4). Seedlings overexpressing miR167 or *arf8* mutant seedlings exhibited a complete loss of N-control over LR emergence, as the ratio remained low (~0.35). Hence, the miR167/ARF8 module acts downstream of N, and more specifically downstream of glutamine or a downstream metabolite (Gifford et al. 2008), in this pathway controlling LR outgrowth (Fig. 3).

MiR167 primary transcript and mature miR167 levels were not found to be considerably affected by N-status in whole-seedling samples using qRT-PCR (Pant et al. 2009) or Illumina small RNA sequencing (P. Nuc and W. Scheible, unpublished). However, different cell types in complex tissues/organs can differ dramatically in their transcriptional programs (Birnbaum et al. 2003); thus, cell-type specific responses, such as N-repression of miR167 in root pericycle cells, can be easily masked.

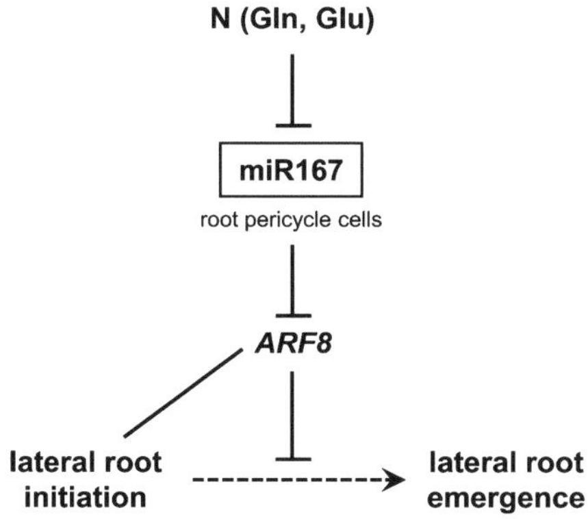

Fig. 3 Nitrogen control of lateral root (LR) outgrowth by miR167 and auxin response factor 8 (ARF8). Organic N sources, such as glutamine (Gln) or glutamate (Glu) repress miR167 expression in root pericycle cells. Accordingly, the expression of miR167-target ARF8 increases, leading to a stimulation of LR initiation in pericycle cells, but repression of LR emergence when N is available. The model is adapted from Gifford et al. (2008). Copyright (2008) National Academy of Sciences, USA

3.2 The miR393/AFB3 N-Regulatory Module Controls Root System Architecture by Integration of External and Internal N Availability

Using a high-throughput 454 sequencing approach for small RNAs, Vidal et al. (2010) have recently described miR393 as induced in roots 2 h after nitrate addition to *A. thaliana* wild-type seedlings that were grown on ammonium as the sole N source. This response of miR393 was absent in nitrate-reductase null mutants, indicating that the induction of miR393 is related to an N-metabolite downstream of nitrate reduction, and indeed treatment with ammonium or glutamate restored the response. The induction of miR393 by N is transient and far less marked compared to for example the response of miR399, miR395, or miR398 to P-, S-, or Cu-limitation, respectively. This probably explains why, similar to miR167, miR393 or *MIRNA393* primary transcripts were also not found as N-regulated in a comparison of N-replete and N-limited *A. thaliana* seedlings (see above).

The target gene of miR393, *AFB3*, encodes an auxin F-box receptor (AFB) and was ~4-fold induced already 1 h after nitrate addition to *A. thaliana* seedlings, but *AFB3* transcript decreased rapidly (2 or 4 h) thereafter, suggesting active transcript degradation by miR393 (Vidal et al. 2010). Consistent with this, (1) miR393 overexpressing lines had diminished *AFB3* levels, (2) lines expressing a miR393-resistant *AFB3* transcript under control of the *AFB3* promoter no longer showed a decrease of *AFB3* transcript in the hours following nitrate induction, and (3) NR null mutants displayed high and stable *AFB3* induction in the hours after nitrate addition (Vidal et al. 2010).

Nitrate addition increased *AFB3* expression and modulated the auxin response preferentially in root tips. Auxin-responsive genes and auxin-related genes not responsive to auxin (e.g., *ARF9*, *ARF18*, or an auxin efflux carrier) were regulated by nitrate, suggesting that nitrate, via the miR393/AFB3 module, modulates auxin signaling at multiple levels. Nitrate and auxin treatment are known to inhibit primary root growth (Walch-Liu and Forde 2008). Consistently, nitrate-treated wild-type plants have shorter primary roots as compared to control (KCl)-treated plants. Nitrate inhibition of primary root growth was also observed in the auxin-receptor mutants *tir1*, *afb1*, or *afb2*, but not in *afb3* mutant or miR393 overexpressing plants, indicating the specific involvement of the auxin-receptor AFB3 in this signal pathway (Vidal et al. 2010). In addition, these authors demonstrated that the miR393/AFB3 module is involved in a pathway that regulates LR growth in response to nitrate.

Nitrate/N-responsive miR393 and miR167 are not the only known miRs involved in regulation of LR growth, but possibly represent pathways to specifically integrate signals derived from N/nitrate. Two other miRs, i.e., miR164 and miR390, were implicated in the perception and modulation of auxin signals, thereby affecting *A. thaliana* LR development (Guo et al. 2005; Marin et al. 2010). Hence, several miRs appear to be at the core of a larger regulatory network that determines LR growth in balance with external and internal signals.

3.3 Potential Roles of Decreased miR169 Expression During N-Limitation

MiR169 is encoded by a large gene family in *A. thaliana* (14 loci) and other plant species that have arisen by tandem gene duplications (http://www.mirbase.org, Li et al. 2008, 2010b). Several phylogenetically related *MIRNA169* primary transcripts (pri-miR169h–n) were found to be strongly downregulated by N-limitation (Hsieh et al. 2009; Pant et al. 2009). The mature miR169 was also less abundant in P-limitation (see Sect. 1.2), predominantly in roots. The targets of miR169 in *A. thaliana* and other plant species are several transcription factor genes from the *HAP2* family (Combier et al. 2006; Fahlgren et al. 2007). The encoded transcription factors are also termed nuclear factor Y-A (NF-YA) subunits. Transcripts of several of these genes, including *NF-YA2*, *NF-YA5*, *NF-YA6*, *NF-YA8*, and *NF-YA10*, increase during N- and P-limitation (Pant et al. 2009; Scheible et al. 2004), thereby showing the opposite response compared to miR169. Furthermore, promoter-GUS analyses revealed colocalization of HAP2 and miR169 in vascular tissues and flower tissues, and miR169 overexpression or suppression of miR169 activity by target mimicry led to opposite changes in HAP2 expression (Chen et al. 2010).

The biological function(s) of miR169 during N- or P-limitation is/are not yet clear. However, it has been reported previously that miR169 influences drought resistance in *A. thaliana* via inhibition of *NF-YA5* (Li et al. 2008). *Nfya5* knockout mutants and plants overexpressing miR169 showed enhanced leaf water loss and were more sensitive to drought stress, whereas NF-YA5 overexpressers show opposite phenotypes (Li et al. 2008). Low expression of miR169 during N limitation could, therefore, contribute to the documented increased drought tolerance of N-limited plants (Lodeiro et al. 2000; Castaings et al. 2008). In legume species, another potential role of miR169 might be related to the development of N-fixing nodules during N-limitation. In barrel medic, the miR169-target gene *HAP2-1* is a key regulator for differentiation of nodule primordia, and miR169 overexpression or knockdown of the miR169-target gene *HAP2-1* leads to a developmental block of nodule development (Combier et al. 2006). Decreased expression of miR169 during N-limitation might, therefore, represent one of possibly several necessary signals to trigger nodule development. In this context, miR169 was found in phloem sap of rapeseed and was also found to decrease strongly in phloem sap during N- and P-limitation (see below), suggesting the intriguing possibility that miR169 also has a role as long-distance signal (for further details and discussion, see Pant et al. 2009).

4 A Role for miR398 and Other miRs in the Regulation of Copper Homeostasis and Oxidative Stress Tolerance

Copper (Cu) is an essential micronutrient for plants as a cofactor for proteins such as plastocyanin (PC), copper/zinc (Cu/Zn) superoxide dismutases (CSD), cytochrome c oxidase, or laccases (Burkhead et al. 2009, Fig. 4). Whereas PC as the most abundant Cu protein is essential, other Cu proteins can be functionally replaced

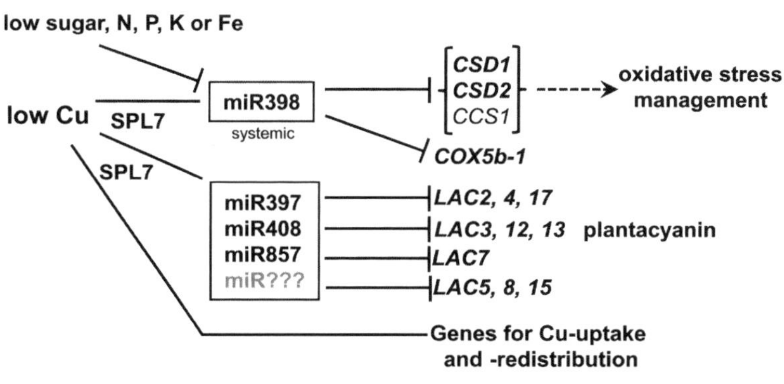

Fig. 4 MicroRNA regulation of copper homeostasis. Low Cu leads to a strong SPL7-dependent induction of miR398, whereas other nutrient limitations repress miR398 expression (see text for more information). During Cu limitation, the transcription factor SPL7 is also required for induction of several other miRs (miR397, miR408, miR857) and a series of genes required for Cu uptake and redistribution. MiR398 represses genes required for production of Cu/Zn superoxide dismutase involved in oxidative stress tolerance, as well as a subunit of the Cu-containing cytochrome c oxidase complex. MiR397, miR408, miR857, and probably another unknown SPL7-dependent miR (miR???) repress several genes encoding Cu-containing laccases and plantacyanin. Together, the miRs inhibit expression of genes encoding nonessential Cu-proteins (shown in bold italics) during Cu-limitation. The model is derived from Abdel-Ghany and Pilon (2008), Abdel-Ghany et al. (2005), Beauclair et al. (2010), Cohu et al. (2009), Hsieh et al. (2009), Pant et al. (2009), and Yamasaki et al. (2007, 2009)

(e.g., CSD by Fe superoxide dismutase) or are dispensable during Cu limitation. MiR398 and three other miRs have evolved as central regulators for these adaptations during Cu limitation in recent years.

Like miR395 and miR399, miR398 was originally computationally predicted and three target genes (i.e., *CSD1*and *CSD2* encoding cytosolic and chloroplastic CSD, respectively, and *At3g15640* encoding cytochrome c oxidase subunit V, COX5b) were confirmed by 5′-RACE analysis (Bonnet et al. 2004; Jones-Rhoades and Bartel 2004). Copper chaperone CCS1 (*CCS1*) is a more recently identified/confirmed target of miR398 (Beauclair et al. 2010; Cohu et al. 2009; Pant et al. 2009). *CCS1* encodes the chaperone required for Cu delivery to CSD and to generate the mature proteins. *CSD1*, *CSD2*, and *CCS1* are regulated by miR398 both by transcript cleavage and translational repression (Beauclair et al. 2010; Dugas and Bartel 2008). Under Cu-limited (<1 μM) conditions, *CSD* and *COX5b-1* transcripts are downregulated (Abdel-Ghany et al. 2005). This downregulation is mediated by miR398 which increases strongly in Cu-limitation (Abdel-Ghany et al. 2005; Yamasaki et al. 2007) (Fig. 4). MiR398, thus, appears to be a key factor in the regulation of Cu homeostasis, by shutting off gene expression of nonessential Cu proteins during Cu-limitation.

Three *MIRNA398* genes are known in *A. thaliana*. MiR398b and c, but not miR398a, are strongly induced in low Cu (Yamasaki et al. 2009). The transcription factor SQUAMOSA promoter binding protein-like7 (SPL7), a close homolog of Cu regulator Crr1 from *C. reinhardtii* (Kropat et al. 2005), has been shown to be essential for this induction. SPL7 binds directly to GTAC *cis* elements in the *MIRNA398* gene promoters in vitro, and these motifs are essential and sufficient for the response to Cu deficiency in vivo (Yamasaki et al. 2009). Under Cu limitation, SPL7 activates the transcription of more genes involved in Cu homeostasis, including Cu transporters (COPT1, COPT2, or ZIP2) and Cu chaperone, and is required for the accumulation of additional Cu-limitation inducible miRs, i.e., miR397, miR408, miR857 (Abdel-Ghany and Pilon 2008; Yamasaki et al. 2009). Interestingly, these miRs degrade transcripts encoding additional Cu proteins such as a series of laccases or plantacyanin (Abdel-Ghany and Pilon 2008) (Fig. 4) that are believed to function for example in lignin polymerization (Bao et al. 1997) or maintenance of cell wall structure and integrity (Ranocha et al. 2002). SPL7, thus, appears to be a central gene regulating the levels of many Cu proteins via microRNAs such as miR397, miR398b, miR398c, miR408, and miR857 (Yamasaki et al. 2009).

Similar to other plant antioxidant systems, CSD has an important role in counteracting stress-induced reactive oxygen species (ROS) that affect many cellular functions by damaging nucleic acids, oxidizing proteins, and causing lipid peroxidation (Foyer et al. 1994). ROS are also a major cause of reduced crop productivity (Apel and Hirt 2004). Accordingly, induction of *CSD1* and *CSD2* in *A. thaliana* in oxidative stress conditions was found (Kliebenstein et al. 1998) and established to be posttranscriptional mediated by downregulation of miR398 (Sunkar et al. 2006), demonstrating the importance of miR398 for oxidative stress tolerance. Moreover, transgenic *A. thaliana* plants overexpressing a miR398-resistant form of *CSD2* were more tolerant to high light, heavy metals, and other oxidative stresses, thus pointing out an approach for improving plant productivity under oxidative stress conditions (Sunkar et al. 2006). Accumulation of ROS in plants is known to be a secondary consequence of abiotic stresses (Apel and Hirt 2004) including nutrient stresses (Shin et al. 2005). Consistently, the expression of miR398 was found to be downregulated during ozone and salt stress (Jagadeeswaran et al. 2009), and miR398a, but not miR398b/c, was downregulated during N, P, K, Fe, or sugar limitation (Hsieh et al. 2009; Pant et al. 2009) (Fig. 4). Vice versa, miR398 and *MIRNA398c* promoter activity were also reported to be induced by sucrose (Dugas and Bartel 2008) independently of physiologically relevant levels of Cu (0–15 μM).

MIRNA398 genes were identified, although at low copy number, in a range of monocotyledonous and dicotyledonous plants, including *A. thaliana*, rice, corn, barrel medic, poplar, grapevine, or cotton (http://www.mirbase.org). In rice, *CSD1*, *CSD2*, and *CCS1* have been recently confirmed as genuine miR398 targets, and conservation of the miR398 binding site in *CCS1* revealed for nine monocotyledonous and dicotyledonous plant species (Li et al. 2010a). Similar to miR399 or miR395, these results suggest deep conservation of miR398 function.

5 Phloem Mobility and Systemic Signaling by Nutrient-Responsive miRs

Until recently miRs were regarded as molecules with local, cell autonomous activity (Alvarez et al. 2006; Tretter et al. 2008). MiR399 represented the first example that questioned this view (Lin et al. 2008; Pant et al. 2008), and further examples, i.e., miR172 and miR395, are meanwhile known (Martin et al. 2009; Buhtz et al. 2010, see below). Reciprocal micrografting experiments initially revealed that a root *pho2* genotype was necessary and sufficient for shoot Pi accumulation (cf. Sect. 1.1), suggesting that PHO2 acts in roots (Bari et al. 2006). However, during P-limitation the *A. thaliana MIRNA399* genes are much (~50-fold) more expressed in shoots than in roots in P-limited *A. thaliana* seedlings, and very high miR399 levels were shown to be required for *PHO2* inhibition (Bari et al. 2006), suggesting that it might be shoot-derived miR399 that leads to inhibition of PHO2 in roots. This hypothesis was confirmed by another set of micrografting experiments that demonstrated that P-replete chimeric plants composed of a wild-type rootstock and a miR399 overexpressing scion have (1) wild-type-like very low or undetectable *MIRNA399* gene expression, but high levels of mature miR399 in their wild-type roots and (2) 5–10-fold reduced root *PHO2* transcript levels, and 3–4-fold increased shoot Pi concentrations, like genotypes with genetic knockout of PHO2 in the root (Pant et al. 2008, also see Lin et al. 2008). Furthermore, miR399 was also shown to be present in phloem sap of P-limited, but not P-sufficient rapeseed and pumpkin plants (Buhtz et al. 2008; Pant et al. 2008). Together, these results showed for the first time systemic control of a biological process, i.e., maintenance of plant Pi homeostasis, by a phloem-mobile miR399 (Fig. 1).

Using high-throughput small RNA sequencing of phloem sap collected from nutrient-limited (–P, –Cu, –N, –S, –Fe) or nutrient-replete *B. napus* plants, a more systematic overview of phloem miRs in general and nutrient-responsive phloem miRs in particular was obtained (Buhtz et al. 2008, 2010; Pant et al. 2009). These studies revealed that a large number of known miRs (e.g., 32 miRs representing 18 families, Buhtz et al. 2008) are contained in the phloem and that several of these phloem miRs are not only present, but respond to changes in growth conditions. The abundance of many of the aforementioned P-, N-, S-, or Cu-responsive miRs changed drastically in phloem sap sampled from P-, N-, S-, or Cu-starved plants. This included a strong increase of miR399, miR827, miR2111 (but not miR778 or miR398) in phloem sap from P-limited plants (Pant et al. 2009), a strong increase of miR395 and miR398 in sap from S- and Cu-starved plants, respectively (Buhtz et al. 2008, 2010), or a decrease of miR169 in sap sampled from P- or N-limited plants (Pant et al. 2009). Moreover, miR399* and miR2111* species were abundant in phloem sap from P-limited plants (Pant et al. 2009). These results show that, besides miR399, other miRs also qualify as candidate long-distance signals during the adaptation of plants to nutrient limitations. The movement of miR395 during S-limitation through graft junctions from wild-type scions to rootstocks of miR processing *hen1-1* mutant, and downregulation of *APS4* target gene transcript has indeed been reported recently (Buhtz et al. 2010), suggesting functional importance of miR395 phloem movement. One situation in which this might be important is when a local shortage of sulfur occurs in leaves. This would lead to local production of miR395 that then moves via the phloem to the roots, where it reduces sulfate assimilation and channels S-flux to the xylem (cf. Sect. 2) and hence would improve S-translocation to the shoot.

The high abundance of miR2111 in phloem sap of P-starved rapeseed plants is interesting in conjunction with the root-specific expression of its target gene (cf. Sect. 1.3) and suggests another systemic regulatory circuitry during P-starvation, analogous to miR399-PHO2 paradigm. The significance of the highly abundant miR399* and miR2111* in phloem sap of P-starved plants is more controversial. A function for these small RNAs in gene regulation seems possible (cf. Sect. 1.3), and intriguing target genes were predicted (Pant et al. 2009, Table 1), but experimental confirmation failed thus far (Hsieh et al. 2009). Another possible function may be that these miR*s assist in long-distance movement of miR399 and miR2111 by forming miR/miR* duplexes in the phloem. Alternatively, these miR*s could somehow regulate the activity of their corresponding miRs via duplex formation (Hsieh et al. 2009).

6 Conclusions

Since their discovery, miRs had a comet-like rise as key regulators, both in plants and animals. The identification of new nutrient-responsive miRs, the elucidation of their biological roles, and the miR field in general are progressing rapidly, as indicated by the fact that much of the cited work in this chapter is very recent. The authors consider it possible that the current knowledge about nutrient-responsive miRs is just the tip of an iceberg. These miRs appear to be at the core of the regulation that occurs during plant adaptation to macro- and micronutrient stress. More and detailed knowledge is particularly expected from (1) the analysis of the newly identified nutrient-responsive miRs, (2) cell-type specific studies, and (3) the global investigation of the inhibitory effect of miRs on translation and protein output, which might uncover a range of unpredicted and unpredictable targets and specific functions for miR isoforms.

References

Abdel-Ghany SE, Pilon M (2008) MicroRNA-mediated systemic downregulation of copper protein expression in response to low copper availability in Arabidopsis. J Biol Chem 283:15932–15945

Abdel-Ghany SE, Müller-Moulé P, Niyogi KK, Pilon M, Shikanai T (2005) Two P-type ATPases are required for copper delivery in *Arabidopsis thaliana* chloroplasts. Plant Cell 17:1233–1251

Adai A, Johnson C, Mlotshwa S, Archer-Evans S, Manocha V, Vance V, Sundaresan V (2005) Computational prediction of miRNAs in *Arabidopsis thaliana*. Genome Res 15:78–91

Allen E, Xie Z, Gustafson AM, Carrington JC (2005) microRNA-directed phasing during trans-acting siRNA biogenesis in plants. Cell 121:207–221

Alvarez JP, Pekker I, Goldshmidt A, Blum E, Amsellem Z, Eshed Y (2006) Endogenous and synthetic microRNAs stimulate simultaneous, efficient, and localized regulation of multiple targets in diverse species. Plant Cell 18:1134–1151

Apel K, Hirt H (2004) Reactive oxygen species: metabolism, oxidative stress, and signal transduction. Annu Rev Plant Biol 55:373–399

Aung K, Lin SI, Wu CC, Huang YT, Su CL, Chiou TJ (2006) Pho2, a Pi overaccumulator, is caused by a nonsense mutation in a microRNA399 target gene. Plant Physiol 141:1000–1011

Bao W, O'malley DM, Whetten R, Sederoff RR (1997) A laccase associated with lignification in loblolly pine xylem. Science 260:672–674

Bari R, Datt Pant B, Stitt M, Scheible WR (2006) PHO2, microRNA399, and PHR1 define a Pi-signaling pathway in plants. Plant Physiol 141:988–999

Beauclair L, Yu A, Bouché N (2010) microRNA-directed cleavage and translational repression of the copper chaperone for superoxide dismutase mRNA in Arabidopsis. Plant J 62:454–462

Birnbaum K, Shasha DE, Wang JY, Jung JW, Lambert GM, Galbraith DW, Benfey PN (2003) A gene expression map of the Arabidopsis root. Science 302:1956–1960

Bonnet E, Wuyts J, Rouzé P, Van de Peer Y (2004) Detection of 91 potential conserved plant microRNAs in *Arabidopsis thaliana* and *Oryza sativa* identifies important target genes. Proc Natl Acad Sci USA 101:11511–11516

Bouché N (2010) New insights into miR398 functions in Arabidopsis. Plant Signal Behav 5(6):684–686

Branscheid A, Sieh D, Pant BD, May P, Devers EA, Elkrog A, Schauser L, Scheible WR, Krajinski F (2010) Expression pattern suggests a role of MiR399 in the regulation of the cellular response to local Pi increase during arbuscular mycorrhizal symbiosis. Mol Plant Microbe Interact 23:915–926

Brodersen P, Voinnet O (2009) Revisiting the principles of microRNA target recognition and mode of action. Nat Rev Mol Cell Biol 10:141–148

Brodersen P, Sakvarelidze-Achard L, Bruun-Rasmussen M, Dunoyer P, Yamamoto YY, Sieburth L, Voinnet O (2008) Widespread translational inhibition by plant miRNAs and siRNAs. Science 320:1185–1190

Buhtz A, Springer F, Chappell L, Baulcombe DC, Kehr J (2008) Identification and characterization of small RNAs from the phloem of *Brassica napus*. Plant J 53:739–749

Buhtz A, Pieritz J, Springer F, Kehr J (2010) Phloem small RNAs, nutrient stress responses, and systemic mobility. BMC Plant Biol 10:64

Burkhead JL, Gogolin Reynolds KA, Abdel-Ghahny SE, Cohu CM, Pilon M (2009) Copper homeostasis. New Phytol 182:799–816

Castaings L, Camargo A, Pocholle D, Gaudon V, Texier Y, Boutet-Mercey S, Taconnat L, Renou JP, Daniel-Vedele F, Fernandez E et al (2008) The nodule inception-like protein 7 modulates nitrate sensing and metabolism in Arabidopsis. Plant J 57:426–435

Chen JW, Chiu TY, Chiang SF, Chen YS, Chiou TJ (2010) Interplay of microRNA169 and HAP2 in response to phosphate deficiency. In: Poster 19017 at the 21st international conference for Arabidopsis research

Chiou TJ, Aung K, Lin SI, Wu CC, Chiang SF, Su CL (2006) Regulation of Pi homeostasis by microRNA in Arabidopsis. Plant Cell 18:412–421

Cohu CM, Abdel-Ghany SE, Gogolin Reynolds KA, Onofrio AM, Bodecker JR, Kimbrel JA, Niyogi KK, Pilon M (2009) Copper delivery by the copper chaperone for chloroplast and cytosolic copper/zinc superoxide dismutases: regulation and unexpected phenotypes in an Arabidopsis mutant. Mol Plant 2:1336–1350

Combier JP, Frugier F, de Billy F, Boualem A, El-Yahyaoui F, Moreau S, Vernié T, Ott T, Gamas P, Crespi M et al (2006) MtHAP2-1 is a key transcriptional regulator of symbiotic nodule development regulated by microRNA169 in *Medicago truncatula*. Genes Dev 20:3084–3088

Crawford NM (1995) Nitrate: nutrient and signal for plant growth. Plant Cell 7:859–868

Delhaize E, Randall PJ (1995) Characterization of a phosphate-accumulator mutant of *Arabidopsis thaliana*. Plant Physiol 107:207–213

Dugas DV, Bartel B (2008) Sucrose induction of Arabidopsis miR398 represses two Cu/Zn superoxide dismutases. Plant Mol Biol 67:403–417

Ebbs ML, Bender J (2006) Locus-specific control of DNA methylation by the Arabidopsis SUVH5 histone methyltransferase. Plant Cell 18:1166–1176

Fahlgren N, Howell MD, Kasschau KD, Chapman EJ, Sullivan CM, Cumbie JS, Givan SA, Law TF, Grant SR, Dangl JL et al (2007) High-throughput sequencing of Arabidopsis microRNAs: evidence for frequent birth and death of MIRNA genes. PLoS ONE 2:e219

Foyer CH, Descourvieres P, Kunert KJ (1994) Protection against oxygen radicals: an important defense mechanism studied in transgenic plants. Plant Cell Environ 17:507–523

Franco-Zorrilla JM, Valli A, Todesco M, Mateos I, Puga MI, Rubio-Somoza I, Leyva A, Weigel D, García JA, Paz-Ares J (2007) Target mimicry provides a new mechanism for regulation of microRNA activity. Nat Genet 39:1033–1037

Friedländer MR, Chen W, Adamidi C, Maaskola J, Einspanier R, Knespel S, Rajewsky N (2008) Discovering microRNAs from deep sequencing data using miRDeep. Nat Biotechnol 26:407–415

Fritz C, Palacios-Rojas N, Feil R, Stitt M (2006) Regulation of secondary metabolism by the carbon-nitrogen status in tobacco: nitrate inhibits large sectors of phenylpropanoid metabolism. Plant J 46:533–548

Fujii H, Chiou TJ, Lin SI, Aung K, Zhu JK (2005) A miRNA involved in Pi-starvation response in Arabidopsis. Curr Biol 15:2038–2043

Gifford ML, Dean A, Gutierrez RA, Coruzzi GM, Birnbaum KD (2008) Cell-specific nitrogen responses mediate developmental plasticity. Proc Natl Acad Sci USA 105:803–808

Guo HS, Xie Q, Fei JF, Chua NH (2005) MicroRNA directs mRNA cleavage of the transcription factor NAC1 to downregulate auxin signals for Arabidopsis lateral root development. Plant Cell 17:1376–1386

Hsieh LC, Lin SI, Shih AC, Chen JW, Lin WY, Tseng CY, Li WH, Chiou TJ (2009) Uncovering small RNA-mediated responses to Pi deficiency in Arabidopsis by deep sequencing. Plant Physiol 151:2120–2132

Huang SQ, Xiang AL, Che LL, Chen S, Li H, Song JB, Yang ZM (2010) A set of miRNAs from *Brassica napus* in response to sulphate deficiency and cadmium stress. Plant Biotechnol J. doi:10.1111/j.1467-7652.2010.00517.x

Jagadeeswaran G, Zheng Y, Li YF, Shukla LI, Matts J, Hoyt P, Macmil SL, Wiley GB, Roe BA, Zhang W, Sunkar R (2009) Cloning and characterization of small RNAs from *Medicago truncatula* reveals four novel legume-specific microRNA families. New Phytol 184:85–98

Jones-Rhoades MW, Bartel D (2004) Computational identification of plant microRNAs and their targets, including a stress-induced miRNA. Mol Cell 18:787–799

Kawashima CG, Yoshimoto N, Maruyama-Nakashita A, Tsuchiya YN, Saito K, Takahashi H, Dalmay T (2009) Sulphur starvation induces the expression of microRNA-395 and one of its target genes but in different cell types. Plant J 57:313–321

Kliebenstein DJ, Monde RA, Last RL (1998) Superoxide dismutase in Arabidopsis: an eclectic enzyme family with disparate regulation and protein localization. Plant Physiol 118:637–650

Kropat J, Tottey S, Birkenbihl RP, Depège N, Huijser P, Merchant S (2005) A regulator of nutritional copper signaling in Chlamydomonas is an SBP domain protein that recognizes the GTAC core of copper response element. Proc Natl Acad Sci USA 102:18730–18735

Krouk G, Crawford NM, Coruzzi GM, Tsay YF (2010) Nitrate signaling: adaptation to fluctuating environments. Curr Opin Plant Biol 13:266–273

Lacombe S, Nagasaki H, Santi C, Duval D, Piegu B, Bangratz M, Breitler JC, Guiderdoni E, Brugidou C, Hirsch J, Cao X, Brice C, Panaud O, Karlowski WM, Sato Y, Echeverria M (2008) Identification of precursor transcripts for 6 novel miRNAs expands the diversity on the genomic organisation and expression of miRNA genes in rice. BMC Plant Biol 8:123

Lelandais-Brière C, Naya L, Sallet E, Calenge F, Frugier F, Hartmann C, Gouzy J, Crespi M (2009) Genome-wide *Medicago truncatula* small RNA analysis revealed novel microRNAs and isoforms differentially regulated in roots and nodules. Plant Cell 21:2780–2796

Lenburg ME, O'Shea EK (2006) Signaling Pi starvation. Trends Biochem Sci 21:383–387

Leustek T (2002) Sulfate metabolism. In: Somerville CR, Meyerowitz EM (eds) The Arabidopsis Book. American Society of Plant Biologists, Rockville

Li WX, Oono Y, Zhu J, He XJ, Wu JM, Iida K, Lu XY, Cui X, Jin H, Zhu JK (2008) The Arabidopsis NFYA5 transcription factor is regulated transcriptionally and posttranscriptionally to promote drought resistance. Plant Cell 20:2238–2251

Li YF, Zheng Y, Addo-Quaye C, Zhang L, Saini A, Jagadeeswaran G, Axtell MJ, Zhang W, Sunkar R (2010a) Transcriptome-wide identification of microRNA targets in rice. Plant J 62:742–759

Li Y, Fu Y, Ji L, Wu C, Zheng C (2010b) Characterization and expression analysis of the Arabidopsis miR169 family. Plant Sci 178:271–280

Liang G, Yang F, Yu D (2010) MicroRNA395 mediates regulation of sulfate accumulation and allocation in *Arabidopsis thaliana*. Plant J 62:1046–1057

Lin SI, Chiou TJ (2008) Long-distance movement and differential targeting of microRNA399s. Plant Signal Behav 3:730–732

Lin SI, Chiang SF, Lin WY, Chen JW, Tseng CY, Wu PC, Chiou TJ (2008) Regulatory network of microRNA399 and PHO2 by systemic signaling. Plant Physiol 147:732–746

Lodeiro AR, González P, Hernández A, Balagué LJ, Favelukes G (2000) Comparison of drought tolerance in nitrogen-fixing and inorganic nitrogen-grown common beans. Plant Sci 154:31–41

Lundmark M, Kørner CJ, Nielsen TH (2010) Global analysis of microRNA in Arabidopsis in response to Pi starvation as studied by locked nucleic acid-based microarrays. Physiol Plant. doi:10.1111/j.1399-3054.2010.01384.x

Marin E, Jouannet V, Herz A, Lokerse AS, Weijers D, Vaucheret H, Nussaume L, Crespi MD, Maizel A (2010) miR390, Arabidopsis TAS3 tasiRNAs, and their AUXIN RESPONSE FACTOR targets define an autoregulatory network quantitatively regulating lateral root growth. Plant Cell 22:1104–1117

Martin A, Adam H, Díaz-Mendoza M, Zurczak M, González-Schain ND, Suárez-López P (2009) Graft-transmissible induction of potato tuberization by the microRNA miR172. Development 136:2873–2881

Maruyama-Nakashita A, Nakamura Y, Tohge T, Saito K, Takahashi H (2006) Arabidopsis SLIM1 is a central transcriptional regulator of plant sulfur response and metabolism. Plant Cell 18:3235–3251

Mi S, Cai T, Hu Y, Chen Y, Hodges E, Ni F, Wu L, Li S, Zhou H, Long C et al (2008) Sorting of small RNAs into Arabidopsis argonaute complexes is directed by the 5# terminal nucleotide. Cell 133:116–127

Morcuende R, Bari R, Gibon Y, Zheng W, Pant BD, Bläsing O, Usadel B, Czechowski T, Udvardi MK, Stitt M et al (2007) Genome-wide reprogramming of metabolism and regulatory networks of Arabidopsis in response to phosphorus. Plant Cell Environ 30:85–112

Okamura K, Phillips MD, Tyler DM, Duan H, Chou YT, Lai EC (2008) The regulatory activity of microRNA* species has substantial influence on microRNA and 3′ UTR evolution. Nat Struct Mol Biol 15:354–363

Pant BD, Buhtz A, Kehr J, Scheible WR (2008) MicroRNA399 is a long-distance signal for the regulation of plant Pi homeostasis. Plant J 53:731–738

Pant BD, Musialak-Lange M, Nuc P, May P, Buhtz A, Kehr J, Walther D, Scheible WR (2009) Identification of nutrient-responsive Arabidopsis and rapeseed microRNAs by comprehensive real-time polymerase chain reaction profiling and small RNA sequencing. Plant Physiol 150:1541–1555

Peng M, Hannam C, Gu H, Bi YM, Rothstein SJ (2007) A mutation in NLA, which encodes a RING-type ubiquitin ligase, disrupts the adaptability of Arabidopsis to nitrogen limitation. Plant J 50:320–337

Peng M, Hudson D, Schofield A, Tsao R, Yang R, Gu H, Bi YM, Rothstein SJ (2008) Adaptation of Arabidopsis to nitrogen limitation involves induction of anthocyanin synthesis which is controlled by the NLA gene. J Exp Bot 59:2933–2944

Poirier Y, Bucher M (2002) Phosphate transport and homeostasis in Arabidopsis. In: Somerville CR, Meyerowitz EM (eds) The Arabidopsis book. American Society of Plant Biologists, Rockville

Ranocha P, Chabannes M, Chamayou S, Danoun S, Jauneau A, Boudet AM, Goffner D (2002) Laccase down-regulation causes alterations in phenolic metabolism and cell wall structure in poplar. Plant Physiol 129:145–155

Rhoades MW, Reinhart BJ, Lim LP, Burge CB, Bartel B, Bartel DP (2002) Prediction of plant microRNA targets. Cell 110:513–520

Rouached H, Arpat AB, Poirier Y (2010) Regulation of phosphate starvation responses in plants: signaling players and cross-talks. Mol Plant 3:288–299

Ru P, Xu L, Ma H, Huang H (2006) Plant fertility defects induced by the enhanced expression of microRNA167. Cell Res 16:457–465

Rubio V, Linhares F, Solano R, Martín AC, Iglesias J, Leyva A, Paz-Ares J (2001) A conserved MYB transcription factor involved in phosphate starvation signaling both in vascular plants and in unicellular algae. Genes Dev 15:2122–2133

Saito K (2004) Sulfur assimilatory metabolism. The long and smelling road. Plant Physiol 136:2443–2450

Scheible WR, Morcuende R, Czechowski T, Fritz C, Osuna D, Palacios-Rojas N, Schindelasch D, Thimm O, Udvardi MK, Stitt M (2004) Genome-wide reprogramming of primary and secondary metabolism, protein synthesis, cellular growth processes, and the regulatory infrastructure of Arabidopsis in response to nitrogen. Plant Physiol 136:2483–2499

Shin R, Berg RH, Schachtman DP (2005) Reactive oxygen species and root hairs in Arabidopsis root response to nitrogen, phosphorus and potassium deficiency. Plant Cell Physiol 46:1350–1357

Stitt M (1999) Nitrate regulation of metabolism and growth. Curr Opin Plant Biol 2:178–186

Sunkar R, Jagadeeswaran G (2008) In silico identification of conserved microRNAs in large number of diverse plant species. BMC Plant Biol 8:37

Sunkar R, Zhu JK (2004) Novel and stress-regulated microRNAs and other small RNAs from Arabidopsis. Plant Cell 16:2001–2019

Sunkar R, Kapoor A, Zhu JK (2006) Posttranscriptional induction of two Cu/Zn superoxide dismutase genes in Arabidopsis is mediated by downregulation of miR398 and important for oxidative stress tolerance. Plant Cell 18:2051–2065

Takahashi H, Yamazaki M, Sasakura N, Watanabe A, Leustek T, Engler JA, Engler G, Van Montagu M, Saito K (1997) Regulation of sulfur assimilation in higher plants: a sulfate transporter induced in sulfate-starved roots plays a central role in *Arabidopsis thaliana*. Proc Natl Acad Sci USA 94:11102–11107

Takahashi H, Watanabe-Takahashi A, Smith FW, Blake-Kalff M, Hawkesford MJ, Saito K (2000) The roles of three functional sulphate transporters involved in uptake and translocation of sulphate in *Arabidopsis thaliana*. Plant J 23:171–182

Tian CE, Muto H, Higuchi K, Matamura T, Tatematsu K, Koshiba T, Yamamoto KT (2004) Disruption and overexpression of auxin response factor 8 gene of Arabidopsis affect hypocotyl elongation and root growth habit, indicating its possible involvement in auxin homeostasis in light condition. Plant J 40:333–343

Todesco M, Rubio-Somoza I, Paz-Ares J, Weigel D (2010) A collection of target mimics for comprehensive analysis of microRNA function in *Arabidopsis thaliana*. PLoS Genet 6:e1001031

Tretter EM, Alvarez JP, Eshed Y, Bowman JL (2008) Activity range of Arabidopsis small RNAs derived from different biogenesis pathways. Plant Physiol 147:58–62

Valdés-López O, Arenas-Huertero C, Ramírez M, Girard L, Sánchez F, Vance CP, Luis Reyes J, Hernández G (2008) Essential role of MYB transcription factor: PvPHR1 and microRNA: PvmiR399 in phosphorus-deficiency signalling in common bean roots. Plant Cell Environ 31:1834–1843

Vidal EA, Araus V, Parry G, Green PJ, Coruzzi GM, Gutierrez RA (2010) Nitrate-responsive miR393/*AFB3* regulatory module controls root system architecture in *Arabidopsis thaliana*. Proc Natl Acad Sci USA 107:4477–4482

Voinnet O (2009) Origin, biogenesis, and activity of plant microRNAs. Cell 136:669–687

Walch-Liu P, Forde BG (2008) Nitrate signalling mediated by the NRT1.1 nitrate transporter antagonises L-glutamate-induced changes in root architecture. Plant J 54:820–828

Wang JW, Schwab R, Czech B, Mica E, Weigel D (2008) Dual effects of miR156-targeted SPL genes and CYP78A5/KLUH on plastochron length and organ size in *Arabidopsis thaliana*. Plant Cell 20:1231–1243

Wang JW, Czech B, Weigel D (2009) miR156-regulated SPL transcription factors define an endogenous flowering pathway in *Arabidopsis thaliana*. Cell 138:738–749

Wu G, Poethig RS (2006) Temporal regulation of shoot development in *Arabidopsis thaliana* by miR156 and its target SPL3. Development 133:3539–3547

Wu MF, Tian Q, Reed JW (2006) Arabidopsis microRNA167 controls patterns of ARF6 and ARF8 expression, and regulates both female and male reproduction. Development 133:4211–4218

Yaeno T, Iba K (2008) BAH1/NLA, a RING-type ubiquitin E3 ligase, regulates the accumulation of salicylic acid and immune responses to *Pseudomonas syringae* DC3000. Plant Physiol 148:1032–1041

Yamasaki H, Abdel-Ghany SE, Cohu CM, Kobayashi Y, Shikanai T, Pilon M (2007) Regulation of copper homeostasis by micro-RNA in Arabidopsis. J Biol Chem 282:16369–16378

Yamasaki H, Hayashi M, Fukazawa M, Kobayashi Y, Shikanai T (2009) SQUAMOSA promoter binding protein-like7 is a central regulator for copper homeostasis in Arabidopsis. Plant Cell 21:347–361

MicroRNA Function in Seed Biology

Wioletta E. Pluskota, Cristina Martínez-Andújar, Ruth C. Martin, and Hiroyuki Nonogaki

Contents

Abstract The involvement of microRNA (miRNA) in various developmental programs in plants, such as leaf and flower development, and the response of plants to biotic and abiotic stresses, has been characterized. Information on the involvement of miRNA in seed biology is limited, but emerging. miRNAs and transcription factors or other genes that are targeted by miRNA are expressed during embryogenesis, seed maturation, imbibition, and seedling establishment, suggesting that the downregulation of target genes by miRNA may play a critical role in seed development and germination. This chapter focuses on the biological function of miRNA and miRNA-targeted genes involved in seed formation, the sensitivity of mature seeds to plant hormones, and postembryonic seedling development.

H. Nonogaki (✉) and C. Martínez-Andújar
Department of Horticulture, Oregon State University, Corvallis, OR 97331, USA
e-mail: hiro.nonogaki@oregonstate.edu

W.E. Pluskota
Department of Plant Physiology and Biotechnology, University of Warmia and Mazury, Oczapowskiego 1A, 10-718 Olsztyn, Poland

R.C. Martin
USDA-ARS, National Forage Seed Production Research Center, Corvallis, OR 97331, USA

V.A. Erdmann and J. Barciszewski (eds.), *Non Coding RNAs in Plants*, RNA Technologies, DOI 10.1007/978-3-642-19454-2_21,

Keywords Development • Dormancy • Embryogenesis • Germination • miRNA • Seed

1 MicroRNAs in Seed Development

Results of massive sequencing of small RNAs (sRNAs) in developing rice seeds (embryo and endosperm) and vegetative tissues (roots, leaves, and seedlings) indicated that many miRNAs are differentially expressed in seed and vegetative tissues (Xue et al. 2009). Differential expression of miRNAs between seeds and other tissues has also been demonstrated using *Arabidopsis* mutants (Zhan and Lukens 2010). Seed developmental programs seem to be controlled by a specific set of miRNAs and their target genes in both monocots and dicots. Their function in early embryogenesis, endosperm development, and seed maturation is described here.

1.1 Embryogenesis

Embryo and endosperm development is initiated upon fertilization, except in some special cases such as apomixis where seeds and fruits develop without fertilization (Koltunow and Grossniklaus 2003). Double fertilization involves the fusion of one of the two haploid sperm nuclei in a pollen tube with a haploid egg cell, which gives rise to the diploid embryo, and the fusion of the second haploid pollen sperm nucleus with two haploid nuclei in the central cell, which differentiates into the triploid endosperm. Recent studies have revealed information about the involvement of miRNAs and their targets in prefertilization events. sRNA involvement in male gametophyte development is described in another chapter. Here, the possible involvement of *AUXIN RESPONSE FACTORs* (*ARFs*), some of which are the targets of miRNAs, in female gametophyte development is examined. In angiosperms, the unique process of egg cell specification requires the asymmetric distribution of auxin within the embryo sac (Pagnussat et al. 2009). Proper accumulation of ARFs, auxin signal transduction proteins, in the embryo sac is also required (Pagnussat et al. 2009). Artificial miRNAs (amiRNA) can be designed to target individual genes or groups of endogenous genes (Schwab et al. 2006). This approach was used to characterize the function of ARFs in the embryo sac. An amiRNA that has the backbone of miR164 and that targets a short sequence largely conserved among nine *ARFs* (*ARF1–8* and *ARF19*) (termed as *ami-ARFa*) was generated and expressed in plants under the control of an embryo sac promoter. The down-regulation of *ARFs* caused by *ami-ARFa* resulted in defective embryo sacs containing multiple egg cells or lacking specification between the egg cell and synergid cells (Pagnussat et al. 2009). These results suggest that miRNA-targeted *ARF*s (*ARF6* and *ARF8*) (Rhoades et al. 2002; Wu et al. 2006), *trans*-acting siRNA (tasiRNA)-targeted *ARF*s (*ARF2*, *ARF3* and *ARF4*) (Allen et al. 2005; Williams et al. 2005), and/or sRNA nontargeted *ARF*s are involved in the specification of the female gametophyte.

Following proper development of the female and male gametophytes and successful fertilization, cell division occurs in the zygote, and divided cells proceed with

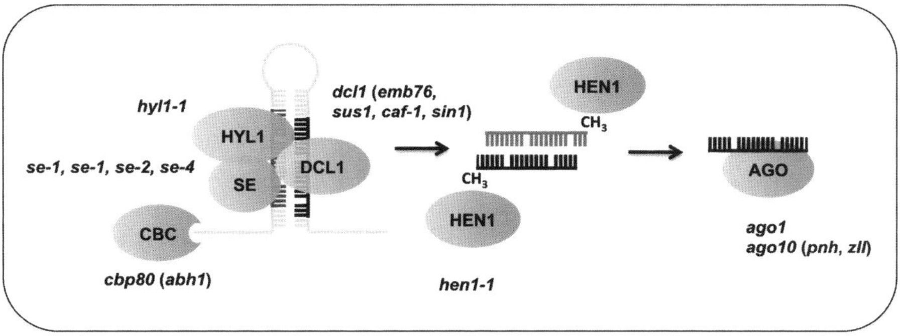

Fig. 1 Schematic representation of the miRNA processing pathway and the key proteins described in the text for miRNA biogenesis-associated embryogenesis mutants. The names of mutants associated with these components that exhibit embryo defects are also shown. See text for full names of proteins and mutants. Only major events during miRNA maturation are shown here to follow the description in the text. For a more comprehensive schematic of this pathway, see review articles (Montgomery and Carrington 2008; Voinnet 2009)

histodifferentiation, which is a major event during early embryogenesis. In animals, miRNAs are involved in major developmental processes such as neurogenesis and cardiovascular development during embryogenesis (Stefani and Slack 2008; Liu and Olson 2010). Strong evidence supporting the involvement of miRNAs in plant embryogenesis is provided from mutants defective in *DICER-LIKE1* (*DCL1*), a homolog of the *Drosophila melanogaster* gene *Dicer* (Golden et al. 2002). Formation of miRNAs requires an RNaseIII domain-containing protein, termed *DICER-1* in animals and *DCL1* in plants, to catalyze the processing of a miRNA precursor with a fold-back structure (Hannon 2002; Xie et al. 2003) (Fig. 1). Interestingly, *DCL1* itself is a target of miR162, indicating that *DCL1* mRNA is subject to negative feedback regulation (Xie et al. 2003). Mutants defective in *DCL1*, which has also been designated with other gene names such as *EMBRYO DEFECTIVE76* (*EMB76*) (Franzmann et al. 1995), *SUSPENSOR1* (*SUS1*), *CARPEL FACTORY* (*CAF*) (Jacobsen et al. 1999), or *SHORT INTEGUMENTS1* (*SIN1*) (Golden et al. 2002), contain low levels of miRNA and are embryo-lethal (Schauer et al. 2002). These results indicate that regulation of gene expression by miRNA is essential for normal embryogenesis in plants.

The most prominent phenotype of *emb76*, a *DCL1*-defective mutant, is observed in the suspensor that supports the embryo and provides nutrients and hormones to the developing embryo. Therefore, this mutant has been renamed as *sus1* (Schwartz et al. 1994). The suspensor, in addition to providing physical and chemical support to the embryo, also plays a critical role in the coordinated operation of the embryo developmental programs. Regardless of miRNA involvement, many embryo-defective mutants have aberrant cell division in the suspensor. For example, the abnormal pattern of cell division in the suspensor cells in *lec1* mutants leads to the formation of a secondary embryo in addition to the authentic embryo originating from the embryo proper (Lotan et al. 1998). This observation provides two important implications: (1) suspensor cells have totipotency to differentiate into an embryo and (2) the embryonic program is strictly suppressed in suspensor cells during normal embryogenesis. The suppression of suspensor proliferation during normal embryogenesis is thought to be maintained by an inhibitory signal from the embryo proper (Schauer et al. 2002). The *emb76/sus1* mutants, which are defective in *DCL1* and hence in miRNA processing, exhibit embryo arrest and fail to suppress proliferation of the suspensor cells (Schwartz et al. 1994), probably because a suppressive signal from the embryo to the suspensor is missing in these mutants.

The phenotypes of other *DCL1*-defective mutants, *caf-1* and *sin1*, indicate that embryo arrest in these mutants is associated with abnormal development of integuments, the maternal tissue from which the testa or seed coat originates (Lang et al. 1994; Golden et al. 2002). In wild-type ovules, the outer integument cell layers totally cover the inner integument that encloses the embryo sac. By contrast, most ovules in *caf-1* and *sin1* mutants show uncoordinated growth of both the inner and outer integuments (Golden et al. 2002). These results suggest that *DCL1* and miRNAs play a role in embryogenesis not only through their function in the zygote but also through gametophytic maternal effects on the development of the embryo (Ray et al. 1996).

The embryo lethality phenotypes of *dcl1* mutants are probably due to the deregulation of multiple target genes from miRNAs. Nearly 50 miRNA target genes are overexpressed in *dcl1* mutant embryos (Nodine and Bartel 2010). *SQUAMOSA PROMOTER-BINDING PROTEIN-LIKE10* (*SPL10*) and *SPL11*, which are targeted by miR156, are derepressed over 100-fold in the *dcl1* mutant embryos. The precise suppression of *SPL10* and *SPL11* by miR156 in the eight-cell-stage embryo seems critical for preventing premature accumulation of transcripts from hundreds of genes normally induced during the maturation phase of embryo development (Nodine and Bartel 2010). miR156 and other *SPLs* targeted by miR156 are also involved in developmental stages after germination of mature seeds.

The phenotypes of multiple *DCL1*-defective mutants indicate the importance of miRNA processing during embryogenesis. For the accurate processing of pri- and pre-miRNA, SERRATE (SE), a C_2H_2-type zinc finger protein (Prigge and Wagner 2001), and HYPONASTIC LEAVES1 (HYL1), a double-stranded RNA-binding protein (Lu and Fedoroff 2000), are required in addition to DCL1 (Montgomery and Carrington 2008) (Fig. 1). SE and HYL1 interact with each other (Lobbes et al. 2006). They also interact with DCL1 in nuclear dicing bodies (Fang and Spector 2007; Fujioka et al. 2007). While SE may also be a component of the large nuclear cap-binding complex (CBC) and play an additional role in splicing of mRNA, both SE and HYL1 are required for proper processing of pri-miRNA (Laubinger et al. 2008). In the mutants defective in *SE* and *HYL1*, pri-miRNA levels are increased, whereas mature miRNA levels are decreased (Yang et al. 2006b). Mutations in these components of the miRNA processing machinery also affect embryo development.

SE is expressed in the meristems of the shoot and root and in the adaxial portion of the cotyledons of torpedo-stage embryos, with less expression in walking-stick-stage embryos and no expression in the mature embryo (Prigge and Wagner 2001). *se-1* mutants develop abnormal embryos similar to *sin1* mutants that are defective in *DCL1* (Ray et al. 1996). *se-2* and *se-3* mutants also show severe embryo defects, with *se-3* exhibiting conditional lethality (Grigg et al. 2005). Mutations of the strong *se-4* allele confer embryo lethality. In this mutant, abnormal development and cell divisions are visible from the heart stage onward. Cotyledon primordia are not recognized by cell shape or arrangement in *se-4* mutant embryos. Irregular cell divisions continue throughout the abnormal embryos during the torpedo stage and result in irregularly shaped mature embryos, which become lethal (Lobbes et al. 2006).

Consistent with the critical role of SE in the general miRNA processing pathway, microarray analysis indicates accumulation of pri-miRNAs and deregulation of miRNA-targeted genes in *se* mutants compared with wild type (Lobbes et al. 2006). The *se* mutant phenotypes probably result from a combination of effects on multiple miRNA targets that are specifically upregulated in this mutant. The involvement of *PHABULOSA* (*PHB*), a *homeodomain leucine zipper* (*HD-Zip*) *III* gene, which is targeted by miR165/miR166 (Reinhart et al. 2002; Tang et al. 2003), in the *se* mutant phenotypes has been well characterized. PHB functions in determining adaxial leaf fate (McConnell et al. 2001). *PHB* is expressed during the globular and later stages of embryogenesis. In *se-3* mutants that exhibit conditional embryo lethality, *PHB* expression in the globular embryo is elevated, and its expression domain is spatially expanded compared to that in wild-type embryos (Grigg et al. 2005). Aberrant expression of *PHB* that is caused by the reduction in both SE function and miR165/miR166 levels is responsible, at least in part, for the embryo defects observed in *se* mutants. Interestingly, a mutation in an *SE*-like gene in animals has also been shown to cause defects in embryo development (Golling et al. 2002; Grigg et al. 2005).

Mutations in HYL1, an interaction partner for SE and DCL1, which is also essential for normal RNA processing (Fig. 1), also affect embryo development. However, unlike *dcl1*, *se-3* and *se-4* mutants, *hyl1* mutants are not embryo-lethal, although they do have a strongly compromised seed set (Vazquez et al. 2004). When *hyl1* mutants are crossed with *se-1* mutants, siliques from the F1 progeny produce abortive seeds due to embryo lethality in *hyl-1 se-1* double mutants (Yang et al. 2006b) (Fig. 2). Thus, proper miRNA processing by the DCL1-SE-HYL1 complex is essential for normal embryogenesis.

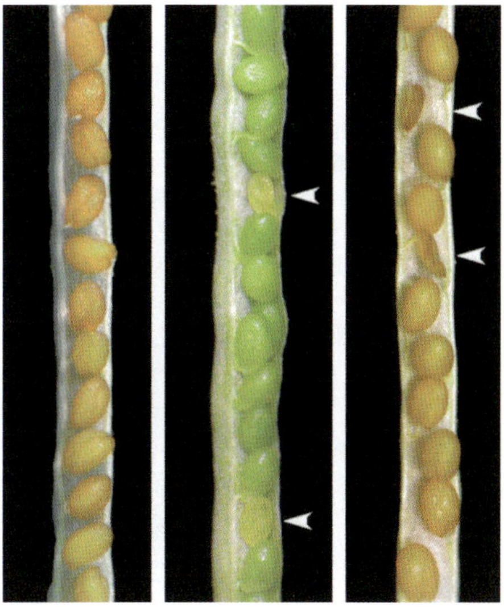

Fig. 2 Embryo lethality observed in miRNA-associated mutants. *Left*, a wild-type silique containing normally developed mature seeds; *middle* and *right*, early and mature stages of siliques from F1 progeny of a cross between and *hyponastic leaves1-1* (*hyl1-1*) and *serrate-1* (*se-1*) mutants, which contain the abortive seeds (*arrowheads*) (Yang et al. 2006b). Copyright Wiley-Blackwell

Following proper processing of pri- and pre-miRNA and the excision of the miRNA/miRNA* duplex, a methyltransferase HUA ENHANCER1 (HEN1) methylates the ribose of the last nucleotide on each strand of the duplex, which is important for miRNA maturation (Chen 2005; Yu et al. 2005) (Fig. 1). Therefore, mutations in *HEN1* hinder miRNA biogenesis. *miR159*, *miR163*, *miR167*, *miR173*, *miR176*, and *miR177* are undetectable in *hen1-1* mutants and in *caf-1*, a *dcl1* mutant. The various developmental defects observed in *hen1-1* mutants are similar to those of *caf/sin1* mutants (Park et al. 2002). Therefore, methylation of the miRNA/miRNA* duplex is also an important step in the maturation of miRNA and is essential for the completion of normal embryogenesis.

Properly processed mature miRNA is loaded as a guide RNA into a complex containing an ARGONAUTE (AGO) protein (Fig. 1). The AGO-miRNA complex functions to suppress transcripts by either irreversible cleavage or translational repression (Chapman and Carrington 2007; Montgomery and Carrington 2008). The fact that AGO itself is targeted by miR168 (Rhoades et al. 2002) represents another negative feedback regulation of the miRNA pathway by a miRNA analogous to the negative feedback regulation of *DCL1* by miR162 (Xie et al. 2003). *Arabidopsis ago1* mutant is not embryo-lethal (Bohmert et al. 1998), which is likely due to partial functional redundancies among the ten AGO family members in this species (Mallory and Vaucheret 2006). When a mutation in *AGO10* (also called *PINHEAD* [*PNH*] or *ZWILLE* [*ZLL*]) is combined with that in *AGO1*, the *ago1 ago10* double mutant is embryo-lethal (Lynn et al. 1999; Mallory and Vaucheret 2006).

As examined above, mutations in the major components essential for miRNA biogenesis, processing, and loading result in severe embryo defects. In these mutants, the phenotypes are detected at relatively early stages of embryogenesis, many of which result in lethality. Through the analysis of those mutants, a specific set of miRNAs involved in embryogenesis, such as miR156, miR162, miR165/miR166, and miR168 have been identified as described above. An *Arabidopsis* mutant *foc*, *floral organs in carpels*, contains a *Ds* transposon insertion in the 3′ regulatory region of *MIR160a*. The level of mature miR160 is reduced in this mutant (Liu et al. 2010). *foc* mutants produces aberrant seeds. During the two- to four-cell stages of embryo proper development, the suspensor cells in *foc* embryos exhibit aberrant longitudinal cell divisions, which result in a double-filed suspensor at this stage and a triple-filed suspensor at the later globular (-like) stage (Fig. 3). The *foc* embryo proper at the transition stage lacks a distinguishable hypophysis, while the heart (-like) stage embryo of this mutant exhibits asymmetric cotyledon lobes (Liu et al. 2010). These findings suggest an essential role of miR160 in the regulation of embryogenesis programs.

MIR160 is strongly expressed in wild-type embryos (Liu et al. 2010). miR160 targets *ARF10*, *ARF16*, and *ARF17* (Rhoades et al. 2002). Consistent with the reduction in miR160 level in *foc*, the accumulation of *ARF10*, *ARF16*, and *ARF17*, is increased in the developing embryos of this mutant. Auxin is a master regulator of embryogenesis (Bowman and Floyd 2008) and many other events during plant development (Chapman and Estelle 2009). During embryogenesis, proper distribution of auxin within, and between, the embryo proper and suspensor is essential for patterning and polarity of the embryo. Auxin transporters such as PIN FORMED1

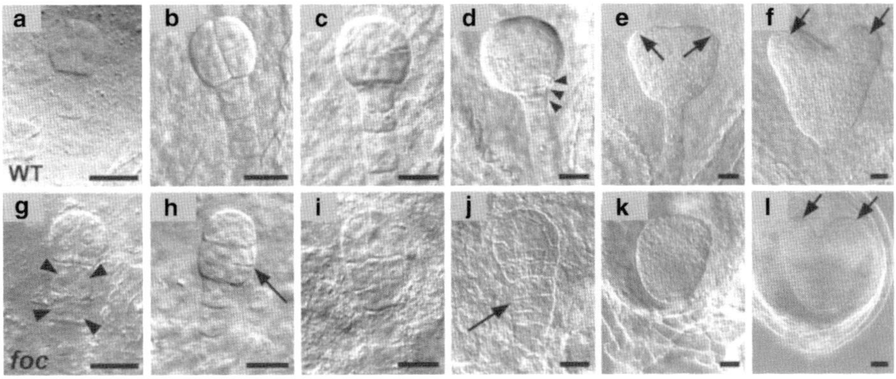

Fig. 3 Embryo defects in *foc* mutants. *Top* and *bottom* rows show wild-type and *foc* embryos, respectively. *Arrowheads* in (**g**) indicate aberrant longitudinal cell divisions in the suspensor cells in *foc*, resulting in a double-filed suspensor. An *arrow* in (**h**) shows abnormal cell divisions in the central and basal domains of *foc* embryo. *Arrowheads* in (**d**) indicate normal hypophysis region in wild-type embryo, which is missing in foc (**j**). An *arrow* in (**j**) indicates a triple-filed suspensor. *Arrows* in (**e**), (**f**) and (**l**) indicate cotyledon buttresses or lobes (Liu et al. 2010). Copyright Wiley-Blackwell

(PIN1) and PIN7, which mediate auxin efflux from cells, regulate auxin distribution in various tissues of the embryo (Bowman and Floyd 2008). ARFs play a critical role in auxin sensing (Guilfoyle and Hagen 2007). *foc* mutants that increases *ARF10*, *ARF16*, and *ARF17* expression during embryo development actually show altered auxin responses (Liu et al. 2010). The defects observed in *foc* mutant embryos are likely caused by the deregulation of these *ARF*s from miR160 due to the reduced level of miRNA160, which resulted in altered auxin sensing in the mutant embryos.

As a consequence of defects during embryo development, *foc* mutant seedlings exhibit abnormal cotyledons (Liu et al. 2010). Seedling defects are also observed in the *mARF17* mutants that express a miR160-resistant version of *ARF17*. miRNA target genes can be deregulated by introducing silent mutations in the region of the target gene complimentary to the miRNA. In this way, the target gene becomes resistant to miRNA and cleavage by miRNA-guided RISC (RNA-Induced Silencing Complex), which causes over accumulation of intact and functional proteins. The disruption of miR160 regulation of *ARF17* in *mARF17* miRNA-resistant mutants results in young seedlings with lobed and extra cotyledons, indicative of trilateral and quadrilateral embryonic symmetry (Mallory et al. 2005). The phenotypes of *mARF17* mutants also provide evidence to support the function of miRNA-targeted *ARFs* in embryo patterning.

Another interesting characteristic of the *foc* mutant is vivipary, i.e., precocious germination of seeds while developing on the maternal plants. Some embryos of *foc* mutants emerge from developing seeds contained in siliques (Liu et al. 2010). As discussed later in seed germination control, deregulation of *ARF10* from miR160 causes hypersensitivity to abscisic acid (ABA) in imbibed seeds (Liu et al. 2007a). ABA generally suppresses vivipary (Kao et al. 1996; Gubler et al. 2005). Therefore, the reduction in miR160 level and the increase in *ARF10* accumulation in *foc* mutants are expected to exert suppressive effects on vivipary. The vivipary phenotype in *foc* mutants may not be through the ARF10 function, but through ARF16 and/or ARF17 function. Alternatively, ARF10 might have different roles in terms of ABA sensitivity control in developing and imbibed seeds.

Mutations in miR164 and their targets also affect embryo patterning and change cotyledon morphology. miR164 targets five members of the NAC-domain (named after *Petunia* NO *APICAL MERISTE* [*NAM*], *Arabidopsis* A*TAF* cDNA clones, and *Arabidopsis* C*UP-SHAPED COTYLEDON2* [*CUC2*]) (Aida et al. 1997) gene family including *NAC1*, *CUC1*, and *CUC2* (Rhoades et al. 2002). Expression of a miR164-resistant version of *CUC1* (termed *5mCUC1*) causes cotyledon orientation defects (Mallory et al. 2004). Plants overexpressing *MIR164* phenocopy *cuc1 cuc2* double mutants (Aida et al. 1997) that exhibit cup-shaped cotyledons, or mimic *cuc1* and *cuc2* single mutants (Aida et al. 1997; Vroemen et al. 2003) that have partially fused cotyledons (Laufs et al. 2004; Mallory et al. 2004). Interaction of *CUCs* and other key regulators of tissue differentiation in the embryo, such as *SHOOT MERISTEMLESS* (*STM*) and *KANDAI*, and their regulation by auxin is summarized elsewhere (Bowman and Floyd 2008).

While miRNAs are the best understood of the sRNAs, a large number of other sRNAs, such as small interfering RNAs (siRNAs), are produced by RNA polymerase IV (PolIV), a homolog of DNA-dependent RNA polymerase II (Mosher et al. 2009). PolIV-dependent siRNAs (p4-siRNAs) are strongly expressed in flowers and young developing siliques that contain seeds at mid-embryogenesis. p4-siRNAs are detected predominantly in the endosperm or seed coat. Expression of p4-siRNAs in developing endosperm is specifically from maternal chromosomes. The uniparental expression of p4-siRNAs suggests a link between genomic imprinting and RNA silencing in plants (Mosher et al. 2009). A more detailed description of p4 siRNA is described in another chapter. However, it should be noted that endosperm differentiation is also an integral part of seed development.

1.2 Seed Development and Maturation

In the previous section, miRNA involvement in cell fate and tissue domain determination during early through mid embryogenesis is described. Here, the function of some miRNAs and their target genes affecting later seed developmental programs including seed maturation are described. While histodifferentiation is a major event during early embryogenesis, cell expansion and elongation, which are important determinants of seed size, are predominant during mid through late embryogenesis. Mutations in *MIR159* genes could change seed size through the action of MYB transcription factors. *MIR159a* and *MIR159b* are expressed in mature seeds (Allen et al. 2007). Loss of function in *MIR159a* or *MIR159b* does not cause any obvious phenotypes. By contrast, *mir159ab* double mutants produce fewer seeds with irregular shape and reduced size compared to wild type (Allen et al. 2007). These results suggest functional redundancy between *MIR159a* and *MIR159b* in terms of seed development. miR159a and miR159b are predicted to target seven transcription factors including two *GAMYB-like* genes, *MYB33* and *MYB65*. Expression of these two genes is elevated in *mir159ab* double mutants. Consistent with the results of expression analysis, triple mutant *mir159ab myb33* and quadruple mutant *mir159ab myb33 myb65* exhibit suppression of the *mir159ab* phenotypes (Allen et al. 2007). These data suggest that deregulation of *MYB33* and *MYB65* from miR159 is predominantly responsible for the *mir159ab* phenotypes. This study demonstrates miRNA function in the determination of seed size and also provides a good example of redundancy within a *MIR* gene family.

Potential involvement of miRNA-targeted and tasiRNA-targeted *ARFs* in female gametophyte development and the role of *ARF10*, *ARF16*, and *ARF17* in embryo patterning are mentioned above. ARFs also seem to be factors controlling seed size in *Arabidopsis*. *ARF2* has been characterized for its function during seed development. *ARF2* is not a miRNA target, but a tasiRNA target (Williams et al. 2005). Mutants with *arf2* T-DNA insertion alleles, *arf2-6*, *arf2-7*, and *arf2-8*, show increased seed size. It has been suggested that ARF2 is a suppressor protein associated with cell growth (Okushima et al. 2005). The function of ARF2 is probably to suppress downstream genes that have a positive effect on cell expansion and seed size increase. The physiological role of this gene might be to prevent excessive growth of individual seeds, which could reduce the number of seeds that the maternal plant can potentially produce.

In rice, *ARF6* and *ARF8* seem to play a role in seed development. Massively parallel signature sequencing was performed using cDNA libraries generated from rice immature seeds 3, 6, 9, and 12 days after anthesis. This study shows that rice (*Oryza sativa*) miRNAs, osa-miR167, osa-miR397, osa-miR398, osa-miR408, osa-miR528, osa-miR1866-3p, and osa-miRc11 are preferentially expressed in seeds (Xue et al. 2009). One of them, osa-miR167 targets *ARF6* and *ARF8* in rice (Yang et al. 2006a). Auxin increases miR167 levels in rice cell culture (Yang et al. 2006a). In rice seeds, the IAA level is approximately 40-fold higher than in other tissues, which is consistent with the high abundance of osa-miR167 in rice seeds (Matsuda et al. 2005; Xue et al. 2009). While ARF2 is a repressor, ARF6 and ARF8 are activators of downstream targets (Hagen and Guilfoyle 2002; Tiwari et al. 2003; Yang et al. 2006a). In rice cell culture, *ARF8* activates *OsGH3-2*, a gene encoding an auxin-conjugating enzyme (Yang et al. 2006a), which catalyzes the conjugation of IAA to different compounds to decrease the cellular concentration of free IAA (Staswick et al. 2002). Delivery of synthetic miR167 into rice cell culture leads to a decrease in *ARF8* and *OsGH3-2* mRNA levels, which is expected to increase cellular free IAA level. This suggests that the auxin-osa-miR167-*ARF8-OsGH3* pathway is involved in a positive feedback loop in which exogenous auxin increases its own cellular level (Yang et al. 2006a). It is possible that osa-miR167, ARF6, and ARF8 are involved in a similar pathway in developing rice seeds. This pathway may function to rapidly increase endogenous free IAA levels in developing seeds.

Rice seed development is also affected by miR172. miR172 targets *APETALLA2* (*AP2*) and *AP2*-like genes in *Arabidopsis* (Aukerman and Sakai 2003). In rice, miR172 is predicted to target five *AP2*-like genes (Lee et al. 2007). miR172 is not present in rice embryo, endosperm, and pericarp 10 days after pollination. When *MIR172b* is overexpressed with the constitutive 35S promoter, *35S:MIR172b* plants produce seeds with reduced weight (Zhu et al. 2009). This phenotype suggests that suppression of *AP2*-like gene(s) has a negative effect on seed weight in rice. This is somewhat counterintuitive to what is known in *Arabidopsis*. In *Arabidopsis*, *ap2* mutants have an increase in seed mass, which is transmitted through an effect on the maternal plant and is most likely due to changes in sugar metabolism (Ohto et al. 2005). However, since mature seeds from rice plants overexpressing *MIR172b* exhibit irregular-shaped seeds including double-grained seeds (Zhu et al. 2009), reduced seed weight is most likely a consequence of severe defects in flower organ development, which are caused by the suppression of *AP2-like* during flower development.

Upon completion of developmental programs, seeds are released from maternal plants. The regulation of *ARF6* and *ARF8* by miR167 is also important for seed dispersal or shattering, which terminates the seed developmental program on the maternal plant (Todesco et al. 2010). The role of miR167 in seed shattering was suggested through studies using a "miRNA target mimicry" technique in *Arabidopsis* (Franco-Zorrilla et al. 2007; Todesco et al. 2010). This technique was developed to address difficulties in characterizing the function of plant miRNAs, many of which have gene families and exhibit redundancy. When an artificial mimic miRNA target that interrupts pairing with the corresponding miRNA family is transformed into plants, this noncleavable mimic miRNA competes with native targets for the miRNA family. The mutation results in sequestration of the miRNA family and deregulation of the native targets. This approach has been applied for miR167 (mimic miR167: *MIM167*). Seeds of *MIM167* plants often do not fill completely and remain attached to the dry siliques. Other mutants such as *shatterproof1* (*shp1*), *shp2*, and 35S:*FRUITFULL* also exhibit non-seed-shattering phenotypes (Liljegren et al. 2000). However, the phenotype in these mutants is caused by a failure in silique dehiscence due to the lack of lignification at the valve margins, which are the sites of dehiscence located at the boundary between the valves and replum (Liljegren et al. 2000). By contrast, dehiscence normally happens in *MIM167* siliques, but seeds often remain attached to the replums of dehiscent siliques (Todesco et al. 2010). A similar phenotype is observed in *seedstick* (*stk*) mutants, which has defects in the funiculus, an umbilical cord-like structure connecting seeds to the maternal plant (Pinyopich et al. 2003). The phenotype of *MIM167* mutants implies that miR167-targeted *ARF6* and/or *ARF8* function is possibly associated with the detachment of seeds from the funiculi. Thus, miRNA is involved in a wide range of events during seed development.

2 MicroRNAs in Seed Dormancy and Germination

Early studies of miRNA processing-related mutants indicated that miRNAs might also be important for seed germination. Mature seeds of the mutants that are defective in SE, HYL1, and CBP80 (also called ABA HYPERSENSITIVE1 or ABH1), a large subunit of the multifunctional nuclear CBC (Fig. 1), which is hypothesized to interact with SE for miRNA processing (Laubinger et al. 2008), were hypersensitive to ABA (Lu and Fedoroff 2000; Bezerra et al. 2004). These findings suggest that proper processing of pri-miRNA into mature forms is important for germination.

Compared to the information available for the function of miRNA in embryogenesis and later seed development, knowledge on miRNA involvement in seed dormancy and germination is still limited. However, functional genomics focusing on a specific set of miRNAs and their target genes expressed during seed germination and postgermination has led to significant discoveries of interesting mechanisms, including feedback regulation of hormone sensing in imbibed seeds, epigenetic control of dormancy and germination, and also potential interaction between separate miRNA pathways during seedling growth. The details of these findings are discussed here.

2.1 Hormone Sensitivity

Many plant miRNAs have been identified by massive sequencing and modern bioinformatic studies (Gustafson et al. 2005). miRNA microarrays and sequencing-by-synthesis approaches enable profiling differentially expressed miRNA genes in various organs and tissues, during certain developmental stages, among distinct mutants or under the effects of some chemicals (Zhang et al. 2008; Fahlgren et al. 2009; Mica et al. 2010). When research on seed germination-associated miRNAs was initiated, only a small number (~28) of plant miRNA species were known. However, a modified slot blot method, which was typically used for screening monoclonal antibodies, was used for miRNA expression analysis. In this method, a sRNA-blotted membrane was prepared from a preparative gel and was hybridized with multiple miRNA probes contained in separate slots that allowed simultaneous examination of the expression of ~30 species of miRNAs (Martin et al. 2005). This approach was utilized to identify several miRNAs that might be involved in seed germination and postgermination (Martin et al. 2006). The function of these and other miRNAs have been further characterized with expression analyses and functional genomics using mutants. miR156, miR159, and miR160 are the major miRNAs associated with seed germination. While miR159 and miR160 are involved in germination in a strict sense (Perino and Côme 1991), which specifies the stages before radicle emergence, miR156 plays a more important role after radicle emergence (postgermination). miR159 and miR160 seem to be major miRNAs controlling seed germination. Both miRNAs affect germination through the modulation of hormonal sensitivity of seeds. miR159 targets *MYB33* and *MYB101*, which are positive regulators of ABA responses, and controls transcript levels of these two *MYB* genes in *Arabidopsis* seeds. Transgenic seeds overexpressing *MIR159* contain lower levels of *MYB33* and *MYB101* transcripts and exhibit reduced sensitivity to ABA compared to wild type (Reyes and Chua 2007). This phenotype is similar to that observed in *myb33* and *myb101* mutant seeds. When the deregulation approach was used for *MYB33*, seeds expressing miR159-resistant mutant *MYB33* (*mMYB33*) were hypersensitive to ABA (Reyes and Chua 2007). Seeds overexpressing the *Turnip mosaic virus* (TuMV) P1/HC-Pro viral silencing suppressor protein, which is known to inhibit miRNA function (Mallory et al. 2002; Kasschau et al. 2003), also show ABA hypersensitivity, suggesting that miRNA function is required to maintain normal ABA sensitivity in imbibed seeds (Reyes and Chua 2007). The accumulation of miR159 is reduced in *abi3* mutants, suggesting that ABI3 acts upstream of *MIR159*, although it is not known whether ABI5, another component important for ABA signal transduction, acts in the same pathway (Reyes and Chua 2007).

While miR159 targets ABA signaling molecules, miR160 targets *ARF10*, *ARF16* and *ARF17* mentioned above (Rhoades et al. 2002). The function of *ARF10* has been characterized in terms of seed germination control. Transgenic plants expressing a miR160-resistant form of *ARF10*, which has silent mutations in the miRNA target site (*mARF10*), exhibited developmental defects such as serrated leaves, curled stems, contorted flowers, and twisted siliques. In addition, *mARF10* mutant seeds and plants were hypersensitive to ABA during sensu stricto germination and postgermination, in a dose-dependent manner (Liu et al. 2007a). This is interesting because the mutation was created in the auxin signal transduction protein but the phenotype observed involved ABA sensitivity. ABA hypersensitivity of *mARF10* seeds was mimicked in wild-type seeds treated with exogenous auxin. These results suggest that auxin-ABA cross talk is present in imbibed seeds, and that the down-regulation of *ARF10* by miR160 is essential for the auxin-ABA cross talk during sensu stricto germination. The increased ABA sensitivity in *mARF10* seeds is expected to antagonize the positive action of gibberellin (GA) on seed germination, such as the elimination of DELLA, a seed germination repressor protein, and the induction of cell wall-loosening factors (e.g., xyloglucan endotransglycosylase) required for cell expansion and the generation of embryo growth potential (Nonogaki 2008) (Fig. 4). The molecular mechanisms underlying the ARF10 modulation of ABA sensitivity in seeds are not known.

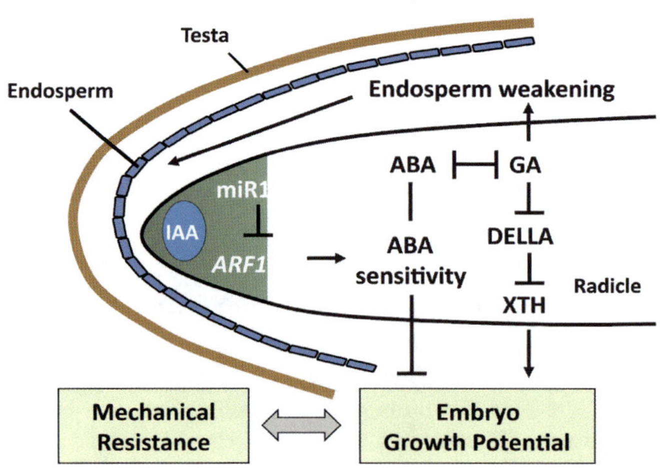

Fig. 4 A model illustrating a role of miR160 and *AUXIN RESPONSE FACTOR10* (*ARF10*) in the regulation of abscisic acid (ABA) sensitivity during seed germination. Seeds and seedlings deregulated of *ARF10* from miR160 are hypersensitive to ABA, which is expected to antagonize gibberellin (GA), a seed germination-promoting hormone. Elimination of DELLA germination repressor proteins and induction of cell wall proteins such as xyloglucan endotransglycosylases (XTHs), which are promoted by GA, can be affected by ARF10 overaccumulation to negatively affect germination in the presence of ABA. Copyright Landes Bioscience

2.2 Epigenetic Regulation

Chromatin remodeling by histone monoubiquitination plays an important role in the induction and/or maintenance of seed dormancy in *Arabidopsis*. Histone monoubiquitination regulates ABA levels in developing seeds through histone H2B monoubiquitination (Liu et al. 2007b; Chinnusamy et al. 2008). Other epigenetic process such as cytosine DNA methylation, a stable epigenetic mark for maintenance of gene silencing, is also integral to ABA-regulated processes including seed dormancy and germination (Chinnusamy et al. 2008; Ortega-Galisteo et al. 2008). *DEMETER-LIKE PROTEIN3* (*DML3*) or *REPRESSOR OF SILENCING1* (*ROS1*)-like is a DNA glycosylase/lyase that catalyzes DNA demethylation (Ortega-Galisteo et al. 2008). *DML3* is a target of miR402 (Sunkar and Zhu 2004; Kim et al. 2010). Overexpression of miR402 promotes seed germination under stress conditions. *dml3* mutant seeds exhibit a similar phenotype to miR402 overexpressors in *Arabidopsis* (Kim et al. 2010). The downregulation of *DML3* by miR402 acts positively on seed germination under stresses. It has been suggested that the negative effect of *DML3* on seed germination is exerted through demethylation (or derepression) of target genes. Targets of DML3 are therefore hypothesized to be germination suppressors (Kim et al. 2010), although the targets have not been identified and changes in methylation status of genes in imbibed seeds have not been well characterized.

3 MicroRNAs in Postgermination

3.1 Hormone and Stress Sensitivity

mARF10 mutants described above for germination also exhibit phenotypes after seed germination. *mARF10* seedlings exhibit bent cotyledons, and their growth at postgerminative stages is also hypersensitive to ABA, which is even more exaggerated than ABA sensitivity observed during seed germination. Growth of *mARF10* seedlings is suppressed by 0.2 μM ABA. The suppression of seedling growth by 0.2 μM ABA is not observed in wild-type or transgenic seedlings expressing nonmutated *ARF10* (Liu et al. 2007a).

Other miRNAs are associated with sensitivity to abiotic stresses at postgerminative stages, which may or may not be related to ABA sensitivity. miR398 targets genes encoding Cu/Zn superoxide dismutases (SOD), cytosolic CSD1 and chloroplastic CSD2 that are important for copper homeostasis (Yamasaki et al. 2007; Abdel-Ghany and Pilon 2008; Beauclair et al. 2010) and detoxification of superoxide radicals (Sunkar et al. 2006). Oxidative stresses reduce miR398 levels, which results in the accumulation of *CSD1* and *CSD2* in *Arabidopsis* seedlings. Expression of miR398-resistant *CSD2* makes transgenic seedlings tolerant to high light, heavy metals, and other oxidative stresses (Sunkar et al. 2006). This type of modification in *Arabidopsis* provides proof of concept for potential miRNA technology for improving stand establishment of plants under stress conditions in agriculture.

In fact, miR398 is also expressed in *Medicago truncatula*, a legume model species (Trindade et al. 2010). In tobacco, overexpression of miR398 negatively affects postgerminative seedling growth (Feng et al. 2010).

3.2 Developmental Phase Transition

A major event during postgermination stages is reserve mobilization. Seeds of many species are able to develop cotyledon-stage seedlings even on water-moistened filter paper. This is because seeds accumulate nutrition in storage tissues, such as the endosperm or cotyledons, during seed development, and these materials are mobilized following radicle emergence to supply nutrition to growing seedlings. When seed storage tissues complete reserve mobilization, seedlings have to become nutritionally independent or autotrophic, which requires development of vegetative or adult leaves. miR156 plays a critical role during this transitional phase. The suppression of *SPL13* by miR156 is essential for normal development of vegetative leaves and autotrophic seedling development. When silent mutations are created in the *SPL13* sequence complementary to the miR156 sequence, the mutant *SPL13* (*mSPL13*) is deregulated from miR156, becomes resistant to it, and overaccumulates in the mutant seedlings (Martin et al. 2010b). Because silent mutations do not change the amino-acid sequences, the mutants overaccumulate functional proteins that cause a delay in the development of the shoot apical meristem (SAM) and/or leaf primordia (Martin et al. 2010a). These results suggest that SPL13 is a negative regulator of vegetative leaf development in the cotyledon-stage seedlings. *SCHNARCHZAPFEN* (*SNZ*), an AP2-like gene, is downregulated specifically in *mSPL13* mutants. Interestingly, *SNZ* is a target of miR172, which suggests that two separate miRNA pathways, miR156-*SPL13* and miR172-*SNZ*, interact. Expression analysis using quantitative reverse transcription (RT)-PCR indicated that *MIR172a* and *MIR172b* are overexpressed in *mSPL13* mutant seedlings, suggesting that *SPL13* positively regulates miR172 to reduce *SNZ* expression and vegetative leaf emergence. There may be miRNA gene regulation cascades in which the miR156 pathway acts upstream of the miR172 pathway during postgerminative stages. While only *SPLs* involved in postgerminative stages are described in this chapter, miR156 and *SPLs* are also involved in the juvenile-to-adult phase transition (Wu et al. 2009), shoot maturation (Shikata et al. 2009), and flowering (Wang et al. 2009). Similar gene regulation cascades are also present in the regulation of juvenile-to-adult phase transition in both monocots (Chuck et al. 2007) and dicots (Wu et al. 2009).

References

Abdel-Ghany SE, Pilon M (2008) MicroRNA-mediated systemic down-regulation of copper protein expression in response to low copper availability in Arabidopsis. J Biol Chem 283: 15932–15945

Aida M, Ishida T, Fukaki H et al (1997) Genes involved in organ separation in Arabidopsis: an analysis of the cup-shaped cotyledon mutant. Plant Cell 9:841–857

Allen E, Xie Z, Gustafson AM et al (2005) microRNA-directed phasing during trans-acting siRNA biogenesis in plants. Cell 121:207–221

Allen RS, Li J, Stahle MI et al (2007) Genetic analysis reveals functional redundancy and the major target genes of the Arabidopsis miR159 family. Proc Natl Acad Sci USA 104: 16371–16376

Aukerman MJ, Sakai H (2003) Regulation of flowering time and floral organ identity by a microRNA and its *APETALA2*-like target genes. Plant Cell 15:2730–2741

Beauclair L, Yu A, Bouchè N (2010) microRNA-directed cleavage and translational repression of the copper chaperone for superoxide dismutase mRNA in Arabidopsis. Plant J 62:454–462

Bezerra IC, Michaels SD, Schomburg FM et al (2004) Lesions in the mRNA cap-binding gene *ABA HYPERSENSITIVE 1* suppress *FRIGIDA*-mediated delayed flowering in *Arabidopsis*. Plant J 40:112–119

Bohmert K, Camus I, Bellini C et al (1998) *AGO1* defines a novel locus of *Arabidopsis* controlling leaf development. EMBO J 17:170–180

Bowman JL, Floyd SK (2008) Patterning and polarity in seed plant shoots. Annu Rev Plant Biol 59:67–88

Chapman EJ, Carrington JC (2007) Specialization and evolution of endogenous small RNA pathways. Nat Rev Genet 8:884–896

Chapman EJ, Estelle M (2009) Mechanism of auxin-regulated gene expression in plants. Annu Rev Genet 43:265–285

Chen X (2005) microRNA biogenesis and function in plants. FEBS Lett 579:5923–5931

Chinnusamy V, Gong Z, Zhu J-K (2008) Abscisic acid-mediated epigenetic processes in plant development and stress responses. J Integr Plant Biol 50:1187–1195

Chuck G, Cigan AM, Saeteurn K et al (2007) The heterochronic maize mutant *Corngrass1* results from overexpression of a tandem microRNA. Nat Genet 39:544–549

Fahlgren N, Sullivan CM, Kasschau KD et al (2009) Computational and analytical framework for small RNA profiling by high-throughput sequencing. RNA 15:992–1002

Fang Y, Spector DL (2007) Identification of nuclear dicing bodies containing proteins for microRNA biogenesis in living Arabidopsis plants. Curr Biol 17:818–823

Feng X-M, Qiao Y, Mao K et al (2010) Ectopic overexpression of *AtmiR398b* gene in tobacco influences seed germination and seedling growth. Plant Cell Tiss Organ Cul 102:53–59

Franco-Zorrilla JM, Valli A, Todesco M et al (2007) Target mimicry provides a new mechanism for regulation of microRNA activity. Nat Genet 39:1033–1037

Franzmann LH, Yoon ES, Meinke DW (1995) Saturating the genetic map of *Arabidopsis thaliana* with embryonic mutations. Plant J 7:341–350

Fujioka Y, Utsumi M, Ohba Y et al (2007) Location of a possible miRNA processing site in SmD3/SmB nuclear bodies in Arabidopsis. Plant Cell Physiol 48:1243–1253

Golden TA, Schauer SE, Lang JD et al (2002) *Short Integuments1/suspensor1/carpel Factory*, a Dicer homolog, is a maternal effect gene required for embryo development in Arabidopsis. Plant Physiol 130:808–822

Golling G, Amsterdam A, Sun Z et al (2002) Insertional mutagenesis in zebrafish rapidly identifies genes essential for early vertebrate development. Nat Genet 31:135–140

Grigg SP, Canales C, Hay A et al (2005) SERRATE coordinates shoot meristem function and leaf axial patterning in *Arabidopsis*. Nature 437:1022–1026

Gubler F, Millar AA, Jacobsen JV (2005) Dormancy release, ABA and pre-harvest sprouting. Curr Opin Plant Biol 8:183–187

Guilfoyle TJ, Hagen G (2007) Auxin response factors. Curr Opin Plant Biol 10:453–460

Gustafson AM, Allen E, Givan S et al (2005) ASRP: the *Arabidopsis* small RNA project database. Nucl Acids Res 33:D637–D640

Hagen G, Guilfoyle T (2002) Auxin-responsive gene expression: genes, promoters and regulatory factors. Plant Mol Biol 49:373–385

Hannon GJ (2002) RNA interference. Nature 418:244–251

Jacobsen SE, Running MP, Meyerowitz EM (1999) Disruption of an RNA helicase/RNAse III gene in Arabidopsis causes unregulated cell division in floral meristems. Development 126:5231–5243

Kao CY, Cocciolone SM, Vasil IK et al (1996) Localization and interaction of the cis-acting elements for abscisic acid, VIVIPAROUS1, and light activation of the C1 gene of maize. Plant Cell 8:1171–1179

Kasschau KD, Xie Z, Allen E et al (2003) P1/HC-Pro, a viral suppressor of RNA silencing, interferes with Arabidopsis development and miRNA unction. Dev Cell 4:205–217

Kim JY, Kwak KJ, Jung HJ et al (2010) MicroRNA402 affects seed germination of *Arabidopsis thaliana* under stress conditions via targeting DEMETER-LIKE Protein3 mRNA. Plant Cell Physiol 51:1079–1083

Koltunow AM, Grossniklaus U (2003) APOMIXIS: a developmental perspective. Annu Rev Plant Biol 54:547–574

Lang JD, Ray S, Ray A (1994) sin1, a mutation affecting female fertility in Arabidopsis, interacts with mod1, its recessive modifier. Genetics 137:1101–1110

Laubinger S, Sachsenberg T, Zeller G et al (2008) Dual roles of the nuclear cap-binding complex and SERRATE in pre-mRNA splicing and microRNA processing in *Arabidopsis thaliana*. Proc Natl Acad Sci USA 105:8795–8800

Laufs P, Peaucelle A, Morin H et al (2004) MicroRNA regulation of the CUC genes is required for boundary size control in Arabidopsis meristems. Development 131:4311–4322

Lee DY, Lee J, Moon S et al (2007) The rice heterochronic gene *SUPERNUMERARY BRACT* regulates the transition from spikelet meristem to floral meristem. Plant J 49:64–78

Liljegren SJ, Ditta GS, Eshed Y et al (2000) *SHATTERPROOF* MADS-box genes control seed dispersal in *Arabidopsis*. Nature 404:766–770

Liu N, Olson EN (2010) MicroRNA regulatory networks in cardiovascular development. Dev Cell 18:510–525

Liu P-P, Montgomery TA, Fahlgren N et al (2007a) Repression of *AUXIN RESPONSE FACTOR10* by microRNA160 is critical for seed germination and post-germination stages. Plant J 52:133–146

Liu Y, Koornneef M, Soppe WJJ (2007b) The absence of histone H2B monoubiquitination in the Arabidopsis hub1 (rdo4) mutant reveals a role for chromatin remodeling in seed dormancy. Plant Cell 19:433–444

Liu X, Huang J, Wang Y et al (2010) The role of *floral organs in carpels*, an Arabidopsis loss-of-function mutation in *MicroRNA160a*, in organogenesis and the mechanism regulating its expression. Plant J 62:416–428

Lobbes D, Rallapalli G, Schmidt DD et al (2006) SERRATE: a new player on the plant microRNA scene. EMBO Rep 7:1052–1058

Lotan T, Ohto M, Yee KM et al (1998) *Arabidopsis* LEAFY COTYLEDON1 is sufficient to induce embryo development in vegetative cells. Cell 93:1195–1205

Lu C, Fedoroff N (2000) A mutation in the Arabidopsis *HYL1* gene encoding a dsRNA binding protein affects responses to abscisic acid, auxin, and cytokinin. Plant Cell 12: 2351–2366

Lynn K, Fernandez A, Aida M et al (1999) The PINHEAD/ZWILLE gene acts pleiotropically in Arabidopsis development and has overlapping functions with the ARGONAUTE1 gene. Development 126:469–481

Mallory AC, Vaucheret H (2006) Functions of microRNAs and related small RNAs in plants. Nat Genet 38:S31–S36

Mallory AC, Reinhart BJ, Bartel D et al (2002) A viral suppressor of RNA silencing differentially regulates the accumulation of short interfering RNAs and micro-RNAs in tobacco. Proc Natl Acad Sci USA 99:15228–15233

Mallory AC, Dugas DV, Bartel DP et al (2004) MicroRNA regulation of NAC-domain targets is required for proper formation and separation of adjacent embryonic, vegetative, and floral organs. Curr Biol 14:1035–1046

Mallory AC, Bartel DP, Bartel B (2005) MicroRNA-directed regulation of Arabidopsis *AUXIN RESPONSE FACTOR17* is essential for proper development and modulates expression of early auxin response genes. Plant Cell 17:1360–1375

Martin RC, Liu PP, Nonogaki H (2005) Simple purification of small RNAs from seeds and efficient detection of multiple microRNAs expressed in Arabidopsis thaliana and tomato (*Lycopersicon esculentum*) seeds. Seed Sci Res 15:319–328

Martin RC, Liu P-P, Nonogaki H (2006) microRNAs in seeds: modified detection techniques and potential applications. Can J Bot 84:189–198

Martin RC, Asahina M, Liu PP et al (2010a) The microRNA156 and microRNA172 gene regulation cascades at post-germinative stages in Arabidopsis. Seed Sci Res 20:79–87

Martin RC, Asahina M, Liu PP et al (2010b) The regulation of post-germinative transition from the cotyledon- to vegetative-leaf stages by microRNA-targeted *SQUAMOSA PROMOTER-BINDING PROTEIN LIKE13* in Arabidopsis. Seed Sci Res 20:89–96

Matsuda F, Miyazawa H, Wakasa K et al (2005) Quantification of indole-3-acetic acid and amino acid conjugates in rice by liquid chromatography-electrospray ionization-tandem mass spectrometry. Biosci Biotechnol Biochem 69:778–783

McConnell JR, Emery J, Eshed Y et al (2001) Role of *PHABULOSA* and *PHAVOLUTA* in determining radial patterning in shoots. Nature 411:709–713

Mica E, Piccolo V, Delledonne M et al (2010) Correction: high throughput approaches reveal splicing of primary microRNA transcripts and tissue specific expression of mature microRNAs in Vitis vinifera. BMC Genomics 11:109

Montgomery TA, Carrington JC (2008) Splicing and dicing with a *SERRATE*d edge. Proc Natl Acad Sci USA 105:8489–8490

Mosher RA, Melnyk CW, Kelly KA et al (2009) Uniparental expression of PolIV-dependent siRNAs in developing endosperm of Arabidopsis. Nature 460:283–286

Nodine M, Bartel D (2010) MicroRNAs prevent precocious gene expression and enable pattern formation during plant embryogenesis. Plant Biol 2010 Abstract #P04005

Nonogaki H (2008) Repression of transcription factors by microRNA during seed germination and postgermination: another level of molecular repression in seeds? Plant Signal Behav 3: 65–67

Ohto MA, Fischer RL, Goldberg RB et al (2005) Control of seed mass by *APETALA2*. Proc Natl Acad Sci USA 102:3123–3128

Okushima Y, Mitina I, Quach HL et al (2005) AUXIN RESPONSE FACTOR 2 (ARF2): a pleiotropic developmental regulator. Plant J 43:29–46

Ortega-Galisteo A, Morales-Ruiz T, Ariza R et al (2008) Arabidopsis DEMETER-LIKE proteins DML2 and DML3 are required for appropriate distribution of DNA methylation marks. Plant Mol Biol 67:671–681

Pagnussat GC, Alandete-Saez M, Bowman JL et al (2009) Auxin-dependent patterning and gamete specification in the *Arabidopsis* female gametophyte. Science 324:1684–1689

Park W, Li J, Song R et al (2002) CARPEL FACTORY, a Dicer homolog, and HEN1, a novel protein, act in microRNA metabolism in *Arabidopsis thaliana*. Curr Biol 12:1484–1495

Perino C, Côme D (1991) Physiological and metabolic study of the germination phases in apple embryo. Seed Sci Technol 19:1–14

Pinyopich A, Ditta GS, Savidge B et al (2003) Assessing the redundancy of MADS-box genes during carpel and ovule development. Nature 424:85–88

Prigge MJ, Wagner DR (2001) The Arabidopsis *SERRATE* gene encodes a zinc-finger protein required for normal shoot development. Plant Cell 13:1263–1280

Ray S, Golden T, Ray A (1996) Maternal effects of the *short integument* mutation on embryo development in Arabidopsis. Dev Biol 180:365–369

Reinhart BJ, Weinstein EG, Rhoades MW et al (2002) MicroRNAs in plants. Genes Dev 16:1616–1626

Reyes JL, Chua N-H (2007) ABA induction of miR159 controls transcript levels of two MYB factors during Arabidopsis seed germination. Plant J 49:592–606

Rhoades MW, Reinhart BJ, Lim LP et al (2002) Prediction of plant microRNA targets. Cell 110:513–520
Schauer SE, Jacobsen SE, Meinke DW et al (2002) DICER-LIKE1: blind men and elephants in Arabidopsis development. Trends Plant Sci 7:487–491
Schwab R, Ossowski S, Riester M et al (2006) Highly specific gene silencing by artificial microRNAs in Arabidopsis. Plant Cell 18:1121–1133
Schwartz BW, Yeung EC, Meinke DW (1994) Disruption of morphogenesis and transformation of the suspensor in abnormal suspensor mutants of Arabidopsis. Development 120:3235–3245
Shikata M, Koyama T, Mitsuda N et al (2009) Arabidopsis SBP-box genes *SPL10*, *SPL11* and *SPL2* control morphological change in association with shoot maturation in the reproductive phase. Plant Cell Physiol 50:2133–2145
Staswick PE, Tiryaki I, Rowe ML (2002) Jasmonate response locus JAR1 and several related Arabidopsis genes encode enzymes of the firefly Luciferase superfamily that show activity on jasmonic, salicylic, and indole-3-acetic acids in an assay for adenylation. Plant Cell 14:1405–1415
Stefani G, Slack FJ (2008) Small non-coding RNAs in animal development. Nat Rev Mol Cell Biol 9:219–230
Sunkar R, Zhu JK (2004) Novel and stress-regulated microRNAs and other small RNAs from Arabidopsis. Plant Cell 16:2001–2019
Sunkar R, Kapoor A, Zhu J-K (2006) Posttranscriptional induction of two Cu/Zn superoxide dismutase genes in Arabidopsis is mediated by downregulation of miR398 and important for oxidative stress tolerance. Plant Cell 18:2051–2065
Tang G, Reinhart BJ, Bartel DP et al (2003) A biochemical framework for RNA silencing in plants. Genes Dev 17:49–63
Tiwari SB, Hagen G, Guilfoyle T (2003) The roles of auxin response factor domains in auxin-responsive transcription. Plant Cell 15:533–543
Todesco M, Rubio-Somoza I, Paz-Ares J et al (2010) A collection of target mimics for comprehensive analysis of microRNA function in *Arabidopsis thaliana*. PLoS Genet 6:e1001031
Trindade I, Capitão C, Dalmay T et al (2010) miR398 and miR408 are up-regulated in response to water deficit in *Medicago truncatula*. Planta 231:705–716
Vazquez F, Gasciolli V, Crete P et al (2004) The nuclear dsRNA binding protein HYL1 is required for microRNA accumulation and plant development, but not posttranscriptional transgene silencing. Curr Biol 14:346–351
Voinnet O (2009) Origin, biogenesis, and activity of plant microRNAs. Cell 136:669–687
Vroemen CW, Mordhorst AP, Albrecht C et al (2003) The *CUP-SHAPED COTYLEDON3* gene is required for boundary and shoot meristem formation in Arabidopsis. Plant Cell 15:1563–1577
Wang J-W, Czech B, Weigel D (2009) miR156-regulated SPL transcription factors define an endogenous flowering pathway in *Arabidopsis thaliana*. Cell 138:738–749
Williams L, Carles CC, Osmont KS et al (2005) A database analysis method identifies an endogenous trans-acting short-interfering RNA that targets the Arabidopsis *ARF2*, *ARF3*, and *ARF4* genes. Proc Natl Acad Sci USA 102:9703–9708
Wu MF, Tian Q, Reed JW (2006) Arabidopsis microRNA167 controls patterns of *ARF6* and *ARF8* expression, and regulates both female and male reproduction. Development 133: 4211–4218
Wu G, Park MY, Conway SR et al (2009) The sequential action of miR156 and miR172 regulates developmental timing in Arabidopsis. Cell 138:750–759
Xie Z, Kasschau KD, Carrington JC (2003) Negative feedback regulation of *Dicer-Like1* in Arabidopsis by microRNA-guided mRNA degradation. Curr Biol 13:784–789
Xue L-J, Zhang J-J, Xue H-W (2009) Characterization and expression profiles of miRNAs in rice seeds. Nucl Acids Res 37:916–930
Yamasaki H, Abdel-Ghany SE, Cohu CM et al (2007) Regulation of copper homeostasis by microRNA in Arabidopsis. J Biol Chem 282:16369–16378

Yang JH, Han SJ, Yoon EK et al (2006a) Evidence of an auxin signal pathway, microRNA167-ARF8-GH3, and its response to exogenous auxin in cultured rice cells. Nucl Acids Res 34:1892–1899

Yang L, Liu Z, Lu F et al (2006b) SERRATE is a novel nuclear regulator in primary microRNA processing in Arabidopsis. Plant J 47:841–850

Yu B, Yang Z, Li J et al (2005) Methylation as a crucial step in plant microRNA biogenesis. Science 307:932–935

Zhan S, Lukens L (2010) Identification of novel miRNAs and miRNA dependent developmental shifts of gene expression in *Arabidopsis thaliana*. PLoS ONE 5:e10157

Zhang J, Zeng R, Chen J et al (2008) Identification of conserved microRNAs and their targets from Solanum lycopersicum Mill. Gene 423:1–7

Zhu QH, Upadhyaya NM, Gubler F et al (2009) Over-expression of miR172 causes loss of spikelet determinacy and floral organ abnormalities in rice (*Oryza sativa*). BMC Plant Biol 9:149

Effects of Virus Infection on Transcriptional Activity of miR164a in Plants

Ariel A. Bazzini and Sebastian Asurmendi

Contents

Abstract MicroRNAs (miRs) are small endogenous RNAs that switch or fine-tune gene expression at the posttranscriptional level. Interestingly, bacteria and viruses can alter miR accumulation and function upon infection and redirect host gene expression. The mechanisms of miR alteration under pathogenic conditions are still unclear with most efforts focused at the posttranscriptional level, leaving the transcriptional one as secondary interest. Here, we discuss the influence of host–pathogen interactions on the transcriptional regulation of miRs. We explore the utilization of GUS reporters to study the timing and localization of miR expression within the context of plant development, as well as after external stimuli such as viral infection. In particular, we focus on the transcriptional regulation of *Arabidopsis thaliana*

A.A. Bazzini (✉) and S. Asurmendi
Instituto de Biotecnología, CICVyA, INTA Castelar, Dr. N. Repetto y Los Reseros s/n, CP 1686 Hurlingham, Buenos Aires, Argentina

Consejo Nacional de Investigaciones Científicas y Técnicas (CONICET), Buenos Aires, Argentina

Department of Genetics, Yale University School of Medicine, New Haven, CT 06510, USA
e-mail: ariel.bazzini@yale.edu

V.A. Erdmann and J. Barciszewski (eds.), *Non Coding RNAs in Plants*, RNA Technologies, DOI 10.1007/978-3-642-19454-2_22,

miR-164a and its response to viral and hormone stimuli. The transcriptional alteration of miR gene expression reveals another important regulatory layer within the host–pathogen interaction and raises the following questions: What *trans*-acting factors regulate miR expression at transcriptional level? How are their activities altered under pathogenic conditions? What is the functional significance of transcriptional alterations of miR within the context of the host–pathogen interaction?

1 miR Biogenesis

MicroRNAs (miRNAs) are an important class of endogenous small silencing RNAs in both plants and animals that regulate gene expression posttranscriptionally. Plant miRs are 20–24-nt RNA molecules that guide RNA-induced silencing complex (RISC) to recognize messenger RNA (mRNA) by partially or fully complementary sequences. This recognition triggers translational repression or mRNA decay (Fig. 1a) (Eulalio et al. 2008; Guo et al. 2010). The primary miRs (pri-miRs) are class II genes, mainly transcribed by RNA polymerase II (pol II) (Lee et al. 2004), contained poly(A) tails and cap structure (Fig. 1) (Aukerman and Sakai 2003; Tam 2001). The pri-miRs fold into imperfect fold-back structures and are first spliced into a stem-loop precursor (pre-miR) and later processed into approximately 21-nt RNA duplex molecules by a Dicer-like enzyme (Fig. 1) (Bartel 2004). One strand of the duplex is selected as the mature miR and further incorporated into the Argonaute (Ago) protein forming the RISC, and the leftover strand, referred as miR*, is degraded (Tomari et al. 2004). miRs are involved in a variety of activities, including plant development, signal transduction, protein degradation, response to environmental stress, and pathogen invasion, among others (Dunoyer and Voinnet 2005; Mallory et al. 2004; Palatnik et al. 2003; Poethig 2009; Ruiz-Ferrer and Voinnet 2009). Expression of some miRs is regulated by hormones (Sempere et al. 2003), enhancers (Brennecke et al. 2003; Sempere et al. 2003), and pathogen infection (Chapman et al. 2004; Dunoyer et al. 2004).

2 Characterization of miR Promoters

As mentioned above, miRs are expressed mainly by pol II, and as such, typically possess *cis*-regulatory sequences 5′-upstream that are responsible for their transcriptional activation. However, sequences other than the basal “promoter” region can also modulate gene expression (Lee et al. 2006), such as “enhancers” and “silencers.” In addition, miR function is also modulated by posttranscriptional regulators. For example, the RNA binding protein Dnd1, protects target mRNAs from miR-mediated repression by preventing the access of the miR to its targets (Kedde et al. 2007). Regarding promoter elements, miR and protein-coding gene promoters have similar structures (Smale 2001). Basically, pol II promoters contain two parts, the core and the upstream elements. The core includes at least two elements, the

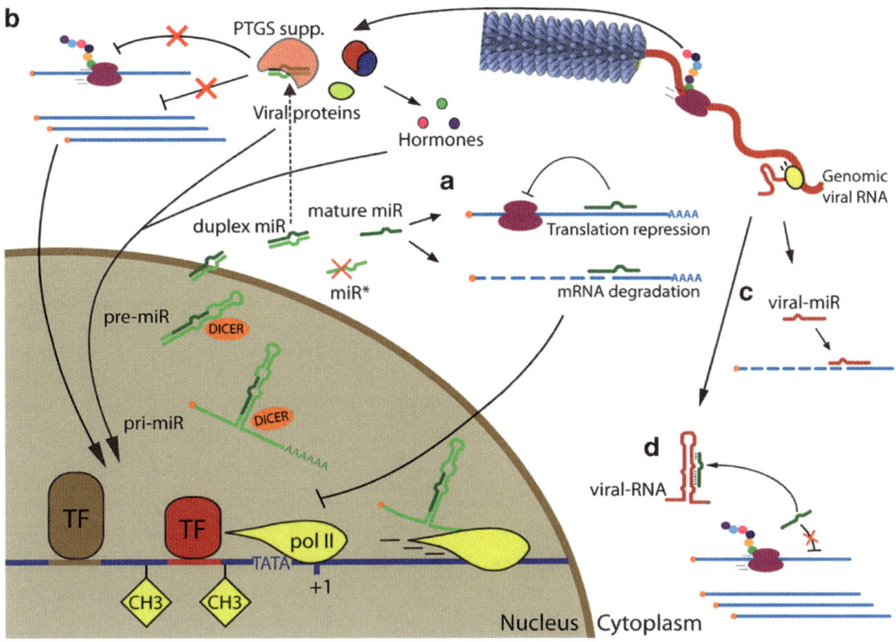

Fig. 1 The biogenesis of microRNAs (miRs) in plant starts with the primary miR (pri-miR) transcription by polymerase II (pol II) regulated by specific transcription factor (TF). The pri-miR is spliced first into precursor of miR (pre-miR) and subsequently into miR duplex involving Dicer protein as the main requirement of the process. The duplex goes to the cytoplasm and the mature miR (*dark green*) is incorporated into a RISC complex while the complementary molecule (miR*) is degraded. (**a**) The most common miRs actions are translational repression and messenger RNA cleavage. (**b**) Viral proteins such as suppressor of posttranscriptional gene silencing (PTGS supp.) interacts with miR pathway component (for example miRs duplex) preventing miR function. Higher level of miR target can some how trigger miR transcription to restore miR target levels. Alternatively, viral proteins could directly interact with the miR promoter or indirectly due to hormone signaling. Any of this alteration might involve alteration of TF, DNA methylation ($-CH_3$) stage or any other chromatin modification. (**c**) Virus can also express miRs (viral-miRs) that recognize and regulate host messenger during infection. (**d**) Viral RNA can also interfere (for example, sequestering) miR function

transcription start site (TSS) and the TATA box, which is usually located at ~30 nucleotides from the TSS, although the TATA box is not always required (Lee et al. 2004). In fact, housekeeping genes as well as genes involved in several developmental pathways are also often regulated by TATA-less promoters (Smale 2001; Weaver 2001). Characteristic sequence motifs have been shown to be conserved in both plants and animals. The core promoters of several miRs have been identified in *Caenorhabditis elegans*, *Homo sapiens*, *Arabidopsis thaliana* (Ath), and *Oryza sativa*, with many similar sequence motifs shared among all the four species, and others specific to individual ones (Zhou et al. 2007).

In silico analysis of promoters sequences can uncover key aspects of promoter function. Several motif-finding algorithms are available such as MEME (Bailey et al. 2009), WordSpy (Wang and Zhang 2006), the *Arabidopsis* Gene Regulatory

Information Server (AGRIS) (Davuluri et al. 2003), the *A. thaliana* Promoter Binding Element Database (AtProbe) (http://exon.cshl.org/cgi-bin/atprobe/atprobe.pl), PlantCARE (Lescot et al. 2002), PLACE (Higo et al. 1999), TRANSFAC (Wingender et al. 2000), PlantProm (Shahmuradov et al. 2003), SoftBerry (http://www.softberry.com), and common query voting (CoVote) (Zhou et al. 2007). The ability to elucidate miR regulatory circuits is aided by an ever-increasing array of genome-wide or large-scale comparison tools, including analyses of *cis*-regulatory elements in miR promoters, protein-coding gene promoters, random genomic sequences, large-scale gene expression profiling (especially miRs and miR targets), and known transcription factor (TF) binding elements. These different sources supply a suitable network for probabilistic modeling of miR regulatory circuits (Joung and Fei 2009).

Based on genome-wide analyses, the size of the miR promoter core varies from gene to species. In general, the core elements within plants are typically located between the first 500 nucleotides upstream of the fold-back miR (in the case of the polycistronic ones, the first fold-back was used). In *C. elegans* and *H. sapiens*, the core promoter regions have also been identified in the first 500 nucleotides, but also between the 500 and 1,000 nucleotides upstream the miR fold-back and even further (Zhou et al. 2007). Even though a computational approach using known motifs is a very useful tool to identify miR promoters and TSS, 5′ rapid amplification of cDNA ends may also be required to advance into the TSS (Xie et al. 2005).

3 miR Origin

Certain miRs are highly conserved among plant families, from mosses to angiosperms (Axtell and Bartel 2005). It has been proposed that at least some miRs evolved de novo through inverted duplication of their future target genes (Allen et al. 2004; Voinnet 2004). These duplications progressively acquired mutations and, through adaptive selection, adopted the "the final" characteristic stem-loop structure. Consequently, the degree of sequence conservation between miR precursors and/or miR targets reflects their evolutionary history following duplication events (Allen et al. 2004; Voinnet 2004). From this evolutionary point of view, few miRs were denominated "young" while others "old." In addition, promoter analyses have also shown sequence conservation between miR and targets, supporting the mentioned evolutionary origin (Wang et al. 2006).

4 Do Pathogen Infections Interfere or Exploit miRs?

There are several observations that suggest an evolutionary connection between pathogen infections and miRs. First, in general, viral and bacterial infections modulate host gene expression in order to facilitate invasion, multiplication, and proliferation (Chapman et al. 2004; Dunoyer and Voinnet 2005; Lakatos et al. 2006;

Navarro et al. 2008; Ruiz-Ferrer and Voinnet 2009; Silhavy and Burgyan 2004). Second, several pathogen infections produce characteristic phenotypes that resemble severe developmental alterations observed after loss or overexpression of miR. Third, several pathogens alter miR pathways at different levels. However, the role and/or mechanism of miR alteration during pathogen infection are still unclear.

Some miR alterations have been explained by a mechanism known as posttranscriptional gene silencing (PTGS), which detects and eliminates homologous double-stranded RNAs (dsRNA) and aberrant or misfolded single-stranded RNAs by small interference RNA (siRNA). In some cases, the dsRNA is generated by an RNA-dependent RNA polymerase (RDR). These RDRs are found in *C. elegans* (Sijen et al. 2001; Smardon et al. 2000), fungi (Cogoni and Macino 1999; Volpe et al. 2002), and *A. thaliana* (Dalmay et al. 2000; Mourrain et al. 2000; Xie et al. 2004). The RDR can also use siRNA to prime the synthesis of the second strand, amplifying the mechanism (Herr and Baulcombe 2004). The dsRNA is cleaved into 21–26 nucleotides RNA (siRNA) by Dicer. In this scenario, the siRNAs provide a very potent defense against invasive nucleic acids, such as those produced during viral infection (Baulcombe 2005; Hamilton and Baulcombe 1999).

However, most viruses have developed a counterdefensive strategy in which viral proteins, referred as PTGS suppressors, block one or more steps of the PTGS pathway (Roth et al. 2004; Vance and Vaucheret 2001), resulting in increased viral replication. Further, recent studies in plants and animals have suggested that viruses not only suppress but also regulate endogenous RNA silencing pathways in more complex ways to manipulate host gene expression on their behalf (Cazalla et al. 2010; Dunoyer and Voinnet 2005; Pfeffer and Voinnet 2006; Silhavy and Burgyan 2004; Voinnet 2005). However, the underlying mechanisms are still unclear. One strategy observed in mammalian viruses involves the recruitment of virally derived miRs to directly downregulate or upregulate host and/or endogenous mRNAs (Fig. 1c). This viral strategy was first described for the Epstein–Barr virus (Pfeffer and Voinnet 2006), and later for several other viruses including *Kaposi's sarcoma-associated herpesvirus*, *human cytomegalovirus*, *mouse gamma herpes virus 68*, *herpes simplex virus 1*, and *simian virus 40* (Dinant et al. 1998; Gupta et al. 2006; Pfeffer et al. 2005; Pfeffer and Voinnet 2006; Samols et al. 2005; Sullivan and Ganem 2005). The existence of virally encoded miRs clearly points to the amazing evolutionary adaptation and capacity of viruses to manipulate host gene expression.

A second mode of action is associated with viral suppressor of PTGS mostly related to plant virus (Fig. 1b). Different authors have demonstrated that viral suppressors of RNA silencing can interfere with miR-mediated regulation of host genes (Chapman et al. 2004; Chen et al. 2004; Dunoyer et al. 2004; Kasschau et al. 2003; Silhavy and Burgyan 2004). For example, transgenic expression of viral suppressors of silencing in plants resulted in changes in miR accumulation and differential host-targeted gene expression (Chapman et al. 2004; Chen et al. 2004; Dunoyer et al. 2004; Kasschau et al. 2003; Silhavy and Burgyan 2004). In addition, many of these transgenic plants showed phenotypes that mimic the viral symptoms. Those studies revealed that viral proteins could interfere with miR pathways, although it remains uncertain whether this is an adaptive strategy or just a side effect due to the shared components of silencing and miR pathways. On the contrary, expression of *Tobacco Mosaic Virus* (TMV) movement and coat proteins with no PTGS suppression activity also interferes with miR accumulation, suggesting that miR alteration would be an adaptive outcome, rather than a side effect (Bazzini et al. 2007). Supporting this adaptive hypothesis, it has been reported that *Turnip Mosaic Virus* infection can specifically induce the accumulation of miR1885, which targets genes involved in plant defense and disease resistance (He et al. 2008). These data clearly suggest an important role for miRs in host–pathogen interactions.

Third, viral infection can also interfere in miR function by sequestering miRs through sequence complementarity (Fig. 1d) (Cazalla et al. 2010). For example, *Herpes virus saimiri* expresses seven viral U-rich non coding RNAs, which constitute potential binding sites for three host-cell miRs. T cells transformed by the virus specifically reduce the abundance of miR-27, with consequent effects on the expression of miR-27 target genes. This viral strategy illustrates the uses of a non coding RNA to manipulate host-cell gene expression via the miR pathway (Cazalla et al. 2010).

Fourth, it is possible (but not likely) that viral infection alters cell energy flow of unrelated pathways, including miR ones (Pfeffer and Voinnet 2006). Nevertheless, this is a theoretical possibility, since no experimental evidence has been reported to date.

In summary, miR pathways can be affected at both transcriptional and posttranscriptional levels (involving miRs processing, accumulation, and activity). Up to now, most studies that have focused on posttranscriptional mechanisms, especially in plants, concern alterations in PTGS suppressor activity (Chapman et al. 2004; Dunoyer et al. 2004; Lakatos et al. 2006). By contrast, we have now uncovered other mechanisms of pathogen-induced miR modulation that occur through the alteration of miR transcription.

5 Virus Infection Affects Transcriptional Activity of miR164a Promoter

The ability of a virus to affect miR expression at transcriptional level was first observed using transgenic plants expressing a reporter gene under the control of the miR164a promoter (Bazzini et al. 2009). miR164 regulates plant development (Baker et al. 2005; Mallory et al. 2004) to establish organ boundaries (Nikovics et al. 2006; Raman et al. 2008; Sieber et al. 2007), leaf senescence (Kim et al. 2009), and response to abiotic stress conditions. miR164 is also involved in hormonal signaling (Guo et al. 2005; Jia et al. 2009) and is conserved among angiosperm plants (Jia et al. 2009). Interestingly, its accumulation increases after infection from different viruses across several different plant species (Bazzini et al. 2007; Tagami et al. 2007). In *A. thaliana*, miR164 is transcribed from three independent loci, miR164a, miR164b, and miR164c, and can potentially target seven members (Gustafson et al. 2005) of the *NAC* TF family such as *CUP-SHAPED COTYLEDON1* and 2 (*CUC1/2*) (Mallory et al. 2004; Raman et al. 2008), *NAC1* (Xie et al. 2000), and *ORE1*

(Kim et al. 2009). The expression patterns of the three miRs164 in *A. thaliana* partially overlap (Sieber et al. 2007), explaining the combination of redundant and nonredundant phenotypes observed in mir164a, b, and/or c gene mutants. For example, *athmir164a* knockout shows an increase in the depth of leaf sinuses (Nikovics et al. 2006), while *ath-mir164a* and *ath-mir164b* knockouts show higher lateral root branching (Guo et al. 2005). *Ath-miR164c* displays defects in carpel fusion (Baker et al. 2005), and the triple mutant (mir164abc) presents complete carpel separation (Sieber et al. 2007).

As mentioned above, in silico analysis is a powerful tool for deciphering promoter topology/function. Examination of approximately 2,500 bases upstream of the mature *A. thaliana* miR164a reveals the putative TSS and several promoter elements (Fig. 2a) (Bazzini et al. 2009) using Plant CARE databases (Lescot et al. 2002). Some of these elements are involved in pathogen defense such as stress response elements, and others are involved in gibberellic, abscisic, salicylic, and jasmonic acids responses (Fig. 2a) (Bazzini et al. 2009). In addition, circadian control and anaerobic drought response motifs are also predicted, as well as 28 enhancer elements and 23 light-responsive related sequences that are not randomly distributed (Fig. 2a) (Bazzini et al. 2009).

The spatial and temporal expression of the miR164a promoter as well as its response after viral infection have been determined in transgenic *A. thaliana* plants carrying the 2,522 bp fragment (−2483 to +39, considering as +1 the TSS) upstream the uidA reporter gene (GUS) (P-miR164) (Bazzini et al. 2009). GUS activity was detected in the entire plant vasculature and in leaf hydathodes (Fig. 2e). In reproductive organs, GUS staining was found in all carpel compound tissues and it was stronger in its vasculature (Fig. 2b, c). GUS expression was also detected in siliques, petals, and stamen vascular tissue and in the septum that separates the lobes of each anther's thecae, whereas no GUS staining was found in the sepals (Fig. 2b, c). In stems, GUS staining was restricted to developing or "young" xylem vessels (Bazzini et al. 2009). Together, these data reveal the precise regulation of miR164 expression at the transcriptional level. Using a similar approach, other groups have shown the specific expression profiles of other miRs. For example, Parizotto et al. (2004) have characterized the expression pattern of *A. thaliana* miR171 driven by an upstream sequence of 1,238 bases. In a similar manner, Vaucheret et al. (2006) have shown GUS expression in meristem, vascular tissues of leaves and roots transcriptionally under control of a 1,339 base-pair fragment of the *A. thaliana* miR168 promoter (position −1358 to −19 relative to the hairpin start).

A temporal profile of miR164 expression has also been examined using the GUS reporter under the control of the miR164a promoter (Fig. 2e) (Bazzini et al. 2009). Peak expression occurs at the transition between 1.12 and 5.1 stages, according to Boyes et al. (2001), while undetectable GUS activity was shown at stage 8 (Fig. 2e). The mentioned peak time point corresponds to just after the switch from vegetative to reproductive growth, when several developmental processes are initiated, including changes in hormone levels. This switch is also relevant within the host–virus interaction, since it coincides with the transient arrest of viral replication (Lunello et al. 2007). Together, these data would implicate miR164a as an important regulator

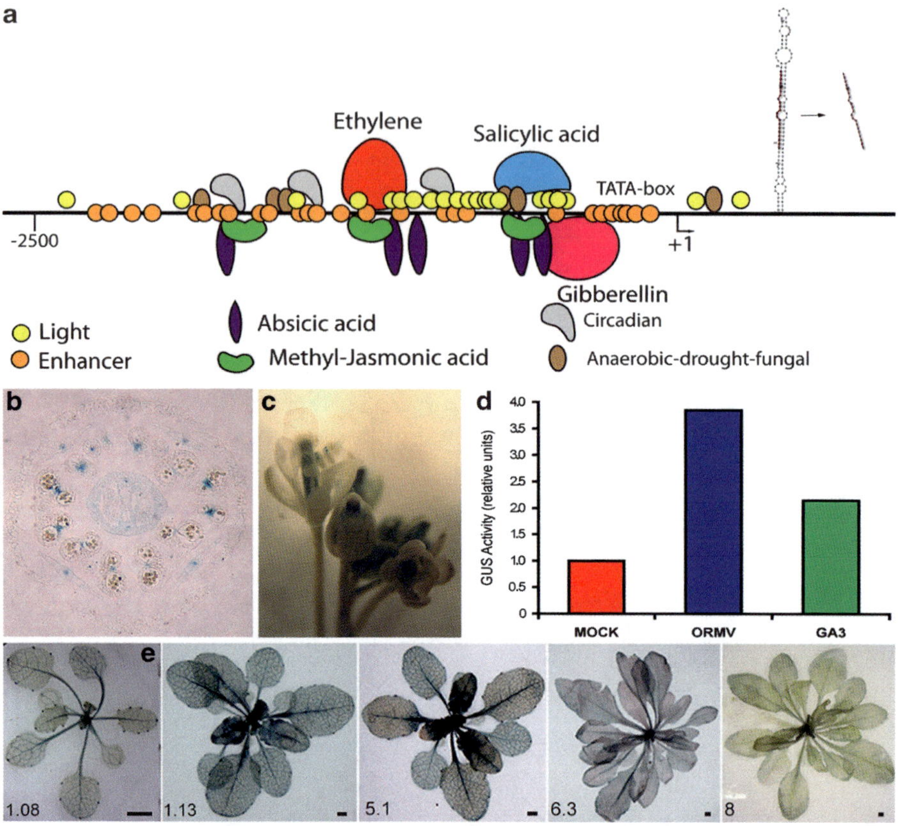

Fig. 2 (**a**) Schematic representation of the regulatory elements distribution along the miR-164a promoter identified in silico. The transcription start site (+1), the stem-loop, and the mature miR are also shown. (**b**) Flower transverse section showing the reporter gene activity controlled by miR164a promoter in the septum that divides both locules of each theca. (**c**) GUS staining of mature and immature transgenic flowers expressing GUS under the miR164a promoter. (**d**) Bar graphics showing the mean GUS activity value measured in transgenic plant expressing GUS under the miR164a promoter after infection with ORMV or treated with gibberellin (GA3). Each treatment had its own control (mock inoculated or mock treated), which were set as 1. (**e**) Time course of GUS activity driven by the miR164a promoter during the development of the transgenic plants from stages 1.08 to stage 8. Bar = 0.5 cm

of developmental transitions in the plant life cycle. The undetectable expression of the miR164a promoter at stage 8 is supported by an inability to detect mature miR164 at this point, as well as the coincident upregulation of the miR-164 target, *ORE1*, a key gene involved in leaf senescence (Kim et al. 2009).

The 2,500 base-pair fragment of the *A. thaliana* miR164 promoter also drives gene expression in monocotyledonous and dicotyledonous plants, such as *Allium cepa*, *Solanum tuberosum*, *Helianthus annuus*, and *Nicotiana benthamiana*, by micro-projectile bombardment or agroinfiltration assays, suggesting the broad conservation of regulatory elements among different species. By contrast, no transcriptional activity is detected in mammalian BHK or insect Sf9 cells transfected with appropriate reporter constructs, suggesting that, at least for miR164 promoter, the regulatory elements themselves, or their distribution/configuration, are not sufficient to induce expression in animals (Bazzini et al. 2009).

Induction of the miR164a promoter by viruses was established by GUS measurement in the transgenic lines infected with *Oilseed rape mosaic virus* (ORMV) (Fig. 2d) and TMV-Cg (Bazzini et al. 2009). Although both viruses infect *A. thaliana*, both differ markedly in the severity of the symptoms they produce on *A. thaliana* plants, from very mild in the case of TMV-Cg to strong in the case of ORMV, even when both viruses are proposed to be strains of the same species of the *Tobamovirus* family (Aguilar et al. 1996; Lartey et al. 1996). Interestingly, the most severe virus (ORMV) significantly increased GUS activity (Fig. 2d). The less severe virus (TMV-Cg) also presented higher mean GUS activity compared to the noninfected samples; however, this finding is not statistically significant (Bazzini et al. 2009). These results indicate that viral infection can interfere, directly or indirectly, with miR expression level (Fig. 1) and support the hypothesis that some viral symptoms are consequences of miR alteration (Cillo et al. 2009).

Interestingly, gibberellin treatment (sprays) also induced GUS activity and mature miR accumulation compared to control water treatment, suggesting that miR164a promoter can also be induced by gibberellin (Fig. 2d). This result is consistent with the presence of several phytohormones-responsive elements within the miR-164a upstream sequence (Fig. 1). Moreover virus infections were reported to alter the concentration of phytohormones such as auxin, gibberellin, and abscisic acid (ABA) (Bari and Jones 2009). Therefore, it makes sense to propose a cross talk between hormone and miR abundance (or vice versa) after virus infection that could potentially alter miR promoter activity (Fig. 1). In fact, recent works has reported a link between miRs, hormones, and pathogen resistance (Navarro et al. 2006, 2008). In addition, the level of pre-miR164a transcripts was also higher after infection, supporting the promoter induction results (Bazzini et al. 2009). Nevertheless, the increased pre-miR164a accumulation could also be explained by a change in the nuclear precursor rate processing. As mentioned before, virus infection increases mature miR levels and miR gene targets (Bazzini et al. 2009; Chapman et al. 2004; Dunoyer et al. 2004). These increments have been previously explained by a stabilization process mediated by PTGS suppressors, which bind and sequester sRNAs (including miRs) as double-strand inactive forms. This binding reduces miR activity (Chapman et al. 2004; Csorba et al. 2007) leading to the miR gene target accumulation (Fig. 1b). Therefore, miR inactivation could be sensed by the plant due to the higher miR target level, consequently inducing miR transcription to restore miR target levels (Fig. 1b), proposing a feedback between target level and miR transcription (Fig. 1).

It is unclear whether viral proteins act directly or indirectly at the miR-164a promoter to influence miR expression. The identification of TFs and chromatin marks such as methylation after virus infection will illuminate the underlying mechanisms of the transcriptional alteration of miR (Fig. 1).

In sum, virus infection as well as hormone treatment can alter miR at transcriptional level. Further study will be required to ascertain the functional relevance of such regulation and its relative significance compared to posttranscriptional mechanisms of miR modulation.

6 miR Promoter Can Be Use to Shut Down miR Expression

Only a few loss-of-function mutants have been described for miRs. And most of miR mutants interrupt key genes of the miR biogenesis pathways. For example, *A. thaliana* dicer mutants present strong abnormal development, clearly demonstrating the roles of miRs in the patterning and formation of the whole plant (Deleris et al. 2006). In animals, dicer mutant embryos showed that zygotically expressed miRs dramatically enhance the efficiency of maternal mRNA clearance in the maternal-to-zygotic transition (MZT), a universal step in animal development (Bushati et al. 2008; Giraldez et al. 2006; Lund et al. 2009). However, all these examples affect the production of all, or most, miRs. The future challenge lies in suppressing the expression of individual miR family members. In animals, knockdown of miRs was done by genetic deletion or by "miRZip," antisense miRs expressed by lentivectors (Lesnik and Antes 2010; Zhao et al. 2007). For example, genetic deletion of miR-1-2 revealed numerous functions in the heart, including regulation of cardiac morphogenesis, electrical conduction, and cell cycle control (Zhao et al. 2007). Alternatively, in plants, few approaches have been described. One of them is referred to as target mimicry, whereby an uncleavable target is used to sequester specific miRs into nonproductive interactions, most likely reducing the availability of the miR to carry out its normal regulatory role (Franco-Zorrilla et al. 2007). A comparable technique has also been utilized in animals (Choi et al. 2007). Another common strategy is the transgenic expression of miR resistant target, by direct mutagenesis into the complementary region recognized with the miR. This particular technique, while effective to establish the role of individual miR-target interactions, does not reveal the full spectrum activity of particular miRs.

An alternative strategy to knock down miR function involves the targeting of miR promoters. Since promoter sequences are responsible for driving gene expression, these same sequences can be targeted through transcriptional silencing to prevent gene expression. Given that only a few loss-of-function mutants have been described, it is likely that the targeting of miR promoters will serve as a productive avenue by which to reveal miR function/action. Vaistij et al. (2010) cleverly demonstrated that it is possible to use RNA interference (RNAi) to suppress miR accumulation. Using hairpin RNAi constructs designed to target both primary miR transcripts and their promoters, they reduced the accumulation of miR163 and miR171a, resulting in a concomitant increase in the levels of particular targets. Intriguingly, this strategy resulted in the DNA methylation of targeted promoter regions. This study demonstrates that knockdown of single miR expression can be accomplished using RNAi directed at promoter regions by transcriptional silencing (Vaistij et al. 2010).

7 Conclusion

This chapter focuses on miR promoter regions, their characteristic structure, their responsiveness to external stimuli, and their potential applications in knockdown strategies. In particular, the example of *A. thaliana* miR164a is used to illustrate the precise temporal and spatial developmental expression pattern of a miR and how it can be modulated after hormone treatments and viral infection. Importantly, the study of *A. thaliana* miR164a reveals an important feature of the host–pathogen interaction: modulation of miR expression at the level of miR transcription. In sum, new important questions emerge involving miR promoter elements. Which miRs are expressed under what specific conditions? Which miRs are differentially regulated at transcriptional level after viral infection? Which *trans*-acting factors regulate miR expression?

Acknowledgments We thank Almasia N.I., Manacorda C.A., Mongelli V.C., Conti G., Maroniche G.A., Rodriguez M.C., Distéfano A.J., Hopp H.E., and del Vas M. for the great teamwork studying the *A. thaliana* miR164 promoter. We especially thank Drs Natalia Almasia and Florencia Del Viso for discussions and comments on this chapter. We owe our thanks to Dr. Carter Takacs for critical reading of this chapter. This research was partially supported by PICT 2008 No. 1094 and PICT 2005 No. 32598, Agencia Nacional de Promocion Científica y Técnológica (ANPCyT), to A.B. and S.A., respectively, and by Instituto Nacional de Tecnología Agropecuaria (INTA, Argentina) PE 3454 to S.A.

References

Aguilar I, Sanchez F, Martin Martin A et al (1996) Nucleotide sequence of Chinese rape mosaic virus (oilseed rape mosaic virus), a crucifer tobamovirus infectious on *Arabidopsis thaliana*. Plant Mol Biol 30:191–197

Allen E, Xie Z, Gustafson AM et al (2004) Evolution of microRNA genes by inverted duplication of target gene sequences in *Arabidopsis thaliana*. Nat Genet 36:1282–1290

Aukerman MJ, Sakai H (2003) Regulation of flowering time and floral organ identity by a MicroRNA and its APETALA2-like target genes. Plant Cell 15:2730–2741

Axtell MJ, Bartel DP (2005) Antiquity of microRNAs and their targets in land plants. Plant Cell 17:1658–1673

Bailey TL, Boden M, Buske FA et al (2009) MEME SUITE: tools for motif discovery and searching. Nucleic Acids Res 37:W202–W208

Baker CC, Sieber P, Wellmer F et al (2005) The early extra petals1 mutant uncovers a role for microRNA miR164c in regulating petal number in Arabidopsis. Curr Biol 15:303–315

Bari R, Jones JD (2009) Role of plant hormones in plant defense responses. Plant Mol Biol 69:473–488

Bartel DP (2004) MicroRNAs: genomics, biogenesis, mechanism, and function. Cell 116:281–297

Baulcombe D (2005) RNA silencing. Trends Biochem Sci 30:290–293

Bazzini AA, Hopp HE, Beachy RN et al (2007) Infection and coaccumulation of tobacco mosaic virus proteins alter microRNA levels, correlating with symptom and plant development. Proc Natl Acad Sci USA 104:12157–12162

Bazzini AA, Almasia NI, Manacorda CA et al (2009) Virus infection elevates transcriptional activity of miR164a promoter in plants. BMC Plant Biol 9:152

Boyes DC, Zayed AM, Ascenzi R et al (2001) Growth stage-based phenotypic analysis of Arabidopsis: a model for high throughput functional genomics in plants. Plant Cell 13:1499–1510

Brennecke J, Hipfner DR, Stark A et al (2003) bantam encodes a developmentally regulated microRNA that controls cell proliferation and regulates the proapoptotic gene hid in Drosophila. Cell 113:25–36

Bushati N, Stark A, Brennecke J et al (2008) Temporal reciprocity of miRNAs and their targets during the maternal-to-zygotic transition in Drosophila. Curr Biol 18:501–506

Cazalla D, Yario T, Steitz J (2010) Down-regulation of a host microRNA by a Herpesvirus saimiri noncoding RNA. Science 328:1563–1566

Chapman EJ, Prokhnevsky AI, Gopinath K et al (2004) Viral RNA silencing suppressors inhibit the microRNA pathway at an intermediate step. Genes Dev 18:1179–1186

Chen J, Li WX, Xie D et al (2004) Viral virulence protein suppresses RNA silencing-mediated defense but upregulates the role of microRNA in host gene expression. Plant Cell 16: 1302–1313

Choi WY, Giraldez AJ, Schier AF (2007) Target protectors reveal dampening and balancing of nodal agonist and antagonist by miR-430. Science 318:271–274

Cillo F, Mascia T, Pasciuto MM et al (2009) Differential effects of mild and severe Cucumber mosaic virus strains in the perturbation of MicroRNA-regulated gene expression in tomato map to the 3′ sequence of RNA 2. Mol Plant Microbe Interact 22:1239–1249

Cogoni C, Macino G (1999) Gene silencing in *Neurospora crassa* requires a protein homologous to RNA-dependent RNA polymerase. Nature 399:166–169

Csorba T, Bovi A, Dalmay T et al (2007) The p122 subunit of Tobacco Mosaic Virus replicase is a potent silencing suppressor and compromises both small interfering RNA- and microRNA-mediated pathways. J Virol 81:11768–11780

Dalmay T, Hamilton A, Rudd S et al (2000) An RNA-dependent RNA polymerase gene in Arabidopsis is required for posttranscriptional gene silencing mediated by a transgene but not by a virus. Cell 101:543–553

Davuluri RV, Sun H, Palaniswamy SK et al (2003) AGRIS: Arabidopsis gene regulatory information server, an information resource of Arabidopsis cis-regulatory elements and transcription factors. BMC Bioinformatics 4:25

Deleris A, Gallego-Bartolome J, Bao J et al (2006) Hierarchical action and inhibition of plant Dicer-like proteins in antiviral defense. Science 313:68–71

Dinant S, Kusiak C, Cailleteau B et al (1998) Field resistance against potato virus Y infection using natural and genetically engineered resistance genes. Eur J Plant Pathol 104:377–382

Dunoyer P, Voinnet O (2005) The complex interplay between plant viruses and host RNA-silencing pathways. Curr Opin Plant Biol 8:415–423

Dunoyer P, Lecellier CH, Parizotto EA et al (2004) Probing the microRNA and small interfering RNA pathways with virus-encoded suppressors of RNA silencing. Plant Cell 16: 1235–1250

Eulalio A, Huntzinger E, Izaurralde E (2008) Getting to the root of miRNA-mediated gene silencing. Cell 132:9–14

Franco-Zorrilla JM, Valli A, Todesco M et al (2007) Target mimicry provides a new mechanism for regulation of microRNA activity. Nat Genet 39:1033–1037

Giraldez AJ, Mishima Y, Rihel J et al (2006) Zebrafish MiR-430 promotes deadenylation and clearance of maternal mRNAs. Science 312:75–79

Guo HS, Xie Q, Fei JF et al (2005) MicroRNA directs mRNA cleavage of the transcription factor NAC1 to downregulate auxin signals for arabidopsis lateral root development. Plant Cell 17:1376–1386

Guo H, Ingolia NT, Weissman JS et al (2010) Mammalian microRNAs predominantly act to decrease target mRNA levels. Nature 466:835–840

Gupta A, Gartner JJ, Sethupathy P et al (2006) Anti-apoptotic function of a microRNA encoded by the HSV-1 latency-associated transcript. Nature 442:82–85

Gustafson AM, Allen E, Givan S et al (2005) ASRP: the Arabidopsis small RNA project database. Nucleic Acids Res 33:D637–D640

Hamilton AJ, Baulcombe DC (1999) A species of small antisense RNA in posttranscriptional gene silencing in plants. Science 286:950–952

He XF, Fang YY, Feng L et al (2008) Characterization of conserved and novel microRNAs and their targets, including a TuMV-induced TIR-NBS-LRR class R gene-derived novel miRNA in Brassica. FEBS Lett 582:2445–2452

Herr AJ, Baulcombe DC (2004) RNA silencing pathways in plants. Cold Spring Harb Symp Quant Biol 69:363–370

Higo K, Ugawa Y, Iwamoto M et al (1999) Plant cis-acting regulatory DNA elements (PLACE) database: 1999. Nucleic Acids Res 27:297–300

Jia X, Mendu V, Tang G (2009) An array platform for identification of stress-responsive microRNAs in plants. Meth Mol Biol 639:253–269

Joung JG, Fei Z (2009) Identification of microRNA regulatory modules in Arabidopsis via a probabilistic graphical model. Bioinformatics 25:387–393

Kasschau KD, Xie Z, Allen E et al (2003) P1/HC-Pro, a viral suppressor of RNA silencing, interferes with Arabidopsis development and miRNA unction. Dev Cell 4:205–217

Kedde M, Strasser MJ, Boldajipour B et al (2007) RNA-binding protein Dnd1 inhibits microRNA access to target mRNA. Cell 131:1273–1286

Kim JH, Woo HR, Kim J et al (2009) Trifurcate feed-forward regulation of age-dependent cell death involving miR164 in Arabidopsis. Science 323:1053–1057

Lakatos L, Csorba T, Pantaleo V et al (2006) Small RNA binding is a common strategy to suppress RNA silencing by several viral suppressors. EMBO J 25:2768–2780

Lartey RT, Voss TC, Melcher U (1996) Tobamovirus evolution: gene overlaps, recombination, and taxonomic implications. Mol Biol Evol 13:1327–1338

Lee Y, Kim M, Han J et al (2004) MicroRNA genes are transcribed by RNA polymerase II. EMBO J 23:4051–4060

Lee JY, Colinas J, Wang JY et al (2006) Transcriptional and posttranscriptional regulation of transcription factor expression in Arabidopsis roots. Proc Natl Acad Sci USA 103:6055–6060

Lescot M, Dehais P, Thijs G et al (2002) PlantCARE, a database of plant cis-acting regulatory elements and a portal to tools for in silico analysis of promoter sequences. Nucleic Acids Res 30:325–327

Lesnik JK, Antes TJ (2010) Permanent knockdown of microRNAs using lentivectors. Biotechniques 48:321–323

Lund E, Liu M, Hartley RS et al (2009) Deadenylation of maternal mRNAs mediated by miR-427 in Xenopus laevis embryos. RNA 15:2351–2363

Lunello P, Mansilla C, Sanchez F et al (2007) A developmentally linked, dramatic, and transient loss of virus from roots of *Arabidopsis thaliana* plants infected by either of two RNA viruses. Mol Plant Microbe Interact 20:1589–1595

Mallory AC, Dugas DV, Bartel DP et al (2004) MicroRNA regulation of NAC-domain targets is required for proper formation and separation of adjacent embryonic, vegetative, and floral organs. Curr Biol 14:1035–1046

Mourrain P, Beclin C, Elmayan T et al (2000) Arabidopsis SGS2 and SGS3 genes are required for posttranscriptional gene silencing and natural virus resistance. Cell 101:533–542

Navarro L, Dunoyer P, Jay F et al (2006) A plant miRNA contributes to antibacterial resistance by repressing auxin signaling. Science 312:436–439

Navarro L, Jay F, Nomura K et al (2008) Suppression of the microRNA pathway by bacterial effector proteins. Science 321:964–967

Nikovics K, Blein T, Peaucelle A et al (2006) The balance between the MIR164A and CUC2 genes controls leaf margin serration in Arabidopsis. Plant Cell 18:2929–2945

Palatnik JF, Allen E, Wu X et al (2003) Control of leaf morphogenesis by microRNAs. Nature 425:257–263

Parizotto EA, Dunoyer P, Rahm N et al (2004) In vivo investigation of the transcription, processing, endonucleolytic activity, and functional relevance of the spatial distribution of a plant miRNA. Genes Dev 18:2237–2242

Pfeffer S, Voinnet O (2006) Viruses, microRNAs and cancer. Oncogene 25:6211–6219

Pfeffer S, Sewer A, Lagos-Quintana M et al (2005) Identification of microRNAs of the herpesvirus family. Nat Methods 2:269–276

Poethig RS (2009) Small RNAs and developmental timing in plants. Curr Opin Genet Dev 19:374–378

Raman S, Greb T, Peaucelle A et al (2008) Interplay of miR164, CUP-SHAPED COTYLEDON genes and LATERAL SUPPRESSOR controls axillary meristem formation in *Arabidopsis thaliana*. Plant J 55:65–76

Roth BM, Pruss GJ, Vance VB (2004) Plant viral suppressors of RNA silencing. Virus Res 102:97–108

Ruiz-Ferrer V, Voinnet O (2009) Roles of plant small RNAs in biotic stress responses. Annu Rev Plant Biol 60:485–510

Samols MA, Hu J, Skalsky RL et al (2005) Cloning and identification of a microRNA cluster within the latency-associated region of Kaposi's sarcoma-associated herpesvirus. J Virol 79:9301–9305

Sempere LF, Sokol NS, Dubrovsky EB et al (2003) Temporal regulation of microRNA expression in *Drosophila melanogaster* mediated by hormonal signals and broad-complex gene activity. Dev Biol 259:9–18

Shahmuradov IA, Gammerman AJ, Hancock JM et al (2003) PlantProm: a database of plant promoter sequences. Nucleic Acids Res 31:114–117

Sieber P, Wellmer F, Gheyselinck J et al (2007) Redundancy and specialization among plant microRNAs: role of the MIR164 family in developmental robustness. Development 134:1051–1060

Sijen T, Fleenor J, Simmer F et al (2001) On the role of RNA amplification in dsRNA-triggered gene silencing. Cell 107:465–476

Silhavy D, Burgyan J (2004) Effects and side-effects of viral RNA silencing suppressors on short RNAs. Trends Plant Sci 9:76–83

Smale ST (2001) Core promoters: active contributors to combinatorial gene regulation. Genes Dev 15:2503–2508

Smardon A, Spoerke JM, Stacey SC et al (2000) EGO-1 is related to RNA-directed RNA polymerase and functions in germ-line development and RNA interference in *C. elegans*. Curr Biol 10:169–178

Sullivan CS, Ganem D (2005) MicroRNAs and viral infection. Mol Cell 20:3–7

Tagami Y, Inaba N, Kutsuna N et al (2007) Specific enrichment of miRNAs in *Arabidopsis thaliana* infected with Tobacco mosaic virus. DNA Res 14:227–233

Tam W (2001) Identification and characterization of human BIC, a gene on chromosome 21 that encodes a noncoding RNA. Gene 274:157–167

Tomari Y, Matranga C, Haley B et al (2004) A protein sensor for siRNA asymmetry. Science 306:1377–1380

Vaistij FE, Elias L, George GL et al (2010) Suppression of microRNA accumulation via RNA interference in *Arabidopsis thaliana*. Plant Mol Biol 73:391–397

Vance V, Vaucheret H (2001) RNA silencing in plants – defense and counterdefense. Science 292:2277–2280

Vaucheret H, Mallory AC, Bartel DP (2006) AGO1 homeostasis entails coexpression of MIR168 and AGO1 and preferential stabilization of miR168 by AGO1. Mol Cell 22:129–136

Voinnet O (2004) Shaping small RNAs in plants by gene duplication. Nat Genet 36:1245–1246

Voinnet O (2005) Induction and suppression of RNA silencing: insights from viral infections. Nat Rev Genet 6:206–220

Volpe TA, Kidner C, Hall IM et al (2002) Regulation of heterochromatic silencing and histone H3 lysine-9 methylation by RNAi. Science 297:1833–1837

Wang G, Zhang W (2006) A steganalysis-based approach to comprehensive identification and characterization of functional regulatory elements. Genome Biol 7:R49

Wang Y, Hindemitt T, Mayer KF (2006) Significant sequence similarities in promoters and precursors of *Arabidopsis thaliana* non-conserved microRNAs. Bioinformatics 22:2585–2589

Weaver RF (2001) Molecular biology. McGraw-Hill, Boston, pp 279–280

Wingender E, Chen X, Hehl R et al (2000) TRANSFAC: an integrated system for gene expression regulation. Nucleic Acids Res 28:316–319

Xie Q, Frugis G, Colgan D et al (2000) Arabidopsis NAC1 transduces auxin signal downstream of TIR1 to promote lateral root development. Genes Dev 14:3024–3036

Xie Z, Johansen LK, Gustafson AM et al (2004) Genetic and functional diversification of small RNA pathways in plants. PLoS Biol 2:E104

Xie Z, Allen E, Fahlgren N et al (2005) Expression of Arabidopsis MIRNA genes. Plant Physiol 138:2145–2154

Zhao Y, Ransom JF, Li A et al (2007) Dysregulation of cardiogenesis, cardiac conduction, and cell cycle in mice lacking miRNA-1-2. Cell 129:303–317

Zhou X, Ruan J, Wang G et al (2007) Characterization and identification of microRNA core promoters in four model species. PLoS Comput Biol 3:e37

Populus Short RNAs

Nathaniel Robert Street

Contents

Abstract As a model species, *Populus* offers opportunities to study processes unique to woody, perennial species, such as seasonal senescence, dormancy and wood production. As a long-lived species, it is also possible that different adaptive survival strategies have been selected for in comparison to annual species. To date, a number of miRNAs have been shown to be differentially expressed or induced in response to abiotic stress and to be involved in processes such as wood development. Although *Populus* has not yet been extensively profiled for short RNAs, the available data has been used to identify a number of phased loci and to characterise association of short RNA and repetitive elements in the genome. There appears to be a hotspot of sRNA production on chromosome 19, which contains the proposed sex-determining locus.

Keywords: Populus • miRNA • short RNA • Sex locus • NBS-LRR

N.R. Street (✉)
Department of Ecology and Environmental Science, Umeå Plant Science Centre, Umeå University, 901 87 Umeå, Sweden
e-mail: nathaniel.street@plantphys.umu.se

V.A. Erdmann and J. Barciszewski (eds.), *Non Coding RNAs in Plants*, RNA Technologies, DOI 10.1007/978-3-642-19454-2_23,

1 Overview

Populus is now firmly established as an important model plant species, especially for studying processes unique to woody, perennial species, such as seasonal senescence, dormancy and wood production (Jansson and Douglas 2007). As a long-lived species, it is also possible that different adaptive survival strategies have been selected for in comparison to annuals. To date, there have been very few studies examining miRNAs in *Populus*, and these have been performed on a range of species. Those studies have uncovered a number of non-conserved miRNA families, and many of these have been shown to be responsive to differential conditions and abiotic perturbations in addition to the already well-defined roles of miRNAs in developmental processes. There remains only a single study examining the total sRNA population (Klevebring et al. 2009), and it is likely that greater sequencing depth will be required to fully characterise the full diversity of different sRNA classes in *Populus*. The studies of Lu et al. (2005, 2008), Jia et al. (2009) and Li et al. (2009) were all performed using traditional cloning and Sanger sequencing approaches, while studies of Barakat et al. (2007) and Klevebring et al. (2009) used 454 pyrosequencing. There are no publications to date making use of the much higher throughput Illumina and SOLiD sequencing technologies, both of which have short RNA protocols available.

For the purposes of this chapter, the dataset of Klevebring et al. (2009) has been updated to the recently released v2 assembly of the *Populus trichocarpa* genome (http://www.phytozome.net) and to miRBase release 15 using identical methodology to that detailed in the original publication.

2 miRNAs

The latest release of miRBase (v15, Griffiths-Jones et al. 2008) contains miRNAs identified in *P. trichocarpa* (234 miRNAs) and *Populus euphratica* (8 miRNAs), representing a total of 49 families. Recently, Zhou et al. (2010) have also described sequencing of miRNAs in *Populus cathayana*. Table 1 provides sequence counts for miRNA families from Viridiplantae precursor sequences for the data of Klevebring et al. (2009).

In general, highly conserved miRNAs, particularly those regulating developmental processes, showed the highest read counts (i.e. expression level), although the relative abundance of families differs between species (Klevebring et al. 2009). This will, at least in part, reflect the underlying biological state of the plants and tissues sampled.

Although some families are classified as specific to *P. euphratica* (Li et al. 2009), this is likely a limitation of sequencing coverage in previous work rather than those families actually being specific to a particular *Populus* species. In addition, imperfect

Table 1 miRNA family expression

Matching miRNAs	Raw read count
miR159	41,653
miR398	11,557
miR167	6,624
miR408	5,989
miR168	2,695
miR1450	2,075
miR475	1,600
miR166	1,222
miR396	1,124
miR172	984
miR156	733
miR1447	659
miR472	573
miR160	567
miR169	334
miR473	289
miR397	288
miR164	285
miR171	232
miR394	131
miR393	125
miR476	117
miR162	91
miR1444	87
miR828	59
miR1511	36
miR395	35
miR319	27
miR399	22
miR858	17
miR156, miR157	14
miR2911	12
miR403	10
miR390	6
miR479	5
miR482	4
miR477	3
miR478	2
miR2111	1
miR827	1

Raw sequence counts of sRNA sequences from Klevebring et al. (2009) with perfect matches to mature sequences of known miRNA families (miRBase v15). Where an sRNA matches multiple families, the family IDs are concatenated definition of mature miRNA sequences often makes comparison between datasets problematic. For example, if sRNA sequence reads from Klevebring et al. (2009) are searched against precursor rather than mature miRNA sequences in miRBase, there are low read count (i.e. lowly expressed) matches to two of the suggested eight *P. euphratica* specific miRNA families. Such issues will need addressing as many *Populus* miRNAs in miRBase lack the currently required evidence for classification as bona fide miRNAs and remain in silico predictions only or lack evidence of miRNA* production (Klevebring et al. 2009; Meyers et al. 2008).

It remains unclear how many novel, non-conserved miRNAs are still to be identified in *Populus*. Klevebring et al. (2009) identified 414 potential miRNAs with only 143 of those matching currently known miRNAs. Although some degree of false positives can be expected in the remaining predictions, at least a percentage of them can be expected to represent bona fide, but as yet, unconfirmed miRNAs. Updated analysis of the same data to the recently released v2 assembly of the *Populus* genome (http://www.phytozome.net) identified 493 predicted miRNAs of which 82 have miRNA* evidence. As was found in Klevebring et al. (2009), the majority of predicted miRNAs that match current miRBase entries lack miRNA* evidence in this, or any, dataset. Certainly, a deeper coverage sequencing dataset covering a range of tissues, developmental states and environmental conditions would be of significant benefit to better define the miRNA population of *Populus*.

Of the few studies examining miRNAs in *Populus*, the majority have considered their role in stress response. Lu et al. (2005) identified miRNAs that exhibit altered expression in response to mechanical stress in woody tissue and including miRNAs that are not conserved in common with *Arabidopsis*. Even for those conserved miRNAs that were examined, species-specific patterns of expression were found, suggesting divergence in their regulatory roles and control. Lu et al. (2008) identified a number of *Populus* specific miRNA families and showed that the majority of miRNAs within these families respond differentially to a range of abiotic stress conditions. In contrast to short-lived annual species, tree species have to survive across multiple years and be able to respond and survive repeated exposure to stress conditions. As such, it is possible that unique evolutionary adaptive strategies and control mechanisms have been selected for in woody species, and the results of Lu et al. (2008) suggest that non-conserved miRNAs may be one such mechanism. Jia et al. (2009) identified a number of miRNAs that showed altered expression in response to UV-B exposure. Interestingly, they performed promoter analysis of those miRNA loci and found that they share many promoter elements in common to their predicted target genes. Li et al. (2009) identified miRNAs in the highly stress tolerant *P. euphratica*. They identified a number of novel miRNAs that had not previously been characterised in *Populus*, a subset of which were shown to be induced by dehydration stress. The genes predicted to be targeted by these miRNAs included a diverse range of functions, again showing how miRNAs can lead to complex remodelling of transcriptional networks. Barakat et al. (2007), Klevebring et al. (2009) and Street et al. (2008) performed in silico analysis of miRNA predicted targets. A common finding, in agreement with results from other species, was that the majority of miRNA targets are involved in developmental processes and expression domain restriction. Street et al. (2009) used a combinatorial approach to predict that miRNA family 396 may be of particular importance in the control of leaf development. However, this remains to be confirmed in planta.

3 Genome-Wide sRNA Distribution

The genomic distribution of sRNAs in *Populus* largely matches that reported for *Arabidopsis* (Kasschau et al. 2007; Rajagopalan et al. 2006) with 24- and 21-nt sRNAs being the most prevalent size class and with these showing significant association to the location of repetitive elements. Table 2 provides an updated version of Table 1 from Klevebring et al. (2009) and shows that fewer sRNA reads align to the new version of the genome. This is possibly an effect of the removal of some of the alternate haplotypes that were previously represented by short scaffolds in the v1 assembly, which would suggest haplotypic differences in sRNA sequences. Another significant difference between the two genome assemblies was the removal of contaminant sequences (largely fungal), and it is possible that some of the sRNAs that previously had valid alignment were not *Populus* sequences.

The pattern of distribution for an updated repeat analysis of the v2 genome assembly confirms that observed in Klevebring et al. (2009), with the majority of collocation between repeats and sRNAs occurring for 24-nt sRNAs with LTR retrotransposon elements. There is also significant collocation of repeats with 21-nt sRNAs and clear dominance of collocation of 21-nt sRNAs with *NBS-LRR* disease-resistance genes. *NBS-LRR* genes have undergone significant expansion in *Populus* (Tuskan et al. 2006), and it is likely that the correspondingly high production of collocating sRNA sequences serves as a mechanism to limit runaway expression of those genes. Based on current knowledge, it is most likely that 24-nt sRNAs associating with repetitive regions trigger silencing via methylation (Chan et al. 2004), but as *Populus* lacks a series of knockout or knock-down lines for the various sRNA biogenesis pathway genes, this remains to be confirmed, as does the biological significance of production of the various classes of sRNA.

Table 2 sRNA sequence read counts

	Total	Non-redundant
Input	901,887	216,767
Filter valid sequence	625,361	146,894
Filter t/rRNA	502,277	129,750
Filter genome	293,224	75,606

Sequence read count overview for the data of Klevebring et al. (2009) updated to v2 *Populus trichocarpa* genome assembly. See Klevebring et al. (2009) for full details of filtering steps

4 TAS and Phased Loci

A number of papers have reported identification of *TAS3* conservation across a diverse range of plant species, including *Populus* (Axtell et al. 2006; Allen et al. 2005). Klevebring et al. (2009) provided the first sequence-based evidence for the active transcription of *TAS3* in *Populus* including complete conservation of the *Arabidopsis thaliana* active targeting siRNAs. However, in both Klevebring et al. (2009) and the reanalysis of their data to the v2 genome assembly, the *TAS3* locus was not an identified phased locus due to very low expression levels in the sequenced sample. There is no evidence for the conservation of the other TAS loci identified in *Arabidopsis* (Klevebring et al. 2009).

A number of predicted phased loci were identified (Table 3) with only one of these being classified as *trans*-acting (i.e. TAS). Similar to *TAS4* in *Arabidopsis*, this locus targets a MYB transcription factor; however, there is no evidence of sequence conservation between these two loci. All other predicted phased loci target genes in *cis* as well as *trans* and are, therefore, not classified as TAS loci. The majority of these loci target *PPR* and *NBS-LRR* disease-resistance genes, both of which have undergone significant expansion in *Populus* (Tuskan et al. 2006). It is likely that the production of phased siRNAs acts as a self-regulatory mechanism to prevent run-away transcription of these gene families. Phased loci targeting *PPR* genes have also been identified in *Arabidopsis* (Howell et al. 2007); however, these are not conserved in *Populus*, suggesting convergent, and evolutionarily young, selection for phased siRNAs targeting similar gene families in both species.

Reanalysis of the Klevebring et al. (2009) data identified a total of 107 potential phased loci: 75 of these are located on scaffolds representing the assembled 19 *Populus* chromosomes (i.e. scaffold_1 through scaffold_19) with the remaining loci located on scaffolds not yet placed within the main assembly. These scaffolds most likely represent alternate haplotype blocks that lie in highly variable regions of the genome. All phased loci are currently uncharacterised, and therefore, a considerable effort is needed to understand the biological role and importance of these loci.

5 An sRNA Sex Locus?

Perhaps the most intriguing observation to arise form the data presented in Klevebring et al. (2009) was the presence of a significantly higher than average production of sRNAs within the paratelomeric region of chromosome 19. This region contains a strikingly high density of *NBS-LRR* genes and has substantially reduced rates of recombination, and there is mounting evidence that the sex-determining locus is within this region (Pakull et al. 2009; Yin et al. 2008). The sequenced *P. trichocarpa* individual was a female, and evidence suggests that females are the heterochromatic sex in *Populus*, although this is still far from certain (Yin et al. 2008). Klevebring et al. (2009) showed that this region also contains a

Table 3 Phased sRNA Loci

Chromosome	Start bp	End bp	No. of sequences	No. of phased sequences	*p*-Value
Scaffold_1	31277100	31277351	7	7	1.39E−10
Scaffold_1	21959293	21959544	27	6	6.77E−04
Scaffold_2	14284366	14284617	30	7	1.49E−04
Scaffold_3	14874292	14874543	7	7	1.39E−10
Scaffold_5	923826	924077	60	11	9.45E−06
Scaffold_5	12360153	12360404	22	10	1.60E−09
Scaffold_5	3440266	3440517	18	9	4.79E−09
Scaffold_5	3449597	3449848	17	8	8.22E−08
Scaffold_5	3444468	3444719	12	7	9.57E−08
Scaffold_5	17094407	17094658	26	6	5.42E−04
Scaffold_5	2904032	2904283	23	6	2.59E−04
Scaffold_5	2696884	2697135	6	5	6.98E−07
Scaffold_5	3186815	3187066	8	5	6.14E−06
Scaffold_6	545507	545758	25	9	1.69E−07
Scaffold_6	22511165	22511416	15	7	7.16E−07
Scaffold_8	16918920	16919171	11	6	1.68E−06
Scaffold_8	426355	426606	25	6	4.29E−04
Scaffold_10	20253496	20253747	29	10	4.23E−08
Scaffold_11	11718639	11718890	9	5	1.34E−05
Scaffold_12	9234209	9234460	20	9	1.58E−08
Scaffold_12	1293206	1293457	23	6	2.59E−04
Scaffold_12	1293330	1293581	23	6	2.59E−04
Scaffold_14	4039293	4039544	23	6	2.59E−04
Scaffold_14	4039417	4039668	23	6	2.59E−04
Scaffold_15	9987207	9987458	14	7	3.93E−07
Scaffold_17	1711316	1711567	23	6	2.59E−04
Scaffold_17	1711440	1711691	23	6	2.59E−04
Scaffold_17	7943362	7943613	23	6	2.59E−04
Scaffold_17	7943486	7943737	23	6	2.59E−04
Scaffold_17	13837410	13837661	8	5	6.14E−06
Scaffold_18	10385054	10385305	27	6	6.77E−04
Scaffold_19	842998	843249	13	7	2.02E−07
Scaffold_19	15352117	15352368	10	5	2.60E−05

Phased loci from chromosome scaffolds (scaffolds 1 through 19) and containing five or more phased sequences. Loci were identified from the data of Klevebring et al. (2009) data updated to v2 of the *P. trichocarpa* genome assembly

phased locus and that the siRNAs have predicted *NBS-LRR* targets within the same region of chromosome 19.

In contrast to the v1 analysis, reanalysis of the data to the v2 genome assembly identified five phased loci on chromosome 19: Four loci are located within the first 1 Mb, including a locus corresponding to that identified in v1 (Fig. 1). Target prediction for these four loci showed near-exclusive targeting of *NBS-LRR* genes, with the vast majority of target genes located on chromosome 19 and scaffold_31.

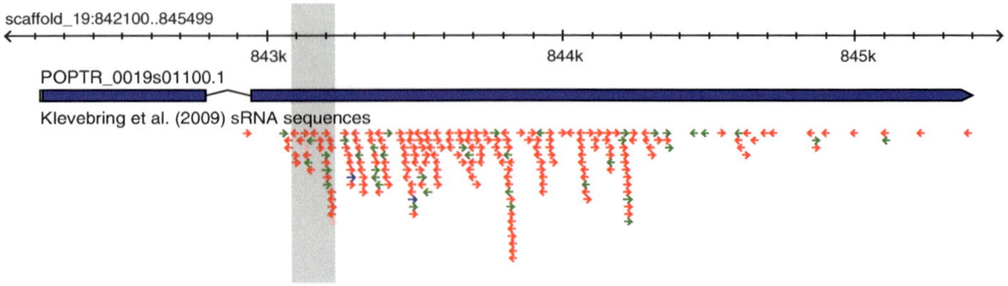

Fig. 1 Example phased sRNA locus. The genomic interval on scaffold_19 from base pair 842579–843578 is shown. This interval includes a predicted phased sRNA locus (identified in *grey*). A predicted protein coding gene model is shown with exons as *solid blue blocks* and introns shown as connecting *black lines*. Mapped short RNA sequences from Klevebring et al. (2009) are shown and sRNA length is indicated by colour: *pink* 16–18 bp; *red* 20–21 bp; *green* 22–23 bp; *blue* 24–25 bp

Four phased loci were also identified on scaffold_31, and one of these was identical to a locus on chromosome 19. This locus almost exclusively targets genes on scaffold_31. Although read counts for nearly all sequences within the loci were low (<5), it appears that alternate haplotypes may produce haplotype-specific phased siRNAs and that these show haplotype divergence for their target *NBS-LRR* genes. As can be seen in Fig. 1, the predicted phased locus covers only a short span of the genomic region producing sRNAs, and it is possible that the phased region could be extended by performing deeper sequencing to identify additional sRNAs that were not present or identified in the Klevebring et al. (2009) dataset. At this region, no sRNA sequences were identified in the dataset of Barakat et al. (2007), indicating the importance of adequate sequencing depth to fully characterise sRNA loci.

The significance of the above-average production of 21- and 24-nt sRNAs within this region of chromosome 19 remains entirely unknown. Similar deviance from the average pattern of sRNA production across the genome has not been reported in other species, but as so few species have been profiled for total sRNA population, no conclusion can be drawn from this. The combination of an over-representation of *NBS-LRR* genes, production of phased sRNA loci targeting those genes and an above-average production of sRNAs make this an interesting genomic region for further consideration and study.

6 Future Perspectives

There is still much to learn about sRNAs and their biological roles in *Populus*, and as a model species, *Populus* offers an opportunity to study the importance of sRNAs in processes and responses that are not present in models such as *Arabidopsis*. There will be much benefit from a considerably more extensive profiling of miRNAs across a range of *Populus* species and across as diverse a sample of tissues, developmental stages and environmental conditions as possible. With the availability of very high throughput sequencing platforms, such as those from Illumina or SOLiD, this is now a highly achievable task in the near-term: Indeed, a single sequencing run on one of these platforms utilising sample bar coding to allow multiplex sequencing would offer significantly greater data than all previous studies combined.

Although it is appealing to complete the discovery of all miRNAs, there is far more to be learnt about the other classes of sRNAs. None of the predicted phased loci so far identified have been functionally characterised. Likewise, the importance of repeat-associated sRNAs remains unknown. A current severe limitation to studying these areas is the lack of mutants for sRNA biogenesis genes. The availability of such material would substantially increase knowledge and would be the first time that the biological role of different sRNA classes has been examined in a long-lived plant species. In light of the findings presented in Yakovlev et al. (2010) regarding a link between epigenetic memory and climatic adaptation in offspring, this could well prove to yield significant advances in our understanding of long-term adaptive mechanisms in tree species.

References

Allen E, Xie Z, Gustafson AM et al (2005) microRNA-directed phasing during *Trans*-acting siRNA biogenesis in plants. Cell 121:207–222

Axtell MJ, Jan C, Rajagopalan R et al (2006) A two-hit trigger for siRNA biogenesis in plants. Cell 127:565–577

Barakat A, Wall PK, Diloreto S et al (2007) Conservation and divergence of microRNAs in *Populus*. BMC Genomics 8:481

Chan SW, Zilberman D, Xie Z et al (2004) RNA silencing genes control de novo DNA methylation. Science 303:1336

Griffiths-Jones S, Harpreet SK, Enright AJ (2008) miRBase: tools for microRNA genomics. Nucleic Acids Res 36:D154–D158

Howell MDD, Fahlgren N, Chapman EJ et al (2007) Genome-wide analysis of the RNA-DEPENDENT RNA POLYMERASE6/DICER-LIKE4 pathway in Arabidopsis reveals dependency on miRNA- and tasiRNA-directed targeting. Plant Cell 19:926–942

Jansson S, Douglas CJ (2007) *Populus*: a model system for plant biology. Annu Rev Plant Biol 58:435–458

Jia X, Ren L, Chen Q-J et al (2009) UV-B-responsive microRNAs in *Populus tremula*. J Plant Physiol 166:2046–2047

Kasschau KD, Fahlgren N, Chapman EJ et al (2007) Genome-Wide profiling and analysis of arabidopsis siRNAs. PLoS Biol 5(3):e57+

Klevebring D, Street NR, Fahlgren N et al (2009) Genome-wide profiling of *Populus* small RNAs. BMC Genomics 10:620

Li B, Yin W, Xia X (2009) Identification of microRNAs and their targets from *Populus euphratica*. Biochem Biophys Res Commun 388:272–277

Lu S, Sun Y-H, Shi R et al (2005) Novel and mechanical stress-responsive microRNAs in *Populus trichocarpa* that are absent from Arabidopsis. Plant Cell 17:2186–2203

Lu S, Sun Y-H, Chiang VL (2008) Stress-responsive microRNAs in *Populus*. Plant J 55:131–151

Meyers BC, Axtell MJ, Bartel B et al (2008) Criteria for annotation of plant microRNAs. Plant Cell 20:3186–3190

Pakull B, Groppe K, Meyer M et al (2009) Genetic linkage mapping in aspen (*Populus tremula* Land *Populus tremuloides* Michx.). Tree Genet Genomics 5:505–515

Rajagopalan R, Vaucheret H, Trejo J et al (2006) A diverse and evolutionarily fluid set of microRNAs in *Arabidopsis thaliana*. Genes Dev 20:3407–3425

Street NR, Sjodin A, Bylesjo M et al. (2008) A cross-species transciptomics approach to identify genes involved in leaf development. BMC Genomics 9:589

Tuskan GA, DiFazio S, Jansson S et al (2006) The genome of black cottonwood *Populus trichocarpa* (Torr& Gray). Science 313:1596–1604

Yakovlev IA, Asante DKA, Fossdal CG et al (2010) Differential gene expression related to an epigenetic memory affecting climatic adaptation in Norway spruce. Plant Sci. doi:10.1016/j.plantsci.2010.07.004

Yin T, Difazio SP, Gunter LE et al (2008) Genome structure and emerging evidence of an incipient sex chromosome in *Populus*. Genome Res 18:422–430

Zhou J, Zhuo R, Liu M et al (2010) Identification and characterization of novel microRNAs from *Populus cathayana* Rehd. Plant Mol Biol Rep 29:242–251

The *Arabidopsis thaliana* Double-Stranded RNA Binding (DRB) Domain Protein Family

Andrew L. Eamens, Shaun J. Curtin, and Peter M. Waterhouse

Contents

Abstract Double-stranded RNA (dsRNA) binding domain proteins (dsRBP) are integral players in small non-protein-coding RNA-mediated regulatory pathways in plants, animals and insects. DICER, a RNase III-like class of endonuclease, requires the assistance of a dsRBP to efficiently and accurately process its dsRNA substrate. In the model dicotyledonous plant species *Arabidopsis thaliana* (*Arabidopsis*), DICER-LIKE1 (DCL1) requires the coordinated action of dsRNA BINDING DOMAIN1 (DRB1) to process microRNAs (miRNAs) from their precursor molecules of partially self-complementary dsRNA. Small-interfering RNA (siRNA) generation on the contrary, from either endogenous or exogenous transcribed perfectly dsRNA substrates, requires the DCL4/DRB4 partnership. The *Arabidopsis* genome encodes four DICER-LIKE (DCL) and five double-stranded RNA binding (DRB) proteins. The DCL protein family has been shown to act in both a redundant and hierarchical manner in small RNA (sRNA) biogenesis. The DCL1/DRB1 and DCL4/DRB4 partnerships for miRNA and siRNA production, respectively, suggested that each DCL requires a corresponding DRB protein for its efficient and accurate processing of its associated sRNA class. This chapter covers our current

P.M. Waterhouse (✉) and A.L. Eamens
School of Molecular Bioscience, University of Sydney, Sydney, NSW 2006, Australia
e-mail: peter.waterhouse@sydney.edu.au

S.J. Curtin
Department of Agronomy and Plant Genetics, University of Minnesota,
1991 Upper Buford Circle, St. Paul, MN 55108, USA

V.A. Erdmann and J. Barciszewski (eds.), *Non Coding RNAs in Plants*,
RNA Technologies, DOI 10.1007/978-3-642-19454-2_24,

knowledge of the *Arabidopsis* DRB family members, DRB1 to DRB5, outlining their individual function(s) and DCL association(s) in the parallel sRNA-mediated RNA silencing pathways of *Arabidopsis*.

Keywords DCL protein • DRB protein • RNA silencing • miRNA • siRNA • sRNA

1 Small RNA-Directed RNA Silencing in Plants

RNA silencing describes a diverse array of mechanisms that regulate gene expression in all organisms with the notable exception of bacteria and the fission yeast *Saccharomyces cerevisiae*. It is generally believed that RNA silencing is an ancient, evolutionarily conserved process, with similar biochemically functioning protein machinery found in both plants and animals (Margis et al. 2006; Sharp 2001). In the plant kingdom, RNA silencing has evolved to an extraordinary level of sophistication, with the parallel RNA silencing pathways associated with (1) regulating the expression of developmentally important genes, (2) controlling transposon replication, (3) maintaining chromatin state and (4) defending against invading viruses. RNA silencing mechanisms interfere with gene expression by either inhibiting transcription initiation or directing sequence specific messenger RNA (mRNA) degradation. More recently in plants, RNA silencing has also been shown to inhibit gene expression via translational repression, but this mechanism of silencing is thought to occur less frequently in plants than in the animal system (Beauclair et al. 2010; Brodersen et al. 2008).

Small non-protein-coding RNAs act as the sequence specificity determinants of RNA silencing. These 20–25 nucleotide (nt) small RNAs (sRNAs) are generated from longer precursor molecules of double-stranded RNA (dsRNA) by the RNase III-like endonuclease DICER (DCR; Bernstein et al. 2001; Gregory et al. 2005). Once processed from its dsRNA precursor, the sRNA is loaded to the *R*NA-*i*nduced *s*ilencing *c*omplex (RISC). RISC contains an ARGONAUTE (AGO) protein family member at its core, forming the catalytic centre that executes a specific RNA silencing mechanism (Baumberger and Baulcombe 2005; Liu et al. 2007; Rand et al. 2005). In the model dicotyledonous plant species *Arabidopsis thaliana* (*Arabidopsis*), five classes of naturally occurring sRNAs have been identified, including the (1) microRNA (miRNA) (Lee and Ambros 2001), (2) small-interfering RNA (siRNA) (Hamilton and Baulcombe 1999), (3) repeat-associated small-interfering RNA (rasiRNA) (Meister and Tuschl 2004), (4) *trans*-acting small-interfering RNA (tasiRNA) (Adenot et al. 2006; Xie et al. 2005) and (5) natural antisense transcript small-interfering RNA (natsiRNA) (Borsani et al. 2005) sRNA classes. In addition to producing endogenous sRNA species, the *Arabidopsis* encoded DICER-LIKE (DCL) proteins are also responsible for the generation of siRNAs from viruses, or transgene-encoded hairpin RNAs (hpRNAs; Deleris et al. 2006; Fusaro et al. 2006).

1.1 The Arabidopsis DICER-LIKE (DCL) Protein Family

As mentioned above, DCR processing of sRNA from its dsRNA substrate is the key initiator of RNA silencing. The endonuclease activity of DCR was first biochemically identified in *Drosophila melanogaster* (*Drosophila*), and its conserved enzymatic activity has since been found in mammals, worms, insects, plants and fungi (Bernstein et al. 2001; Schauer et al. 2002; Golden et al. 2002). In *Arabidopsis*, DCL1 was the first homologue of the *Drosophila* DCR to be characterised (Schauer et al. 2002). The *Arabidopsis* genome encodes three additional DCL protein family members, termed DCL2, DCL3 and DCL4, respectively. All four DCL enzymes have the same arrangement of functional motifs found in the *Drosophila* DCR and produce the different sRNA size classes that mediate the individual initiation steps of the functionally diverse RNA silencing pathways of *Arabidopsis* (Finnegan et al. 2003; Margis et al. 2006; Vaucheret 2006).

Of the four DCL proteins encoded by the *Arabidopsis* genome, miRNA accumulation has been shown to almost exclusively rely on the action of DCL1 (Bouché et al. 2006; Park et al. 2002; Reinhart et al. 2002). The nuclear-localised DCL1 recognises long molecules of partially self-complementary non coding RNA that have the ability to fold back onto themselves to form dsRNA stem-loop structures. These primary-miRNA (pri-miRNA) transcripts are sequentially cleaved by DCL1 to liberate predominantly 21-nt miRNAs, which are in turn used as sequence-specificity guides by RISC to silence the expression of cognate mRNAs via mRNA cleavage (Kurihara and Watanabe 2004; Xie et al. 2004). DCL2 has a similar functional domain arrangement to DCL1, except that this protein has one less dsRNA-binding motif (dsRBM). The overlapping 3′ ends of *Arabidopsis* transcripts convergently transcribed from opposite DNA strands are initially processed by DCL2. This DCL2-generated 24-nt natsiRNA then sets the phasing for DCL1 processed 21-nt secondary natsiRNAs which direct cleavage-mediated silencing of complementary transcripts (Borsani et al. 2005). In addition to producing natsiRNAs, DCL2 is responsible for the generation of 22-nt secondary siRNAs as part of the plant's defence response mechanism to suppress virus replication (Bouché et al. 2006; Deleris et al. 2006). DCL3 is also responsible for processing secondary siRNAs from virus or plant-transcribed viral dsRNA. DCL3 does not appear to participate in the biogenesis of siRNAs from RNA viruses; however, it has been shown to produce 24-nt siRNAs from DNA viruses, which replicate in the nucleus (Akbergenov et al. 2006; Xie et al. 2004). DCL3 is also responsible for the production of the 24-nt size class of sRNA associated with RNA-directed DNA methylation (RdDM). In *Arabidopsis dcl3* plants (plants that lack DCL3 activity), the loss of these 24-nt rasiRNAs is associated with transposon reactivation, heterochromatin remodelling, and the increased expression of other classes of non-protein-coding repeat sequences (Chan et al. 2004; Xie et al. 2004).

Like DCL1, DCL4 processes sRNAs of the 21-nt size class. *TAS* loci produce non coding RNA transcripts which are targeted for cleavage by DCL1-generated miRNAs. This cleavage event identifies the *Tas* transcript for dsRNA synthesis by a plant encoded RNA-dependent RNA polymerase (RDR; Allen et al. 2005). The dsRNA transcript is then processed into phased 21-nt tasiRNAs by the endonuclease activity of DCL4, and this sequentially processed plant-specific class of sRNA are subsequently loaded by RISC and used as guides to direct RNA silencing of complementary mRNAs (Adenot et al. 2006; Xie et al. 2005). In addition to tasiRNA biogenesis, DCL4 is also required for the generation of the 21-nt size class of exogenous siRNAs derived either from replicating viruses or from the introduction of a hpRNA-encoding transgene (Bouché et al. 2006; Curtin et al. 2008; Fusaro et al. 2006). Although each of the four *Arabidopsis* DCLs appear to play a specific role in one or more of the parallel RNA silencing pathways, studies have shown that their functions are partially redundant: that is, when the activity of a particular DCL is lost, its preferred dsRNA substrate can be processed by one of the other active DCLs, giving rise to a siRNA profile typical of the activity of the substituting DCL (Xie et al. 2005; Gasciolli et al. 2005). Furthermore, and in addition to their functional redundancy, when processing viral RNAs into siRNAs, the *Arabidopsis* DCL family shows a functional hierarchy as follows: DCL4 > DCL2 > DCL3 > DCL1 (Deleris et al. 2006; Ding and Voinnet 2007; Dunoyer et al. 2007; Fusaro et al. 2006).

To efficiently and accurately process their respective dsRNA substrates into sRNAs, DCL1 and DCL4 have been shown to require the assistance of two members of a second dsRNA interacting protein family, the *Arabidopsis* *d*s*R*NA *B*INDING DOMAIN (DRB) protein family. DCL1 requires the assistance of DRB1 to process miRNAs from their precursor molecules of imperfectly dsRNA, and DCL4 requires the coordinated action of DRB4 to efficiently generate siRNAs from perfectly dsRNA (Adenot et al. 2006; Kurihara et al. 2006; Vazquez et al. 2004). However, besides the well-characterised DCL1/DRB1 and DCL4/DRB4 partnerships, requirement of the coordinated action of the three remaining members of the *Arabidopsis* DRB family, namely, DRB2, DRB3 and DRB5, by either DCL2 or DCL3 (or DCL1 or DCL4), for the efficient and accurate processing of their respective dsRNA substrates is yet to be experimentally validated.

1.2 The Arabidopsis dsRNA BINDING DOMAIN (DRB) Protein Family

Double-stranded RNA binding domain proteins (dsRBPs) have been identified in both eukaryotes and prokaryotes and are involved in regulating cellular signalling events as well as RNA processing, translation and degradation (Fedoroff 2002). The dsRBPs shown to be involved in miRNA biogenesis or other RNA silencing-related processes in *Drosophila*, humans and nematodes include R3D1 (now termed Laquacious), R2D2, PACT, TRBP and RDE4 (Chendrimada et al. 2005; Jiang et al. 2005; Lee et al. 2006; Liu et al. 2003; Tabara et al. 2002), and each of these dsRBPs contains either two or three dsRBMs. X-ray crystallographic and NMR analyses have revealed that the dsRBM of these RNA silencing-related dsRBPs allows for the protein's interaction with dsRNA by specifically binding to one side of, but not entirely around, its dsRNA substrate (Tian et al. 2004). Their role in RNA silencing in *Arabidopsis* was highlighted by Han et al. (2004) with the discovery of a distinct 36-kDa protein that co-fractionated with dsRNA-processing activity. This protein was identified as HYPONASTIC LEAVES1 (HYL1; referred to as DRB1 from here on) and was shown to possess two, 70 amino-acid residue dsRBMs in its N-terminal half. Four additional proteins were subsequently identified in *Arabidopsis* with high sequence homology to DRB1, designated DRB2 through to DRB5, respectively (Fig. 1a). The protein structures of all five DRB family members resemble that of the dsRBPs, R2D2 and RDE4 with two adjacent dsRBMs in their amino-terminal halves, no dsRNA binding motifs in their carboxyl-termini and sizes of 35–40 kDa (Curtin et al. 2008). To date, dsRBM searches of the *Arabidopsis* proteome have failed to identify any proteins with the three adjacent dsRBM arrangement of the *Drosophila* and human miRNA-specific dsRBPs R3D1 and PACT.

Figure 1b illustrates that at the amino-acid level, DRB2, DRB3 and DRB5 form a single phylogenetic cluster and that this cluster is more closely related to DRB4 than with DRB1. This relationship is also reflected at the genomic level. All five *Arabidopsis DRB* genes consist of three exons, separated by two introns and flanked by a 5′ and 3′ untranslated region (Fig. 1c). However, the first exon of *DRB1* does not encode this protein's first dsRBM as it does in the DRB gene loci *DRB2*, *DRB3*, *DRB4* and *DRB5*. The first two exons encode for the two N-terminal dsRBMs of these four family members, suggesting that *DRB2*, *DRB3* and *DRB5* may have evolved from *DRB4*. The dsRBMs of the DRB protein family members not only allow for their interaction with dsRNA substrates but also function as a protein-protein interaction domain. Using northwestern blot analysis, Hiraguri et al. (2005) showed that DCL1 preferentially interacts with DRB1 over the other four DRB family members, whereas DCL4 exclusively interacts with DRB4. This is consistent with subsequent demonstrations that in *Arabidopsis*, DRB1 and DRB4 are required by DCL1 and DCL4 for efficient and accurate miRNA and tasiRNA biogenesis, respectively (Adenot et al. 2006; Kurihara and Watanabe 2004; Vazquez et al. 2004; Xie et al. 2005). Taken together these analyses also suggested that each individual DCL protein requires the assistance of a specific DRB cofactor for its normal functioning in plants (Hiraguri et al. 2005; Vaucheret et al. 2006). In order to investigate the different roles of the DRBs in the *Arabidopsis* RNA silencing pathways, T-DNA insertions in each family member were identified in either the SALK or GABI *Arabidopsis* collections. These putative knockout mutants were analysed by PCR-based genotyping using allele-specific primers to identify homozygous lines, and null mutants for DRB1 through to DRB5 (*drb1*, *drb2*, *drb3*, *drb4* and *drb5*) were confirmed by quantitative reverse-transcriptase PCR (qRT-PCR) (Curtin et al. 2008). In the subsequent sections of this chapter, we outline what is currently known of the involvement of each of the DRB family members in the parallel sRNA-directed RNA silencing pathways of *Arabidopsis*.

a

Family Member	AGI Locus Identifier	Mutant Allele Characterised	Protein Length (amino acids)	Molecular Weight	Protein Accession No
DRB1	*At1g09700*	*hyl1-2* (SALK_064863)	420	45547	NP_563850
DRB2	*At2g28380*	*drb2-1* (GABI_348A09)	435	47441	NP_565672
DRB3	*At3g26932*	*drb3-1* (SALK_003331)	360	40074	NP_189329
DRB4	*At3g62800*	*drb4-1* (SALK_000736)	356	38416	NP_191839
DRB5	*At5g41070*	*drb5-1* (SALK_031307)	394	43636	NP_198923

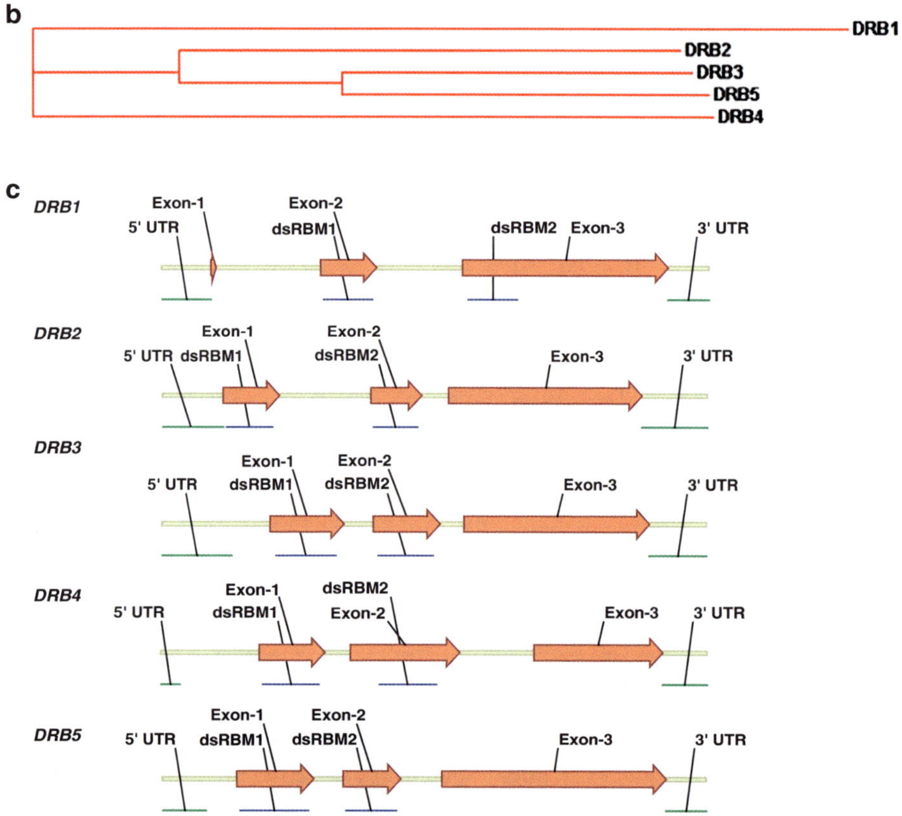

Fig. 1 The *Arabidopsis* dsRNA BINDING DOMAIN (DRB) protein family. (**a**) Table listing the *Arabidopsis* gene identifier (AGI) numbers for individual DRB family members, the mutant alleles used by our research group as well as the protein lengths, molecular weights and accession numbers. (**b**) Phylogenetic tree constructed from the full-length amino-acid sequences of the five DRB family members. As indicated by the gene structures provided below in (**c**), DRBs 2, 3 and 5 are more closely related to DRB4, than to DRB1. Drawing is not to scale. (**c**) Schematic of the gene structure of DRB family members. 5′ and 3′ UTRs are represented by the *thick green lines*, gene exons by *orange arrows* and dsRBMs by the *thin blue lines*. Drawings are not to scale

1.2.1 DRB1

The *hyl1* mutant was originally identified in a developmental screen (Lu and Fedoroff 2000), and subsequent molecular and biochemical analyses have since identified DRB1 (HYL1) as a key machinery component in the initiation step of both the miRNA and closely linked tasiRNA biogenesis pathways (Han et al. 2004; Vazquez et al. 2004; Curtin et al. 2008). The *drb1* pleiotropic phenotype is primarily characterised by yellowish-green upwardly curled (hyponasty) rosette leaves (Fig. 2a). This plant line also exhibits reduced organ size and overall growth, increased lateral organ formation and late flowering with markedly reduced fertility (Lu and Fedoroff 2000; Meins et al. 2005). Overall, the *drb1* phenotype closely resembles that of hypomorphic *dcl1-9* plants (Bouché et al. 2006), to suggest that the severe developmental phenotypes expressed by these two plant lines are due to disruption of the RNA silencing pathways, namely, the miRNA biogenesis pathway. Consistent with the fact that it is an important cofactor of DCL1, which is presumably required in every cell of the plant, either fusion of the *DRB1* promoter region to the β-glucuronidase (*GUS*) reporter gene, or qRT-PCR analysis of the *Drb1* transcript itself, showed that DRB1 is constitutively expressed in all parts of the plant (Fig. 2b; Curtin et al. 2008; Yu et al. 2005).

As outlined for DRB1, the identification of protein factors involved in miRNA biogenesis in *Arabidopsis* stem from the characterisation of phenotypic mutant lines originally isolated in developmental screens and subsequently shown to be defective in wild-type miRNA accumulation (Vaucheret et al. 2004). The accurate processing of pri-miRNA transcripts in specialised nuclear compartments, termed nuclear Dicing-bodies relies on the coordinated action of three dsRNA-interacting proteins: DCL1, DRB1 and SERRATE (SE; Dong et al. 2008; Fujioka et al. 2007; Yang et al. 2006). This initial cleavage event produces the precursor miRNA (pre-miRNA), a smaller intermediate dsRNA stem-loop molecule, which is again cleaved by the combined action of DCL1/DRB1, resulting in the liberation of the miRNA/miRNA* duplex from the stem-loop sequence (Kurihara and Watanabe 2004; Vazquez et al. 2004). The sRNA-specific methyltransferase HUA ENHANCER1 (HEN1) methylates the 3′ dinucleotide overhangs of each duplex strand to protect these and all other sRNA classes from subsequent degradation (Boutet et al. 2003; Kurihara et al. 2006; Park et al. 2002).

We recently demonstrated an additional role specified by DRB1 in miRNA biogenesis (Eamens et al. 2009). The majority of *Arabidopsis* miRNAs identified to date indicate preferential selection of the miRNA guide strand over the opposite duplex strand, the miRNA* passenger strand, for incorporation into RISC. In *drb1* plants, the preferential selection of the miRNA guide strand is lost, leading to an equivalent accumulation of both duplex strands through reduced accumulation of the miRNA guide strand and an inversely proportionate increase in the accumulation of the miRNA* passenger strand (Fig. 3a). We, therefore, proposed a model where DRB1 is functioning in an analogous fashion to the *Drosophila* dsRBP R2D2. In *Drosophila* siRNA biogenesis, DCR2 is capable of processing dsRNA templates in the absence of R2D2; however, it requires R2D2 to form the *R*ISC-*l*oading

Fig. 2 Phenotypes of *drb* mutant plants and the expression patterns of DRB family members. (**a**) Comparison of the phenotypes expressed by T-DNA insertion knockout mutant lines, *drb1*, *drb2*, *drb3*, *drb4* and *drb5*, to wild-type (Columbia-0; Col-0). (**b**) Histochemical staining of plants transformed with a GUS reporter gene expressed under the control of the *DRB1*, *DRB2*, *DRB3*, *DRB4* and *DRB5* promoter sequences

complex (RLC). The RLC heterodimer asymmetrically loads siRNA/siRNA* duplexes to the AGO family member AGO2 for passenger strand degradation. R2D2 binds to the more thermodynamically stable end of the duplex (the duplex end with tighter dsRNA base-pairing), with DCR2 binding to the other end of the duplex so

Fig. 3 (continued) (**f**) SiRNA processing of *Tomato spotted wilt virus* (TSWV) transcripts is DRB4-dependent. (**g**) SiRNA processing of the hpPDS transgene closely mirrors that of viral transcripts. (**h**) DRB proteins are not involved in silencing the expression of repeat elements via RdDM

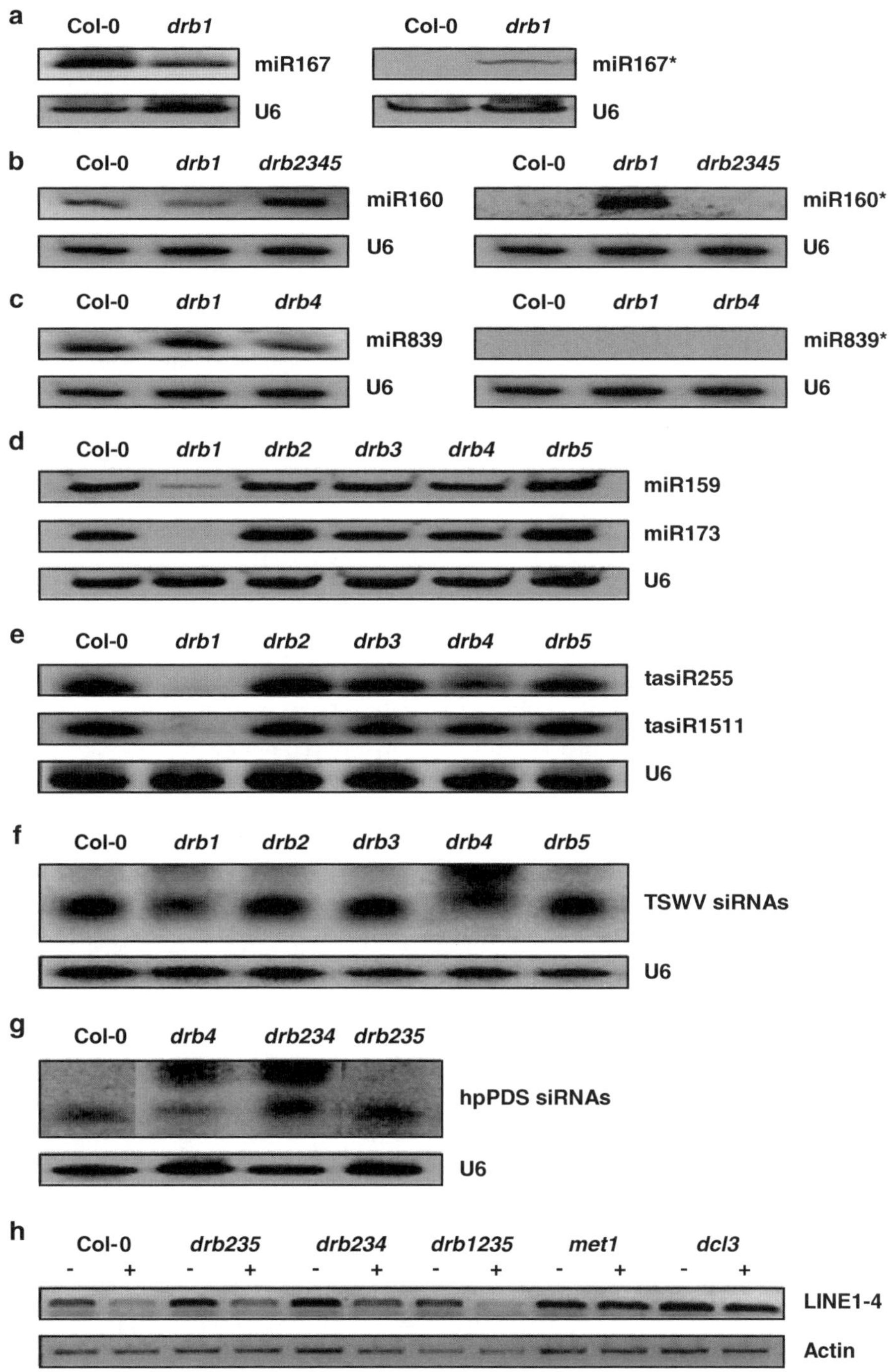

Fig. 3 Small RNA accumulation in *drb* mutant plant lines. (**a**) MiR167 guide and passenger strand accumulation in *drb1*. (**b**) Enhanced miRNA guide strand selection in the quadruple mutant *drb2345*. (**c**) Duplex strand accumulation of the DRB4-dependent miRNA, miR839. (**d**) MiRNA accumulation in the individual *drb* mutant lines. (**e**) TasiRNA accumulation in the *drb* mutants

as to orientate the passenger strand for AGO2-catalysed cleavage. The intact duplex strand, the siRNA guide strand, is then used by RISC to direct silencing of complementary mRNA targets via AGO2-catalysed cleavage (Tomari et al. 2004; Matranga et al. 2005; Rand et al. 2005). MiRNA/miRNA* accumulation profiles in the *drb1* mutant (Fig. 3a) demonstrates that miRNA duplexes are loaded without polarity selection, and that DRB1, like R2D2 in *Drosophila*, appears to be responsible for strand selection by directionally loading miRNA/miRNA* duplexes onto AGO1-catalysed RISC for miRNA* passenger strand degradation (Eamens et al. 2009). The biased selection of the miRNA guide strand is enhanced in the *Arabidopsis* mutant plant line *drb2345*, where of the five DRB family members, only DRB1 is active (Fig. 3b). This suggests that in wild-type plants DRB1 is in competition with one of the other DRBs for interaction with DCL1 and/or miRNA precursor transcripts. DRB2, 3, 4 and 5 may not be able to assist DCL1 in miRNA precursor transcript processing, and/or in directing miRNA duplex strand selection, which in wild-type plants has a repressive effect on miRNA guide strand accumulation.

Following duplex processing, the activity of the nuclear transport protein HASTY (HST) has also been shown to be a requirement for the efficient biogenesis of certain miRNA families (Park et al. 2005). However, the exact role specified by this protein in plant miRNA biogenesis remains unclear, as several miRNAs accumulate to wild-type levels in the *hst* mutant. In the cytoplasm, AGO1-catalysed RISC uses the loaded miRNA as a sequence specificity guide to silence the expression of complementary target transcripts via Slicer-directed cleavage (Baumberger and Baulcombe 2005). Figure 3e shows that in addition to its requirement for normal miRNA accumulation (Fig. 3d), DRB1 is also required for the generation of the closely related class of sRNA, the tasiRNAs, with the accumulation of both sRNA classes severely depleted in *drb1* plants. The sRNA northern blotting data presented in Fig. 3d, e also show that although another DRB family member in addition to DRB1 appears to compete with this dsRBP for interaction with DCL1 and/or miRNA precursor transcripts (Fig. 3b), the association between DRBs 2, 3, 4 and 5 and miRNA/tasiRNA expression appears to be sRNA-specific. Unlike the global reductions to miRNA and tasiRNA accumulation associated with the *drb1* mutation, sRNA accumulation is variable in each individual knockout mutant plant line (*drb2*, *drb3*, *drb4* and *drb5*).

1.2.2 DRB2

All five DRB protein family members contain two adjacent dsRBM in their N-terminal half (Curtin et al. 2008). DRB1 preferentially interacts with DCL1 over the three other *Arabidopsis* DCLs through one of its dsRBMs (Hiraguri et al. 2005). The presence of such motifs in DRBs 2, 3, 4 and 5 gives these proteins the potential to also interact with DCL1 and/or its dsRNA substrates. Using northwestern blot analysis, Hiraguri et al. (2005) showed that DRB2 and DRB5 also bind to DCL1 in vitro, but unlike DRB1, do not preferentially interact with this DCL protein over DCL2, 3 or 4. However, demonstration of this protein-protein interaction in vitro identifies these two DRBs as strong competitors to partner DCL1 in the miRNA biogenesis pathway. The *drb2* mutant is characterised by changes in its rosette leaf morphology. Compared to wild-type plants, *drb2* rosette leaves are darker in colour due to an increased production of anthocyanins. Its leaves are also flatter, ovoid and have predominantly serrated margins on pronouncedly longer petioles (Fig. 2a; Curtin et al. 2008). On closer inspection, via scanning electron microscopy, *drb2* rosette leaves show stomata abnormalities, including aborted stomata and irregular stomatal patterning (data not shown). Phenotypic analyses also revealed that compared to wild-type plants, the roots of *drb2* plants have significantly more root hairs. Complementation of *drb2* with a *Drb2* transgene, under the control of the constitutive *Cauliflower mosaic virus* 35S promoter (35S-P), generated transformant lines that expressed a fully reverted wild-type phenotype, confirming that the developmental abnormalities expressed by this plant line were a result of the *drb2* mutation.

Wild-type plants transformed with a putative 1.6 kb *DRB2* promoter sequence driving the expression of GUS, displayed reporter gene activity in the pollen, funicular tissue of maturing seed and in the testa of germinating seeds (Fig. 2b). GUS expression was also observed in clusters of flowers following fertilisation, intimating that the expression of *Drb2* may be initiated at this developmental time point (Curtin et al. 2008). Curiously, GUS expression was not observed in the same tissues that appear to be affected by the loss of *Drb2* expression. No GUS expression was observed in either the rosette leaf or root tissue of plants transformed with the existing DRB2pro::GUS construct (Curtin et al. 2008). We are currently developing a new series of plant expression vectors containing larger *DRB2* 5′ flanking sequences to assess whether these vectors will return expanded and/or expected patterns of reporter gene expression.

As observed in *drb2* plants, mutations to the expression of either *DRB1*, *DRB4*, *MIR* genes and miRNA-target transcripts have been shown to often result in alterations to leaf shape, morphology, serration and curvature (epinasty or hyponasty) (Mallory et al. 2004; Vaucheret et al. 2004). Northern blotting and qRT-PCR analyses were, therefore, used to assess miRNA and tasiRNA accumulation as well as miRNA target transcript expression in *drb2*. Figure 3d, e suggest that although *drb2* plants express a miRNA-defective-like phenotype, sRNA accumulation appeared to be at approximate wild-type levels for the individual sRNA classes assessed. Analysis of miRNA-target gene expression in *drb2* plants via qRT-PCR also suggested that this DRB family member is not involved in a subsequent effector step of either miRNA or tasiRNA-directed RNA silencing (data not shown). It is worth noting that these analyses were conducted on RNA extracted from 4 week old whole plant samples. We are currently repeating these analyses on RNA extracted from specific tissues where *Drb2* is expressed (this will be based on the GUS expression data obtained from the new series of DRB2pro::GUS plant expression vectors), and using more sensitive sRNA detection techniques such as deep sequencing and miRNA-specific stem-loop qRT-PCR. These alternate approaches are expected to lead to the establishment of specific relationships between DRB2 and the sRNA-directed RNA silencing pathways of *Arabidopsis*.

1.2.3 DRB3 and DRB5

Unlike the *drb1*, *drb2* and *drb4* mutant lines, *drb3* and *drb5* plants do not display distinct developmental phenotypes. Both plant lines are essentially wild-type in appearance; however, they do grow at a faster rate and show an overall increased mature plant size compared to wild-type (Fig. 2a). In the seedlings of plants transformed with the DRB3pro::GUS construct, reporter gene activity is concentrated in the shoot apical meristem (SAM) region and surrounding tissue, extending into the petioles of rosette leaves. As these plants mature, the GUS expressional domain extends into most tissues, abate at much lower levels than its expression in the SAM (Fig. 2b; Curtin et al. 2008). GUS expression was also localised to the SAM region and rosette leaf petioles of plants transformed with the DRB5pro::GUS plant expression vector. The reporter gene expressional domain did not extend into the surrounding tissue in maturing DRB5pro::GUS plants, as was observed for DRB3pro::GUS lines. However, DRB5 promoter-driven GUS activity was detected in the buds of immature flowers (Fig. 2b). Taken together, the reporter gene data for DRB3 and DRB5 demonstrated that the expression of these two highly similar DRB family members is both tissue-specific, and overlapping.

The expression of a similar wild-type phenotype and the tightly overlapping expression of GUS in DRB3 and DRB5 promoter-driven reporter gene lines suggested that these two family members could be functionally redundant. Northern blotting was, therefore, used to assess sRNA accumulation in *drb3* and *drb5* mutants. The miRNA and tasiRNA sRNA classes were shown to accumulate to wild-type levels in both mutant backgrounds to suggest that neither DRB plays a role in the biogenesis of the specific sRNAs analysed (Fig. 3d, e). As outlined for DRB2, additional analyses are currently underway to assess miRNA and tasiRNA accumulation in the specific tissues where these two DRB proteins are expressed, namely the SAM region. Virus-derived siRNA synthesis was also assessed by northern blotting in *drb3* and *drb5* plants. Hiraguri et al. (2005) demonstrated that DRB5 shows an equal affinity for interaction with all four *Arabidopsis* DCL proteins, identifying this DRB and the closely related DRB3 as possibly functioning in the sRNA-directed viral defence pathway of *Arabidopsis*. Recent reports have shown that DCL4 is responsible for generating the vast majority of 21-nt siRNAs originating from replicating viral RNAs. Virus-specific siRNAs of the 22 and 24-nt size class are also detected in virus-infected tissues as a result of DCL2 and DCL3 processing of virus-derived dsRNA transcripts (Deleris et al. 2006; Fusaro et al. 2006). Figure 3f shows that *drb3* and *drb5* returned a similar sRNA accumulation profile to that of wild-type plants. Furthermore, hpRNAs are processed in a similar DCL hierarchical fashion to viral RNAs (Fusaro et al. 2006). The introduction of a hpRNA transgene against PHYTOENE DESATURASE (PDS; hpPDS; Wesley et al. 2001) further indicated that DRB3 and DRB5 are not required for the processing of either class of exogenous dsRNA by DCLs 2, 3 and 4 (Fig. 3g; Curtin et al. 2008).

In *Drosophila*, the RNA silencing protein machinery AGO3, AUBERGINE (AUB) and PIWI are involved in rasiRNA processing and the retro-transposon defence pathway. The expression of these three genes is strictly restricted to

Drosophila's germ line cells (Brennecke et al. 2007). Promoter::GUS fusion results for DRBs 2, 3 and 5 revealed that these three family members are most highly expressed in germline and apical meristem tissues. These tissue-specific expression patterns raised the possibility that DRB2, DRB3 and DRB5, either individually or in combination with one another, may play a similar role in suppressing retro-element activity in *Arabidopsis*, either in a orthologous manner to that of the *Drosophila* AGO3/AUB/PIWI-directed rasiRNA pathway, or via the plant-specific DCL3-catalysed RdDM pathway. We, therefore, analysed the activity of five well-characterised retro-elements in our suite of combinational *drb* mutant backgrounds. Knockout mutant lines (*met1* and *dcl3*) of two well-characterised RdDM proteins, namely, METHYLTRANSFERASE1 (MET) and DCL3, were also included in these analyses as positive controls for plants defective in repressing retro-element expression via RdDM. If any of the five DRB family members were required for DCL3 processing of its preferred dsRNA templates in the initiation phase of RdDM, then we would have expected to see PCR products at higher intensities equivalent of those detected in the two RdDM-defective control lines, *met1* and *dcl3*. Figure 3h shows that for the *Arabidopsis* LINE1-4 retro-transposon, all three *drb* combination mutants analysed, including the *drb235*, *drb234* and *drb1235* mutant backgrounds returned methylation-sensitive PCR products at similar intensities to those amplified from wild-type plants (Curtin et al. 2008). The same banding pattern was observed for all repeat elements analysed, indicating that repeat element activity and the methylation status of such elements is at wild-type levels in each of the *drb* mutant backgrounds. Furthermore, this demonstrated that none of the DRB proteins are involved in the sRNA-directed RdDM pathway of *Arabidopsis* and that DCL3 does not require a corresponding DRB for its wild-type processing of its dsRNA template.

1.2.4 DRB4

Figure 2a illustrates that the rosette leaves of *drb4* plants are typically a light green to yellowish colour, as well as being longer and narrower (lanceolate) than those of wild-type plants. The *drb4* mutant also grows more slowly than wild-type but displays an accelerated juvenile-to-adult phase change. However, the overall size and/or leaf colouration of individual *drb4* lines varies greatly, and this partially penetrative "reduced vigour" phenotype has also been observed in the *dcl4-2* mutant background (Adenot et al. 2006; Nakazawa et al. 2007; Xie et al. 2005). In DRB4pro::GUS plants, reporter gene activity was concentrated, or even restricted to the vascular tissue (Curtin et al. 2008). DRB4 promoter-driven GUS expression was, however, also observed in the root and SAMs and developing anthers of some transformant lines (Fig. 2b). The vascular-associated expression of DRB4 is not unexpected. DRB4 has previously been shown to exclusively interact with DCL4, and this DCL is the primary DCL protein responsible for the processing of viral-derived dsRNAs (Hiraguri et al. 2005; Deleris et al. 2006). Deleris et al. (2006) went on to establish that DCL4 exerts its antiviral behaviour by degrading viral RNAs as they are unloaded from the phloem. DRB4 expression supports such a model and suggests that DRB4 itself, may be acting as a surveillance protein, identifying viral dsRNAs (and/or other long-distance systemic silencing signals) and channelling such molecules to DCL4 for their coordinated processing. The exact expressional domain of DCL4 is not currently known, however, its requirement for tasiRNA biogenesis (Adenot et al. 2006) suggests that it would be expressed throughout the plant. Its expression would, therefore, be expected to overlap that of DRB4. Figure 3f outlines the importance of DRB4 in efficient processing of specific viral dsRNAs. *Tomato spotted wilt virus* (TSWV) was used to infect wild-type *Arabidopsis* and *drb* single mutants. In wild-type plants, the TSWV-specific siRNAs are predominantly of the 21-nt size class, a siRNA accumulation profile that is mirrored in TSWV-infected *drb1*, *drb2*, *drb3* and *drb5* plants. In *drb4* plants, the 21-nt siRNA class is completely lost, replaced by greatly elevated levels of DCL2- and DCL3-generated 22- and 24-nt TSWV-specific siRNAs respectively. The concomitant substitution of the 21-nt size class with those of DCL2 and DCL3-generated size classes was also observed in all *drb4*-containing mutant plant lines transformed with the hpPDS vector (Fig. 3g; Curtin et al. 2008). All together, these results demonstrate that DRB4 cooperation is an absolute requirement by DCL4 for its efficient and accurate processing of dsRNA. The results also confirm the exclusivity of the DCL4/DRB4 partnership in siRNA biogenesis in *Arabidopsis*, and as observed for DCL3 action in the RdDM pathway, DCL2 does not appear to require the assistance of a DRB family member to generate its 22-nt size class of viral-specific siRNAs.

TasiRNAs are an endogenous class of plant-specific sRNA processed from *TAS* loci (Adenot et al. 2006; Allen et al. 2005). The non-protein-coding *Tas* transcripts are initially cleaved by DCL1/DRB1/AGO-generated miRNAs, namely miR173 (*Tas1* and *Tas2*), miR390 (*Tas3*) and miR828 (*Tas4*; Allen et al. 2005; Rajagopalan et al. 2006; Xie et al. 2005). MiRNA cleavage of the *Tas* transcript identifies it as a template for dsRNA synthesis via the combined action of RDR6 and SUPPRESSOR OF GENE SILENCING3 (SGS3; Adenot et al. 2006). As described for siRNA processing from viral- and hpRNA-derived dsRNA, the DCL4/DRB4 partnership processes this perfectly dsRNA molecule into sequential 21-nt tasiRNAs, a process which initiates at the miRNA cleavage site. These sRNAs are then loaded to AGO1-catalysed RISC to direct the cleavage of cognate mRNAs in *trans* (Gasciolli et al. 2005; Xie et al. 2005). Figure 3e shows that in accordance with their established requirement of DCL1 and DCL4 activity, tasiRNA levels are reduced in the *drb1* and *drb4* mutant backgrounds, but remain at approximate wild-type levels in *drb2*, *drb3* and *drb5* plants. The northern blotting data presented in Fig. 3d also confirms that DRB4, and hence DCL4, are not involved in the biogenesis of the majority of plant miRNAs.

Two endogenous miRNAs however, namely miR822 and miR839, do rely on the DCL4/DRB4 partnership for their biogenesis (Rajagopalan et al. 2006). Figure 3c shows that the accumulation of miR839 is reduced in drb4 plants and not the established miRNA-deficient plant line drb1, where this DRB4-dependent miRNA accumulates to its wild-type level. Furthermore, the accumulation of the opposite duplex strand, the miR839* passenger strand, remained unchanged, staying below the level of northern blotting detection-sensitivity in all three plant lines analysed (wild-type, *drb1* and *drb4*). The retention of miRNA guide strand selection in *drb4* plants suggested that unlike DRB1's role in miRNA biogenesis, DRB4 is not involved in the preferential selection of sRNA duplex strands (Eamens et al. 2009). This result was confirmed by bioinformatic analysis of the 21-nt siRNA species generated in a wild-type *Arabidopsis* plant transformed with a hpRNA targeting the 5′ portion of the GREEN FLOURESCENT PROTEIN (GFP) transgene (hpGFP). These analyses revealed that there was no uniform bias for the retention of the siRNA/siRNA* duplex strand with a weaker 5′ terminal dsRNA base-pairing, as shown for the majority of DCL1/DRB1-generated miRNA/miRNA* duplexes detected in the same sRNA sequencing set. These results demonstrate that DRB4 is required by DCL4 for efficient and accurate processing of its dsRNA substrates, however, DRB4 does not direct sRNA duplex strand selection (Eamens et al. 2009).

1.2.5 The *drb235* Triple Mutant

Comparative analysis, at both the DNA and amino-acid level, of DRB family members DRB2, DRB3 and DRB5 revealed that these three proteins are highly similar. Furthermore, their almost identical amino acid sequences suggest that they may be functionally redundant. We, therefore, produced the *drb235* triple mutant by standard genetic crossing (Curtin et al. 2008). Compared with either single (*drb2*, *drb3* and *drb5*) or double (*drb23*, *drb25* and *drb35*) mutant lines, *drb235* plants display a developmentally severe phenotype (Fig. 4a). The miRNA-like pleiotropic *drb2* phenotype was readily expressed in the double mutants, *drb23* and *drb25*. Both double mutant lines displayed higher degrees of leaf margin serration, increased anthocyanin production and rosette leaves that were flatter and more ovular than those of *drb2* plants (Fig. 2a). As described for single mutants of DRB3 and DRB5, *drb35* plants were essentially wild-type in appearance, except that they grew at a faster rate and displayed an overall increase in size compared to wild-type, a phenotypic consequence associated with both the *drb3* and *drb5* mutations (Fig. 4a). The *drb235* triple mutant, however, expresses the most developmentally dramatic phenotype of all the *drb* combination mutants. The rosette leaves of *drb235* plants are pale green in colour with obvious dark-green venation throughout the vascular tissue. In addition, *drb235* rosette leaves loose the *drb2*-assocaited ovulate shape, taking on a *drb4*-like lanceolate appearance. Their margins are highly serrated and are drastically downturned at their tips, forming a more compact rosette. Interestingly, instead of exhibiting the increased vigour displayed by the *drb3*, *drb5* and *drb35* backgrounds, *drb235* plants are much smaller than wild-type and have highly serrated leaves, and fascinated stems and inflorescences (Fig. 4a). The severe phenotype of the *drb235* triple mutant suggests that these three highly similar DRBs share a redundant function in plant development.

Analysis of reporter gene expression in wild-type plants transformed with the DRB2, DRB3 and DRB5 putative promoter sequences demonstrated that the

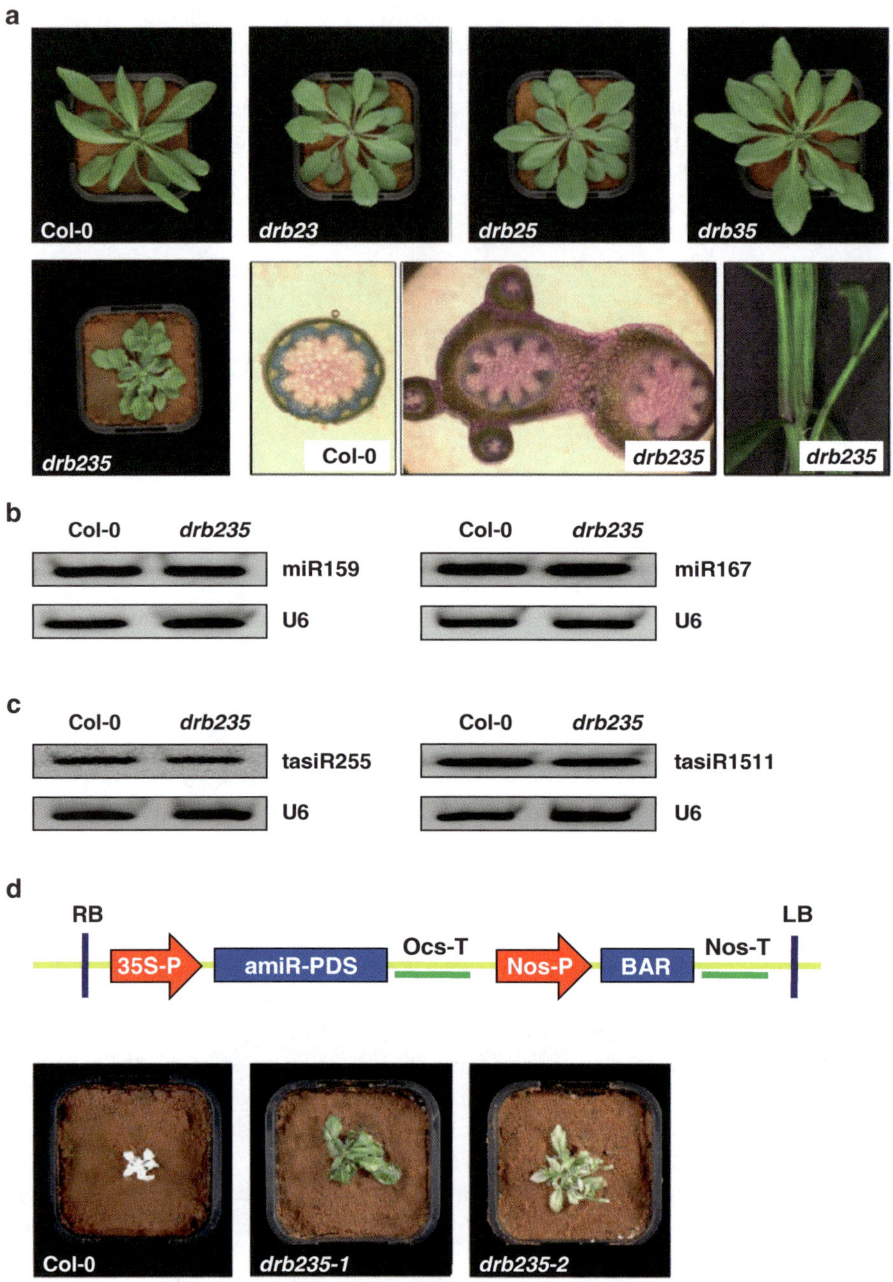

Fig. 4 Phenotype and sRNA accumulation in the *drb235* triple mutant. (**a**) Phenotypes expressed by plant lines *drb23*, *drb25*, *drb35* and *drb235* compared to wild-type (Col-0). Cross section of the fused inflorescences of the *drb235* triple mutant. (**b**) miR159 and miR167 accumulation in *drb235* whole plants. (**c**) tasiRNA accumulation (tasiR255 and tasi1511) in the *drb235* triple mutant. (**d**) Photobleaching of wild-type (Col-0) and *drb235* plants expressing a PDS-targeting amiRNA (amiR-PDS)

expressional domains of DRB3 and DRB5 overlap in the SAM region (Fig. 2b; Curtin et al. 2008). DRB2 promoter-driven GUS expression did not appear to be localised to the same tissue, with GUS activity detected in the pollen, funicular tissue and testa of maturing and germinating seeds, respectively. However, microarray analysis of global gene expression in *Arabidopsis* indicates that in addition to its expression in pollen and seed, *Drb2* is also highly expressed in the SAM (Winter et al. 2007; http://www.bar.utoronto.ca/efp/cgi-bin/efpWeb.cgi). The pleiotropic miRNA-deficient phenotype expressed by the *drb235* triple mutant, in combination with the shared expressional domains for DRBs 2, 3 and 5, suggests not only that these three proteins are functionally redundant but also that they are involved in one of the sRNA-directed pathways of *Arabidopsis*. We, therefore, assessed miRNA and tasiRNA accumulation in this mutant background for comparison to wild-type. Figure 4b, c show that the accumulation of neither class of endogenous sRNA appears to be affected in *drb235* plants. The total RNA used for these sRNA accumulation analyses was extracted from whole plant tissues of 4-week-old plants, and not from the tissues where these three DRB proteins appear to be expressed, namely, the SAM region. As mentioned previously, in addition to producing a new series of DRB2 promoter-driven reporter gene plant expression vectors, we are currently analyzing the expression of endogenous species of sRNA in tissues where DRB2, DRB3 and DRB5 are expressed by the most sensitive methods currently available for such analyses, including sRNA deep sequencing and sRNA-specific stem-loop qRT-PCR.

The germ-line-localised expression of DRBs 2, 3 and 5 indicates that they may be involved in some aspect of development. The interaction between DRB2 and DRB5 with DCL1 (Hiraguri et al. 2005) identifies them as potential candidates for involvement in DCL1-mediated RNA silencing, such as substituting for DRB1 activity in specific tissues, mediating the biogenesis or action of a subset of miRNAs that we have not assayed for. We, therefore, produced an artificial miRNA (amiRNA) plant expression vector (Eamens et al. 2009), based on the pri-miRNA sequence of the well characterised *Arabidopsis* miRNA, miR159b (Allen et al. 2007), and designed to silence the expression of PDS (amiR-PDS). Expression of the amiR-PDS vector in wild-type plants resulted in the production of primary transformants that are completely white in colour due to their absolute lack of chlorophyll (Fig. 4d). Wild-type/amiR-PDS plants are not viable outside of tissue culture, as they are defective in photosynthesis. The amiR-PDS vector was also introduced into the *drb235* background. *Drb235*/amiR-PDS primary transformants displayed a variegated photobleaching pattern that to some extent appeared to be restricted to the vascular tissue and margins of rosette leaves. Furthermore, these amiRNA lines were capable of surviving outside of tissue culture. Seed was collected from *drb235*/amiR-PDS plants and germinated on selection media. The photobleaching pattern expressed by this generation was primarily restricted to the cotyledons and rosette leaf petioles (data not shown). The remainder of the maturing plant's tissues are wild-type in appearance and colouration (with respect to the untransformed *drb235* phenotype). The expression of photobleached plant material in this generation of *drb235*/amiR-PDS transformants mirrors the patterns of GUS activity observed in

DRB3pro::GUS and DRB5pro::GUS plant lines. This suggests that DRB2, DRB3 and/or DRB5 are indeed involved in DCL1-mediated RNA silencing in specific tissues. Although our original molecular analyses (Curtin et al. 2008) suggested that, contrary to expectation, DRBs 2, 3 are not essential for redundant functions in the miRNA, tasiRNA, natsiRNA, hcsiRNA or viral-specific siRNA-mediated RNA silencing pathways, the miRNA-deficient phenotype expressed by the *drb235* triple mutant, in combination with the amiR-PDS data, strongly indicate that they are associated with DCL1-mediated RNA silencing. It is hoped that closer experimental analyses in specific tissues where these three DRB family members are expressed will reveal the involvement of DRBs 2, 3 and/or 5 in modulating the temporal and/or spatial expression of endogenous sRNA species, namely, the miRNA and tasiRNA classes, to ensure normal plant morphogenesis.

2 Concluding Remarks

For the majority of plant miRNAs characterised to date, DCL1 has been shown to require the assistance of DRB1 for their accurate and efficient processing (Kurihara et al. 2006; Vazquez et al. 2004; Vaucheret et al. 2004). We have demonstrated an additional role specified by DRB1 in the miRNA biogenesis pathway, directing the preferential selection, and subsequent incorporation into RISC, of the miRNA/miRNA* duplex strand with a lower thermodynamic stability (Eamens et al. 2009). Interestingly, miRNAs, tasiRNAs and siRNAs processed by the DCL4/DRB4 partnership do not appear to be influence by the same terminal stability rules as those governing DCL1/DRB1-catalysed sRNA duplex processing. In *Arabidopsis*, DCL4/DRB4 generation of siRNAs from either an introduced hpRNA or a replicating virus closely mirror one another (Fusaro et al. 2006). It may, therefore, be desirable for the plant's survival to process siRNAs without strand selection, generating a biologically diverse pool of siRNAs that would target both strands of a replicating virus for sRNA-directed RNA silencing. Conversely, strand selection may have been retained by the DCL1/DRB1-catalysed miRNA biogenesis pathway to ensure that a single, specific sRNA silencing signal is produced at a given developmental time point to ensure that gene expression is normally regulated.

Curiously, analysis of siRNA accumulation profiles in each of the single and/or *drb* combination mutant backgrounds for viral, repeat-associated and hpRNA transgene-derived siRNAs revealed that DCL2 and DCL3 do not require the coordinated action of a corresponding DRB family member for the efficient production of their specific sRNA size class. Although Hiraguri et al. (2005) have shown physical interactions between individual DCLs and DRBs through the adjacent dsRBMs in the N-terminal halves of the DRB proteins in vitro, the cellular localization of each of the five DRB family members is not currently known. The cellular and/or sub-cellular localization of these RNA silencing machinery proteins may not allow for their biological interaction in planta. DCL2 and DCL3 may have, therefore, evolved to efficiently process their preferred dsRNA substrate in the absence of the activity of a DRB family member.

Our original molecular analyses assessing the production of both endogenous and exogenous sRNA species in individual and *drb* combination mutant lines suggested that besides DRB1 and DRB4, the remaining three members of this protein family are not required for sRNA-mediated RNA silencing in *Arabidopsis* (Curtin et al. 2008). However, subsequent analysis of miRNA duplex strand selection in another one of our suite of T-DNA insertion knockout mutant lines, the *drb2345* quadruple mutant, revealed that the preferential selection of the miRNA duplex strand with a weaker 5′ terminal stability was enhanced in this plant line where only DRB1 is active (Eamens et al. 2009). This suggests that in wild-type plants, one of the other four DRB family members is competing with DRB1 for either interaction with its preferred DCL partner, DCL1, or for interaction with its dsRNA substrate, miRNA precursor transcripts. The pleiotropic miRNA-defective phenotype expressed by the *drb235* triple mutant identified these three closely related family members as potential functional competitors to DRB1 in either the miRNA or closely related tasiRNA biogenesis pathway. The expression of an amiRNA plant expression vector targeting PDS for RNA silencing in this plant line, and the observed tissue-restricted silencing offered by this vector, further identified DRBs 2, 3 and/or 5 as potentially playing a role in miRNA biogenesis or miRNA-regulated gene expression. However, further experimental characterisation of this five member protein family is required to definitively establish the individual roles specified by DRB1, DRB2, DRB3, DRB4 and DRB5 in the parallel sRNA-mediated RNA silencing pathways of *Arabidopsis*.

References

Adenot X, Elmayan T, Lauressergues D et al (2006) DRB4-dependent TAS3 trans-acting siRNAs control leaf morphology through AGO7. Curr Biol 16:927–932

Akbergenov R, Si-Ammour A, Blevins T et al (2006) Molecular characterization of geminivirus-derived small RNAs in different plant species. Nucl Acids Res 34:462–471

Allen E, Xie Z, Gustafson AM et al (2005) microRNA-directed phasing during trans-acting siRNA biogenesis in plants. Cell 121:207–221

Allen RS, Li J, Stahle MI et al (2007) Genetic analysis reveals functional redundancy and the major target genes of the Arabidopsis miR159 family. Proc Natl Acad Sci USA 104:16371–16376

Baumberger N, Baulcombe DC (2005) Arabidopsis ARGONAUTE1 is an RNA Slicer that selectively recruits microRNAs and short interfering RNAs. Proc Natl Acad Sci USA 102:11928–11933

Beauclair L, Yu A, Bouché N (2010) microRNA-directed cleavage and translational repression of the copper chaperone for superoxide dismutase mRNA in Arabidopsis. Plant J 62:454–462

Bernstein E, Caudy AA, Hammond SM et al (2001) Role for a bidentate ribonuclease in the initiation step of RNA interference. Nature 409:363–366

Borsani O, Zhu J, Verslues PE et al (2005) Endogenous siRNAs derived from a pair of natural cis-antisense transcripts regulate salt tolerance in Arabidopsis. Cell 123:1279–1291

Bouché N, Lauressergues D, Gasciolli V et al (2006) An antagonistic function for Arabidopsis DCL2 in development and a new function for DCL4 in generating viral siRNAs. EMBO J 25:3347–3356

Boutet S, Vazquez F, Liu J et al (2003) Arabidopsis HEN1: a genetic link between endogenous miRNA controlling development and siRNA controlling transgene silencing and virus resistance. Curr Biol 13:843–848

Brennecke J, Aravin AA, Stark A et al (2007) Discrete small RNA-generating loci as master regulators of transposon activity in Drosophila. Cell 128:1089–1103

Brodersen P, Sakvarelidze-Achard L, Bruun-Rasmussen M et al (2008) Widespread translational inhibition by plant miRNAs and siRNAs. Science 320:1185–1190

Chan SW, Zilberman D, Xie Z et al (2004) RNA silencing genes control de novo DNA methylation. Science 303:1336

Chendrimada TP, Gregory RI, Kumaraswamy E et al (2005) TRBP recruits the Dicer complex to Ago2 for microRNA processing and gene silencing. Nature 436:740–744

Curtin SJ, Watson JM, Smith NA et al (2008) The roles of plant dsRNA-binding proteins in RNA-like pathways. FEBS Lett 582:2753–2760

Deleris A, Gallego-Bartolome J, Bao J et al (2006) Hierarchical action and inhibition of plant dicer-like proteins in antiviral defense. Science 313:68–71

Ding SW, Voinnet O (2007) Antiviral immunity directed by small RNAs. Cell 130:413–426

Dong Z, Han MH, Fedoroff N (2008) The RNA-binding proteins HYL1 and SE promote accurate in vitro processing of pri-miRNA by DCL1. Proc Natl Acad Sci USA 105:9970–9975

Dunoyer P, Himber C, Ruiz-Ferrer V et al (2007) Intra- and intercellular RNA interference in *Arabidopsis thaliana* requires components of the microRNA and heterochromatic silencing pathways. Nat Genet 39:848–856

Eamens AL, Smith NA, Curtin SJ et al (2009) The *Arabidopsis thaliana* double-stranded RNA binding protein DRB1 directs guide strand selection from microRNA duplexes. RNA 15:2219–2235

Fedoroff NV (2002) RNA-binding proteins in plants: the tip of an iceberg? Curr Opin Plant Biol 5:452–459

Finnegan EJ, Margis R, Waterhouse PM (2003) Posttranscriptional gene silencing is not compromised in the Arabidopsis CARPEL FACTORY (DICER-LIKE1) mutant, a homolog of Dicer-1 from Drosophila. Curr Biol 13:236–240

Fujioka Y, Utsumi M, Ohba Y et al (2007) Location of a possible miRNA processing site in SmD3/SmB nuclear bodies in Arabidopsis. Plant Cell Physiol 48:1243–1253

Fusaro AF, Matthew L, Smith NA et al (2006) RNAi-inducing hairpin RNAs in plants act through the viral defence pathway. EMBO Rep 7:1168–1175

Gasciolli V, Mallory AC, Bartel DP et al (2005) Partially redundant functions of Arabidopsis DICER-like enzymes and a role for DCL4 in producing trans-acting siRNAs. Curr Biol 15:1494–1500

Golden TA, Schauer SE, Lang JD et al (2002) SHORT INTEGUMENTS1/SUSPENSOR1/CARPEL FACTORY, a Dicer Homolog, is a maternal effect gene required for embryo development in Arabidopsis. Plant Physiol 130:808–822

Gregory RI, Chendrimada TC, Cooch N et al (2005) Human RISC couple microRNA biogenesis and post-transcriptional gene silencing. Cell 123:631–640

Hamilton AJ, Baulcombe DC (1999) A species of small antisense RNA in post-transcriptional gene silencing in plants. Science 286:950–952

Han M-H, Goud S, Song L et al (2004) The Arabidopsis double stranded RNA-binding protein HYL1 plays a role in microRNA-mediated gene regulation. Proc Natl Acad Sci USA 101:1093–1098

Hiraguri A, Itoh R, Kondo N et al (2005) Specific interactions between Dicer-like proteins and HYL1/DRB-family dsRNA-binding proteins in *Arabidopsis thaliana*. Plant Mol Biol 57:173–188

Jiang F, Ye X, Liu X et al (2005) Dicer-1 and R3D1-L catalyze microRNA maturation in Drosophila. Genes Dev 19:1674–1679

Kurihara Y, Watanabe Y (2004) Arabidopsis micro-RNA biogenesis through DICER-LIKE1 protein functions. Proc Natl Acad Sci USA 101:12753–12758

Kurihara Y, Takashi Y, Watanabe Y (2006) The interaction between DCL1 and HYL1 is important for efficient and precise processing of pri-miRNA in plant microRNA biogenesis. RNA 12:206–212

Lee RC, Ambros V (2001) An extensive class of small RNAs in *Caenorhabditis elegans*. Science 294:862–864

Lee Y, Hur I, Park SY et al (2006) The role of PACT in the RNA silencing pathway. EMBO J 25:522–532
Liu Q, Rand TA, Kalidas S et al (2003) R2D2, a bridge between the initiation and effector steps of the Drosophila RNAi pathway. Science 301:1921–1925
Liu X, Park JK, Jiang F et al (2007) Dicer-1, but not Loquacious, is critical for assembly of miRNA-induced silencing complexes. RNA 13:2324–2329
Lu C, Fedoroff N (2000) A Mutation in the Arabidopsis HYL1 gene encoding a dsRNA binding protein affects responses to abscisic acid, auxin, and cytokinin. Plant Cell 12:2351–2366
Mallory AC, Dugas DV, Bartel DP et al (2004) Micro-RNA regulation of NAC-domain targets is required for proper formation and separation of adjacent embryonic, vegetative and floral organs. Curr Biol 14:1035–1046
Margis R, Fusaro AF, Smith NA et al (2006) The evolution and diversification of Dicers in plants. FEBS Lett 580:2442–2450
Matranga C, Tomari Y, Shin C et al (2005) Passenger-strand cleavage facilitates assembly of siRNA into Ago2-containing RNAi enzyme complexes. Cell 123:607–620
Meins F, Si-Ammour A, Blevins T (2005) RNA silencing systems and their relevance to plant development. Ann Rev Cell Dev Biol 21:297–318
Meister G, Tuschl T (2004) Mechanisms of gene silencing by double-stranded RNA. Nature 431:343–349
Nakazawa Y, Hiraguri A, Moriyama H et al (2007) The dsRNA-binding protein DRB4 interacts with the Dicer-like protein DCL4 in vivo and functions in the trans-acting siRNA pathway. Plant Mol Biol 63:777–785
Park W, Li J, Song R et al (2002) CARPEL FACTORY, a Dicer homolog, and HEN1, a novel protein, act in microRNA metabolism in *Arabidopsis thaliana*. Curr Biol 12:1484–1495
Park MY, Wu G, Gonzalez-Sulser A et al (2005) Nuclear processing and export of microRNAs in Arabidopsis. Proc Natl Acad Sci USA 102:3691–3696
Rajagopalan R, Vaucheret H, Trejo J et al (2006) A diverse and evolutionarily fluid set of microRNAs in *Arabidopsis thaliana*. Genes Dev 20:3407–3425
Rand TA, Petersen S, Du F et al (2005) Argonaute2 cleaves the anti-guide strand of siRNA during RISC activation. Cell 123:621–629
Reinhart BJ, Weinstein EG, Rhoades M et al (2002) MicroRNAs in plants. Genes Dev 16:1616–1626
Schauer SE, Jacobsen SE, Meinke DW et al (2002) DICER-LIKE1: blind men and elephants in Arabidopsis development. Trends Plant Sci 7:487–491
Sharp PA (2001) RNA interference. Genes Dev 15:485–490
Tabara H, Yigit E, Siomi H et al (2002) The dsRNA binding protein RDE-4 interacts with RDE-1, DCR-1, and a DExH-box helicase to direct RNAi in *C. elegans*. Cell 109:861–871
Tian B, Bevilacqua PC, Diegelman-Parente A et al (2004) The double-stranded-RNA-binding motif: interference and much more. Nat Rev Mol Cell Biol 5:1013–1023
Tomari Y, Matranga C, Haley B et al (2004) A protein sensor for siRNA asymmetry. Science 306:1377–1380
Vaucheret H (2006) Post-transcriptional small RNA pathways in plants: mechanisms and regulations. Genes Dev 20:759–771
Vaucheret H, Vazquez F, Crété P et al (2004) The action of ARGONAUTE1 in the miRNA pathway and its regulation by the miRNA pathway are crucial for plant development. Genes Dev 18:1187–1197
Vaucheret H, Mallory AC, Bartel DP (2006) Ago1 homeostasis entails coexpression of Mir168 and Ago1 and preferential stabilization of Mir168 by Ago1. Mol Cell 22:129–136
Vazquez F, Vaucheret H, Rajagopalan R et al (2004) Endogenous trans-acting siRNAs regulate the accumulation of Arabidopsis mRNAs. Mol Cell 16:69–79
Wesley SV, Helliwell CA, Smith NA et al (2001) Construct design for efficient, effective and high-throughput gene silencing in plants. Plant J 27:581–590
Winter D, Vinegar B, Nahal H et al (2007) An "Electronic Fluorescent Pictograph" browser for exploring and analyzing large-scale biological data sets. PLoS One 2:e718

Xie Z, Johansen LK, Gustafson AM et al (2004) Genetic and functional diversification of small RNA pathways in plants. PLoS Biol 2:e104

Xie Z, Allen E, Wilken A et al (2005) DICER-LIKE4 functions in trans-acting siRNA biogenesis and vegetative phase change. Proc Natl Acad Sci USA 102:12984–12989

Yang L, Liu Z, Lu F et al (2006) SERRATE is a novel nuclear regulator in primary microRNA processing in Arabidopsis. Plant J 47:841–850

Yu L, Yu X, Shen R et al (2005) HYL1 gene maintains venation and polarity of leaves. Planta 221:231–242

Mutagenesis by Transitive RNAi

Katherine A. Petsch, Chonglie Ma, Michael J. Scanlon, and Richard A. Jorgensen

Contents

Abstract Transitive RNAi is a posttranscriptional mechanism of gene silencing that is based on the phenomenon of "transitivity." This term refers to the spreading of silencing outside of the initial target sequence and is associated with transgene-induced posttranscriptional gene silencing (PTGS). Transitive RNAi is triggered by placing an inverted repeat sequence immediately 3′ of the sense transgene that is to be targeted. Placement of the inverted repeat in this region is thought to increase the efficiency by which RDR6 initiates copying of the transgene to generate double-stranded RNA (dsRNA). In a proof-of-concept approach, we showed that select subsets of genes can be manipulated with transitive RNAi in a high-throughput forward mutagenesis approach (Plant J 61:873–882, 2010). Laser microdissection of

M.J. Scanlon(✉) and K.A. Petsch
Department of Plant Biology, Cornell University, Ithaca, NY 14853, USA
e-mail: mjs298@cornell.edu

C. Ma
Department of Plant Sciences, University of Arizona, Tucson, AZ 85721-0036, USA

R.A. Jorgensen
Department of Plant Sciences, University of Arizona, Tucson, AZ 85721-0036, USA

BIO5 Institute, University of Arizona, Tucson, AZ 85721-0240, USA

V.A. Erdmann and J. Barciszewski (eds.), *Non Coding RNAs in Plants*, RNA Technologies, DOI 10.1007/978-3-642-19454-2_25,

Arabidopsis mesophyll cells and en masse cloning of the resulting cDNA libraries into transitive RNAi vectors demonstrated that approximately 15% of genes in the pilot study could generate visible phenotypes, resulting in photosynthetic defects. The capacity for transitive RNAi to silence multiple members of gene family members demonstrated the utility of this approach for forward mutagenesis of redundant gene functions. Targeted silencing of a focused population of gene transcripts by transitive RNAi provides an efficient and complementary approach to procedures that target the entire genome. The ability of RNAi to target closely related genes holds promise for its use in forward mutagenesis of polyploid plants, which exhibit high levels of genetic redundancy. Advantages of transitive RNAi as a forward genetic approach, as well as potential drawbacks to this method, are discussed.

Keywords *Arabidopsis* • Laser microdissection • Mutagenesis • Transitive RNAi

1 RNAi as a Forward Mutagenesis Tool

The application of RNAi as a method to downregulate the expression of a single gene and gene paralogs is a frequently used tool for functional analyses in plants (Fu et al. 2007; Hilson et al. 2004; McGinnis et al. 2005, 2007; Small 2007; Tang et al. 2003). Double-stranded RNA (dsRNA), typically generated from either an inverted repeat of the target sequence (IR-RNAi) or a viral-vector-derived RNA (VIGS), acts as the trigger for posttranscriptional gene silencing (PTGS) of homologous sequences (Becker and Lange 2010; Bernacki et al. 2008; Waterhouse and Helliwell 2003; Watson et al. 2005). Although IR-RNAi and virus induced gene silencing (VIGS) function to efficiently downregulate gene expression of a particular target sequence/s, both methods are of limited use for high-throughput mutagenesis on a global scale. The key drawback to IR-RNAi relates to the cloning strategy. Sense and antisense copies of a genic segment must be individually cloned into the vector of interest, a strategy that is not suitable for high-throughput analyses involving a large population of target transcripts. Although VIGS does permit the generation of large, comprehensive genic libraries in viral vectors, this approach is limited to cells and tissues that are amenable to viral infection and replication. This inability to precisely control the penetrance of gene silencing could be particularly disadvantageous in assessing gene function in a developmental context.

Another form of RNAi, termed sense RNAi, is triggered by high levels of translatable "sense" transcripts (Napoli et al. 1990; Que et al. 1997). However, the frequency at which gene silencing is triggered by this method is significantly lower than both IR-RNAi and VIGS. Sense RNAi requires RDR6, an RNA-dependent RNA polymerase (RDRP) that is believed to recognize abundant or aberrant transcripts and catalyze the biosynthesis of a complementary RNA strand (Curaba and Chen 2008; Dalmay et al. 2000). Subsequently, the resultant dsRNA acts as a trigger for RNA interference. RDR6 acts in a "transitive" manner, i.e., silencing spreads from the site of initiation of synthesis of copy RNA (generally in the 3′ UTR) to sequences upstream of the sense transcript and from internal sites as well (Lipardi et al. 2001; Petersen and Albrechtsen 2005; Sijen et al. 2001; Vaistij et al. 2002).

A more efficient modification of sense RNAi is termed transitive RNAi, during which an inverted repeat sequence is placed in the 3′ UTR of the construct (Brummell et al. 2003). This 3′ inverted repeat is thought to increase the efficiency by which RDR6 initiates copying of the sense transcript to produce dsRNA. Moreover, the siRNA cleavage products of the inverted repeat region may act as primers for subsequent RDRP activity. Significantly, when full-length coding sequences are utilized upstream of a 3′ UTR inverted repeat, the efficiency of this method as a gene silencing tool approaches that of IR-RNAi (Brummell et al. 2003).

2 Overview of the Transitive RNAi Approach

Here, we review a novel approach for a forward mutagenic screen that targets specific populations of genes for silencing using transitive RNAi (Fig. 1; Petsch et al. 2010). A key aspect of this strategy is the ability to target a focused population of genes for mutagenesis. Toward this end, we utilized laser microdissection to generate a select population of transcripts from *Arabidopsis* mesophyll cells, although any method that enables the isolation of RNA from specific cell and/or tissue populations

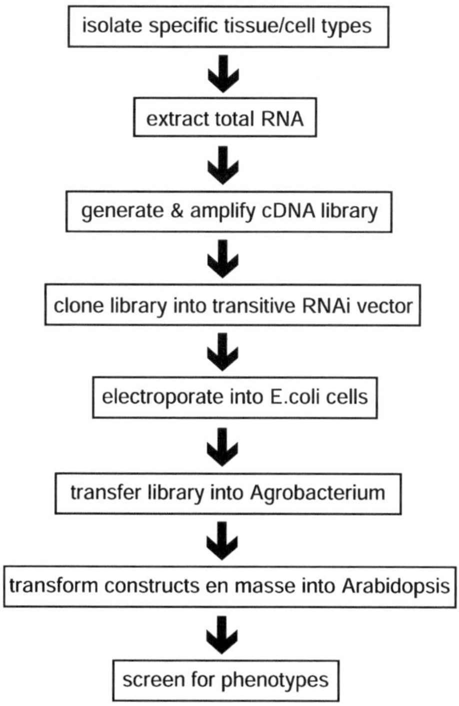

Fig. 1 Overview of the transitive RNAi approach (from Petsch et al. 2010)

could be employed. Following tissue/cell-specific collection, the total RNA is isolated, reverse-transcribed, and PCR-amplified to generate a near full-length cDNA library. The addition of specific restriction enzyme sites during the PCR amplification process allows for the directional cloning of the library into the transitive RNAi vector. Subsequently, electroporation into electrocompetent *E. coli* cells and mating into *Agrobacterium* are performed, allowing for en masse transformation into *Arabidopsis*. Selection of transgenic plants is then a relatively straightforward process that can be performed on growth media supplemented with the selective agent or by spraying the selective agent on plants grown in soil.

3 Laser Capture Microdissection and cDNA Library Construction

In our initial effort to test the efficacy of transitive RNAi for tissue-specific forward mutagenesis on a global scale, we chose to target photosynthetic phenotypes and mesophyll cells (Petsch et al. 2010). Our rationale in choosing mesophyll cells is that photosynthetic phenotypes are induced by mutation in a variety of individual genes and are easily recognized in a large-scale screen. Thus, the likelihood of identifying a mutant phenotype is very high. Additionally, because photosynthesis occurs predominantly in a specialized cell type, the mesophyll cell, it provides a straightforward target for laser microdissection.

To investigate the potential of this approach, two mesophyll cell-specific cDNA libraries were generated via laser microdissection of 2-week-old *Arabidopsis* leaves. For each library, an area of approximately 500,000 μm^2 of mesophyll cells (8–10 leaf sections at a thickness of 10 μm) was microdissected, after which the extracted RNA transcripts were converted into cDNA and amplified to generate either the sense or antisense mesophyll libraries. Bidirectional (i.e., sense and antisense) libraries were constructed because our preliminary data indicated that some genes are more efficiently silenced when cloned in one orientation vs. the other. Random clones from each of the bidirectional mesophyll tissue libraries were selected to assess the quality and genic diversity represented therein. Sequencing of the transcripts revealed a size range from less than 100 bp to over 1 kb, the majority of which were 5′ truncated. Although four out of twenty random clones were comprised of chloroplast DNA, the remaining 16 clones included a diverse array of cDNAs, indicating that no one particular mesophyll transcript is significantly overrepresented in either of the libraries (Petsch et al. 2010).

The sense and antisense mesophyll cDNA libraries were directionally cloned into the transitive RNAi vector, pSR486, using *Pme*I and *AscI* or *Asi*SI and *Bst*Z17I for the sense and antisense libraries, respectively. Key elements of this vector are illustrated in Fig. 2, and the latter also features the SHUTR method originally described by Brummell et al. (2003). Two expression cassettes are located within the transfer-DNA (T-DNA) borders. One cassette induces transitive RNAi and consists of a tandem CaMV 35S promoter, a multiple cloning site (MCS), and an inverted

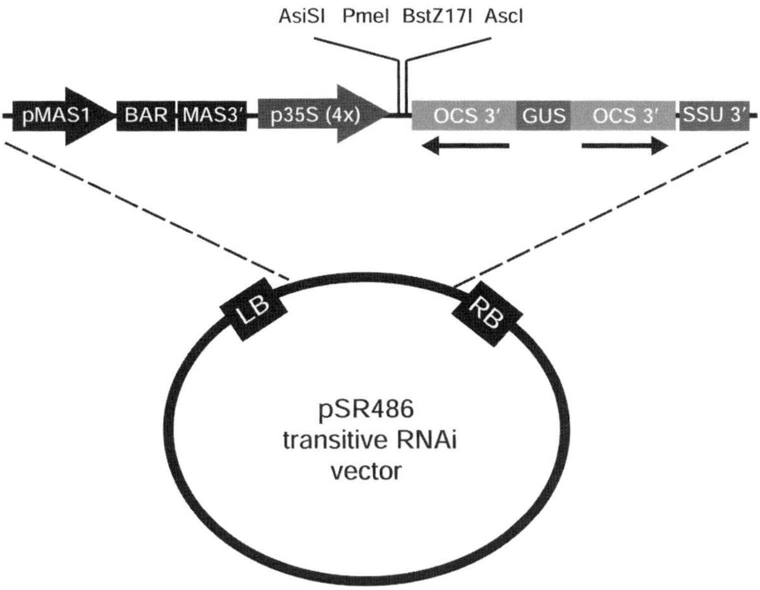

Fig. 2 The pSR486 transitive RNAi vector. Mesophyll cell library cDNAs were cloned directly into the multiple cloning site (MCS) in either the sense or antisense orientation using *Pme*I and *AscI*, or *Asi*SI and *Bst*Z17I restriction enzymes, respectively. The inverted repeat of the OCS 3′ UTR is positioned immediately downstream of the MCS and contains a fragment of the *GUS* gene as a spacer. LB, left border; RB, right border; pMAS1′, mannopine synthase promoter; *BAR*, marker gene conferring resistance to glufosinate ammonium; MAS3′, mannopine synthase terminator; p35S (4x), tandem CaMV 35S promoter; OCS 3′-UTR, octopine synthase 3′-untranslated region; *GUS* spacer, fragment of the *GUSA* gene; SSU 3′, ribulose bisphosphate carboxylase (rbcS) small subunit terminator

repeat of the octopine synthase 3′-untranslated region (OCS 3′-UTR). The second expression cassette contains the *BAR* gene, a selectable marker conferring herbicide resistance. Once cloned into the transitive RNAi vector, the mesophyll libraries were subsequently transformed en masse into *Agrobacterium tumefaciens* and used to infect wild-type *Arabidopsis* (ecotype Columbia (Col-4)).

4 Identification of Transitive RNAi Phenotypes

Seed from the transformed plants was screened on growth media supplemented with 1.5% sucrose and a selective agent (glufosinate ammonium) to ascertain transformants. Five hundred and eleven and 486 transformants were identified in the sense and antisense mesophyll cell library screens, respectively. Of these plants, 76 (14.8%) from the sense library screen and 73 (15%) from the antisense library screen displayed macroscopic phenotypes (Petsch et al. 2010). A sample of representative

Fig. 3 Representative phenotypes generated from the mesophyll cell libraries. First-generation transformants produced from sense- (*upper*) and antisense-orientated (*lower*) cDNAs in the transitive RNAi vector are labeled with their construct number, and the corresponding wild-type (WT) plants are shown on the *left*. Locus IDs for each of the transgenes identified are as follows: 82, AT4G12800; 84, AT5G64040; 86, AT3G03630; 164, AT1G03130; 32, AT1G64520; 42, AT3G01480; 110, AT1G55670; 70, AT1G67090 (modified from Petsch et al. 2010)

phenotypes from each of the mesophyll libraries is reproduced in Fig. 3; the corresponding transgenes that were isolated from these plants are listed in the figure caption. Several of the observed phenotypes (plants 32, 42, and 110) deviated to some degree from previously described mutations in the transgene, which may be attributed to the combined effects of silencing multiple paralogs within a gene family (Hamilton et al. 1998; Sanders et al. 2002), the level of gene downregulation caused by the transitive RNAi construct, or alternatively a positional effect from the transgene insertion site.

The identity of the corresponding transgenes from selected phenotypic and non-phenotypic transgenic plants was determined by PCR using primers that flank the MCS. As expected, a considerable proportion of transgenes are predicted to encode gene products localized to the chloroplast, organelles that comprise a significant portion of leaf mesophyll tissue. Accordingly, the majority of mutants identified in each library exhibited photosynthetic phenotypes (i.e., chlorotic or pale-green leaves), whereas a number of mutants also exhibited growth or developmental abnormalities (Petsch et al. 2010).

5 Validation of Transitive RNAi Phenotypes by Retransformation

Confirmation that several of the observed phenotypes were in fact correlated with silencing of the homologous endogene was achieved by selecting 11 (6 sense and 5 antisense) of the PCR-amplified transgenes (Table 1) and cloning them back into the transitive RNAi vector for retransformation into *Arabidopsis*. Ten of the 11 transgene

Table 1 Description of transgenes used for retransformation into wild-type plants

cDNA library	Plant no.	Locus ID	Gene description	Transgene size (bp)	Transgene sequence
Mesophyll cell (sense)	84	AT5G64040	Subunit of photosystem I located entirely in the thylakoid lumen (*PSI-N*)	480	P; 3′ end
	88	AT2G27260	Similar to hydroxyproline rich glycoprotein (HRGP) family protein	256	P; 3′ end
	97	AT4G12800	Subunit L of photosystem I reaction center (*PSAL*)	213	P; 3′ end
	95	AT5G66570	Extrinsic subunit of photosystem II (*PSBO1*)	314	P; 3′ end
	86	AT3G03630	*O*-acetylserine (thiol) lyase (*CS26*)	287	P; 3′ end
	62	AT3G49190	Condensation domain-containing protein	352	P; 3′ end
Mesophyll cell (anti-sense)	73	AT5G38420	RuBisCO small subunit 2B (*RBCS-2B*)	231	P; 3′ end
	32	AT1G64520	Putative 26S proteasome regulatory subunit (*RPN12*)	969	Full-length
	76	AT3G21055	Photosystem II 5 kD protein subunit PSII-T (*PSBTN*)	172	P; 3′ end
	42	AT3G01480	Chloroplast cyclophilin functioning in the assembly and maintenance of photosystem II (PSII) supercomplexes (*CYP38*)	348	P; 3′ end
	88	AT1G12900	Glyceraldehyde-3-phosphate dehydrogenase (*GAPA-2*)	151	P; 3′ end

P, partial length sequence; 3′ end, sequence is 5′-truncated and biased toward the 3′ region of the cDNA; full-length, sequence covers the full-length of the coding sequence. From Petsch et al. (2010)

constructs tested by retransformation were able to recapitulate the original phenotype; however, the phenotypic severity observed in first-generation retransformants was variable (Petsch et al. 2010). This disparity in phenotype severity observed in separate transformant plants containing identical transgenes presumably reflected innate variations in the extent of endogene silencing induced in individual lines, either by the efficiency of the silencing system or as a result of the transgene insertion site. As an additional validation that the phenotypes were due to transitive RNAi of the selected transgene, quantitative RT-PCR analyses of the ten first-generation transformants were performed and revealed downregulation of the corresponding endogenous transcript ranging from 0.11 to 0.60 of wild-type levels (Fig. 4).

The sole transgene in retransformed plants that was unable to reproduce the original phenotype (plant #62) is predicted to encode a condensation domain-containing protein of unknown function (Petsch et al. 2010). Retransformants containing this transgene were pale green in color and did not induce the distinctive hyponastic leaf phenotype of the initial mutant transformant. Thermal asymmetric interlaced (TAIL)

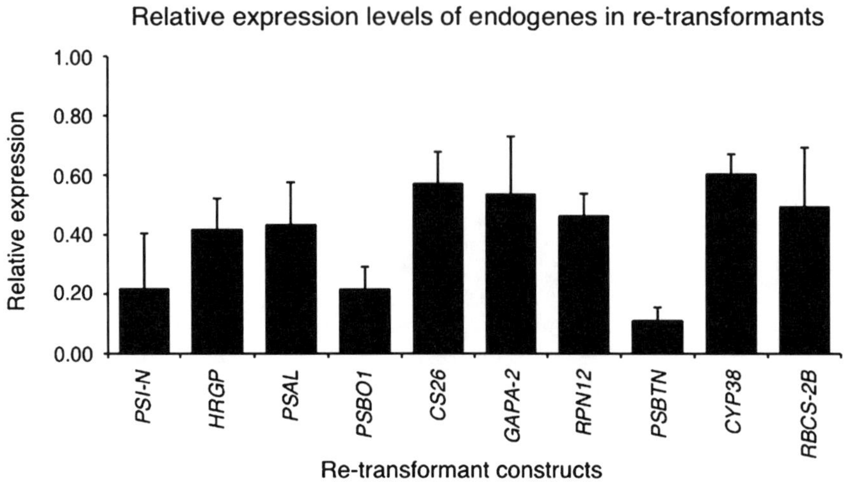

Fig. 4 Relative expression levels of the homologous endogenes in retransformants. Endogenous transcript levels were quantified in wild-type and pooled first-generation retransformed plants using quantitative RT-PCR. The graph shows the relative values, with respect to wild-type plants, of each of the endogenous genes. Values shown are means ± SE (from Petsch et al. 2010)

PCR on genomic DNA extracted from progeny of plant #62 was used to determine the genomic location of the T-DNA. Sequencing of the tertiary PCR product revealed that the T-DNA harboring the condensation domain-containing protein transgene had inserted ~1.5 kb downstream of the *ASYMMETRIC LEAVES 2* (*AS2*) locus. Previously published *AS2* overexpression phenotypes (Nakazawa et al. 2003) very closely resemble those of plant #62, and qRT-PCR analyses revealed upregulation of *AS2* transcripts in plant #62 and its mutant progeny. Overall, these data suggest that the mutation in plant #62 is an activation-tagged allele of *AS2* and does not result from transitive RNAi.

6 Transitive RNAi of Gene Paralogs

Because transitive RNAi is a homology-based method of gene silencing, we sought to determine whether closely related gene paralogs are also subject to transitive RNAi. To address this question, several transgenic lines produced from retransformant construct #73, which contains a 3′ fragment of the RuBisCO small subunit 2B (*RBCS-2B*) gene, were used to examine the expression levels of two closely related gene family members, *RBCS-1B* and *RBCS-3B* (Petsch et al. 2010). The RuBisCO small subunit 2B gene shares 76 and 68% identity, respectively to *RBCS-1B* and *RBCS-3B*. Notably, the first ~60 bp of the *RBCS-2B* transgene shares nearly 100% identity to both *RBCS-1B* and *RBCS-3B*. Quantitative RT-PCR analyses of all three

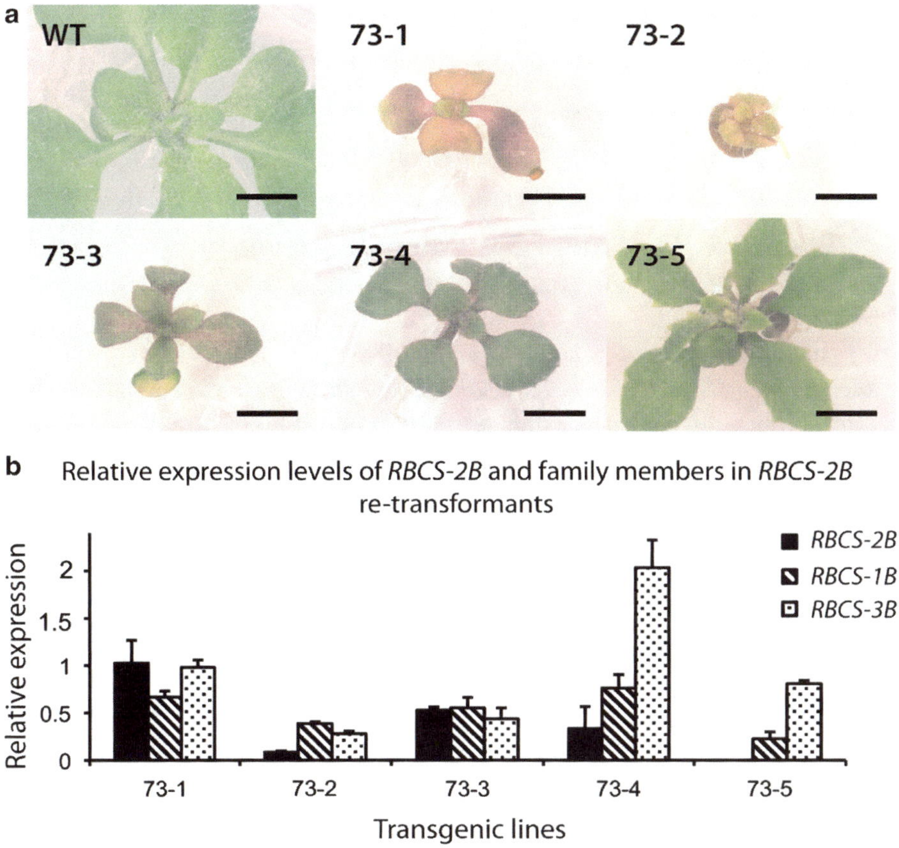

Fig. 5 *RBCS-2B* retransformants. (**a**) Phenotypes generated from *RBCS-2B* retransformation construct #73. Individual transgenic lines are designated 73-1 to 73-5, and a comparative wild-type plant is designated "WT." Scale bars = 0.5 cm. (**b**) Relative expression levels of *RBCS-2B*, *RBCS-1B*, and *RBCS-3B* endogenes in *RBCS-2B* retransformants. Endogenous transcript levels were quantified in wild-type and first-generation retransformed plants using quantitative RT-PCR. The graph shows the relative values, with respect to wild-type plants, for each of the endogenous genes. Values shown are means generated from technical replicates ± SE (modified from Petsch et al. 2010)

endogene paralogs revealed that at least one gene family member was downregulated in each transgenic line (Fig. 5). Strong downregulation of all three endogene paralogs corresponded to a more severe phenotype (Fig. 5, 73-2), whereas lesser levels of downregulation, or downregulation of one or two paralogs alone, corresponded to weaker phenotypes (Fig. 5, 73-1, 73-3 to 73-5). The reduction of only *RBCS-1B* and not *RBCS-2B* in one of the transgenics (73-1), despite the relatively chlorotic phenotype, could be the result of silencing of another closely related gene paralog, *RBCS1A*. This gene paralog also has close homology to *RBCS-2B* and could be responsible for the chlorotic phenotype in combination with *RBCS-1B*. Surprisingly, one transgenic line (73-4) exhibited an ~2-fold upregulation of *RBCS-3B*, even though both *RBCS-2B* and *RBCS-1B* showed lower expression levels relative to wild-type plants (Fig. 5, 73-4). This increase in gene expression may reflect a compensatory mechanism initiated by the plant in response to the reduced transcript levels of the other two paralogs.

7 Transitive RNAi and Laser Microdissection as a Forward Genetics Approach

Our combined approach of using transitive RNAi and laser microdissection as a forward mutagenesis tool offers at least two key advantages that circumvent some of the limitations associated with previous protocols for forward mutagenesis (Petsch et al. 2010). Laser microdissection enables selective populations of genes to be targeted for mutagenesis through the ability to collect specific tissue or cell types, and transitive RNAi provides the potential to overcome genetic redundancy due to gene or whole-genome duplications. Certainly, laser microdissection is not the only means by which to generate a focused population of genes. Several other related techniques, including mechanical microdissection, micropipetting, fluorescence-activated cell sorting (FACS), and RNA subtraction hybridization, could also be applied to produce selective populations of transcripts (Nelson et al. 2008). The transitive RNAi approach has a broad range of possible applications, including the identification of genes affecting particular traits or biological processes in a given location (e.g., cell or tissue type), response to a stimulus (e.g., pathogen, stress, hormone, etc.), or developmental time. However, the success of transitive RNAi in any application is reliant upon both the means by which a specific population of transcripts is acquired and the quality of the experimental design.

In comparison to other methods of forward mutagenesis, i.e., chemical and insertional, the combination of laser microdissection and transitive RNAi offers a degree of precision with respect to the mutagenesis of target genes. This results in a reduction of time and expense that is often required for larger forward mutagenic screens. The increase in efficiency is particularly valuable for laborious screens, especially those requiring replicate, quantitative assays of multiple segregating individuals carrying the same RNAi transgene, e.g., screens for pest tolerance.

One particular advantage of the transitive RNAi approach is the ability for closely related genes to be simultaneously silenced with a single target gene. This was clearly demonstrated with the *RBCS-2B* gene (Fig. 5; Petsch et al. 2010), although the extent of gene silencing and the number of gene paralogs affected was variable in different *RBCS-2B* transgenic lines. Extending these findings, it seems likely that other gene paralogs are also susceptible to silencing by our laser microdissection/transitive RNAi protocol; however, this is dependent on the target fragment used and the degree of sequence identity with other gene family members. The capacity to silence gene paralogs is not possible with the more conventional methods of forward mutagenesis, including chemical and insertional, reflecting the novelty of this forward mutagenesis approach.

8 Library Construction and Methodology Considerations

Based on our preliminary findings, both sense and antisense-oriented libraries are capable of triggering gene silencing with roughly equal efficiencies (Petsch et al. 2010). However, this does not preclude the possibility that certain genes may be more effectively silenced in one orientation vs. the other. In order to minimize the effects of any directional bias that might exist, it would be beneficial to utilize non-directional libraries in future studies. A further consideration with respect to the transitive RNAi approach is that a small percentage of phenotypes may result from activation tagging as conferred by the 4x35S promoter of the transitive RNAi vector. Although not an adverse occurrence in terms of forward mutagenesis, the identification of an activation-tagged allele in our mutant screen underscores the need to perform quantitative RT-PCR analyses or transgene retransformations to determine the probable cause of an observed mutant phenotype. Additional parameters to be aware of include the combination of binary vector and *Agrobacterium* strain to minimize multiple transgene insertion events. To this end, we chose to employ a pRK290-based binary vector in combination with *Agrobacterium* stain LBA4404, which gives approximately 65% of single-copy integrants (N. Doetsch, unpublished). Use of the same vector with *Agrobacterium* strain GV3101 gives a higher plant transformation rate; however, only 27% of single-copy integrants were observed. Thus, to obtain a higher proportion of single-copy insertion events, a trade-off with transformation efficiency is necessary. Lastly, normalization of the cDNA library may be an important factor to consider for certain applications to increase the abundance of low-abundance transcripts and remove overrepresented highly abundant transcripts.

9 Conclusions

Forward mutagenesis by transitive RNAi provides an alternative and also complementary approach to the more classical methods of mutagenesis. Importantly, this approach offers several key advantages with respect to the cloning strategy and identification of the causative transgene. In contrast to other methods of random mutagenesis, identifying the transgene that is contributing to the phenotype is straightforward and requires a single genomic PCR reaction utilizing primers designed to the cloning vector. Generation of the cDNA library itself is also a relatively simple process that does not require any prior knowledge of the transcript sequence and can be performed using 5′-truncated cDNAs, which are sufficient to trigger effective silencing. Furthermore, the combination of transitive RNAi and laser microdissection greatly enhances the power of this approach by targeting only a small subset of genes, rather than the entire genic content.

Forward mutagenesis by transitive RNAi has particular appeal for use in crop plants, which have high loads of genetic redundancy due to polyploidy and paleopolyploidy. The homology-based posttranscriptional silencing mechanism of transitive RNAi enables the potential knockdown of closely related gene paralogs, a feature that is not possible using traditional protocols for random mutagenesis.

References

Becker A, Lange M (2010) VIGS – genomics goes functional. Trends Plant Sci 15:1–4

Bernacki S, Tuttle JR, Muangsan N et al (2008) Extending functional genomics: VIGS for model and crop plants. In: Kahl G, Meksem K (eds) The handbook of plant functional genomics: concepts and protocols. Wiley-VCH, Weinheim, pp 227–249

Brummell DA, Balint-Kurti PJ, Harpster MH et al (2003) Inverted repeat of a heterologous 3′-untranslated region for high-efficiency, high-throughput gene silencing. Plant J 33:793–800

Curaba JL, Chen XM (2008) Biochemical activities of *Arabidopsis* RNA-dependent RNA polymerase 6. J Biol Chem 283:3059–3066

Dalmay T, Hamilton A, Rudd S et al (2000) An RNA-dependent RNA polymerase gene in *Arabidopsis* is required for posttranscriptional gene silencing mediated by a transgene but not by a virus. Cell 101:543–553

Fu DL, Uauy C, Blechl A et al (2007) RNA interference for wheat functional gene analysis. Transgenic Res 16:689–701

Hamilton AJ, Brown S, Han YH et al (1998) A transgene with repeated DNA causes high frequency, post-transcriptional suppression of ACC-oxidase gene expression in tomato. Plant J 15:737–746

Hilson P, Allemeersch J, Altmann T et al (2004) Versatile gene-specific sequence tags for *Arabidopsis* functional genomics: transcript profiling and reverse genetics applications. Genome Res 14:2176–2189

Lipardi C, Wei Q, Paterson BM (2001) RNAi as random degradative PCR: siRNA primers convert mRNA into dsRNAs that are degraded to generate new siRNAs. Cell 107:297–307

McGinnis K, Chandler V, Cone K et al (2005) Transgene-induced RNA interference as a tool for plant functional genomics. Meth Enzymol 392:1–24

McGinnis K, Murphy N, Carlson AR et al (2007) Assessing the efficiency of RNA interference for maize functional genomics. Plant Physiol 143(4):1441–1451

Nakazawa M, Ichikawa T, Ishikawa A et al (2003) Activation tagging, a novel tool to dissect the functions of a gene family. Plant J 34:741–750

Napoli C, Lemieux C, Jorgensen R (1990) Introduction of a chimeric chalcone synthase gene into petunia results in reversible co-suppression of homologous genes *in trans*. Plant Cell 2:279–289

Nelson T, Gandotra N, Tausta SL (2008) Plant cell types: Reporting and sampling with new technologies. Curr Opin Plant Biol 11:567–573

Petersen BO, Albrechtsen M (2005) Evidence implying only unprimed RdRP activity during transitive gene silencing in plants. Plant Mol Biol 58:575–583

Petsch KA, Ma C, Scanlon MJ et al (2010) Targeted forward mutagenesis by transitive RNAi. Plant J 61:873–882

Que QD, Wang HY, English JJ et al (1997) The frequency and degree of cosuppression by sense chalcone synthase transgenes are dependent on transgene promoter strength and are reduced by premature nonsense codons in the transgene coding sequence. Plant Cell 9:1357–1368

Sanders M, Maddelein W, Depicker A et al (2002) An active role for endogenous β-1,3-glucanase genes in transgene-mediated co-suppression in tobacco. EMBO J 21:5824–5832

Sijen T, Fleenor J, Simmer F et al (2001) On the role of RNA amplification in dsRNA-triggered gene silencing. Cell 107:465–476

Small I (2007) RNAi for revealing and engineering plant gene functions. Curr Opin Biotechnol 18:148–153

Tang GL, Reinhart BJ, Bartel DP et al (2003) A biochemical framework for RNA silencing in plants. Genes Dev 17:49–63

Vaistij FE, Jones L, Baulcombe DC (2002) Spreading of RNA targeting and DNA methylation in RNA silencing requires transcription of the target gene and a putative RNA-dependent RNA polymerase. Plant Cell 14:857–867

Waterhouse PM, Helliwell CA (2003) Exploring plant genomes by RNA-induced gene silencing. Nat Rev Genet 4:29–38

Watson JM, Fusaro AF, Wang MB et al (2005) RNA silencing platforms in plants. FEBS Lett 579:5982–5987

Pol IV-Dependent siRNAs in Plants

Rebecca A. Mosher

Contents

R.A. Mosher (✉)
The School of Plant Sciences, The University of Arizona,
P.O. Box 210036, Tucson, AZ 85721-0036, USA
e-mail: rmosher@cals.arizona.edu

V.A. Erdmann and J. Barciszewski (eds.), *Non Coding RNAs in Plants*,
RNA Technologies, DOI 10.1007/978-3-642-19454-2_26,

Abstract In plants, the most abundant class of small RNAs are 24-nucleotide short interfering (si)RNA. These siRNAs are produced at thousands of discrete genomic locations through the action of the plant-specific DNA-dependent RNA polymerase IV (Pol IV). Pol IV-dependent siRNAs catalyze repressive DNA methylation on transposable elements and other repetitive sequences, but might trigger diverse chromatin modifications at distinct genomic locations, such as DNA demethylation or histone modification. Pol IV-dependent siRNAs are expressed abundantly, and sometimes exclusively, in the developing endosperm, where they are produced from only the maternal chromosomes. The biological role of Pol IV-dependent siRNAs is unclear, but might involve interaction between different genomes or alleles, or stabilizing and buffering the genome from genetic and epigenetic modifications.

Keywords RNA polymerase IV • RNA polymerase V • Transcriptional gene silencing • DNA methylation • Heterochromatin

1 Introduction

In the 1990s, a convergence of research in plants, flies, and worms led to the discovery of RNA-induced gene silencing, an ancient system that cleaves double-stranded (ds)RNA into small molecules with the capacity to silence complementary sequences (Ruvkun 2008). In various eukaryotes, this system evolved to inactivate transcription, suppress translation, or eliminate DNA, affecting widespread biological systems, including development, disease, stress responses, and genome maintenance (Ding and Voinnet 2007; Sunkar et al. 2007; Stefani and Slack 2008; Moazed 2009). Although each organism offers a unique perspective, flowering plants are particularly suited to studying RNA silencing due to amplification and diversification of RNA silencing components (Xie et al. 2004).

Of particular interest to plant scientists is the prevalence of 24 nucleotide (nt) short interfering (si)RNAs capable of transcriptional gene silencing. Unlike their smaller cousins the microRNAs, 24 nt siRNAs direct the formation of silent heterochromatin with epigenetic potential – in other words, the silent state can be propagated in the absence of the initiating signal. Production and action of this class of siRNAs is also interesting because it involves two plant-specific DNA-dependent RNA polymerases with specialized roles in RNA silencing (Pikaard et al. 2008). In this chapter, I discuss the synthesis, expression, molecular function, and biological roles of these small silencers.

2 Components of the Pol IV/V Pathway

The first indication of a pathway specialized for transcriptional gene silencing was the observation that plants produce at least two sizes of small RNAs that are independently affected by viral suppressors of silencing or genetic mutations and that exhibit differential systemic movement (Hamilton et al. 2002). The shorter (21 nt) siRNAs were associated with RNA degradation, while the longer (24 nt) siRNAs were correlated with DNA methylation and transcriptional silencing (Hamilton et al. 2002). Forward and reverse genetic screens eventually uncovered members of the RNA-dependent RNA polymerase (RDR), Dicer-like (DCL), and Argonaute (AGO) families that were specialized for synthesis and function of the different sizes of small RNA (Dalmay et al. 2000; Zilberman et al. 2003; Xie et al. 2004). Next-generation sequencing technologies underlined the importance of the long class of small RNAs – in flowering plants, 24 nt RNA species vastly outnumber shorter small RNA species, both in total number and in complexity (Henderson et al. 2006; Dolgosheina et al. 2008; Zhu et al. 2008).

In addition to specialized biosynthesis enzymes and a distinct silencing mechanism, the 24 nt siRNAs are unique because they require two plant-specific DNA-dependent RNA polymerases (Pikaard et al. 2008). Originally named RNA Polymerase IVa and IVb, these enzymes were later renamed Polymerase IV (Pol IV) and Polymerase V (Pol V) due to their nonoverlapping molecular roles in the biosynthesis and action of 24 nt siRNAs (Wierzbicki et al. 2008). Nearly all of the 24 nt siRNAs in *Arabidopsis* require Pol IV, and this class is, therefore, known as Pol IV-dependent (p4)-siRNAs (Zhang et al. 2007; Mosher et al. 2008) (Table 1).

Table 1 Components required for the synthesis and action of Pol IV-dependent siRNAs

Gene (alias)	Details	References
NRPD1 (NRPD1a, RPD1a, SDE4)	Component of DNA-dependent RNA polymerase IV	Herr et al. (2005), Onodera et al. (2005), Pontier et al. (2005), Zhang et al. (2007), and Mosher et al. (2008)
NRPE1 (NRPD1b, RPD1b, DRD3)	D component of NA-dependent RNA polymerase V	Kanno et al. (2005) and Pontier et al. (2005)
NRPD2/E2 (NRPD2a, RPD2a, DRD2)	Component of DNA-dependent RNA polymerase IV and V	Herr et al. (2005), Kanno et al. (2005), Onodera et al. (2005), and Pontier et al. (2005)
NRPD4/E4 (RDM2)	Component of DNA-dependent RNA polymerase IV and V	He et al. (2009a) and Ream et al. (2009)
NRPE5	Component of DNA-dependent RNA polymerase V	Huang et al. (2009), Lahmy et al. (2009), and Ream et al. (2009)
NRPE7	Component of DNA-dependent RNA polymerase V	Ream et al. (2009)

(continued)

Table 1 (continued)

Gene (alias)	Details	References
NRPD7	Component of DNA-dependent RNA polymerase IV	Ream et al. (2009)
NRPD9/E9	Component of DNA-dependent RNA polymerase IV and V	Ream et al. (2009)
RPB1	Component of DNA-dependent RNA polymerase II	Zheng et al. (2009)
RDR2	RNA-dependent RNA polymerase	Xie et al. (2004), Lu et al. (2006), and Kasschau et al. (2007)
DCL3	Dicer endonuclease	Xie et al. (2004) and Kasschau et al. (2007)
HEN1	siRNA methyltransferase	Li et al. (2005) and Yu et al. (2005)
AGO4	Argonaute protein	Zilberman et al. (2003) and Havecker et al. (2010)
AGO6	Argonaute protein	Zheng et al. (2007) and Havecker et al. (2010)
AGO9	Argonaute protein	Havecker et al. (2010) and Olmedo-Monfil et al. (2010)
CLSY1	Similar to SWI/SNF nucleosome remodeling proteins; possibly needed for Pol IV transcription	Smith et al. (2007)
DRD1	Similar to SWI/SNF nucleosome remodeling proteins; component of DDR complex; required for Pol V transcription	Kanno et al. (2004) and Law et al. (2010)
DMS3 (IDN1)	Component of DDR complex; required for Pol V transcription	Ausin et al. (2009), Kanno et al. (2010), and Law et al. (2010)
RDM1	Component of DDR complex; required for Pol V transcription	Gao et al. (2010) and Law et al. (2010)
KTF1 (SPT5, RDM3)	Contains AGO-hook motif; required for Pol V transcription	Bies-Etheve et al. (2009), He et al. (2009c), and Huang et al. (2009)
IDN2 (RDM12)	Required for RdDM	Ausin et al. (2009) and Zheng et al. (2010)
RDM4 (DMS4)	Associated with Pol V and Pol II	He et al. (2009b) and Kanno et al. (2010)
DRM2	De novo DNA methyltransferase	Cao and Jacobsen (2002)
SUVH2	SRA methyl-DNA binding domain; required for RdDM	Johnson et al. (2007)
SUVH9	SRA methyl-DNA binding domain; required for RdDM	Johnson et al. (2007)
Maize genes		
Rmr6 (ZmNrpd1)	Component of DNA-dependent RNA polymerase IV	Erhard et al. (2009)
Mop1 (ZmRdr2)	RNA-dependent RNA polymerase	Alleman et al. (2006)

(continued)

Table 1 (continued)

Gene (alias)	Details	References
Rmr1	SWI/SNF nucleosome remodeling protein	Hale et al. (2007)
Mop2 (Rmr7/ZmNrpd2)	Component of DNA-dependent RNA polymerase IV	Sidorenko et al. (2009) and Stonaker et al. (2009)

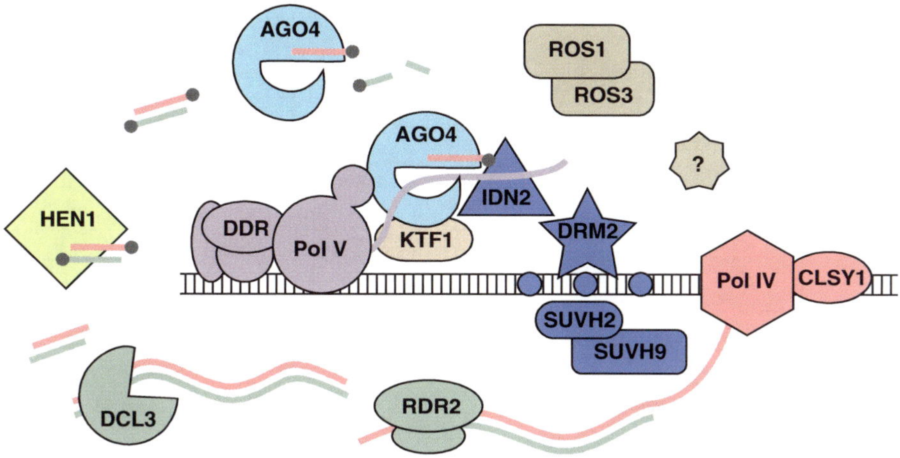

Fig. 1 The Pol IV/V pathway. RNA polymerase IV (Pol IV, *pink hexagon*) is hypothesized to initiate p4-siRNA production by transcribing DNA with the assistance of the SWI/SNF nucleosome remodeling protein CLSY1 (*pink oval*). Pol IV transcripts are substrates for RNA-dependent RNA polymerase 2 (RDR2, *green ovals*), which generates double-stranded (ds)RNA that is cleaved by Dicer-like ribonuclease 3 (DCL3, *green* pac-man). DCL3 chops dsRNA into 24-nucleotide siRNA duplexes that are methylated by HUA ENHANCER 1 (HEN1, *yellow*). ARGONAUTE 4 (AGO4, *light blue*) binds the siRNA duplexes and cleaves one strand, leaving one p4-siRNA "loaded" in the enzyme. AGO4 physically associates with the carboxy-terminal domain of RNA Polymerase V (Pol V, *purple joined circles*) while the loaded p4-siRNA anneals to complementary sequences transcribed by Pol V. Pol V transcription is aided by the DDR complex (*purple ovals*) and the SPT5-family member KTF1 (*orange oval*). With the assistance of INVOLVED IN DE NOVO 2 (IDN2, *blue triangle*) AGO4 recruits DOMAINS REARRANGED METHYLTRANSFERASE 2 (DRM2, *blue star*) to methylate DNA (*blue circles*). DNA methylation also requires the Su(var)3–9 homologs SUVH2 and SUVH9 (*blue rectangles*). At some loci other complexes capable of chromatin modification might be recruited by AGO4, including REPRESSOR OF SILENCING 1 and 3 (ROS1, ROS3, *brown rectangles*) or unknown enzymes (*brown star*)

2.1 *Biosynthesis of p4-siRNAs*

Like the canonical eukaryotic RNA polymerases, Pol IV and Pol V are multisubunit enzymes and mutations in either the largest or second-largest subunits of Pol IV (encoded by *NPRD1* and *NPRD2*, respectively) eliminate production of p4-siRNAs (Ream et al. 2009). *In vitro* transcriptional activity has yet to be detected for Pol IV, but its *in vivo* function requires the Metal A and Metal B sites known to catalyze transcription in other RNA polymerases, indicating that Pol IV produces a transcript *in vivo* (Fig. 1) (Haag et al. 2009). The SWI/SNF nucleosome remodeling protein

CLASSY1 (CLSY1) is required for p4-siRNA production at many genomic loci and might recruit Pol IV to chromatin (Smith et al. 2007). The characteristics of a Pol IV transcript (e.g., 5′ and 3′ end modifications) are also unknown, but might signal to the cell that the transcript is "aberrant" and thereby recruit RNA-DEPENDENT RNA POLYMERASE 2 (RDR2). Although *Arabidopsis* contains six RDRs, RDR2 functions nonredundantly to generate dsRNA from a Pol IV transcript (Fig. 1) (Xie et al. 2004; Lu et al. 2006; Kasschau et al. 2007).

DICER-LIKE 3 (DCL3) is the primary Dicer endonuclease for cleavage of RDR2-generated dsRNA and generates 24 nt small RNA duplexes with characteristic 2 nt overhangs at the 3′ end (Fig. 1) (Xie et al. 2004; Lu et al. 2006; Kasschau et al. 2007). In the absence of DCL3, other Dicer family members generate 21–22 nt siRNAs from RDR2 products (Xie et al. 2004; Kasschau et al. 2007; Mosher et al. 2008). These smaller siRNAs do not accumulate to the same level as 24 nt p4-siRNAs, perhaps due to feedback between p4-siRNA action and synthesis (discussed below) or due to reduced affinity between RDR2 products and other Dicers. Because the molecular phenotype of *dcl3* is similar but weaker than the *rdr2* phenotype, it is possible that these smaller p4-siRNAs retain their function (Xie et al. 2004). After Dicer cleavage from long dsRNA, p4-siRNAs are methylated at the 2′ hydroxyl by the siRNA methyltransferase HUA ENHANCER 1 (HEN1) (Fig. 1) (Xie et al. 2004; Yu et al. 2005). This modification stabilizes p4-siRNAs and microRNAs by inhibiting terminal uridylation, which triggers degradation of the small RNA (Li et al. 2005).

2.2 Downstream Factors in the Pol IV/V Pathway

Small RNAs do not function alone, but instead bind to AGO proteins and guide these effectors to RNA transcripts. AGO proteins bind the small RNA duplex and "slice" one strand, leaving a single-stranded small RNA bound to the AGO and ready to hybridize with complementary sequences (Fig. 1) (Matranga et al. 2005). *Arabidopsis* contains ten AGO proteins, and AGO4 binds the majority of p4-siRNAs (Morel et al. 2002). AGO6 and AGO9 also bind p4-siRNAs and are partially redundant with AGO4 at some loci (Zheng et al. 2007; Havecker et al. 2010). How these three AGOs interact and compete to facilitate p4-siRNA activity is unclear, but tissue-specific expression might account for many of the differences between AGO activities (Havecker et al. 2010).

P4-siRNAs direct AGO4 to nascent transcripts generated by Pol V (Fig. 1) (Wierzbicki et al. 2009). Several proteins are required for Pol V transcription including DEFECTIVE IN RNA-DIRECTED DNA METHYLATION 1 (DRD1), a SWI/SNF nucleosome remodeling protein, DEFECTIVE MERISTEM SILENCING 3 (DMS3), an SMC-hinge domain protein, and RNA-DIRECTED DNA METHYLATION 1 (RDM1), a single-stranded methyl-DNA binding protein (Kanno et al. 2004, 2008; Wierzbicki et al. 2008, 2009; Gao et al. 2010; Law et al. 2010). These proteins form the DDR complex and physically associate with Pol V, perhaps creating a permissive chromatin structure to recruit or activate Pol V transcription (Law et al. 2010). RDM1 is also associated with Pol II and it is unclear whether Pol II transcripts are also targeted by AGOs or whether Pol II enables transcription by Pol V (Zheng et al. 2009; Gao et al. 2010). Pol II and Pol V action are also linked by REQUIRED FOR DNA METHYLATION 4 (RDM4) (He et al. 2009b; Kanno et al. 2010). RDM4 is a transcriptional regulator conserved throughout eukaryotes and interacts with both Pol II and Pol V. Loss of RDM4 function causes pleiotropic phenotypes including reduction in Pol V transcription (He et al. 2009b; Kanno et al. 2010).

In addition to RNA–RNA interactions between p4-siRNAs and nascent transcripts, there are protein–protein interactions between AGOs and Pol V. AGO4, AGO6, and AGO9 physically associate with Pol V through a reiterated GW-motif in their carboxy terminal domains known as an "AGO-hook" (Li et al. 2006; Till et al. 2007; Havecker et al. 2010). KOW-DOMAIN TRANSCRIPTION FACTOR 1 (KTF1) also contains an AGO-hook and associates with both AGO4 and Pol V (Bies-Etheve et al. 2009; He et al. 2009c; Huang et al. 2009). KTF1 has similarity to the SPT5-family of transcription elongation factors, but loss-of-function mutations do not affect accumulation of Pol V transcripts, indicating that KTF1 functions downstream of Pol V transcription, perhaps in stabilizing the AGO4/Pol V complex (He et al. 2009c).

Once localized to chromatin through protein and RNA interactions, AGO4 directly or indirectly recruits enzymes capable of DNA methylation or other chromatin modification (discussed below). Surprisingly, AGO4 "slicing" activity is required at some, but not at all p4-siRNA loci, indicating that AGO4 performs both catalytic and noncatalytic roles in silencing (Qi et al. 2006).

2.3 *The Pol IV/V Pathway in Maize*

The Pol IV/V pathway is best characterized in *Arabidopsis*, but the discovery of several mutations in maize has greatly expanded our knowledge of this pathway. Maize contains a single ortholog of the largest subunit of Pol IV (*NRPD1*) and three nonredundant genes encoding the second-largest subunit (*NRPD2*) (Erhard et al. 2009; Sidorenko et al. 2009; Stonaker et al. 2009). Maize also contains an *RDR2* ortholog, and at least one SWI/SNF nucleosome remodeling protein similar to *CLSY1* and *DRD1* (Alleman et al. 2006; Hale et al. 2007). Mutations in these genes were identified in genetic screens for loss of paramutation (see Sect. 7.6), but each mutation is also deficient in p4-siRNA production. Interestingly, loss of the Pol IV largest subunit has a stronger phenotype than any of the other mutations, indicating that Pol IV might have functions in maize beyond p4-siRNA production, perhaps by associating with multiple isoforms of NRPD2 (Hale et al. 2009).

2.4 *Pol IV-Independent 24 nt siRNAs*

Although deep sequencing of small RNA populations indicates that nearly all 24 nt siRNAs are Pol IV dependent, there are also Pol IV-independent 24 nt siRNAs that arise from DCL3 cleavage of RDR2-independent dsRNA. For instance, transgenes engineered to produce RNA hairpins generate all sizes of siRNAs, including Pol IV-independent 24 nt siRNAs (Kanno et al. 2005; Daxinger et al. 2009). Similarly, when endogenous inverted repeats are transcribed, they generate long dsRNA hairpins that are cleaved by multiple Dicers to produce 21, 22, and 24 nt siRNAs (Zhang et al. 2007; Dunoyer et al. 2010a). Pol IV-independent 24 nt siRNAs associate with AGO4 and cause RNA-directed DNA methylation (RdDM) in *cis* and *trans* (Zilberman et al. 2004; Daxinger et al. 2009; Dunoyer et al. 2010a). Accumulation of DNA methylation requires Pol V, but it is unclear if Pol V is required to initiate RdDM from Pol IV-independent 24 nt siRNAs, or if these siRNAs trigger *bona fide* p4-siRNA production (Kanno et al. 2005).

2.5 A Special Case of Pol IV-Dependence: nat-siRNAs

In addition to large numbers of 24 nt p4-siRNAs, a small number of siRNAs from natural antisense (NAT) gene pairs also require Pol IV (Borsani et al. 2005; Katiyar-Agarwal et al. 2006, 2007). These nat-siRNAs range in size from 22 to 40 nt and are proposed to posttranscriptionally silence a member of the NAT gene pair. Nat-siRNAs are produced after biotic or abiotic stress triggers convergent Pol II transcription at NAT gene pairs, producing dsRNA that is cleaved by DCL1, DCL2, or DCL4. Initial nat-siRNA generation can then trigger further dsRNA production through RDR6. Pol IV, but not RDR2 or DCL3, is also required for nat-siRNA biosynthesis, although the role of Pol IV in this process unclear. Because of the unconventional genetic requirements for nat-siRNA production, they are considered distinct from canonical p4-siRNAs.

3 P4-siRNA Producing Loci

High-throughput short-read sequencing technologies have revolutionized the study of small RNAs, allowing researchers to sequence millions of small RNA molecules from end to end. In flowering plants, the vast majority of unique small RNA sequences are 24 nt p4-siRNAs, many of which are present at very low frequency (<1 per million) (Henderson et al. 2006; Nobuta et al. 2008). Because p4-siRNAs are cleaved at random from precursor transcripts, a 500-bp locus has the potential to generate almost 1,000 unique p4-siRNA sequences, most of which cannot be related to each other until mapped onto a genome. The extreme complexity of the p4-siRNA population makes it necessary to consider p4-siRNA loci, rather than p4-siRNA sequences (Lu et al. 2006; Zhang et al. 2007; Mosher et al. 2008). P4-siRNAs are produced from thousands of distinct genomic regions, covering a minimum of 1% of the *Arabidopsis* genome and a higher fraction of transposable element-rich genomes such as maize and wheat (Mosher et al. 2008; Nobuta et al. 2008; Cantu et al. 2010).

3.1 *Repetitive Sequences*

Many p4-siRNAs match the genome multiple times indicating that repetitive elements are an abundant source of p4-siRNAs (Kasschau et al. 2007; Zhang et al. 2007; Mosher et al. 2008; Nobuta et al. 2008; Cantu et al. 2010). Both tandem and dispersed repeats produce p4-siRNAs, including pericentromeric heterochromatin, rDNA arrays, and transposable elements (Kasschau et al. 2007; Zhang et al. 2007; Mosher et al. 2008; Nobuta et al. 2008; Cantu et al. 2010). RNA silencing systems in diverse organisms target repetitive DNA, leading to the hypothesis that p4-siRNAs are part of an ancient defense against invasive genetic elements (Plasterk 2002). This hypothesis suggests that Pol IV transcribes repetitive DNA and the resulting p4-siRNAs recruit AGO4 in *cis* and *trans* to silence all copies of the repeated sequence. The genome-wide identification of p4-siRNA loci through parallel short-read sequencing supports this model, as over half of the genomic regions producing p4-siRNAs share homology with TEs (Zhang et al. 2007; Mosher et al. 2008). Furthermore, when the tobacco *TNT1* retrotransposon is introduced into *Arabidopsis*, it generates p4-siRNAs only when present in many copies (Perez-Hormaeche et al. 2008).

While it is undeniable that many repetitive elements generate p4-siRNAs, it is unclear whether *all* repetitive elements generate p4-siRNAs. The nature of repetitive loci makes it difficult to determine if every copy of a repeat produce p4-siRNAs or if only a subset of copies produce p4-siRNAs that perfectly match other copies of the repeat. Because of the high complexity of the p4-siRNA population, informative molecules that overlap the single-nucleotide differences between repeat copies are rare. As sequencing depth increases and additional small RNA sequencing data become available, researchers will be better able to address this question.

3.2 *Low Copy and Unique Sequences*

P4-siRNAs are also produced from unique genomic regions (Mosher et al. 2008; Wierzbicki et al. 2008). Some of these regions, such as *FLOWERING WAGENINGEN* (*FWA*) and the intergenic locus *siRNA02*, contain internal repeat structures that are not present elsewhere in the genome, but others contain strictly nonrepetitive sequence (Xie et al. 2004; Chan et al. 2006; Mosher et al. 2008). P4-siRNAs target repressive DNA methylation to some of these loci, indicating they are functionally similar to repetitive p4-siRNA loci (Mosher et al. 2008; Wierzbicki et al. 2008). It is intriguing to speculate that these unique, intergenic regions of p4-siRNA production might be the precursors of repetitive p4-siRNA loci by forming safe havens for mobile genetic elements. Insertion of a TE into active chromatin might be deleterious and selected against through evolution. Insertion of a TE into a region of repressed chromatin would be neutral and could become fixed in a population.

3.3 P4-siRNA Loci in Maize

Deep sequencing of small RNA populations in wild type and the maize *RDR2* mutant (*mop1*) demonstrates that p4-siRNAs are also produced from repetitive regions of the maize genome (Nobuta et al. 2008). Interestingly, many p4-siRNA loci in maize generate both 24 and 22 nt siRNAs. Production of 22 nt siRNA is prominent at high copy elements such as LTR retrotransposons, but is also found at low-copy DNA-based TEs. Synthesis of 24 nt siRNAs requires Pol IV, RDR2, and a DRD1/CLSY homolog, and therefore, these molecules appear to be canonical p4-siRNAs (Hale et al. 2007; Nobuta et al. 2008; Erhard et al. 2009). The 22 nt siRNAs are RDR2-independent, but might require Pol IV or other p4-siRNA biosynthesis enzymes (Hale et al. 2007; Nobuta et al. 2008; Stonaker et al. 2009). Production of two distinct siRNA species from p4-siRNA loci is not a general feature of cereals because abundant 22 nt siRNAs are not found in barley, sorghum, wheat, or rice (Nobuta et al. 2008).

4 Regulation of p4-siRNA Expression

Although we have a significant understanding of p4-siRNA synthesis downstream of Pol IV transcription, we known surprisingly little about initiation and regulation of p4-siRNA expression. For example, it is unclear what general transcription factors are required for Pol IV activity or whether specific transcription factors recruit Pol IV in response to environmental or developmental cues. Pol IV transcripts, RDR2-generated dsRNA, and unmethylated p4-siRNAs do not accumulate in *Arabidopsis*, suggesting that Pol IV transcription is a critical and rate-limiting step in p4-siRNA biosynthesis.

Recent analysis of tissue-specific accumulation of p4-siRNAs in *Arabidopsis* has determined that many p4-siRNA loci are expressed ubiquitously, while some loci accumulate specifically in the female gametophyte and developing endosperm (Mosher et al. 2009). However, this study only analyzed p4-siRNAs expressed in floral tissue and might, therefore, have missed loci expressed specifically in other tissues or under specific environmental conditions. The presence of at least two distinct expression patterns for p4-siRNAs indicates that there are tissue-specific factors controlling Pol IV transcription or p4-siRNA stability. Undoubtedly, the unraveling of p4-siRNA transcriptional control will be a fruitful area for future research as high-throughput small RNA sequencing increases in ease and popularity.

4.1 Feedback Reinforces p4-siRNA Expression

An interesting aspect of p4-siRNA expression is feedback between p4-siRNA function and synthesis – at some loci p4-siRNAs do not accumulate in *ago4* or *pol v* mutants (Zilberman et al. 2004; Kanno et al. 2005; Pontier et al. 2005; Qi et al. 2006;

Havecker et al. 2010). In fission yeast, a stable feedback loop exists whereby AGO1 cleavage of nascent transcripts recruits an RDR complex that generates further siRNAs (Moazed 2009). This is an attractive model for p4-siRNA feedback; however, p4-siRNA accumulation is also disrupted in methyltransferase mutants, indicating that Pol V transcripts do not contribute directly to p4-siRNA production (Zilberman et al. 2004; Pontier et al. 2005; Johnson et al. 2008). Instead, DNA methylation targeted by AGO4 and Pol V is required to recruit Pol IV and initiate further p4-siRNA production.

Feedback between DNA methylation and p4-siRNA synthesis does not occur at all p4-siRNA-generating loci and can vary in degree from causing a slight increase in p4-siRNA expression to being required for all p4-siRNA accumulation (Pontier et al. 2005; Mosher et al. 2008). The extent of feedback is unrelated to the level of DNA methylation directed by p4-siRNAs – some loci lose DNA methylation and retain p4-siRNA expression, while other loci lose p4-siRNA expression without changes in DNA methylation (Mosher et al. 2008). P4-siRNAs can direct varied chromatin modifications (see Sect. 6.2), and these might work in combination to generate the diversity of feedback responses observed at p4-siRNA loci (Mosher et al. 2008).

Feedback between p4-siRNA action and synthesis is consistent with the model of p4-siRNAs as agents to maintain asymmetric DNA methylation (Law and Jacobsen 2010). Because CHH methylation (where H = A, T, or C) is not passed to both strands during DNA replication, it must be repeatedly targeted by *de novo* methyltransferases. Similarly, modified histones are randomly distributed to newly replicated DNA and interspersed with unmodified nucleosomes. P4-siRNAs produced before S phase might be the signal to ensure that the correct methylation patterns are deposited on both sister chromatids. Correctly modified chromatin can then recruit Pol IV for production of further p4-siRNAs in preparation for the next round of replication. However, if p4-siRNA-directed chromatin modifications are erased, a p4-siRNA independent pathway exists to faithfully reestablish p4-siRNA production although details of this mechanism are unknown (Zhang et al. 2007).

4.2 Genomic Imprinting of p4-siRNAs

Feedback between p4-siRNA action and synthesis might explain the unique p4-siRNA expression pattern detected in *Arabidopsis* endosperm. Although this tissue is a product of fertilization and contains chromosomes from both maternal and paternal parents, only p4-siRNAs matching maternal chromosomes accumulate (Mosher et al. 2009). By contrast, transacting siRNAs are expressed from both chromosomes. Because p4-siRNAs levels rise dramatically after fertilization, maternal-specific p4-siRNAs cannot be a result of maternal carry-over, but rather must result from parent-of-origin-dependent gene expression, or genetic imprinting. Maternal-specific p4-siRNA expression in the endosperm requires expression of p4-siRNAs in the female gametophyte, indicating that p4-siRNA action on maternal chromosomes before fertilization feeds back to recruit Pol IV after fertilization (Mosher et al. 2009).

Maternal-specific expression of p4-siRNAs correlates with reduced DNA methylation of maternal chromosomes in the developing endosperm (Gehring et al. 2009; Hsieh et al. 2009), hinting that in the female gametophyte p4-siRNAs might direct DNA demethylation.

5 Localization of p4-siRNAs and the Pol IV/V Pathway Machinery

The Pol IV/V pathway is sometimes described as a "nuclear silencing" pathway due to its role in modifying DNA and chromatin. Transient expression in *Nicotiana benthamiana* of fluorescently labeled AGO4 and DCL3 demonstrated that these proteins accumulate exclusively in the nucleus and supported the model of the Pol IV/V pathway as strictly nuclear (Xie et al. 2004). Subsequent analysis focused solely on the nucleus and described distinct subnuclear localization of all Pol IV/V pathway components (detailed below) (Li et al. 2006; Pontes et al. 2006). However several recent lines of evidence indicate that p4-siRNAs are capable of cell-to-cell movement, indicating that a cytoplasmic phase must occur.

5.1 Subnuclear Localization

When viewed by immunofluorescence, Pol IV accumulates in the nucleus in a punctate pattern including localization at the 5S rDNA repeat loci and the nucleolus organizer regions (Pontes et al. 2006). Other speckles in the nucleus might represent genomic regions with a high concentration of p4-siRNA generating loci. Immunoflorescent signals for Pol V colocalize with Pol IV at these nucleoplasmic foci, and chromatin immunoprecipitation indicates that Pol V and AGO4 are present at p4-siRNA-generating genomic regions (Pontes et al. 2006; Wierzbicki et al. 2008, 2009). These data are consistent with a model where Pol IV and Pol V both transcribe p4-siRNA loci and AGO4 associates with nascent Pol V transcripts (Wierzbicki et al. 2008, 2009).

Surprisingly, immunoflorescent analysis of other proteins within the Pol IV/V pathway indicates that transit between subnuclear compartments is important for p4-siRNA synthesis or maturation (Li et al. 2006; Pontes et al. 2006). In addition to somewhat diffuse nucleoplasmic localization, RDR2 displays a distinctive crescent-shaped accumulation around the nucleolus and a "nucleolar dot" within the nucleolus (Pontes et al. 2006). Similarly, DCL3, AGO4, and Pol V exhibit scattered nucleoplasmic fluorescence and a strong signal within the nucleolus (Li et al. 2006; Pontes et al. 2006). Accumulation of RDR2 and DCL3 is lost in Pol IV mutants, and DCL3 accumulation is also disrupted by an *rdr2* mutation. AGO4 and Pol V consistently display immunofluorescence throughout the nucleus, but do not accumulate in the nucleolar dot in *rdr2* or *dcl3* mutant backgrounds, indicating that accumulation at this site is dependent on p4-siRNA production (Pontes et al. 2006).

Within the nucleolar dot AGO4 colocalizes with SmD3, a marker for Cajal bodies (Li et al. 2006). Cajal bodies are sites of RNA processing and might also be the site for siRNA production, maturation, and loading into AGO effector complexes (Li et al. 2006; Pontes et al. 2006). More recent evidence has indicated that AGO4 and Pol V also form nucleolar bodies that are distinct from Cajal bodies (Li et al. 2008). The role of these "AB" bodies in p4-siRNA biogenesis or function is unknown.

5.2 Cytoplasmic Localization

Evidence that silencing moves systemically in plants and animals predates the discovery of small RNAs, yet until recently the mobile signal was unknown (Voinnet and Baulcombe 1997; Dunoyer et al. 2010a, b; Molnar et al. 2010). Grafting in *Arabidopsis* indicates that all sizes of small RNA are mobile and capable of function at distal tissues (Dunoyer et al. 2010a, b; Molnar et al. 2010). Interestingly, in grafts between wild-type scions and Pol IV mutant rootstocks, nearly wild-type levels of p4-siRNAs were detected in roots, indicating that mobile p4-siRNAs account for a large proportion of p4-siRNAs in a wild-type cell (Molnar et al. 2010). This long-distance movement occurs through the plant vasculature system and is more efficient from photosynthetic source to sink, indicating that p4-siRNAs accumulate in the cytoplasm of phloem companion cells (Dunoyer et al. 2010b; Molnar et al. 2010).

Further evidence for cytoplasmic accumulation of p4-siRNAs is found in the developing seed. Analysis of methylation patterns in the two products of fertilization – the embryo and the endosperm – indicated that p4-siRNAs produced in the endosperm target DNA methylation in the embryo (Mosher and Melnyk 2010). TEs that generate p4-siRNAs are extensively demethylated by DEMETER (DME) demethylase in the central cell and endosperm (Gehring et al. 2009; Hsieh et al. 2009). In the embryo these sites display enhanced CHH methylation, the hallmark of p4-siRNA-directed DNA methylation. Because p4-siRNA expression is specific to the endosperm, these data indicate that demethylation of the central cell causes endosperm expression of p4-siRNAs, which then move into the developing embryo and hypermethylate TEs (Mosher and Melnyk 2010). Similar to phloem loading, p4-siRNA movement between endosperm and embryo must involve cytoplasmic localization, although direct evidence demonstrating this is currently lacking.

An outstanding question in cell-to-cell and long-distance movement of p4-siRNAs is whether the small RNA molecules are bound to AGOs. SiRNA duplexes are small enough to pass through nuclear pores and plasmodesmata, but AGO proteins would require active transport. There is evidence that 21 nt siRNA duplexes are more mobile than AGO-bound siRNAs when bombarded into *Arabidopsis* cells, but there is also evidence that endogenous p4-siRNAs are mobile while bound to AGO9 (Dunoyer et al. 2010b; Olmedo-Monfil et al. 2010). AGO9 accumulates in the cytoplasm of somatic cells surrounding the female gametophyte, and loss of AGO9 causes reactivation of TEs within the gametophyte (Olmedo-Monfil et al. 2010). These data indicate that p4-siRNAs from somatic tissue are capable of silencing

TEs in the gametophyte, perhaps by cell-to-cell movement in association with AGO9 (Olmedo-Monfil et al. 2010).

6 Molecular Functions of p4-siRNAs

Pol IV and p4-siRNAs trigger *de novo* methylation at asymmetric cytosines and are associated with other molecular processes, including DNA demethylation and histone modification. It is unclear whether p4-siRNAs target a single enzymatic activity that is modulated by locus-specific factors or if p4-siRNAs partner with multiple enzymes to target different chromatin modifications at each locus. As high-throughput sequencing technologies develop and it becomes easier to generate genome-wide methylation data from various mutants, our understanding of p4-siRNA-directed chromatin modification will undoubtedly improve.

6.1 RNA-Directed DNA Methylation

DNA methylation induced by dsRNA (RdDM) was first observed in the presence of a viral dsRNA and was later identified as a general feature of 24 nt siRNAs (Wassenegger et al. 1994; Mette et al. 2000). RdDM is characterized by methylation of asymmetric cytosines (CHH, where H = A, T, or C) by DOMAINS REARRANGED METHYLTRANSFERASE 2 (DRM2) (Law and Jacobsen 2010). Methylation in this context is "*de novo*" because it must be reestablished after each cell division. P4-siRNAs produced from methylated DNA before replication can act after replication to faithfully reproduce DNA methylation patterns (Law and Jacobsen 2010).

RdDM occurs when p4-siRNAs, or 24 nt Pol IV-independent siRNAs generated from transcribed inverted repeats, bind to AGO4 and associate with nascent Pol V transcripts (Zilberman et al. 2004; Kanno et al. 2008; Dunoyer et al. 2010a). AGO4 recruits the *de novo* methyltransferase DRM2 to methylate cytosines in the CHH context (where H = A, T, or C), the hallmark of RdDM (Fig. 1) (Law and Jacobsen 2010). DRM2 activity requires INVOLVED IN DE NOVO 2 (IDN2), a protein with an XS/XH RNA-binding domain that might recognize the duplex formed by the siRNA/Pol V transcript (Ausin et al. 2009). RdDM also requires two homologs of the *Drosophila* histone methyltransferase Su(var)3–9, SUVH2 and SUVH9. These proteins bind methylated DNA through their SRA domains and assist DRM2 in an unknown way (Johnson et al. 2008).

Many genomic regions undergo RdDM, including TEs, rDNA repeats, and genic sequences, and loss of methylation causes release of transcriptional gene silencing at some regions (Hamilton et al. 2002; Zilberman et al. 2003; Xie et al. 2004). Each RdDM region generates p4-siRNAs that function in *cis*; however, transgene systems demonstrate the capability of 24 nt siRNAs to directed RdDM in *trans*, making it possible that one gene can influence the methylation of other alleles or homologs (Mette et al. 2000; Zhang et al. 2007; Mosher et al. 2008).

Genome-wide analysis of methylation and siRNA accumulation demonstrates that p4-siRNA-producing loci are highly enriched for regions of DNA methylation, indicating that RdDM is a common function of p4-siRNAs (Henderson et al. 2006; Kasschau et al. 2007; Mosher et al. 2008). However, not all p4-siRNA regions exhibit high DNA methylation, and some regions are methylated independently of the Pol IV/V pathway (Mosher et al. 2008).

6.2 Chromatin Modification and Remodeling

P4-siRNAs are also associated with changes in histone methylation, either directly or as a consequence of changes in DNA methylation. Plants with mutations that disrupt p4-siRNA biosynthesis or action have decreased histone H3 lysine 9 dimethylation (H3K9me2) at some p4-siRNA loci (Zilberman et al. 2003; Xie et al. 2004; Wierzbicki et al. 2008). Methylation of H3K9 is the canonical mark for silent heterochromatin and is the chromatin modification targeted by siRNAs in *Schizosaccharomyces pombe* (Moazed 2009). Monomethylation of histone H3 lysine 27 (H3K27me1), another mark of silent chromatin, is also decreased in the absence of p4-siRNAs at some loci (Huettel et al. 2006; Wierzbicki et al. 2009). Concurrent with loss of silent chromatin marks, increases in histone acetylation and H3 lysine 4 methylation are detected when the Pol IV/V pathway is disrupted (Zilberman et al. 2003; Xie et al. 2004; Huettel et al. 2006). Broad changes in heterochromatin are also detected cytologically in the absence of p4-siRNAs. Pol IV mutations display disassociation of nucleolar organizing regions, decondensation of chromocenters, and dispersal of heterochromatic foci (Onodera et al. 2005).

An underexplored chromatin change that can be directed by p4-siRNAs is nucleosome remodeling. Tandem repeats upstream of the maize *b1* locus generate p4-siRNAs capable of transcriptional gene silencing (Sidorenko et al. 2009). Sensitivity to DNase I differs between active and silent alleles at this locus, indicating that nucleosome remodeling might be triggered by p4-siRNAs (Stam et al. 2002; Arteaga-Vazquez and Chandler 2010). Similar changes in nucleosome positioning have yet to be investigated in *Arabidopsis*, but the involvement of two SWI/SNF nucleosome remodeling proteins makes it possible that nucleosome movement is an aspect of p4-siRNA function (Kanno et al. 2004; Smith et al. 2007).

It is unclear whether chromatin modifications are the cause or the consequence of RdDM at p4-siRNA loci. Interplay between DNA and histone methylation is well established for CHG methylation and H3K9 dimethylation and a similar feedback system might occur at p4-siRNA loci (Law and Jacobsen 2010). However, RdDM-independent chromatin modification is also detected at some p4-siRNA loci. These regions have no detectable changes in DNA methylation in the presence of p4-siRNAs, yet the actions of both Pol IV and Pol V are required for wild-type accumulation of p4-siRNAs (Mosher et al. 2008). This indicates that a histone modification or nucleosome rearrangement occurs in response to p4-siRNA action at these loci and feeds back to recruit Pol IV for further p4-siRNA synthesis. Importantly, this modification occurs in the absence of RdDM, indicating that p4-siRNAs are capable of direct chromatin modification.

6.3 DNA Demethylation

Interestingly, in addition to DNA and histone methylation, p4-siRNAs are also associated with DNA demethylation. P4-siRNAs indirectly affect DNA demethylation by upregulating the DNA glycosylase/demethylase *REPRESSOR OF SILENCING 1* (*ROS1*) (Huettel et al. 2006; Penterman et al. 2007). ROS1 and its homologs DEMETER-LIKE 2 (DML2) and DEMETER-LIKE 3 (DML3) excise 5-methyl cytosine at hundreds of genomic sites, including some undergoing RdDM. ROS1 activity might be upregulated by the Pol IV/V pathway to generate negative feedback and dampen RdDM activity (Penterman et al. 2007). A fourth member of the ROS1 family, DME, is expressed specifically in the central cell of the female gametophyte, where it demethylates p4-siRNA-producing regions (Gehring et al. 2009; Hsieh et al. 2009).

It is unclear how DNA glycosylases/demethylases in the ROS1 family are targeted to specific genomic regions, but ROS1 activity requires the RNA-binding protein REPRESSOR OF SILENCING 3 (ROS3) (Zheng et al. 2008). ROS1 and ROS3 physically interact, and ROS3 binds 24 nt siRNAs, hinting that ROS1 and related demethylases might be targeted to genomic regions by p4-siRNAs (Zheng et al. 2008). Further evidence for p4-siRNA-directed DNA demethylation is found at intergenic loci that exhibit increased DNA methylation when either Pol IV or Pol V functions are lost (Pontier et al. 2005; Mosher et al. 2008). This methylation pattern is in contrast to other p4-siRNA loci that display no change or loss of methylation in similar genetic backgrounds.

7 Biological Functions of p4-siRNAs

Although mutations in the Pol IV/V pathway eliminate the majority of plant small RNAs, these mutations have no overt phenotypes in *Arabidopsis*, making it difficult to determine the biological role of p4-siRNAs. Careful molecular characterization, detailed analysis of specific tissues, and study of other plant species indicate several biological systems that require p4-siRNAs. Some of these processes, such as TE silencing, control of gene expression, and remethylation of DNA, are associated with stabilizing and buffering the genome; other processes, including rDNA silencing, paramutation, and genetic imprinting, involve the interaction between different genomes or alleles.

7.1 Silencing of Transposable Elements

Transposable elements (TEs) were the first endogenous sources of p4-siRNAs to be discovered, leading to the hypothesis that the Pol IV/V pathway evolved to defend the genome from mobile elements (Hamilton et al. 2002). Introduction of the tobacco

TNT1 retrotransposon into *Arabidopsis* supports this hypothesis, as *TNT1* proliferation attracts Pol IV and silences the element through RdDM (Perez-Hormaeche et al. 2008). Recent deep sequencing of small RNAs also supports the genome-defense model by identifying thousands of TEs that generate p4-siRNAs (Kasschau et al. 2007; Zhang et al. 2007; Mosher et al. 2008). All classes of TE produce p4-siRNAs, including LTR and non-LTR retrotransposons, DNA transposons, and helitrons, yet very few elements reactivate transcription when p4-siRNAs and corresponding CHH methylation are lost (Kanno et al. 2005; Huettel et al. 2006; Mosher et al. 2008; Yokthongwattana et al. 2010). By contrast, many TEs are derepressed in the *decreased dna methylation 1* (*ddm1*) mutant, indicating that *DDM1* and not Pol IV is the primary defense against mobile elements (Lippman et al. 2004).

It is unclear why p4-siRNAs are produced from so many silent TEs. Pol IV might represent a second line of defense against TE mobility, or p4-siRNAs might be critical to suppress TE activity in a few specific cells, such as the germ line. AGO9 and its associated p4-siRNAs accumulate specifically in the somatic cells surrounding the megaspore mother cell (MMC). Loss of AGO9 causes reactivation of some TE reporters in the MMC; however, *ago9* mutants do not display phenotypes associated with TE mobility (Havecker et al. 2010; Olmedo-Monfil et al. 2010).

The role of p4-siRNAs in TE silencing is also studied in maize, an organism with a significantly higher TE content. *MuDR* is an autonomous maize LTR retrotransposon of the *Mutator* (*Mu*) type encoding two proteins – *mudrA*, a transposase, and *mudrB*, a protein required for genome insertion (Lisch et al. 2002). *MuDR* is silenced by *MuKiller* (*MuK*), a *MuDR* insertion carrying an inverted repeat that generates siRNAs and targets DNA methylation at *MuDR* (Slotkin et al. 2005). *Mop1* (*RDR2*) is required to maintain *MuK*-triggered methylation of *MuDR* but does not affect methylation of all TEs in the maize genome (Lisch et al. 2002). Consistent with this finding, many *Mu*-like TEs are upregulated in *mop1*, while a large number of other TE families are downregulated (Jia et al. 2009).

Interestingly, demethylation of *MuDR* occurs immediately after introduction into the *mop1* background, but transcriptional reactivation of *mudrA* does not occur for several generations, indicating that *Mop1* has two distinct roles in *Mu* element silencing – one affecting DNA methylation and one affecting transcriptional repression (Woodhouse et al. 2006). A third distinct pathway must also exist to repress *MuDR* activity because *mudrB*, an enzyme critical for TE mobility, remains silent in the *mop1* background (Woodhouse et al. 2006). Similarly, *Rmr1* (*DRD1*-like) is required for both initiation and maintenance of *MuK*-triggered methylation, but *rmr1* mutations do not induce TE mobility even after ten generations of inbreeding (Hale et al. 2009).

7.2 *Regulating Gene Expression*

Because the *Arabidopsis* genome is extremely compact, most p4-siRNA loci, including those overlapping TEs, fall within potential promoter regions and might impact gene expression. However loss of p4-siRNAs is associated with altered proximal gene expression in only a few cases, including upregulation of the floral regulator *FWA* and downregulation of the glycosylase/demethylase *ROS1* (Pontier et al. 2005; Huettel et al. 2006; Kasschau et al. 2007; Penterman et al. 2007). Upregulation of *FWA* probably causes the subtle delay in flowering observed in Pol IV and Pol V mutants grown in short days (Pontier et al. 2005; Ream et al. 2009). Delayed flowering is also observed in maize mutants defective for p4-siRNA biosynthesis, indicating that p4-siRNA regulation of flowering could be a distantly conserved trait (Arteaga-Vazquez and Chandler 2010).

Although there are very few *Arabidopsis* genes misexpressed after loss of p4-siRNAs, many genes are misregulated in maize *mop1* (*rdr2*) mutants (Jia et al. 2009). Of 6,000 genes analyzed, 24% were misexpressed in the shoot apical meristem of *mop1* mutants. Strikingly, two thirds of these genes were downregulated, indicating that p4-siRNAs might play a role in gene activation rather than gene silencing (Jia et al. 2009). However, chromatin modifying enzymes were significantly overrepresented among all misregulated genes, suggesting that some of the detected changes in gene expression are indirect (Jia et al. 2009). Misregulation of chromatin modification enzymes in the meristem might also account for the pleiotropic phenotypes observed in maize Pol IV/V pathway mutations (Arteaga-Vazquez and Chandler 2010). A similar tissue-specific analysis in *Arabidopsis* might discover a greater role for the Pol IV/V pathway in gene expression.

7.3 Remethylation of DNA

Another biological process to which p4-siRNAs have been attributed is remethylation of DNA after methylation loss (Teixeira et al. 2009). Because the majority of DNA methylation is symmetric and maintained by DNA methyltransferases that recognize hemimethylated sites, most methylation is not established *de novo* (Law and Jacobsen 2010). Loss of symmetric DNA methylation through *dna methyltransferase 1* (*met1*) or *decreased dna methylation 1* (*ddm1*) mutations can, therefore, generate genetically stable changes in methylation that result in mutant alleles known as epialleles (Kankel et al. 2003). However, not all genomic regions will generate stable epialleles. After a temporary loss of DDM1 function, loci that produce p4-siRNAs slowly regain DNA methylation over several generations (Teixeira et al. 2009). This remethylation requires Pol IV/V pathway components and occurs at all cytosine contexts, not only the CHH methylation traditionally associated with RdDM (Teixeira et al. 2009). By faithfully reestablishing methylation patterns, the Pol IV/V pathway might buffer the genome from temporary changes in DNA methylation activity.

An interesting aspect of p4-siRNA-directed remethylation is that methylation is reestablished gradually over several generations (Teixeira et al. 2009). Because many p4-siRNAs are expressed specifically in the developing endosperm, remethylation might be triggered by p4-siRNAs that move from endosperm to embryo. Indeed, hypermethylation of TE sequences is detected in the embryo compared to vegetative tissues, indicating that RdDM is elevated during embryogenesis (Hsieh et al. 2009). P4-siRNA loci might be demethylated in the central cell to increase p4-siRNA expression in the endosperm and enhance p4-siRNA mediated methylation in the embryo (Hsieh et al. 2009). Originally conceived as a method to ensure suppression of TEs in the embryo, this process would also aid remethylation of the genome.

7.4 Genomic Imprinting

Demethylation of maternal p4-siRNA loci in the endosperm also has implications regarding genomic imprinting, or parent-of-origin dependent gene expression. Genomic imprinting occurs only in placental mammals and flowering plants, where it is primarily restricted to the developing endosperm (Garnier et al. 2008). Binary genomic imprinting, where one parent's allele is completely silent, is a rare expression pattern known to occur at 11 genes in *Arabidopsis*, with an estimated 50 genes imprinted genome-wide (Gehring et al. 2006; Fitz Gerald et al. 2009).

Although p4-siRNAs are rarely produced from genic sequences, all known imprinted genes produce p4-siRNAs (Gehring et al. 2006). In some cases, such as the maternal-expressed *FWA* and *MEDEA*, the imprinted gene contains a repeated sequence that generates p4-siRNAs (Chan et al. 2006). In other imprinted genes, such as *MATERNALLY EXPRESSED PAB C-TERMINAL* (*MPC*), p4-siRNAs are produced from unique sequence elements (Tiwari et al. 2008). Interestingly, p4-siRNA expression is also imprinted, with accumulation in the endosperm occurring specifically from maternal chromosomes (Mosher et al. 2009). In *Arabidopsis*, demethylation of maternal alleles by DME glycosylase/demethylase is required for correct expression of both paternally and maternally expressed imprinted genes (Garnier et al. 2008). Recent genome-wide methylation profiling of endosperm demonstrates that p4-siRNA-producing loci are also maternally demethylated by DME (Gehring et al. 2009; Hsieh et al. 2009).

The coincident demethylation of p4-siRNA loci and demethylation-dependent imprinting of the few genes overlapping p4-siRNAs seem to indicate a causal relationship (Mosher 2010). Because expression of p4-siRNAs in the central cell is required for expression of p4-siRNAs in the endosperm, it is possible that p4-siRNAs direct DME demethylation in the central cell to mark maternal chromosomes in the endosperm (Mosher et al. 2009). However, Pol IV pathway mutations do not display the phenotypes associated with misregulation of imprinted genes. It is, therefore, more plausible that p4-siRNA imprinting occurs in parallel with imprinting of protein-coding genes (Mosher 2010). Because many imprinted genes affect growth and development of the endosperm and embryo, evolution of genomic imprinting is believed to arise from parental conflict over allocation of scarce maternal resources among half-sibling progeny (Garnier et al. 2008). Whether the Pol IV/V pathway has a subtle affect on seed development is yet to be determined.

7.5 rDNA Silencing

P4-siRNAs are also associated with selectively silencing the rDNA repeats of one parent to maintain the correct rRNA dosage in interspecific hybrids (Preuss et al. 2008). This process is called nucleolar dominance and is distinguished from genomic imprinting because the parental origin of the alleles does not influence silencing – rather, the rDNA array of one species is preferentially silenced in hybrids (Tucker et al. 2010). In the allotetraploid *Arabidopsis suecica*, rDNA from *Arabidopsis thaliana* is silenced, while rDNA from *Arabidopsis arenosa* is active (Tucker et al. 2010). Nucleolar dominance is associated with repressive DNA methylation and histone deacetylation, and in *A. suecica* this process requires components of the Pol IV/V pathway, including *RDR2*, *DCL3*, and *DRM2* (Preuss et al. 2008). RNAi-mediated knockdown of the Pol IV/V pathway in *A. suecica* triggers accumulation of rRNA from both species, but the resulting plants are phenotypically normal.

A subset of rDNA genes are also silenced in nonhybrid species to regulate rRNA dosage throughout development (Tucker et al. 2010). In *Arabidopsis*, Pol IV immunolocalizes at rDNA, p4-siRNAs are generated from the intergenic and transcribed spacers of rDNA arrays, and DRM2 methylates a segment of the rDNA repeat (Pontes et al. 2006; Preuss et al. 2008). However, RdDM is not sufficient for silencing at these regions – the histone deacetylase HDA6 is also required. Loss of *hda6* causes spurious transcription of normally silent repeats, even in the presence of copious siRNA production and RdDM (Earley et al. 2010).

7.6 Paramutation

The biological process most clearly regulated by Pol IV and p4-siRNAs is paramutation, the epigenetic conversion of one allele state into another (Arteaga-Vazquez and Chandler 2010). At the *b1* locus in maize, there are high- and low-expressing alleles. Although high expression is generally dominant over low expression, some low-expression alleles (termed paramutagenic) appear dominant over high-expression alleles. Strikingly, when combined in an F1, paramutagenic alleles convert high-expression (paramutable) alleles into low expression, paramutagenic alleles. The newly converted paramutagenic allele is indistinguishable from the original paramutagenic allele and is able to convert other paramutable alleles in subsequent generations (Arteaga-Vazquez and Chandler 2010). Genetic mapping of the region required for paramutation at *b1* identified a set of repeats 100 kb upstream of the start of the *b1* coding sequence (*b1TR*) (Stam et al. 2002). Both paramutagenic and paramutable alleles have seven copies of the *b1TR*, which function both for interallele communication and as an enhancer of transcription at *b1* (Arteaga-Vazquez and Chandler 2010).

Genetic screens for paramutation defects at *b1* and a second paramutable locus, *Pl1*, uncovered a number of components of the Pol IV/V pathway. The first Pol IV pathway mutation identified in maize was *Mediator of paramutation 1* (*Mop1*), which encodes an ortholog of *RDR2* (Lisch et al. 2002; Alleman et al. 2006). Genetic screens have also uncovered a SWI/SNF nucleosome remodeling protein with similarity to *DRD1* and *CLSY*, called *Required to maintain repression 1* (*Rmr1*) (Hale et al. 2007). *Rmr6* encodes the largest subunit of Pol IV and *Rmr7/Mop2* is one of three isoforms of the second-largest subunit of Pol IV (Erhard et al. 2009; Sidorenko et al. 2009; Stonaker et al. 2009).

These mutations suggest the obvious model that p4-siRNAs are produced from the silent, methylated paramutagenic allele and operate *in trans* to methylate and silence the paramutable allele. Once methylated, this allele also becomes a substrate for Pol IV and gains the ability to paramutate other alleles (Arteaga-Vazquez and Chandler 2010). However, this model is not supported by recent data. Although there are subtle differences in DNA methylation and nucleosome occupancy between paramutagenic and paramutable alleles at *b1TR*, these repeats are transcribed equally in paramutagenic and paramutable alleles, and both alleles produce equivalent levels of p4-siRNAs (Stam et al. 2002; Alleman et al. 2006; Arteaga-Vazquez et al. 2010). A role for the Pol IV/V pathway and p4-siRNAs in paramutation is undeniable, but the details of that involvement remain to be uncovered.

8 Evolution of Transcriptional Gene Silencing

In *S. pombe*, siRNAs direct heterochromatinization of centromeric repeats, indicating that transcriptional silencing of repetitive DNA might be an ancient form of RNA silencing. Fission yeast Pol II transcribes centromeric repeats to initiate siRNA production and generates scaffold transcripts that bind siRNA/AGO complexes in addition to generating mRNAs (Moazed 2009). Duplication of several Pol II subunits in plants allowed these roles to be split among Pol II, Pol IV, and Pol V (Pikaard et al. 2008). This duplication and subfunctionalization began early in the evolution of terrestrial plants and was accelerated during the evolution of flowering (Luo and Hall 2007; Mosher 2010).

8.1 Evolution of Pol IV and Pol V

Pol IV and Pol V are closely related to Pol II in the catalytic region, but diverge significantly in the C-terminal domain (CTD), a region of Pol II associated with cotranscriptional processes such as 5′ capping, intron splicing, and termination (Phatnani and Greenleaf 2006; Luo and Hall 2007). The Pol II CTD is composed of a heptapeptide repeat that is heavily phosphorylated to modulate activity. The CTDs of Pol IV and V are unrelated to Pol II at the sequence level; however, the Pol V CTD contains ten tandem copies of a sixteen amino-acid sequence that can be extensively phosphorylated, suggesting convergent evolution with Pol II

(Pontier et al. 2005). Both Pol IV and Pol V CTDs contain the DEFECTIVE IN CHLOROPLAST AND LEAVES (DeCL) domain, a sequence associated with rRNA processing in chloroplasts (Bellaoui et al. 2003).

Pol IV first appeared in the plant lineage through duplication of the largest subunit of Pol II (*NRPB1*) to create *NRPD1* (Luo and Hall 2007). *NRPD1* is absent from most algal species but present in the green algae *Charales*, the ancestors of land plants (Lewis and McCourt 2004). *NRPD2*, the second-largest Pol IV subunit, arose from its Pol II homolog in basal land plants and is present in mosses and liverworts (Luo and Hall 2007). Several smaller Pol II subunits also duplicated and specialized for Pol IV function, including *NRPD4*, *NRPD7*, and *NRPD9* (He et al. 2009a; Ream et al. 2009). *NRPE1*, the largest subunit of Pol V, is not found in basal plants, but is present in angiosperms and arose through a duplication of *NRPD1* (Luo and Hall 2007). Pol V also contains the unique subunits *NRPE5* and *NRPE7*, and shares *NRPE4* and *NRPE9* with Pol IV (He et al. 2009a; Huang et al. 2009; Lahmy et al. 2009; Ream et al. 2009). Pol IV and V subunits continue to duplicate and subfunctionalize in flowering plants. Rice contains two copies each of *NRPD1* and *NRPE1* and maize contains three nonredundant copies of *NRPD2* (Luo and Hall 2007; Sidorenko et al. 2009; Stonaker et al. 2009).

8.2 Transcriptional Gene Silencing Basal Plants

While p4-siRNAs are abundant in many angiosperms, 24 nt siRNAs are a minor component of nonflowering plant transcriptomes (Axtell et al. 2007; Dolgosheina et al. 2008; Morin et al. 2008). Consistent with this, Pol V subunits have not been detected outside of angiosperms, although data for conifers are lacking (Luo and Hall 2007). Interestingly, the DeCL domain in the CTD of Pol IV is a recent addition to the enzyme and is only found in species that also contain Pol V, indicating that the Pol IV/V pathway might have different functions in basal and flowering plants (Luo and Hall 2007).

The moss *Physcomitrella patens* generates primarily 21 nt siRNAs with a small number of regions producing a mixture of 21–24 nt siRNA (Cho et al. 2008). These regions are enriched for TEs and exhibit DNA methylation, making them likely targets of transcriptional silencing via RdDM. Mutations in *PpDCL3* eliminate production of 23–24 nt siRNAs from these loci but do not affect DNA methylation or transcriptional silencing at most regions (Cho et al. 2008). However, two families of LTR retrotransposons are derepressed in *Ppdcl3*, indicating that small RNA-mediated transcriptional silencing occurs to some degree in moss (Cho et al. 2008). Similarly, although 24 nt siRNAs are rare in the conifer *Pinus contorta*, siRNAs of this length exist and match repetitive elements of the rice genome (Morin et al. 2008). This surprising result indicates that some targets of small RNA-mediated silencing might be distantly conserved.

9 Conclusions

In the last decade, we have uncovered an invisible ocean of small molecules in eukaryotes and learned that plant cells are awash in 24 nt siRNAs that are capable of transcriptional gene silencing. These molecules are linked to numerous processes including the repression of repetitive DNA, maintenance of the genome, and interactions between alleles or genomes, but their biological role is still unclear. Flowering plants have evolved two additional DNA-dependent RNA polymerases that specialize in the production and action of p4-siRNAs, further highlighting the importance of these regulators.

References

Alleman M, Sidorenko L, McGinnis KM et al (2006) An RNA-dependent RNA polymerase is required for paramutation in maize. Nature 442:295–298

Arteaga-Vazquez MA, Chandler VL (2010) Paramutation in maize: RNA mediated trans-generational gene silencing. Curr Opin Genet Dev 20:156–163

Arteaga-Vazquez M, Sidorenko L, Rabanal FA et al (2010) RNA-mediated trans-communication can establish paramutation at the b1 locus in maize. Proc Natl Acad Sci USA 107: 12986–12991

Ausin I, Mockler TC, Chory J et al (2009) IDN1 and IDN2 are required for de novo DNA methylation in *Arabidopsis thaliana*. Nat Struct Mol Biol 16:1325–1327

Axtell MJ, Snyder JA, Bartel DP (2007) Common functions for diverse small RNAs of land plants. Plant Cell 19:1750–1769

Bellaoui M, Keddie JS, Gruissem W (2003) DCL is a plant-specific protein required for plastid ribosomal RNA processing and embryo development. Plant Mol Biol 53:531–543

Bies-Etheve N, Pontier D, Lahmy S et al (2009) RNA-directed DNA methylation requires an AGO4-interacting member of the SPT5 elongation factor family. EMBO Rep 10:649–654

Borsani O, Zhu J, Verslues PE, Sunkar R, Zhu J-K (2005) Endogenous siRNAs derived from a pair of natural *cis*-antisense transcripts regulate salt tolerance in *Arabidopsis*. Cell 123:1279–1291

Cantu D, Vanzetti LS, Sumner A et al (2010) Small RNAs, DNA methylation and transposable elements in wheat. BMC Genomics 11:408

Cao XF, Jacobsen SE (2002) Role of the Arabidopsis DRM methyltransferases in de novo DNA methylation and gene silencing. Curr Biol 12:1138–1144

Chan SW-L, Zhang X, Bernatavichute YV et al (2006) Two-step recruitment of RNA-directed DNA methylation to tandem repeats. PLoS Biol 4:1923–1933

Cho SH, Addo-Quaye C, Coruh C et al (2008) *Physcomitrella patens* DCL3 is required for 22–24 nt siRNA accumulation, suppression of retrotransposon-derived transcripts, and normal development. PLoS Genet 4:e1000314

Dalmay T, Hamilton AJ, Rudd S et al (2000) An RNA-dependent RNA polymerase gene in *Arabidopsis* is required for posttranscriptional gene silencing mediated by a transgene but not by a virus. Cell 101:543–553

Daxinger L, Kanno T, Bucher E et al (2009) A stepwise pathway for biogenesis of 24-nt secondary siRNAs and spreading of DNA methylation. EMBO J 28:48–57

Ding SW, Voinnet O (2007) Antiviral immunity directed by small RNAs. Cell 130:413–426

Dolgosheina EV, Morin RD, Aksay G et al (2008) Conifers have a unique small RNA silencing signature. RNA 14:1508–1515

Dunoyer P, Brosnan CA, Schott G et al (2010a) An endogenous, systemic RNAi pathway in plants. EMBO J 29:1699–1712

Dunoyer P, Schott G, Himber C et al (2010b) Small RNA duplexes function as mobile silencing signals between plant cells. Science 328:912–916

Earley KW, Pontvianne F, Wierzbicki AT et al (2010) Mechanisms of HDA6-mediated rRNA gene silencing: suppression of intergenic Pol II transcription and differential effects on maintenance versus siRNA-directed cytosine methylation. Gene Dev 24:1119–1132

Erhard KF, Stonaker JL, Parkinson SE et al (2009) RNA polymerase IV functions in paramutation in *Zea mays*. Science 323:1201–1205

Fitz Gerald JN, Hui PS, Berger F (2009) Polycomb group-dependent imprinting of the actin regulator AtFH5 regulates morphogenesis in *Arabidopsis thaliana*. Development 136:3399–3404

Gao ZH, Liu HL, Daxinger L et al (2010) An RNA polymerase II- and AGO4-associated protein acts in RNA-directed DNA methylation. Nature 465:106–109

Garnier O, Laoueille-Duprat S, Spillane C (2008) Genomic imprinting in plants. Epigenetics 3:14–20

Gehring M, Huh JH, Hsieh TF et al (2006) DEMETER DNA glycosylase establishes MEDEA polycomb gene self-imprinting by allele-specific demethylation. Cell 124:495–506

Gehring M, Bubb KL, Henikoff S (2009) Extensive demethylation of repetitive elements during seed development underlies gene imprinting. Science 324:1447–1451

Haag JR, Pontes O, Pikaard CS (2009) Metal A and metal B sites of nuclear RNA polymerases Pol IV and Pol V are required for siRNA-dependent DNA methylation and gene silencing. PLoS ONE 4:e4110

Hale CJ, Stonaker JL, Gross SM et al (2007) A novel snf2 protein maintains trans-generational regulatory states established by paramutation in maize. PLoS Biol 5:2156–2165

Hale CJ, Erhard KF Jr, Lisch D et al (2009) Production and processing of siRNA precursor transcripts from the highly repetitive maize genome. PLoS Genet 5:e1000598

Hamilton AJ, Voinnet O, Chappell L et al (2002) Two classes of short interfering RNA in RNA silencing. EMBO J 21:4671–4679

Havecker ER, Wallbridge LM, Hardcastle TJ et al (2010) The Arabidopsis RNA-directed DNA methylation argonautes functionally diverge based on their expression and interaction with target loci. Plant Cell 22:321–334

He XJ, Hsu YF, Pontes O et al (2009a) NRPD4, a protein related to the RPB4 subunit of RNA polymerase II, is a component of RNA polymerases IV and V and is required for RNA-directed DNA methylation. Gene Dev 23:318–330

He XJ, Hsu YF, Zhu S et al (2009b) A conserved transcriptional regulator is required for RNA-directed DNA methylation and plant development. Gene Dev 23:2717–2722

He XJ, Hsu YF, Zhu SH et al (2009c) An Effector of RNA-Directed DNA methylation in Arabidopsis Is an ARGONAUTE 4-and RNA-binding protein. Cell 137:498–508

Henderson IR, Zhang XY, Lu C et al (2006) Dissecting *Arabidopsis thaliana* DICER function in small RNA processing, gene silencing and DNA methylation patterning. Nat Genet 38:721–725

Herr AJ, Jensen MB, Dalmay T et al (2005) RNA polymerase IV directs silencing of endogenous DNA. Science 308:118–120

Hsieh TF, Ibarra CA, Silva P et al (2009) Genome-wide demethylation of *Arabidopsis* endosperm. Science 324:1451–1454

Huang LF, Jones AME, Searle I et al (2009) An atypical RNA polymerase involved in RNA silencing shares small subunits with RNA polymerase II. Nat Struct Mol Biol 16:91–93

Huettel B, Kanno T, Daxinger L et al (2006) Endogenous targets of RNA-directed DNA methylation and Pol IV in *Arabidopsis*. EMBO J 25:2828–2836

Jia Y, Lisch DR, Ohtsu K et al (2009) Loss of RNA-dependent RNA polymerase 2 (RDR2) function causes widespread and unexpected changes in the expression of transposons, genes, and 24-nt small RNAs. PLoS Genet 5:e1000737

Johnson LM, Bostick M, Zhang X et al (2007) The SRA methyl-cytosine-binding domain links DNA and histone methylation. Curr Biol 17:379–384

Johnson LM, Law JA, Khattar A et al (2008) SRA-domain proteins required for DRM2-mediated de novo DNA methylation. PLoS Genet 4:e1000280

Kankel MW, Ramsey DE, Stokes TL et al (2003) Arabidopsis MET1 cytosine methyltransferase mutants. Genetics 163:1109–1122

Kanno T, Mette MF, Kreil DP et al (2004) Involvement of putative SNF2 chromatin remodeling protein DRD1 in RNA-directed DNA methylation. Curr Biol 14:801–805

Kanno T, Huettel B, Mette MF et al (2005) Atypical RNA polymerase subunits required for RNA-directed DNA methylation. Nat Genet 37:761–765

Kanno T, Bucher E, Daxinger L et al (2008) A structural-maintenance-of-chromosomes hinge domain-containing protein is required for RNA-directed DNA methylation. Nat Genet 40:670–675

Kanno T, Bucher E, Daxinger L et al (2010) RNA-directed DNA methylation and plant development require an IWR1-type transcription factor. EMBO Rep 11:65–71

Kasschau KD, Fahlgren N, Chapman EJ et al (2007) Genome-wide profiling and analysis of *Arabidopsis* siRNAs. PLoS Biol 5:0479–0493

Katiyar-Agarwal S, Morgan R, Dahlbeck D et al (2006) A pathogen-inducible endogenous siRNA in plant immunity. Proc Natl Acad Sci USA 103:18002–18007

Katiyar-Agarwal S, Gao S, Vivian-Smith A et al (2007) A novel class of bacteria-induced small RNAs in Arabidopsis. Gene Dev 21:3123–3134

Lahmy S, Pontier D, Cavel E et al (2009) PolV(PolIVb) function in RNA-directed DNA methylation requires the conserved active site and an additional plant-specific subunit. Proc Natl Acad Sci USA 106:941–946

Law JA, Jacobsen SE (2010) Establishing, maintaining and modifying DNA methylation patterns in plants and animals. Nat Rev Genet 11:204–220

Law JA, Ausin I, Johnson LM et al (2010) A protein complex required for polymerase V transcripts and RNA-directed DNA methylation in Arabidopsis. Curr Biol 20:951–956

Lewis LA, McCourt RM (2004) Green algae and the origin of land plants. Am J Bot 91:1535–1556

Li JJ, Yang ZY, Yu B et al (2005) Methylation protects miRNAs and siRNAs from a 3′-end uridylation activity in Arabidopsis. Curr Biol 15:1501–1507

Li CF, Pontes O, El-Shami M et al (2006) An ARGONAUTE4-containing nuclear processing center colocalized with cajal bodies in *Arabidopsis thaliana*. Cell 126:93–106

Li CF, Henderson IR, Song L et al (2008) Dynamic regulation of ARGONAUTE4 within multiple nuclear bodies in *Arabidopsis thaliana*. PLoS Genet 4:e27

Lippman Z, Gendrel AV, Black M et al (2004) Role of transposable elements in heterochromatin and epigenetic control. Nature 430:471–476

Lisch D, Carey CC, Dorweiler JE et al (2002) A mutation that prevents paramutation in maize also reverses mutator transposon methylation and silencing. Proc Natl Acad Sci USA 99:6130–6135

Lu C, Kulkarni K, Souret FF et al (2006) MicroRNAs and other small RNAs enriched in the Arabidopsis RNA-dependent RNA polymerase-2 mutant. Genome Res 16:1276–1288

Luo J, Hall BD (2007) A multistep process gave rise to RNA polymerase IV of land plants. J Mol Evol 64:101–112

Matranga C, Tomari Y, Shin C et al (2005) Passenger-strand cleavage facilitates assembly of siRNA into Ago2-containing RNAi enzyme complexes. Cell 123:607–620

Mette MF, Aufsatz W, van der Winden J et al (2000) Transcriptional silencing and promoter methylation triggered by double-stranded RNA. EMBO J 19:5194–5201

Moazed D (2009) Small RNAs in transcriptional gene silencing and genome defence. Nature 457:413–420

Molnar A, Melnyk CW, Bassett A et al (2010) Small silencing RNAs in plants are mobile and direct epigenetic modification in recipient cells. Science 328:872–875

Morel JB, Godon C, Mourrain P et al (2002) Fertile hypomorphic ARGONAUTE (ago1) mutants impaired in post-transcriptional gene silencing and virus resistance. Plant Cell 14:629–639

Morin RD, Aksay G, Dolgosheina E et al (2008) Comparative analysis of the small RNA transcriptomes of *Pinus contorta* and *Oryza sativa*. Genome Res 18:571–584

Mosher RA (2010) Maternal control of Pol IV-dependent siRNAs in Arabidopsis endosperm. New Phytol 186:358–364

Mosher RA, Melnyk CW (2010) siRNAs and DNA methylation: seedy epigenetics. Trends Plant Sci 15:204–210

Mosher RA, Schwach F, Studholme D et al (2008) PolIVb influences RNA-directed DNA-methylation independently of its role in siRNA biogenesis. Proc Natl Acad Sci USA 105:3145–3150

Mosher RA, Melnyk CW, Kelly KA et al (2009) Uniparental expression of PolIV-dependent siRNAs in developing endosperm of *Arabidopsis*. Nature 460:283–286

Nobuta K, Lu C, Shrivastava R et al (2008) Distinct size distribution of endogeneous siRNAs in maize: evidence from deep sequencing in the mop1-1 mutant. Proc Natl Acad Sci USA 105:14958–14963

Olmedo-Monfil V, Duran-Figueroa N, Arteaga-Vazquez M et al (2010) Control of female gamete formation by a small RNA pathway in Arabidopsis. Nature 464:628–632

Onodera Y, Haag JR, Ream T et al (2005) Plant nuclear RNA polymerase IV mediates siRNA and DNA methylation-dependent heterochromatin formation. Cell 120:613–622

Penterman J, Uzawa R, Fischer RL (2007) Genetic interactions between DNA demethylation and methylation in Arabidopsis ([OA]). Plant Physiol 145:1549–1557

Perez-Hormaeche J, Potet F, Beauclair L et al (2008) Invasion of the *Arabidopsis* genome by the tobacco retrotransposon Tnt1 is controlled by reversible transcriptional gene silencing. Plant Physiol 147:1264–1278

Phatnani HP, Greenleaf AL (2006) Phosphorylation and functions of the RNA polymerase IICTD. Gene Dev 20:2922–2936

Pikaard CS, Haag JR, Ream T et al (2008) Roles of RNA polymerase IV in gene silencing. Trends Plant Sci 13:390–397

Plasterk RHA (2002) RNA silencing: the genome's immune system. Science 296:1263–1265

Pontes O, Li CF, Nunes PC et al (2006) The Arabidopsis chromatin-modifying nuclear siRNA pathway involves a nucleolar RNA processing center. Cell 126:79–92

Pontier D, Yahubyan G, Vega D et al (2005) Reinforcement of silencing at transposons and highly repeated sequences requires the concerted action of two distinct RNA polymerases IV in *Arabidopsis*. Gene Dev 19:2030–2040

Preuss SB, Costa-Nunes P, Tucker S et al (2008) Multimegabase silencing in nucleolar dominance involves siRNA-directed DNA methylation and specific methylcytosine-binding proteins. Mol Cell 32:673–684

Qi YJ, He XY, Wang XJ et al (2006) Distinct catalytic and non-catalytic roles of ARGONAUTE4 in RNA-directed DNA methylation. Nature 443:1008–1012

Ream TS, Haag JR, Wierzbicki AT et al (2009) Subunit compositions of the RNA-silencing enzymes Pol IV and Pol V reveal their origins as specialized forms of RNA polymerase II. Mol Cell 33:192–203

Ruvkun G (2008) The perfect storm of tiny RNAs. Nat Med 14:1041–1045

Sidorenko L, Dorweiler JE, Cigan AM et al (2009) A dominant mutation in mediator of paramutation2, one of three second-largest subunits of a plant-specific RNA polymerase, disrupts multiple siRNA silencing processes. PLoS Genet 5:e1000725

Slotkin RK, Freeling M, Lisch D (2005) Heritable transposon silencing initiated by a naturally occurring transposon inverted duplication. Nat Genet 37:641–644

Smith LM, Pontes O, Searle L et al (2007) An SNF2 protein associated with nuclear RNA silencing and the spread of a silencing signal between cells in Arabidopsis. Plant Cell 19:1507–1521

Stam M, Belele C, Dorweiler JE et al (2002) Differential chromatin structure within a tandem array 100 kb upstream of the maize b1 locus is associated with paramutation. Gene Dev 16:1906–1918

Stefani G, Slack FJ (2008) Small non-coding RNAs in animal development. Nat Rev Mol Cell Biol 9:219–230

Stonaker JL, Lim JP, Erhard KF et al (2009) Diversity of Pol IV function is defined by mutations at the maize *rmr7* locus. PLoS Genet 5:e1000706

Sunkar R, Chinnusamy V, Zhu JH et al (2007) Small RNAs as big players in plant abiotic stress responses and nutrient deprivation. Trends Plant Sci 12:301–309

Teixeira FK, Heredia F, Sarazin A et al (2009) A role for RNAi in the selective correction of DNA methylation defects. Science 323:1600–1604

Till S, Lejeune E, Thermann R et al (2007) A conserved motif in Argonaute-interacting proteins mediates functional interactions through the Argonaute PIWI domain. Nat Struct Mol Biol 14:897–903

Tiwari S, Schulz R, Ikeda Y et al (2008) MATERNALLY EXPRESSED PAB C-TERMINAL, a novel imprinted gene in *Arabidopsis*, encodes the conserved C-terminal domain of polyadenylate binding proteins. Plant Cell 20:2387–2398

Tucker S, Vitins A, Pikaard CS (2010) Nucleolar dominance and ribosomal RNA gene silencing. Curr Opin Cell Biol 22:351–356

Voinnet O, Baulcombe DC (1997) Systemic signalling in gene silencing. Nature 389:553

Wassenegger M, Heimes S, Riedel L et al (1994) Rna-directed de-novo methylation of genomic sequences in plants. Cell 76:567–576

Wierzbicki AT, Haag JR, Pikaard CS (2008) Noncoding transcription by RNA polymerase Pol IVb/Pol V mediates transcriptional silencing of overlapping and adjacent genes. Cell 135:635–648

Wierzbicki AT, Ream TS, Haag JR et al (2009) RNA polymerase V transcription guides ARGONAUTE4 to chromatin. Nat Genet 41:630–634

Woodhouse MR, Freeling M, Lisch D (2006) Initiation, establishment, and maintenance of heritable *MuDR* transposon silencing in maize are mediated by distinct factors. PLoS Biol 4:339

Xie Z, Johansen LK, Gustafson AM et al (2004) Genetic and functional diversification of small RNA pathways in plants. PLoS Biol 2:642–652

Yokthongwattana C, Bucher E, Caikovski M et al (2010) MOM1 and Pol-IV/V interactions regulate the intensity and specificity of transcriptional gene silencing. EMBO J 29:340–351

Yu B, Yang Z, Li J et al (2005) Methylation as a crucial step in plant microRNA biogenesis. Science 307:932–935

Zhang X, Henderson IR, Lu C et al (2007) Role of RNA polymerase IV in plant small RNA metabolism. Proc Natl Acad Sci USA 104:4536–4541

Zheng X, Zhu J, Kapoor A et al (2007) Role of Arabidopsis AGO6 in siRNA accumulation, DNA methylation and transcriptional gene silencing. EMBO J 26:1–11

Zheng X, Pontes O, Zhu J et al (2008) ROS3 is an RNA-binding protein required for DNA demethylation in Arabidopsis. Nature 455:1259–1262

Zheng B, Wang Z, Li S et al (2009) Intergenic transcription by RNA polymerase II coordinates Pol IV and Pol V in siRNA-directed transcriptional gene silencing in Arabidopsis. Gene Dev 23:2850–2860

Zheng ZM, Xing Y, He XJ et al (2010) An SGS3-like protein functions in RNA-directed DNA methylation and transcriptional gene silencing in Arabidopsis. Plant J 62:92–99

Zhu QH, Spriggs A, Matthew L et al (2008) A diverse set of microRNAs and microRNA-like small RNAs in developing rice grains. Genome Res 18:1456–1465

Zilberman D, Cao X, Jacobsen SE (2003) *ARGONAUTE4* control of locus specific siRNA accumulation and DNA and histone methylation. Science 299:716–719

Zilberman D, Cao X, Johansen LK et al (2004) Role of *Arabidopsis ARGONAUTE4* in RNA-directed DNA methylation triggered by inverted repeats. Curr Biol 14:1214–1220

RNAi-Mediated Protection Against *Citrus Tristeza Virus* in Transgenic *Citrus* Plants

Nuria Soler, Carmen Fagoaga, Sinda Chiibi, Carmelo López, Pedro Moreno, Luis Navarro, Ricardo Flores, and Leandro Peña

Contents

Abstract RNA silencing is a sequence-specific mechanism of inhibition of gene expression evolutionarily conserved in most eukaryotes. RNA interference (RNAi), a technology based on the use of double-stranded RNA (dsRNA) to trigger RNA silencing, can be achieved in plants by genetic transformation with sense and antisense cDNAs derived from target viral sequences separated by an intron (intron-hairpin constructs). Upon transcription, the resulting hairpin RNA transcript usually acts as a strong inducer of RNA silencing. This strategy has been widely used to produce virus-resistant transgenic plants. *Citrus tristeza virus* (CTV) (genus *Closterovirus*, family *Closteroviridae*) is the causal agent of the most devastating viral diseases of citrus trees in the world. It only infects phloem-associated tissues of *Citrus* species and relatives within the family Rutaceae. CTV is one of the largest and most complex plant RNA viruses, with a single-stranded, plus-sense RNA genome of 19.3 kb, organized in 12 open reading frames (ORFs), potentially coding for at least 17 polypeptides, and two 5′ and 3′ unstranslated regions (UTRs). Replication and expression of the genomic RNA results in more than 30 different

L. Peña (✉), N. Soler, C. Fagoaga, S. Chiibi, P. Moreno, and L. Navarro
Centro de Protecci6n Vegetal y Biotecnología, Instituto Valenciano de Investigaciones Agrarias (IVIA), Carretera Moncada-Náquera, Km 4.5, 46113 Moncada, Valencia, Spain
e-mail: lpenya@ivia.es

C. López and R. Flores
Instituto de Biología Molecular y Celular de Plantas (CSIC-UPV), Universidad Politécnica, Avenida de los Naranjos, 46022 Valencia, Spain

V.A. Erdmann and J. Barciszewski (eds.), *Non Coding RNAs in Plants*, RNA Technologies, DOI 10.1007/978-3-642-19454-2_27,

plus and minus RNA species, as well as their corresponding dsRNA forms. Concomitantly, citrus hosts have developed a strong antiviral response through RNA silencing, as inferred from the high level of virus-derived siRNAs observed in infected tissues. As a counterdefense, CTV encodes at least three silencing suppressor proteins acting intracellularly and/or intercellularly to overcome antiviral defense. Under these circumstances, searching for RNAi-induced resistance against CTV in transgenic citrus plants becomes a real challenge. We have used intron-hairpin constructs targeting several viral regions, with our present interest focusing on one or the three CTV genes encoding silencing suppressors, or on conserved domains important for viral replication and encapsidation.

Keywords Closterovirus • CTV • Intron-hairpin • RNA silencing • Silencing suppressor

1 Citrus and Tristeza

Citrus is the most economically important fruit tree crop worldwide, with more than 110 million tons in 2009 produced in more than 7.6 million Ha (FAO 2010). Besides the genus *Citrus* that belongs to the family *Rutaceae*, subfamily *Aurantioideae*, citrus crops include two other genera of economic importance: *Poncirus* and *Fortunella*. There are only three true *Citrus* species: citron (*C. medica* L.), mandarin (*C. reticulata* Blanco), and pummelo (*C. grandis* (L.) Osb.). Since the three ancestral species reproduce only sexually and are original from the same geographical area, Southeast Asia, several hybridizations among these species followed by frequent somatic mutations generated the major citrus types of economic importance, including sweet oranges (*C. sinensis* (L.) Osb.), mandarins (*C. deliciosa* Ten., *C. tangerina* Hort. Ex Tan., *C. clementina* Hort. Ex Tan., *C. nobilis* André non Lour., *C. unshiu* (Mak.) Marc., etc.), lemons (*C. limon* (L.) Burm. f.), and limes (*C. aurantifolia* (Christm.) Swing.; *C. latifolia* Tan.; *C. limonia* (L.) Osb.). Grapefruit is a much more recent type, first described in Barbados in 1750, and originated from a natural hybridization between pummelo and sweet orange probably followed by introgression with pummelo (Nicolosi et al. 2000).

Another important citrus genotype is sour orange (*C. aurantium* L.). Its use marked the origin of modern citriculture around mid-nineteenth century when bud grafting scion varieties onto sour orange rootstock became a universal practice, mainly due to its resistance to *Phytophthora* spp., but also because of its excellent agronomic attributes, particularly its capacity to induce high fruit yield and quality, and its adaptability to all soils. However, the massive use of sour orange was the basis of the dramatic outcome of several tristeza epidemics that in the last 80 years have caused the death of more than 100 million sweet orange, mandarin, and grapefruit scion varieties propagated on this rootstock in Argentina, Brazil, California, Florida, Israel, Venezuela, and Spain. Moreover, the disease keeps spreading into new areas, either by propagation of infected buds or by different aphid species, mainly *Toxoptera citricida* (Kildaky) and *Aphis gossypii* (Glover). This situation forced the progressive replacement of sour orange by tristeza-resistant or tolerant rootstocks, which do not perform as well as sour orange.

Tristeza is a bud union disease and refers to the decline of most scion types propagated on sour orange or lemon rootstocks. Its causal agent is *Citrus tristeza virus* (CTV), a member of the genus *Closterovirus*, family *Closteroviridae*. In nature, CTV infects only citrus species and relatives within the family *Rutaceae*, subfamily *Aurantioideae*, and within these hosts it invades only phloem-associated tissues. Indeed, tristeza decline results from virus-induced necrosis of the rootstock phloem just below the bud union.

Additionally, virulent CTV isolates cause stem pitting on some sweet orange, grapefruit, and lime scion varieties regardless of the rootstock, resulting in reduced vigor, yield, and fruit quality. A third syndrome observed by biological indexing but rarely in the field, is characterized by stunting, small pale or yellow leaves, reduced root system, and sometimes complete growth cessation of sour orange, grapefruit, or lemon seedlings (Moreno et al. 2008).

2 CTV Genome

CTV virions are long flexuous particles (2,000 × 11 nm) formed by two coat proteins (CPs) that encapsidate a plus-sense single-stranded RNA (ssRNA) of approximately 19.3 kb organized in 12 open reading frames (ORFs) potentially encoding at least 17 protein products and two 5′ and 3′ unstranslated regions (UTRs) of 107 and 273 nt, respectively (Karasev et al. 1995). The 5′-proximal ORF 1a encodes a polyprotein containing two papain-like protease domains, plus methyltransferase-like and helicase-like domains. ORF 1b encodes a putative RNA-dependent RNA polymerase that it is thought to be expressed by a +1 frameshift mechanism (Karasev et al. 1995). The ten ORFs located in the 3′ moiety of the genome are expressed through a set of 3′-co-terminal subgenomic (sg) mRNAs (Hilf et al. 1995), and they encode the CPs of 25 and 27 kDa (p25 and p27), which encapsidate about 97 and 3% of the genome, respectively, and proteins p33, p6, p65, p61, p18, p13, p20, and p23 (Pappu et al. 1994; Karasev et al. 1995). Both CPs, together with p65 and p61, are involved in virion assembly (Satyanarayana et al. 2000). Additionally, p27 has been shown to initiate encapsidation of the genomic RNA from its 5′ end (Satyanarayana et al. 2004). The p20 protein accumulates in amorphous inclusion bodies of CTV-infected cells (Gowda et al. 2000). The small hydrophobic p6 may operate as a membrane anchor (Satyanarayana et al. 2000) and its homologue in the *Beet yellows virus* (BYV), also of the genus *Closterovirus*, is a movement protein (Peremyslov et al. 2004). The protein p23, a RNA binding protein with a Zn finger domain (López et al. 2000), which regulates the asymmetrical accumulation of the plus and minus strands during RNA replication (Satyanarayana et al. 2002), has no homologue counterpart in other closteroviruses, and likely it is the determinant of the seedling yellows syndrome (Albiach-Martí et al. 2010). When ectopically expressed in transgenic citrus plants, p23 induces aberrations resembling CTV symptoms (Ghorbel et al. 2001). Moreover, p23, p20, and p25 act as RNA silencing suppressors in *Nicotiana tabacum* and *N. benthamiana* plants, being p25 intercellular, p23 intracellular, and p20 both inter- and intracellular silencing suppressors (Lu et al. 2004). The function of p33, p13, and p18 remains unknown.

3 Resistance to CTV

Breeding for resistance to CTV in scion varieties has been largely unsuccessful, mainly due to the complex reproductive biology of citrus. Most genotypes are facultative apomictic, which means that adventitious embryos are generated directly from maternal nucellar cells, precluding the development of the less vigorous zygotic embryos. Although this is the basis for propagation of citrus rootstocks, apomixis seriously limits the recovery of sexual progeny populations in breeding programs. Some important genotypes have total or partial pollen and/or ovule sterility and cannot be used as parents in breeding programs, and there are many cases of cross- and self-incompatibility. Additionally, citrus have a long juvenile period, and most species need at least 5 years to start flowering in subtropical areas, and usually several years more to achieve fully mature characteristics. Citrus types have high heterozygosity, and there is a lack of basic knowledge about how the most important horticultural traits are inherited some of which, as those related to fruit quality and maturity time, show quantitative inheritance. All these features together with their large plant size have greatly impeded genetic improvement of citrus through conventional breeding.

The only successful results from breeding for CTV resistance come from the first recorded artificial hybridization, carried out by Swingle and Webber in Florida in 1893 in relation to disease problems. Since a severe freeze destroyed most of the seedlings, they decided to use the cold-hardy relative *Poncirus trifoliata* as a parent in crosses aimed at incorporating higher cold tolerance to *Citrus* scions. None of the progeny trees combined cold hardiness with good fruit quality. However, the Carrizo and Troyer citranges (sweet orange × *P. trifoliata*) and the Swingle citrumelo (grapefruit × *P. trifoliata*) hybrid rootstocks resulting from these crosses are widely used by the most important citrus industries due to their tolerance to CTV-induced decline.

Resistance to CTV in *P. trifoliata* has been attributed to a single dominant locus (*Ctv*), which has been thoroughly characterized and mapped (Yoshida 1985, 1993; Gmitter et al. 1996; Mestre et al. 1997; Fang et al. 1998). Because of the complex genetics of citrus, it is extremely difficult to introduce this resistance locus into citrus varieties by conventional breeding. However, its cloning has been attempted in several laboratories (Deng et al. 2001; Yang et al. 2003). A BAC library developed from "Pomeroy" *P. trifoliata*, homozygous for *Ctv*, was used for a 1.2-Mb genome walk spanning the region between *Ctv*-flanking markers. Sequencing of a set of four overlapping BAC clones in this region, using shotgun sequencing, and resolution of their ends by sequencing of additional BAC clones further localized *Ctv* to a 282-kb region comprising 22 predicted genes (Yang et al. 2003). Sequence analysis of the *Ctv* locus in this region identified 61 simple sequence repeats (SSRs) that were used to further narrow down the locus in the *Poncirus* genome to 121 kb, comprising ten genes. Each of these genes has been individually cloned in *Agrobacterium*-based binary vectors and used to transform susceptible grapefruit varieties in Erik Mirkov's laboratory (Rai 2006). The transgenic lines expressing any of the ten candidate genes were susceptible to CTV infection, suggesting that more than one gene in the locus is involved in resistance to CTV or that the role of other genomic loci has been overlooked.

In general, citrus genotypes are hosts for CTV, but there is a wide diversity in their response against viral infection, which is strain-dependent. While Mexican lime is a symptomatic host of most CTV strains, which show systemic infection and reach relatively high virus titers, only the aggressive strains induce symptoms in sweet orange and grapefruit systemic hosts. On the contrary, most mandarins are nonsymptomatic but systemic hosts, and sour oranges and lemons hardly tolerate virus spread and show very low CTV titer. Five quantitative trait loci (QTLs) have been associated with the partial resistance of sour orange to certain CTV strains (Asins et al. 2004). There are also species, as pummelo, in which the resistant/susceptible response depends on specific cultivar/strain combinations (Garnsey et al. 1987), with a single dominant gene for resistance to CTV called *Ctv2* having been mapped in "Chandler" pummelo (Fang and Roose 1999).

Nowadays, the only possibility to protect susceptible commercial varieties from severe CTV isolates is classical cross protection with mild CTV strains. This approach has prevented the low yield and small-sized fruits of "Pera" sweet orange in Sao Paulo, Brazil (Costa and Müller 1980) and "Marsh" grapefruit in South Africa (van Vuuren et al. 1993). In both cases, protection was based on the search for field CTV isolates causing asymptomatic infection in the citrus cultivar of interest, and the use of budwood from those trees to propagate preinoculated plants for new plantings. The mild strain would then protect the new plants against infections with upcoming severe CTV strains. Considering that "Pera" sweet orange is the main variety of the citrus industry in Sao Paulo, one of the largest in the world, and that all new nursery "Pera" plants are infected with a mild isolate as part of the commercial production system, it is clear that in the last 40 years cross protection has been a tool of paramount importance for the success of the Brazilian citriculture. However, attempts to apply the same strategy in other citrus areas, such as Australia, Japan, or Florida, have failed because protection was highly dependent on the citrus scion variety, the prevailing CTV strains and the environmental conditions, with the afforded protection being only temporary.

4 CTV and RNA Silencing

In recent times, RNA silencing has arisen as a mechanism that explains many cases of genetic plant defence against viral infections and cross protection between closely related virus strains (Covey et al. 1997; Ratcliff et al. 1997, 1999). RNA silencing is induced by double-stranded RNA (dsRNA) or highly structured ssRNA and results in sequence-specific ssRNA degradation through generation of 21–24 nt short interfering RNAs (siRNAs) by RNaseIII-like enzymes called Dicer (Bernstein et al. 2001). The siRNAs are loaded into an RNA-induced silencing complex (RISC) and one strand of the siRNA is degraded, while the other primes the Argonaute (AGO)-containing RISC active complex that then targets for cleavage of ssRNA sharing sequence similarity with the inducing dsRNA (Hammond et al. 2000). Accumulation of high levels of dsRNA derived from subgenomic RNAs, a characteristic feature of CTV replication (Hilf et al. 1995), could trigger RNA silencing. Moreover, Fagoaga et al. (2006) and Ruiz-Ruiz et al. (2011) have observed high accumulation of viral-specific siRNAs in CTV-infected Mexican lime plants, indicating a strong natural RNA silencing-mediated antiviral response. It is then tempting to speculate that in cross protection the siRNAs generated by the mild CTV isolate could prevent subsequent infections by severe strains through targeting and degradation of their highly homologous genomic and subgenomic RNAs. Even within highly divergent CTV variants, it is easy to find large portions along the CTV genome with more than 24-nt identical.

In principle, pathogen-derived resistance (PDR) would be based on the same mechanism as cross protection, but it could represent a better and more predictable strategy to achieve durable resistance to CTV in citrus. In PDR, introduction and expression in plants of pathogen genes in a dysfunctional form, in excess, or at the wrong developmental stage, could interfere with the pathogen life cycle having minimal effects on the host, and providing resistance to infection (Sanford and Johnston 1985). Since the first demonstration of virus-derived resistance in transgenic plants by using the CP gene of *Tobacco mosaic virus* (TMV) (Abel et al. 1986), this strategy has been proved to be widely applicable. Two of the most successful examples of CP-mediated protection against viruses in plants refer to fruit tree species, namely, the "SunUp" transgenic papaya resistant to *Papaya ringspot virus* (PRSV), which is commercialized in USA since 1999 (Gonsalves 1998; Ming et al. 2008), and the "Honeysweet" transgenic plum resistant to *Plum pox virus* (PPV), which has been approved for commercial release in USA in 2009 (Marshall 2010). In both cases, transgenic plants were generated with the aim of overexpressing the CP transgene, thus getting an ectopic overaccumulation of the corresponding protein that would reencapsidate the challenging virus soon after initiating infection. However, only unique transgenic events with several CP transgene insertions showed strong resistance to the challenging viruses (Scorza et al. 1994; Ravelonandro et al. 1997), particularly in field trial assays (Hily et al. 2004). Molecular analysis of these events revealed very low transgene mRNA levels and undetectable CP accumulation (Scorza et al. 1994). More detailed analyses showed constitutive transgene methylation and transgene-derived siRNA accumulation (Scorza et al. 2001; Hily et al. 2005). Since all these features are characteristic of RNA silencing, it is clear today that random integration of several foreign DNA copies during genetic transformation was responsible for the resistance in those transgenic lines that never showed transgene CP accumulation but expressed strong RNA silencing.

To generate CP-mediated resistance to CTV in transgenic citrus, we incorporated into Mexican lime a transgene derived from the p25 CP gene from severe and mild

CTV strains, with more than 40 independent transgenic lines being produced. Mexican lime was chosen as a citrus model because it is very sensitive to CTV and the potential resistance could be easily tested by evaluating leaf cupping, vein clearing, and stem pitting symptoms in the greenhouse within a few months after challenging. When eight to ten propagations of each transgenic line were graft- and aphid-inoculated with CTV, two types of response to viral challenge were observed: most lines developed CTV symptoms similar to those of the nontransgenic controls, but six of the 40 lines exhibited resistance against the virus. Resistance consisted of a fraction of the propagations, ranging from 10 to 33%, that were immune to CTV, with the rest showing a significant delay in virus accumulation and symptom onset in at least three consecutive flushes (about 1 year) after inoculation (Domínguez et al. 2002a). These results were reproduced with four of the six transgenic lines in an additional challenge experiment in which propagations were again graft-inoculated with CTV (Fig. 1). Since several transgenic lines showed complex T-DNA insertions and undetectable p25 accumulation, but consistent partial resistance, an RNA silencing mechanism was proposed to explain the protection against CTV. In a new set of experiments, Mexican lime plants were transformed with untranslatable versions of the p25 gene, but, in general, the protection achieved was rapidly overcome by the challenging virus (Domínguez et al. 2002b).

Fig. 1 Response to graft inoculation with a severe CTV isolate exhibited by representative Mexican lime plants transformed with the *p25* CP transgene (*left*) and with an empty vector construct (*right*). One year after challenging, the *p25*-transgenic plant shows mild vein clearing symptoms, while the control plant is affected by intense leaf distortion and vein corking

The 3′-terminal gene of CTV codes for p23, which is an RNA-binding protein (López et al. 2000) involved in regulating the balance of plus and minus RNA strands during replication (Satyanarayana et al. 2002). Considering its regulatory role, we decided to explore whether overexpression of this protein in transgenic citrus could interfere with CTV replication and provide resistance. More than 50 transgenic lines of Mexican lime were generated carrying the *p23* gene or a truncated version thereof. Unexpectedly, constitutive expression of p23 induced phenotypic aberrations resembling symptoms incited by CTV in nontransgenic Mexican lime, whereas transgenic plants expressing the p23-truncated version were normal. The onset of CTV-like symptoms in p23-transgenic plants was associated with the accumulation of p23, and its level paralleled symptom intensity (Ghorbel et al. 2001). Overexpression of p23 in other CTV-susceptible citrus genotypes, including sweet and sour orange, and the CTV-resistant *P. trifoliata*, also led to CTV-like symptoms that were not visible when these plants were transformed with a truncated p23 version (Fagoaga et al. 2005). Altogether, these results indicate that p23 is an important CTV pathogenicity determinant that interferes with plant development in *Citrus* species and relatives. In the course of the experiments to incorporate *p23* into Mexican lime, three out of 60 lines carrying the *p23* gene of the severe strain CTV T36, and two out of 20 lines carrying *p23* from the mild strain CTV T317, were visually normal and developed as controls transformed with the empty vector or nontransformed. These five lines displayed characteristics typical of RNA silencing: multiple copies and methylation of the silenced transgene, low levels of the corresponding mRNA, and accumulation of *p23*-specific siRNAs. When propagations of these silenced lines were graft- or aphid-inoculated with CTV, some were immune, since they neither expressed symptoms nor accumulated virions or viral RNA. Other propagations were moderately resistant because they showed delayed expression of leaf symptom and attenuated stem pitting compared to the controls. The susceptible propagations showed normal symptom expression and elevated virus titer, as the empty-vector controls (Fagoaga et al. 2006).

A characteristic of the *p25* and *p23* transgene-mediated RNA silencing is that vegetative propagations from the same transgenic line showed different responses against CTV, with some propagations being immune and others susceptible to viral challenge. This variable response among clonal transformants carrying viral-derived transgenes indicates that factors other than the genetic background of the transgenic plant, such as environmental conditions or the developmental stage, play a key role in RNA-mediated resistance.

5 RNAi Against CTV

To further enhance RNA silencing against CTV, a new set of constructs was designed from a highly conserved region (>90% homology) comprising part of the *p23* gene and the 3′-UTR, which is critical for recognition by the replicase complex. Mexican lime plants were transformed with the 3′-terminal 549 nucleotides of the CTV genome in sense, antisense, and intron-hairpin formats. Intron-hairpin constructs are strong inducers of RNA interference against plant viruses (Smith et al. 2000) because, upon transcription, they generate a dsRNA molecule that acts as a highly efficient trigger for RNA silencing, leading to cleavage and degradation of target complementary viral and transgene-derived RNAs (Fire et al. 1998).

After challenge by graft inoculation, propagations from all sense, antisense, and empty-vector transgenic lines were susceptible to CTV, except for a single sense line with a complex transgene integration pattern that showed transgene-derived siRNAs in association with low levels of the transgene-derived transcript. By contrast, nine of the 30 intron-hairpin lines showed CTV resistance, with 9–56% of the propagations (depending on the line) remaining uninfected after graft inoculation and the others being susceptible. As indicated above, factors other than the genetic background of the transgenic plant, including differences in the physiological and ontological stage of individual propagations, may be critical for the efficiency of RNA silencing-mediated resistance in clonal plants. Resistance was always associated with the presence of transgene-derived siRNAs, but their level in different sense and intron-hairpin transformants was variable irrespective of the response to CTV infection. Empty-vector infected controls also accumulated high levels of siRNAs from the viral 3′-UTR, most likely derived from genomic and subgenomic dsRNAs. Indeed, CTV-infected plants accumulated 1–2 orders of magnitude more siRNAs than noninfected intron-hairpin transformants, indicating that the virus has evolved very efficient counterdefense strategies based on expression of three different silencing suppressor proteins (Lu et al. 2004), probably targeting different components of the RNA silencing pathway.

In intron-hairpin lines with single transgene integration, CTV resistance was correlated with low accumulation of the transgene-derived transcript rather than with high accumulation of transgene-derived siRNAs (López et al. 2010). As resistance could not be predicted by high transgene-derived siRNA levels, our results suggest that only a fraction of the transgene-derived siRNAs, perhaps those resulting from HEN1-mediated methylation (Yang et al. 2006; Yu et al. 2005) and/or those programming RISC (Omarov et al. 2007; Pantaleo et al. 2007), are competent for RNA silencing, with the other fraction being quickly degraded, as proposed to occur with most virus-derived siRNAs in infected plants (Qu and Morris 2005). Besides, rather than blocking the biogenesis of siRNAs, CTV silencing suppressors would prevent their loading into the AGO-containing RISC complex or proper functioning of another downstream step in the RNA silencing pathway. The p23 intracellular suppressor has characteristics in common with protein 2b of cucumoviruses and protein P0 of poleroviruses. All the three are pathogenicity factors that induce developmental aberrations when overexpressed in transgenic plants, which are reminiscent of the phenotypes of plants affected in the miRNA pathway (Fagoaga et al. 2005; Lewsey et al. 2007; Bortolamiol et al. 2007). As 2b and P0 target members of the Argonaute family, 2b by binding AGO1 directly to prevent the RISC complex from cleaving its target RNA (Zhang et al. 2006), and P0 by targeting members of the Argonaute family for degradation (Bortolamiol et al. 2007; Baumberger et al. 2007), CTV p23 might act at the same level. Recent results have shown that miRNA168, which targets AGO1, is upregulated by CTV infection in Mexican lime and other citrus genotypes (Ruiz-Ruiz et al. 2011). It will be worth testing whether p23 is actually targeting AGO1.

The success of RNA interference (RNAi) against CTV would depend on whether transgene-induced RNA silencing can substantially attenuate or block virus gene expression and, more specifically, the accumulation of three silencing suppressor proteins (López et al. 2010). To achieve this aim, we have designed a transformation vector carrying a cassette comprising the complete untranslatable versions of genes *p25*, *p20*, and *p23* plus the 3′-UTR in sense and antisense configurations, and separated by the piv2 intron of the potato (*Solanum tuberosum*) gene *st-ls1* under the control of CaMV 35S promoter and the *nopaline synthase* terminator. This construct has been used to transform Mexican lime, and the resulting transgenic lines have been propagated and challenged by graft inoculation with severe CTV strains under greenhouse conditions. Preliminary results indicate that this strategy may provide the best level of resistance against CTV achieved so far in this host (Soler et al. unpublished results) (Fig. 2).

Another RNAi construct has been designed to target sequences proximal to the 5′-UTR, as the subgenomic dsRNAs derived from this region are much less abundant than those from the 3′-half of the genome (Moreno et al. 2008) and as the siRNAs derived from the 5′-UTR accumulate at much lower levels than those from the 3′-UTR (Ruiz-Ruiz et al. 2011). However, CTV strains show similarities as low as 44% in the 5′-UTR (Gowda et al. 2003), with the corresponding sequences having been classified into three types (I, II, and III) based on intragroup sequence identity higher than 88% (López et al. 1998). Consequently, for a construct potentially silencing all known CTV strains, one fragment of at least 50 nt and 100%

Fig. 2 Response to graft inoculation with the clonal *Citrus tristeza virus* (CTV)-T36 strain of shoots from Mexican limes transformed with an empty vector control (*right*) and an intron-hairpin *p25-p20-p23*+3′-UTR (ihp) construct (*left*). The control shoot shows leaf epinasty and yellowing, while the ihp transgenic shoot remains symptomless

intragroup sequence identity was chosen for each group. The first fragment (nucleotides 25–85 from group I) comprises two stem-loop structures within the 5′-UTR that are required for virus replication (Gowda et al. 2003). The second fragment (nucleotides 105–522 from group II) covers the ORF 1a translation initiation and the first part of the papain-like protease PRO I domain. The third fragment (nucleotides 1531–1604 from group III) comprises part of the PRO I and PRO II domains of the polyprotein 1a. The three regions were RT-PCR amplified and cloned as a fragment of 588 nt that was then subcloned in a transformation plasmid in sense and antisense configurations, separated by the intron of the *pyruvate orthophosphate dikinase* gene from *Flaveria trinervia*, under the control of the 35S CaMV promoter and the *octopine synthase* terminator. This construct has been used to transform Mexican lime plants, and several transgenic lines are currently being propagated to be challenged by graft inoculation with severe CTV strains under greenhouse conditions (Chiibi et al. unpublished results).

In summary, these studies show that RNAi can be extended to CTV in its natural hosts. Whether transgenic citrus plants expressing CTV-derived sequences could be an efficient alternative to cross protection for controlling in the field CTV strains inducing stem pitting remains to be tested. So far, only partial protection to CTV has been achieved in greenhouse experiments with transgenic Mexican lime, but it should be mentioned that this experimental host allows CTV to reach very high titers compared with other citrus species, particularly sour orange, in which CTV is essentially unable to move cell-to-cell (Folimonova et al. 2008). With the aim of developing sour orange rootstocks resistant to the tristeza syndrome, we have transferred to this genotype those constructs providing some level of protection to CTV in Mexican lime. Transgenic sour orange lines carrying *p25*, *p23*, and several intron-hairpin constructs are currently being tested for resistance to decline in field trials performed in a cooperative project (with Catalina Anderson at the Estación experimental INTA-Concordia (Argentina)), in an area where the brown citrus aphid (*T. citricida*) vector and severe CTV strains are prevalent.

Acknowledgments This research is currently being supported by grants AGL2009-08052 from the Ministerio de Ciencia e Innovación, and Prometeo/2008/121 from the Generalitat Valenciana.

References

Abel PP, Nelson RS, De B et al (1986) Delay of disease development in transgenic plants that express the tobacco mosaic virus coat protein gene. Science 232:738–743

Albiach-Martí MR, Robertson C, Gowda S et al (2010) The pathogenicity determinant of *Citrus tristeza virus* causing the seedling yellows syndrome maps at the 3′-terminal region of the viral genome. Mol Plant Pathol 11:55–67

Asins MJ, Bernet GP, Ruiz C et al (2004) QTL analysis of citrus tristeza virus-citradia interaction. Theor Appl Genet 108:603–611

Baumberger N, Tsai CH, Lie M et al (2007) The Polerovirus silencing suppressor P0 targets ARGONAUTE proteins for degradation. Curr Biol 17:1609–1614

Bernstein E, Caudy AA, Hammond SM et al (2001) Role for a bidentate ribonuclease in the initiation step of RNA interference. Nature 409:363–366

Bortolamiol D, Pazhouhandeh M, Marrocco K et al (2007) The Polerovirus F box protein P0 targets ARGONAUTE1 to suppress RNA silencing. Curr Biol 17:1615–1621

Costa AS, Müller GW (1980) Tristeza control by cross protection: a U.S.-Brazil cooperative success. Plant Dis 64:538–541

Covey SN, Al-Kaff N, Lángara A et al (1997) Plants combat infection by gene silencing. Nature (London) 385:781–782

Deng Z, Huang S, Ling P et al (2001) Fine genetic mapping and BAC contig development for the citrus tristeza virus resistance gene locus in *Poncirus trifoliata* (Raf.). Mol Genet Genomics 265:739–747

Domínguez A, Hermoso de Mendoza A, Guerri J et al (2002a) Pathogen-derived resistance to *Citrus tristeza virus* (CTV) in transgenic Mexican lime (*Citrus aurantifolia* (Christ.) Swing.) plants expressing its p25 coat protein gene. Mol Breed 10:1–10

Domínguez A, Fagoaga C, Navarro L et al (2002b) Constitutive expression of untranslatable versions of the p25 coat protein gene of *Citrus tristeza virus* (CTV) in transgenic Mexican lime plants does not confer resistance to the virus. In: Duran-Vila N, Milne RG, da Graça JV (eds) Proceedings of the 15th conference international organization of citrus virologists. IOCV, Riverside, pp 341–344

Fagoaga C, López C, Moreno P et al (2005) Viral-like symptoms induced by the ectopic expression of the *p23* gene of *Citrus tristeza virus* are citrus specific and do not correlate with the pathogenicity of the virus strain. Mol Plant Microbe Interact 18:435–445

Fagoaga C, López C, Hermoso de Mendoza A et al (2006) Post-transcriptional gene silencing of the p23 silencing suppressor of *Citrus tristeza virus* confers resistance to the virus in transgenic Mexican lime. Plant Mol Biol 60:153–165

Fang DQ, Roose ML (1999) A novel gene conferring citrus tristeza virus resistance in *Citrus maxima* (Burm.) Merrill. HortScience 34:334–335

Fang DQ, Federici CT, Roose ML (1998) A high-resolution linkage map of the citrus tristeza virus resistance gene region in *Poncirus trifoliata* (L.) raf. Genetics 150:883–890

FAO (2010). http://www.fao.org/es/esc/common/ecg/243/es/bull2006.pdf

Fire A, Xu S, Montgomery MK et al (1998) Potent and specific genetic interference by double-stranded RNA in *Caenorhabditis elegans*. Nature 391:806–811

Folimonova SY, Folimonov AS, Tatineni S et al (2008) *Citrus tristeza virus*: survival at the edge of the movement continuum. J Virol 82:6546–6556

Garnsey SM, Barrett HC, Hutchison DJ (1987) Identification of citrus tristeza virus resistance in citrus relatives and its potential applications. Phytophylactica 19:187–191

Ghorbel R, López C, Fagoaga C et al (2001) Transgenic citrus plants expressing the citrus tristeza virus p23 protein exhibit viral-like symptoms. Mol Plant Pathol 2:27–36

Gmitter FG, Xiao SY, Huang S et al (1996) A localized linkage map of the citrus tristeza virus resistance gene region. Theor Appl Genet 92:688–695

Gonsalves D (1998) Control of *Papaya ringspot vir*us in papaya: a case study. Annu Rev Phytopathol 36:415–437

Gowda S, Satyanarayana T, Davis CL et al (2000) The *p20* gene product of *Citrus tristeza virus* accumulates in the amorphous inclusion bodies. Virology 274:246–254

Gowda S, Satyanarayana T, Ayllón MA et al (2003) The conserved structures of the 5′ nontranslated region of *Citrus tristeza* virus are involved in replication and virion assembly. Virology 317:50–64

Hammond SM, Bernstein E, Beach D et al (2000) An RNA-directed nuclease mediates post-transcriptional gene silencing in Drosophila cells. Nature 404:293–296

Hilf ME, Karasev AV, Pappu HR et al (1995) Characterization of citrus tristeza virus subgenomic RNAs in infected tissue. Virology 208:576–582

Hily JM, Scorza R, Malinowski T et al (2004) Stability of gene silencing-based resistance to *Plum pox virus* in transgenic plum (*Prunus domestica* L.) under field conditions. Transgenic Res 13:427–436

Hily JM, Scorza R, Webb K et al (2005) Accumulation of the long class of siRNA is associated with resistance to *Plum pox virus* in a transgenic woody perennial plum tree. Mol Plant Microbe Interact 18:794–799

Karasev AV, Boyko VP, Gowda S et al (1995) Complete sequence of the citrus tristeza virus RNA genome. Virology 208:511–520

Lewsey M, Robertson FC, Canto T et al (2007) Selective targeting of miRNA-regulated plant development by a viral counter-silencing protein. Plant J 50:240–252

López C, Ayllón MA, Navas-Castillo J et al (1998) Molecular variability of the 5′- and 3′-terminal regions of citrus tristeza virus RNA. Phytopathology 88:685–691

López C, Navas-Castillo J, Gowda S et al (2000) The 23-kDa protein coded by the 3′-terminal gene of *Citrus tristeza virus* is an RNA-binding protein. Virology 269:462–470

López C, Cervera M, Fagoaga C et al (2010) Accumulation of transgene-derived siRNAs is not sufficient for RNAi-mediated protection against *Citrus tristeza virus* in transgenic Mexican lime. Mol Plant Pathol 11:33–41

Lu R, Folimonov A, Shintaku M et al (2004) Three distinct suppressors of RNA silencing encoded by a 20-kb viral RNA genome. Proc Natl Acad Sci USA 101:15742–15747

Marshall A (2010) 2nd-generation GM traits progress. Nat Biotechnol 28:306

Mestre PF, Asins MJ, Carbonell EA et al (1997) New gene(s) involved in the resistance of *Poncirus trifoliata* (L.) Raf. to *Citrus tristeza virus*. Theor Appl Genet 95:691–695

Ming R, Hou S, Feng Y et al (2008) The draft genome of the transgenic tropical fruit tree papaya (*Carica papaya* Linnaeus). Nature 452:991–996

Moreno P, Ambrós S, Albiach-Martí MR et al (2008) *Citrus tristeza virus*: a pathogen that changed the course of the citrus industry. Mol Plant Pathol 9:251–268

Nicolosi E, Deng ZN, Gentile A et al (2000) Citrus phylogeny and genetic origin of important species as investigated by molecular markers. Theor Appl Genet 100:1155–1166

Omarov RT, Ciomperlik JJ, Scholthof HB (2007) RNAi-associated ssRNA-specific ribonucleases in Tombusvirus P19 mutant-infected plants and evidence for a discrete siRNA-containing effector complex. Proc Natl Acad Sci USA 104:1714–1719

Pantaleo V, Szittya G, Burgyan J (2007) Molecular bases of viral RNA targeting by viral small interfering RNA-programmed RISC. J Virol 81:3797–3806

Pappu HR, Karasev AV, Anderson EJ et al (1994) Nucleotide sequence and organization of eight 3′ open reading frames of the citrus tristeza closterovirus genome. Virology 199:35–46

Peremyslov VV, Pan YW, Dolja VV (2004) Movement protein of a closterovirus is a type III integral transmembrane protein localized to the endoplasmic reticulum. J Virol 78:3704–3709

Qu F, Morris TJ (2005) Suppressors of RNA silencing encoded by plant viruses and their role in viral infections. FEBS Lett 579:5958–5964

Rai M (2006) Refinement of the Citrus tristeza virus resistance gene (*Ctv*) positional map in *Poncirus trifoliata* and generation of transgenic grapefruit (*Citrus paradisi*) plant lines with candidate resistance genes in this region. Plant Mol Biol 61:399–414

Ratcliff F, Harrison BD, Baulcombe DC (1997) A similarity between viral defense and gene silencing in plants. Science 276:1558–1560

Ratcliff FG, MacFarlane SA, Baulcombe DC (1999) Gene silencing without DNA. RNA-mediated cross-protection between viruses. Plant Cell 11:1207–1216

Ravelonandro M, Scorza R, Bachelier JC et al (1997) Resistance of transgenic *Prunus domestica* to plum pox virus infection. Plant Dis 81:1231–1235

Ruiz-Ruiz S, Navarro B, Gisel A et al (2011) Citrus tristeza virus infection induces the accumulation of viral small RNAs (21–24-nt) mapping preferentially at the 3'-terminal region of the genomic RNA and affects the small RNA profile of sensitive hosts. Plant Mol Biol 75:607–619

Sanford JC, Johnston SA (1985) The concept of parasite-derived resistance – deriving resistance genes from the parasites own genome. J Theor Biol 113:395–405

Satyanarayana T, Gowda S, Mawassi M et al (2000) Closterovirus encoded HSP70 homolog and p61 in addition to both coat proteins function in efficient virion assembly. Virology 278:253–265

Satyanarayana T, Gowda S, Ayllón MA et al (2002) The p23 protein of *Citrus tristeza virus* controls asymmetrical RNA accumulation. J Virol 76:473–483

Satyanarayana T, Gowda S, Ayllón MA et al (2004) Closterovirus bipolar virion: evidence for initiation of assembly by minor coat protein and its restriction to the genomic RNA 5′ region. Proc Natl Acad Sci USA 101:799–804

Scorza R, Ravelonandro M, Callahan AM et al (1994) Transgenic plums (*Prunus domestica* L.) express the plum pox virus coat protein gene. Plant Cell Rep 14:18–22

Scorza R, Callahan A, Levy L et al (2001) Post-transcriptional gene silencing in plum pox virus resistant transgenic European plum containing the plum pox potyvirus coat protein gene. Transgenic Res 10:201–209

Smith NA, Singh SP, Wang MB et al (2000) Total silencing by intron-spliced hairpin RNAs. Nature 407:319–320

Van Vuuren SP, Collins RP, da Graça JV (1993) Evaluation of Citrus tristeza virus isolates for cross protection of grapefruit in South-Africa. Plant Dis 77:24–28

Yang ZN, Ye XR, Molina J et al (2003) Sequence analysis of a 282-kilobase region surrounding the citrus tristeza virus resistance gene (*Ctv*) locus in *Poncirus trifoliata* L. Raf. Plant Physiol 131:482–492

Yang Z, Ebright YW, Yu B et al (2006) HEN1 recognizes 21-24 nt small RNA duplexes and deposits a methyl group onto the 2′ OH of the 3′ terminal nucleotide. Nucleic Acids Res 34:667–675

Yoshida Y (1985) Inheritance of susceptibility to *Citrus tristeza virus* in trifoliate orange. Bull Fruit Tree Res Stn 12:17–25

Yoshida T (1993) Inheritance of immunity to *Citrus tristeza virus* of trifoliate orange in some citrus intergeneric hybrids. Bull Fruit Tree Res Stn 25:33–43

Yu B, Yang Z, Li J et al (2005) Methylation as a crucial step in plant microRNA biogenesis. Science 307:932–935

Zhang X, Yuan YR, Pei Y et al (2006) Cucumber mosaic virus-encoded 2b suppressor inhibits Arabidopsis Argonaute1 cleavage activity to counter plant defense. Genes Dev 20:3255–3268

Small RNAs for Crop Improvement: Applications and Considerations for Ecological Risk Assessments

Carol Auer

Contents

Abstract The world faces many challenges in the development of crop systems that produce more food, fiber, and biofuels while minimizing environmental impacts. Today, the central dogma of "one gene, one protein" has been supplanted by a realization that small RNA molecules have profound effects in plants and can be engineered to create novel crop traits. Crop improvement using small RNAs has potential in two broad areas: (1) modifying metabolic and biochemical pathways and (2) silencing genes in pest organisms to provide plant protection. In the USA, genetically engineered (GE) crops have been grown since 1992 with RNA-mediated

C. Auer (✉)
Department of Plant Science, University of Connecticut, Storrs, CT 06269, USA
e-mail: carol.auer@uconn.edu

V.A. Erdmann and J. Barciszewski (eds.), *Non Coding RNAs in Plants*,
RNA Technologies, DOI 10.1007/978-3-642-19454-2_28,

virus resistance and a few other traits. At present, proof-of-concept experiments have demonstrated potential for insect resistance, nematode resistance, increased nutritional value, decreased human allergens, better postharvest quality, new flower colors, and other traits. As more RNA-mediated traits are invented, regulatory agencies are faced with the task of assessing the safety of these traits for humans and the environment. In general, it appears that current ecological risk assessment (ERA) frameworks will be suitable for RNA-mediated traits. However, there are some new challenges in predicting risks prior to experimental field trials or commercial crop production. Targeted research is needed to answer key questions regarding the following: (1) environmental persistence and abundance of artificial small RNAs, (2) off-target effects within GE crops, (3) negative effects on nontarget organisms, (4) effects of genetic mutations and polymorphisms in crops and other organisms, (5) uncertainty in risk analysis, and (6) technologies to rapidly identify and quantify GE crops and foods containing artificial small RNAs.

Keywords Antisense • Biotechnology • Ecological risk assessment • Genetic engineering • Posttranslational gene suppression • RNA interference

1 Introduction

Every country faces contemporary challenges in their quest for efficient and economical production of human food, animal feed, fiber, and biofuels. Challenges for crop production include attacks from a shifting palette of insects and diseases, limitations on essential inputs (e.g., water, fuel, nitrogen fertilizer), and demands for agricultural products (e.g., palm oil) that can have negative environmental impacts. Changes in world climate present another challenge for production systems (Godfray et al. 2010). Perhaps most important, the world's population and energy consumption are expected to continue increasing in the coming years. Thus, there is an urgent need for strategies that will create food security while mitigating negative environmental impacts (Godfray et al. 2010). One biotechnology strategy that is playing a role in meeting these challenges is the engineering of artificial small RNAs and RNA interference (RNAi) in plants.

Crop improvement has long been a cornerstone of agricultural science, and breeding programs at government institutions, universities, and seed companies have transformed agronomic and horticultural crops. This transformative effect can be seen in the breeding programs that created F_1 corn hybrids after World War II, and in the research that led to the Green Revolution. Plant biotechnology continues to receive considerable investment because of its potential to deliver the next generation of crops. The government of China announced the investment of $3.5 billion into development of GE crops and animals (Stone 2008). In some countries, the adoption of GE crops is trending upward, but other nations have largely rejected biotechnology (http://www.isaaa.org/). Some promoters of biotechnology believe that the next generation of GE crops will be more successful because they will directly benefit consumers and/or the environment (Martino-Catt and Sachs 2008).

Potential direct benefits include improved nutritional value through modifications to proteins, carbohydrates, lipids, vitamins, minerals, nutrients, allergens, toxins, and antinutrient compounds (Newell-McGloughlin 2008). GE crops could become "functional foods" with higher concentrations of healthful biochemicals (e.g., antioxidants). Other experts predict that biotechnology will improve pest resistance, drought tolerance, salinity tolerance, nitrogen use efficiency, high temperature tolerance, nitrogen fixation, and photosynthetic efficiency (Godfray et al. 2010).

The central dogma of "one gene, one protein" has been supplanted by a realization that non coding DNAs and their RNA molecules have profound effects in cells and organisms without translation into proteins. In plants, RNAi can be broadly defined as a group of processes that synthesize short RNA molecules to manipulate complementary RNA or DNA sequences (Table 1). Two classes of small RNAs,

Table 1 Definitions for terminology used in this chapter

Term	Definition
Cisgenic cisgenesis	Crops modified through modern molecular methods to express genetic elements (promoter, coding region, flanking regions, terminators) from the same species or a closely related, sexually compatible plant (Schouten et al. 2006; Schouten and Jacobsen 2008). DNA sequences are inserted in their normal orientation and represent a complete copy of the natural gene
Conventional crop	Cultivars, varieties, F1 hybrids, or land races that have been improved by crossing sexually compatible parents and selecting superior progeny for cultivation. Conventional crops include those with novel phenotypes produced through mutation-assisted breeding programs
Ecological risk assessment (ERA)	The process or framework through which future risks (negative impacts) to the environment are estimated based on expert knowledge and hypothesis-driven research
Genetically engineered crop	Crops expressing new traits through modern molecular methods and the insertion of DNA sequences into nuclear or chloroplast genomes. DNA sequences may be from the same plant species, other plant species, distantly related organisms (e.g., bacteria), or synthetic genes
Hairpin dsRNA	RNA molecules in a stem-loop structure where two inverted repeat sequences (stem) are joined at one end by a loop region. This is processed to yield a RNA duplex and then a short interfering RNA that targets mRNA cleavage. Hairpin dsRNA have shown good stability and function in GE plants
Host-delivered RNAi (HD-RNAi)	Plants modified through modern molecular methods to express artificial small RNAs that silence essential genes in pests such as viruses, nematodes, insects, or other pathogens. The attacking organism experiences gene silencing after consuming or contacting plants containing small RNAs
Intragenic intragenesis	Crops modified through modern molecular methods using genetic elements (e.g., promoters, genes, terminating sequences) from the same species or closely related, sexually compatible donor species with the possibility of rearrangement and/or recombination of genetic elements (Rommens 2007). Nonplant DNA, such as marker genes, is excluded. Intragenesis could be used to produce RNA-mediated traits
RNA-mediated trait	Phenotype produced through modern molecular methods through the expression and action of artificial small RNAs. The term is inclusive of RNA interference (RNAi), antisense technology, cosuppression events, and posttranslational gene suppression
Small RNAs	Any RNA macromolecule made by plant cells that participates in RNAi pathways

microRNAs and small interfering RNAs, act as sequence-specific repressors of gene expression by (1) altering transcription through DNA and/or histone methylation, (2) cleaving targeted mRNAs, or (3) inhibiting mRNA translation (Ghildiyal and Zamore 2009; Eamens et al. 2008). Functional small RNAs generally contain ~20–24 ribonucleotides, although a larger class has been reported with ~30–40 ribonucleotides (Eamens et al. 2008). Review articles and thousands of research reports have been published (see reviews in Ghildiyal and Zamore 2009; Parrott et al. 2010; Auer and Frederick 2009; Eamens et al. 2008; Hebert et al. 2008; Verma 2008; Dunoyer and Voinnet 2008; Vaucheret 2006; Baulcombe 2004; Kusaba 2004). Today, it is understood that small RNAs control the expression of developmentally regulated genes, repress repetitive genomic elements, and provide virus resistance in plants (Eamens et al. 2008). In at least one case, conventional breeding has selected a crop trait under the control of small RNAs. The soybean yellow seed coat trait has been attributed to gene silencing in the anthocyanin biosynthesis pathway (Tuteja et al. 2004).

From a historical perspective, our knowledge about small RNAs has emerged from three areas of plant research: (1) early experiments to produce transgenic plants, (2) studies of virus resistance, and (3) direct investigation of RNA-mediated phenomena (Ghildiyal and Zamore 2009; Eamens et al. 2008). Research questions first emerged in the late 1980s and the early 1990s when unexpected effects were noticed in transgenic plants. In some situations, plants overexpressing recombinant genes showed a silencing or downregulation of the encoded protein instead of the expected increase in protein production (Sheehy et al. 1988; Napoli et al. 1990). An unexpected inhibition of protein synthesis could also be seen when the recombinant DNA sequence was inserted backward. In 1988, scientists reported success in downregulating polygalacturonase (PG) enzyme activity by 70–90% in tomato fruits (Sheehy et al. 1988). The term "antisense technology" was defined as gene silencing from insertion of DNA sequence in reverse orientation. In 1990, researchers reported that transgenic petunia plants overexpressing chalcone synthase genes had unexpected reductions in enzyme activity and changes in flower color (Napoli et al. 1990). These effects became known as "gene silencing," "posttranscriptional gene silencing (PTGS)," "cosuppression," or "homology-dependent gene silencing" (Eamens et al. 2008; Vaucheret 2006; Tuteja et al. 2004). At about the same time, researchers found that expression of viral coat protein genes in plants could confer resistance to the viral pathogen from which the coat protein gene was taken (Lindbo and Dougherty 2005; Baulcombe 2004). In retrospect, we now know that these scientists were studying a powerful mechanism found in all eukaryotes.

The power of small RNAs is widely recognized today and being applied to crops in ways that could not have been imagined 20 years ago. The USA has led in the regulation and commercialization of GE crops containing RNA-mediated traits (Table 2). The first GE crop in USA, the Flavr Savr tomato, carried an RNA-mediated trait that modified fruit ripening (Sanders and Hiatt 2005). Small RNAs were used to create the first GE virus-resistant papaya (Chiang et al. 2001). Since 1992, RNA-mediated traits have been approved in tomato, potato, squash, soybean, papaya, and tobacco, but only virus-resistant squash and papaya are cultivated today.

Table 2 Genetically modified (GE) crops approved in USA for commercial production with novel traits derived from antisense technology, posttranscriptional gene silencing, or RNAi as of August 2010

Crop	Trait	Applicant	Date of approval	Mechanism of action
Apple	Postharvest quality	Okanagan specialty fruits	Pending	Reduced browning due to silencing of polyphenol oxidase in fruit tissues
Papaya	Virus resistance	University of Florida	Pending	Viral coat protein gene for PRSV virus
Plum tree	Virus resistance	USDA-ARS	Pending	Viral coat protein gene for plum pox virus
Tobacco	Product quality	Vector tobacco	2001	Nicotine reduced through silencing of gene in biosynthetic pathway
Potato	Virus resistance	Monsanto	2000	Viral coat protein gene for PLRV stacked with Bt insect resistance
Potato	Virus resistance	Monsanto	1998	Viral coat protein gene for PVY stacked with Bt insect resistance
Potato	Virus resistance	Monsanto	1998	Viral coat protein gene for PVY stacked with Bt insect resistance
Potato	Virus resistance	Monsanto	1998	Genes for PLRV virus resistance stacked with Bt insect resistance
Soybean	Oil quality	DuPont	1997	Increased oleic acid content from silencing of GEFAD2-1 gene
Tomato Flavr Savr	Fruit quality	Calgene	1996	Silencing of polygalacturonase gene involved in fruit ripening
Papaya	Virus resistance	Cornell University	1996	Viral coat protein gene for PRSV
Squash	Virus resistance	Asgrow	1996	Viral coat protein genes for CMV, WMMV2 and ZYMV
Tomato Flavr Savr	Fruit quality	Calgene	1995	Silencing of polygalacturonase gene
Tomato Flavr Savr	Fruit quality	Calgene	1995	Silencing of polygalacturonase gene
Tomato	Fruit quality	Zeneca/Petroseed	1995	Silencing of polygalacturonase gene
Tomato Flavr Savr	Fruit quality	Calgene	1994	Silencing of polygalacturonase gene
Tomato	Fruit quality	DNA Plant Tech	1995	Silencing of amino cyclopropane carboxylic acid synthase (ACCS) involved in ethylene biosynthesis
Tomato Flavr Savr	Fruit quality	Calgene	1994	Silencing of polygalacturonase gene
Squash	Virus resistance	Upjohn	1994	Viral coat proteins MWV2 and ZYMV
Tomato Flavr Savr	Fruit quality	Calgene	1992	Silencing of polygalacturonase gene

All information obtained from public documents (http://usbiotechreg.nbii.gov/ and http://www.aphis.usda.gov/brs/not_reg.html)

Applications have been filed with the US government seeking commercialization of RNA-mediated traits in apple trees (decreased fruit browning) and plum trees (plum pox virus resistance). These applications are likely to increase because RNA-mediated traits can provide advantages such as the following: (1) avoiding the introduction of novel proteins that could have toxic or allergenic properties, (2) insertion of a single DNA sequence can silence an entire gene family, and (3) promoters can target gene silencing to specific tissues and organs.

This chapter is written with two goals in mind. First, it aims to provide an updated review of RNA-mediated traits in two broad classes: (1) modifications to plant metabolic and biochemical pathways, and (2) silencing genes in pest organisms to provide protection. Research in model plants or in the early "proof-of-concept" stage has generally been excluded from this review. It is important to note that biotechnology companies rarely publish their cutting-edge discoveries. Thus, it is inevitable that open-access research described by scientists in universities and other institutions is reported.

The second major goal of this chapter is to provide an introduction to predictive ecological risk assessment (ERA) with respect to GE crops expressing RNA-mediated traits. In general, plant scientists have little familiarity with the process of risk assessment and its role in regulatory review. Thus, this section provides an overview of ERA and identifies six areas that need special attention and more information to support risk assessment of RNA-mediated traits. The chapter concludes with a section on future applications and challenges for RNA-mediated traits.

2 RNA-Mediated Traits in Crops

Because small RNAs are both powerful and ubiquitous in the plant kingdom, they could theoretically be used to modify an array of traits in any agronomic or horticultural crop. This section divides crop improvement into five categories: (1) improving nutritional value, (2) removing antinutritional compounds or allergens, (3) altering secondary metabolites, (4) modifying postharvest traits and products, and (5) modifying flower color.

2.1 Improving Nutritional Value

Plant breeders have tried for many years to improve the nutritional content of plants fed to humans and livestock. Various groups have used RNAi to increase the essential amino acid lysine in maize (*Zea mays*). Houmard et al. (2007) decreased the catabolic enzyme lysine-ketoglutarate reductase/saccharophine dehydrogenase (ZLRK/SDH) in maize using an endosperm-specific promoter. This confined RNAi action to the seed and avoided abnormal plant growth due to excessive lysine concentrations in vegetative tissues. Other approaches to increasing lysine have focused on suppression of the dihydrodipicolinate synthase (DHPS) gene or a maize zein storage protein (reviewed in Verma 2008). Wheat (*Triticum*) that contained higher amounts of fiber and digestion-resistant starch could benefit human health, so RNAi was used to create high-fiber wheat by downregulating two starch branching enzymes (SBEIIa and SBEIIb) in wheat endosperm (Regina et al. 2006). Wheat harvested from these GE plants had increased amylose content that was shown to be beneficial to large-bowel function in rats. GE tomato plants with increased carotenoid and flavonoid phytonutrients were created through suppression of the photomorphogenesis regulatory gene *DET1* (Davuluri et al. 2005). This is an example of cross-talk between two pathways that seem to have little in common (light signaling and secondary metabolites) (Dixon 2005). Some people with kidney disease and restricted protein consumption could benefit from low-protein rice. A rice mutant called *Low glutelin content1* was shown to have less glutelin protein due to small RNAs suppressing a multigene family involved in glutelin synthesis (Kusaba et al. 2003).

2.2 Removing Antinutritional Compounds and Allergens

Although breeding has reduced many undesirable compounds in conventional food and feed crops, small RNAs can also be employed. Cassava (*Manihot esculenta*) is a major source of calories in sub-Saharan Africa, but the crop contains potentially toxic levels of the cyanogenic compound linamarin. Siritunga and Sayre (2003) reduced linamarin in cassava by downregulation of the cytochrome P450 gene that catalyzes the first step in linamarin biosynthesis. The reduced synthesis of linamarin in leaves was associated with up to 99% reduction in the harvested root crop. Cotton (*Gossypium*) seeds were modified by RNAi to reduce a toxic secondary metabolite and convert the seeds into a potential source of protein (Sunilkumar et al. 2006). Conventional cotton seeds and cotton seed oil contain a toxic terpenoid molecule called gossypol, making them useless as human food. RNAi was used to interrupt gossypol synthesis during seed development by downregulating the cadinene synthase gene using a seed-specific promoter. Normal terpenoid levels occurred in cotton leaves to protect against insect attack.

Small RNAs can eliminate proteins in legumes and other crops that produce allergic reactions in some people. For example, many processed foods contain soybeans (*Glycine max*), presenting a serious threat to people allergic to soybean seed storage proteins. GE soybeans were generated with suppression of the *Gly m Bd 30 K* gene and it was shown that the allergenic protein could be suppressed while seed morphology and overall protein composition remained unchanged (Herman et al. 2003). While many proof-of-concept reports show that protein allergens can be reduced in foods, the future of anti-allergenic crops is uncertain because the cost of crop development must be weighed against the benefits for a small percentage of the human population.

2.3 Secondary Metabolites

Plant secondary metabolites are the source of many pharmaceutical compounds and commonly used stimulants (e.g., nicotine, caffeine). An interesting example of RNA-mediated crop improvement is the reduction of nicotine in tobacco leaves used for cigarettes (Table 2). Vector 21–41 tobacco was engineered through an antisense approach to silence the quinolinic acid phosphoribosyltransferase (QPTase) gene. Since QPTase is a key enzyme in the biosynthetic pathway for nicotine and related alkaloids, the overall effect was a reduction in the levels of nicotine, nor-nicotine, and total alkaloids in the leaves (Federal Register 67, (232) page 71929, 2002). Coffee is an economically valuable crop, but some people are adversely affected by the natural stimulant caffeine. Ogita et al. (2003, 2004) silenced the gene for one of the N-methyltransferase enzymes involved in caffeine biosynthesis in coffee plants (*Coffea canephora*). Silencing the theobromine synthase gene (*CaMXMT1*) reduced caffeine 50–70% in young leaves. In opium poppy (*Papaver somniferum*), RNAi was used to silence a multigene family coding for the codeinone reductase enzyme (Allen et al. 2004). This allowed a precursor to the alkaloid biosynthesis pathway to accumulate while the downstream products of morphine and codeine were almost eliminated. The authors suggested that these GE poppies could produce novel compounds for pharmaceutical production without being used for production of illegal drugs.

2.4 Modifying Postharvest Traits and Products

One of the earliest RNA-mediated traits was the reduction of PG activity to slow ripening and softening in tomatoes. In 1988, Calgene scientists reported their success in using antisense technology to downregulate PG activity by 70–90% in tomato fruits (Sheehy et al. 1988). Shortly thereafter, several biotechnology companies received approval in the USA for GE tomatoes with delayed ripening (Table 2); the best known example is the Flavr Savr tomato (Sanders and Hiatt 2005). The browning of fruit tissues due to polyphenol oxidase activity is another trait that has been studied for decades. Recent patents suggest that silencing of polyphenol oxidase(s) can preserve fruit and juice quality for food processors and consumers (US Provisional Patent Application No. 61/031,821, February 27, 2008). An application for commercialization of GE apples with reduced browning has been submitted to US regulators (Table 2). The company believes that their technology could be applied to a range of pome and stone fruits (http://www.okspecialtyfruits.com/mb-current-products-available.php).

About 25% of potato crops are put into long-term storage, a process that requires maintaining tuber dormancy. Antisense expression of a potato gene (G1-1) created GE potato lines with longer tuber dormancy (Marmiroli et al. 2000). In a field trial, antisense G1-1 potato plants performed similar to controls. There is demand for high-amylose (unbranched linear) starch for various food and industrial products, so antisense technology was used to suppress two starch branching enzymes (SBE A and B) in potatoes, resulting in higher amylose (unbranched) starch content (Schwall et al. 2000). Antisense technology may also be used to suppress amylose starch in GE potatoes for industrial products (Hofvander et al. US Patent #6,784,338 B1, 2004).

2.5 *Flower Color*

One of the earliest reports of antisense technology involved suppressing genes involved in synthesis of petunia flower pigments (Napoli et al. 1990). More recently, researchers have stacked two different approaches to create a blue-hued rose using silencing of dihydroflavonol 4-reductase and expression of two recombinant genes for delphinidin synthesis (Katsumoto et al. 2007). RNAi suppressed three anthocyanin biosynthesis genes in gentian flowers (*Gentiana* hybrid "Albireo") to produce a light blue to white petal color (Nakatsuka et al. 2008). Since many of the genes for pigment biosynthesis have been cloned, it is likely that RNAi could be used to modify flower color in many ornamental species.

3 Host-Delivered RNAi (HD-RNAi) for Crop Protection

Crops can be engineered to silence essential genes in organisms attacking them, thus providing protection to the GE crop. This approach to plant protection has been called "Host-Delivered RNAi" (HD-RNAi) (Table 1). This section reviews HD-RNAi mechanisms that could provide protection from viruses, bacteria, fungi, nematodes, and insects.

3.1 *Plant Viruses*

Viruses are a serious problem and conventional crop breeding programs have sought to incorporate virus resistance genes for decades (Kang et al. 2005; Baulcombe 2004). Plants appear to have two major pathways for virus resistance: resistance genes and proteins (R genes), and RNAi. While there is thought to be functional overlap between these two pathways, the cross-talk and interactions are not well understood (Soosaar et al. 2005). The discovery of RNAi is closely interwoven with the invention of transgenic plants and research on virus resistance (Lindbo and Dougherty 2005; Baulcombe 2004; Tepfer 2002; Waterhouse et al. 1998). Transgenic tobacco plants expressing different versions of the tobacco etch virus (TEV) coat protein provided the earliest example of engineered virus resistance (reviewed in Lindbo and Dougherty 2005). It was shown that high TEV resistance and low mRNA concentrations were produced by a RNA-mediated response in the cell cytoplasm (Lindbo and Dougherty 2005). Another noteworthy breakthrough came from Waterhouse et al. (1998) where viral coat protein transgenes inserted in the sense and antisense directions allowed formation of hairpin dsRNA in tobacco. Virus resistance and gene silencing through an RNA-dependent RNA polymerase (RdRp) system produced a self-perpetuating, sequence-specific degradation of targeted mRNA. Many groups have generated virus resistant plants under laboratory conditions and two virus-resistant GE crops (papaya and squash) are currently grown in the USA (Table 2).

Single-stranded DNA geminiviruses cause significant damage to crops (e.g., cassava, tomato), but small RNAs have more difficulty stopping these pathogens (Shepherd et al. 2009; Levy et al. 2008; Lucioli et al. 2008; Vance and Vaucheret 2001). Resistance to tomato yellow leaf curl geminivirus (TYLCV) was demonstrated in transgenic tomatoes expressing a truncated sense or antisense version of the replication associated protein (*Rep*) gene (Polston and Hiebert 2006, US patent #20100095402). Fuentes et al. (2006) reported that gene silencing using the TYLCV replication associated protein (C1) gene produced resistance in tomato. In cassava, resistance to African cassava mosaic virus (ACMV) was demonstrated using hairpin dsRNA with homology to the viral replication-associated sequence *Rep/AC1* (Vanderschuren et al. 2009). Barriers to stable geminivirus control have been attributed to the following: (1) mixed populations of viral pathogens are common and might circumvent the sequence homology necessary for RNAi, (2) viruses might encode proteins that interfere with RNAi, (3) geminiviruses might evolve rapidly through mutation, recombination, and pseudorecombination, and these sequence changes could block RNAi, and (4) high concentrations of small RNAs may be needed to stop virus replication (Vanderschuren et al. 2009; Shepherd et al. 2009; Lucioli et al. 2008).

3.2 Bacterial and Fungal Pathogens

There is relatively little information about how plants might use small RNAs to protect themselves from bacterial and fungal pathogens. At present, there is some evidence to suggest that small RNAs change their expression during pathogen attack and subsequently regulate the expression of genes involved in disease resistance pathways (Jin 2008; Navarro et al. 2008). It has been suggested that small RNAs may act to silence negative regulator molecules in the plant cell under normal circumstances, but allow the quick upregulation of genes that are needed when pathogens attack (Jin 2008). Escobar et al. (2001) generated resistance to crown gall disease (*Agrobacterium tumefaciens*) in *Arabidopsis* using expression of dsRNA for two bacterial genes (*iaaM* and *ipt*). In some cases, they were able to decrease tumor production to nearly zero, suggesting that resistance to crown gall disease could be engineered in trees and ornamental woody plants.

3.3 Nematodes

Plant-parasitic nematodes (unsegmented roundworms) are a significant pest of many crops, with global damage estimates up to $125 billion per year (Fuller et al. 2008; Gheysen and Vanholme 2007; Fairbairn et al. 2007). The majority of the damage is done by sedentary endoparasitic nematodes of the Tylenchoidea superfamily, especially the root-knot nematodes (*Meloidogyne* species) and cyst nematodes (*Heterodera* and *Globodera* species) (Fuller et al. 2008; Fairbairn et al. 2007). Chemical controls and crop rotation are typically used to protect crops, but rotation is of limited use when the host plant range is large and nematicides (e.g., methyl bromate) have unacceptable environmental impacts (e.g., ozone depletion, toxicity). Conventional breeding programs have generally been unable to develop effective resistance (Fuller et al. 2008), providing a strong rationale for control through biotechnology.

Several landmark studies showed that HD-RNAi can silence essential genes in nematodes after they absorb or consume artificial small RNAs (Yadav et al. 2006; Huang et al. 2006). Yadav et al. (2006) showed that tobacco plants expressed hairpin dsRNA targeting two *Meloidogyne* (root knot) nematode genes had more than 95% resistance to *Meloidogyne incognita*. Nematodes surviving exposure to the transgenic tobacco roots had developmental problems and lacked transcripts for the targeted genes. Huang et al. (2006) showed that *Arabidopsis* plants expressing hairpin dsRNA for a gene involved in plant–parasite interaction (*16D10*) suppressed formation of root galls by *Meloidogyne* nematodes and reduced egg production. The *Arabidopsis* plants showed some resistance to four economically important species of *Meloidogyne*. Steeves et al. (2006) showed that GE soybeans expressing the *Meloidogyne* major sperm promoter (MSP) gene reduced nematode eggs and reproductive potential. However, there is no clear evidence of efficacy under field conditions. In situations where a range of nematode species are attacking the crop, it is not clear whether control can be accomplished with a single artificial small RNA. The recent announcement of the genome sequence for the plant parasitic nematode *Meloidogyne hapla* may lead to rapid advances in understanding host–parasite interactions and designing HD-RNAi (Oppermana et al. 2008).

3.4 Insects

Small RNAs have been used to investigate the basic biology of *Drosophila* and other insects (Price and Gatehouse 2008; Gordon and Waterhouse 2007). Many of these studies injected nanogram concentrations of dsRNA into the insect hemocoel (the space between organs filled with hemolymph fluids) because insects, mollusks, and vertebrates appear to lack the RdRP enzyme that replicates small RNAs for systemic RNAi action (Huvenne and Smagghe 2010; Gordon and Waterhouse 2007; Gatehouse 2008; Price and Gatehouse 2008). Initially, it was thought that insect pests would have to continuously contact or consume small RNAs for effective gene silencing. Today, the pesticidal properties of artificial small RNAs have been demonstrated in laboratory experiments using seven insect orders and various target genes (Huvenne and Smagghe 2010).

Two landmark studies in 2007 demonstrated HD-RNAi insect resistance, suggesting a potential alternative to Bt crops (Baum et al. 2007; Mao et al. 2007). Baum et al. (2007) showed that silencing of a vacuolar ATPase gene (V-type ATPase A gene) expressed in the midgut of Western Corn Rootworm (WCR) led to larvae mortality and stunted growth. A diet containing dsRNA for the V-type ATPase A generated systemic gene silencing in larvae within 24 h. The authors created GE corn plants expressing dsRNA for WCR V-type ATPase A and showed reduced WCR feeding damage. The HD-RNAi concept was extended to tests with three types of dsRNA (V-ATPase A, V-ATPase E and B-tubulin) and three beetle pests: Southern corn rootworm, Colorado potato beetle, and cotton boll weevil. All three dsRNA treatments were able to kill Southern corn rootworm and Colorado potato beetle larvae, but higher concentrations were required in the artificial diet. Cotton boll weevils appeared to be insensitive to the dsRNA treatments even with orthologous boll weevil genes.

Plants have evolved to synthesize many different toxins, and, in turn, insects have developed enzymes to detoxify plant tissues (Gordon and Waterhouse 2007). Mao et al. (2007) used HD-RNAi to alter the sensitivity of cotton bollworms to the natural toxin gossypol using a cytochrome P450 monooxygenase gene important to bollworm larval growth with expression in the midgut cells. The CYP6AE14 gene was found to be causally related to cotton bollworm tolerance of gossypol, a toxic secondary metabolite produced naturally in cotton plants. GE tobacco and Arabidopsis plants producing CYP6AE14 dsRNA were fed to larvae, effectively decreasing mRNA for CYP6AE14, stunting growth, and increasing sensitivity to gossypol. Using Arabidopsis mutants lacking three of the four Dicer genes, longer dsRNA molecules were made, increasing gene silencing in larvae. Demonstrations of pesticidal properties are needed under field conditions with natural insect populations.

4 Crop Biotechnology Lexicon

The lexicon of plant biotechnology has expanded with the goal of differentiating between GE crops based upon the original source of the DNA sequences (e.g., bacterial, plant, synthetic) (Table 1). This discussion became public when research groups suggested that the source of the DNA should be explicitly referenced to promote consumer acceptance and rapid regulatory approval. Nielsen (2003) suggested five different terms that would convey the similarity of the inserted genetic elements to changes available through conventional breeding. For example, "xenogenic" would refer to GE organisms with laboratory-designed synthetic genes while "intragenic" would refer to modifications using genes from the same genome. Strauss (2003) presented the term "genomics-guided transgenes" (GGT) and suggested that GGTs were likely to carry less risk due to the source of the inserted DNA sequences and warrant less regulatory review and data before experimental field trials. Subsequently, Schouten and colleagues defined "cisgenic" plants as those engineered using native genetic sequences (promoter, coding sequences, terminators) from sexually compatible plants in their normal orientation. A cisgenic approach could transfer a disease-resistance gene from a wild apple species into an elite apple cultivar, respecting natural species barriers similar to conventional crop breeding. Rommens (2007) reintroduced the term "intragenic" to designate the insertion of plant-derived DNA using only native genetic elements, but allowing the rearrangement and reorientation of the genetic elements (e.g., antisense orientation). The intragenic approach is compatible with the creation of RNA-mediated traits

In response, Giddings (2006) argued that there is no defensible difference between cisgenic and transgenic plants based on risk-associated criteria. He reiterated the idea that phylogenetic distance between the DNA donor and the host was not an indicator of risk in the final transgenic organism. Instead, the emphasis should be on the novel traits and expression patterns in the GE crop (Giddings 2006).

It is worth remembering that plants are complex organisms with many natural compounds for protection (e.g., toxins, antinutrients), flexible cross-talk at the molecular and physiological levels, and intricate responses to biotic and abiotic stresses. While many toxic and antinutritional compounds have been minimized through conventional breeding, some crops (e.g., cassava, potato, canola) still express alkaloids, glucosinolates, lectins, and other chemicals toxic to mammals. Anti-nutritional compounds in legumes (e.g., soybean lectins) and oilseeds (e.g., glucosinolates in canola) are known to vary in concentration depending on cultivar genotype. Recognition of these naturally occurring chemicals supports the position that evolutionary distance between DNA donor and host does not necessarily correlate with potential risk to humans (e.g., toxicity, allergenicity) or the environment. Modifying crops through artificial small RNAs could conceivably produce crops and food products that were less nutritious or even harmful. Molecular cross-talk has been demonstrated in tomato where silencing a gene controlling photomorphogenesis altered the synthesis of secondary metabolites (Dixon 2005; Davuluri et al. 2005). Gene silencing can unexpectedly reduce crop yield and change plant phenotypes (Schwall et al. 2000). RNAi has produced unexplained pollen lethality in *Arabidopsis* (Xing and Zachgo 2007). In these experiments, RNAi constructs downregulating the AGAMOUS-LIKE18 (AGL18) MADS-box gene produced T1 progeny where 21% had a pollen lethality phenotype. The authors offered no single explanation for these observations, but positional effects from DNA insertion could produce rearrangements, deletions could interfere with the estimated 3,500 genes involved in pollen formation, or RNAi constructs could generate "off-target" effects by silencing genes important to pollen development. As a strategy for crop improvement, RNA-mediated traits deserve risk assessment on a case-by-case basis to ensure adequate protection for humans, livestock, and the environment. Risk assessment frameworks and research have the opportunity to build public confidence in biotechnology. Adding more words to the biotechnology lexicon could yield confusion rather than clarification.

5 Implications for Predictive Ecological Risk Assessment

5.1 Introduction to ERA

Predictive ERAs have become an established component of the regulatory process for GE crops in many countries (Craig et al. 2008). Many papers and conferences have debated the utility of ERA frameworks and the best practices for implementing them (Auer 2008; Nickson 2008; Raybould 2007; Andow and Zwalen 2006; Hayes 2004; Conner et al. 2003). In general, regulatory agencies utilize frameworks that combine a case-by-case approach, overall weight of evidence, and research data to evaluate the potential ecological risks from a GE crop. Despite some acknowledged limitations, ERA frameworks offer many benefits including identifying potential hazards, creating a framework for gathering relevant data, directing new research to answer specific questions about future risks, guiding the proactive development of risk management strategies, creating harmonization between countries, and providing a common language for discussion among regulators, companies, stakeholders, and citizens.

Predictive risk assessment is the process through which future risks (harms, negative impacts) are estimated based on current knowledge and hypothesis-driven scientific research (Hayes 2004). Risk assessment frameworks typically involve logical steps of problem formulation, identification of potential hazards, identification of exposure pathways, risk characterization, prediction of the severity of harm (negligible, low, moderate, high), and an expression of uncertainty. The classic definition of an ecological risk is a negative impact that is the product of a hazard (a defined adverse impact) and an exposure (a mechanism or route by which the hazard is experienced) (Hayes 2004). This classic definition can sometimes be difficult to apply to a specific GE crop, but it is still widely used as a framework for characterizing potential risks. Mechanisms of exposure may involve interaction with the GE plants themselves, the protein and/or biochemical product of the transgene, sexually compatible plants that receive the transgene, and transgenic propagules or plant parts (e.g., pollen, seeds, rhizomes, bulbs). Examples of potential hazards include: unintended effects on nontarget species (e.g., beneficial insects), the creation of problematic weeds or invasives, harm to threatened and endangered species, and reduction in overall biodiversity.

Figure 1 provides a conceptual model for risk characterization for a HD-RNAi trait in a GE tree. Exposure pathways and environmental hazards are noted as potential elements in an ERA. If no novel proteins (e.g., marker genes) are expressed in the GE tree, detailed knowledge of inserted genetic sequences and flanking DNA regions would be required to conduct polymerase chain reaction (PCR) for monitoring, segregation, and to identify preservation of the GE tree and its products. If the tree is grafted and only the rootstock has been engineered for synthesis of artificial small RNAs, systemic gene silencing (transitivity) could generate trait expression (e.g., insect resistance, virus resistance) in the stems, leaves, or fruits (Eamens et al. 2008; Dunoyer and Voinnet 2008). In this case, reverse transcriptase PCR (RT-PCR) could be necessary to directly amplify the artificial small RNAs extracted from above-ground plant tissues (see discussion below on tracking crops).

Environmental risks are evaluated with regard to specific ERA end points that deserve protection and are relevant to the specific crop. Risk assessment end points can be expressed from the level of individuals (e.g., one individual of an endangered species) to higher organizational levels such as populations, communities, ecosystems

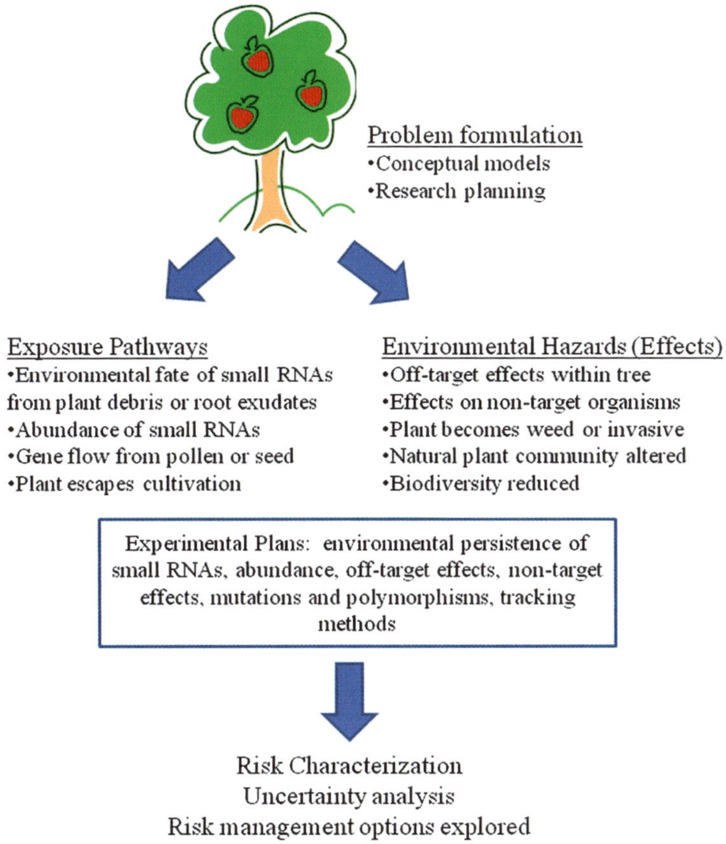

Fig. 1 Conceptual model for the characterization of ecological risks from a GE tree expressing artificial small RNAs

and landscapes (Hayes 2004). Well-developed predictive ERAs will consider the spatial areas in which an impact would occur, the period of time during which the risk would be experienced, the reversibility of the hazards, and the severity of harm to valued risk assessment end points. In practice, if the predictive ERA procedure identifies some potential risks, analysts and regulators often explore management tools or stewardship practices to prevent harm (Fig. 1).

It is important to distinguish between different types of risks with regard to regulatory authority. For example, the long-term effectiveness of a HD-RNAi virus-resistant crop could be reduced by rapid viral mutations and a high degree of polymorphism. Assuming that this did not generate a direct or indirect risk to the environment or farming systems, the biotechnology company would be expected to evaluate this problem (e.g., limits to product longevity and efficacy) with regard to commercialization (see discussion in Tepfer 2002).

5.2 *RNA-Mediated Traits: Special Considerations for Ecological Risk*

In general, it appears that ecological risk from RNA-mediated traits and HD-RNAi crops can be analyzed using the current ERA framework that has been developed for other GE crops. In fact, crops with RNA-mediated traits have already been approved in the USA, suggesting a degree of familiarity and experience (Table 2). Some of the questions about potential ecological risks are familiar to regulators and stakeholders. What are the potential hazards and exposure pathways for this GE crop? Are there likely to be significant effects on nontarget organisms (e.g., beneficial insects), communities, and ecosystems? Could gene flow to native or naturalized relatives occur and create significant environmental harms? Could new weeds or invasive species be produced? However, there are also differences between RNA-mediated traits and GE crops expressing novel proteins.

Six questions have been formulated to identify areas deserving special attention and research for credible ERA (Auer and Frederick 2009). The goal here is to be comprehensive in consideration of potential risks rather than to present detailed evidence about the potential severity or likelihood of specific risks. The six questions are as follows: (1) What "off-target" effects could occur within the crop or closely related plants? (2) What nontarget organisms could be negatively affected? (3) How persistent are small RNA molecules in terrestrial and aquatic environments? (4) What will be the effect of mutations and polymorphisms in the GE crop and in organisms that consume the crop? (5) What tools will be available for rapid detection and tracking of small RNAs, GE crops, and derived food products? and (6) How should uncertainty be expressed in these risk assessments?

5.2.1 Off-Target Effects Within GE Plants

One small RNA might silence a whole family of genes and mRNAs, thus raising concerns about "off-target" gene silencing that causes unintended physiological changes, pleiotropic phenotypes, and/or environmental hazards. Off-target effects could also arise from "transitive silencing" when RNA-dependent RNA polymerase (RdRp) amplifies the RNAi signal throughout the plant producing unintended effects in other tissues and organs (Eamens et al. 2008). Experiments in bacteria have shown molecular cross-talk where off-target mRNA binding decreased expression of genes other than the target (Leonard et al. 2008). Unexpected cross-talk was seen in tomato fruits when downregulation of the *DET1* gene (photomorphogenesis developmental transcription factor) increased activity in secondary metabolism pathways (Davuluri et al. 2005). RNAi can produce unexpected effects on plant reproduction, such as reduced pollen viability, even when other aspects of plant growth appear to be normal (Xing and Zachgo 2007).

Scientists producing HD-RNAi nematode-resistant tobacco using a nematode transcription factor gene examined a genomic database and stated that they found no homology between the nematode gene and tobacco genome (Fairbairn et al. 2007). This suggested a lower probability of off-target effects in the GE tobacco. Such in silico approaches to predicting off-target effects are important, but they will be limited by the availability of genomes and transcriptomes for crop plants. Research groups should be encouraged to study the probability of off-target effects in GE crops to address these questions.

5.2.2 Effects on Nontarget Organisms

Small RNAs constitute a highly conserved mechanism for controlling gene expression in eukaryotes. Thus, nontarget organisms could be harmed if they absorbed or consumed small RNAs with sequence homology that silenced critical genes (Whangbo and Hunter 2008; Small 2007). Nontarget effects have been studied with regard to food and feed safety. A study of rice showed that 4,759 human genome sequences and 270 human transcriptome sequences showed perfect complementarity to one of the 285,864 unique small RNAs in rice. Complementarity was also detected in cow, chicken, pig, and mouse genomes and transcriptomes. However, the authors presented multiple lines of evidence suggesting that humans and other mammals have no significant dietary exposure to intact RNAs due to their digestive systems and other factors. A history of safe consumption by humans and other mammals was presented to suggest that exposure to artificial small RNAs is unlikely to cause nontarget effects in mammals (Ivashuta et al. 2009; Parrott et al. 2010). It was proposed that RNA molecules be considered "generally recognized as safe" (GRAS) components of food and feed.

In contrast to mammals, invertebrates (e.g., nematodes, insects) can absorb and react to small RNAs. This suggests that testing might be important to predict potential impacts on susceptible nontarget organisms (e.g., beneficial insects). HD-RNAi experiments based on insect housekeeping genes (e.g., vacuolar ATPases and Β-tubulin, Baum et al. 2007) produced small RNAs that negatively impacted more than one insect species. Genomic databases and well-designed laboratory feeding studies could prove useful in characterizing the likelihood of negative effects on nontarget organisms such as beneficial insects and casual herbivores. However, the lack of genomic information for many nontarget species is likely to be a limitation.

5.2.3 Environmental Persistence and Abundance of Small RNAs

If nontarget organisms could be harmed by artificial small RNAs, then it will be important to characterize RNA persistence and abundance in various ecosystems. Very little is known about the persistence of extracellular small RNAs (also called environmental RNA), although there is some information about their natural function in communication, symbiotic relationships, and other processes (Whangbo and Hunter 2008; Vlassov et al. 2007). Extracellular DNA has been found in aquatic and terrestrial environments that can persist for months or even years despite the presence of nuclease enzymes (Vlassov et al. 2007; Stotzky 2000). Absorption of DNA into complex organic molecules in the soil is believed to provide protection from nuclease degradation. However, very little research has been done on the persistence of RNAs in different ecosystems (Vlassov et al. 2007). Bacterial biofilms are known to contain a complex mixture of molecules including single-stranded RNA. Extracellular RNA persists in the blood plasma of mammals. In one medical study, a single-stranded RNA virus survived in animal blood samples stored on filter paper at 32°C for 3 months (Michaud et al. 2007). Research should be conducted to determine if environmental persistence is affected by the ribonucleotide sequence; some RNAs could be more stable or resistant to degradation. A study on HD-RNAi insect resistance suggested that difference in effectiveness could be due to RNA length and persistence in the plant–pest system (Baum et al. 2007).

Concentration or abundance of small RNAs is another important factor in characterizing risk. A study of soybean seeds showed that the total pool of naturally occurring small RNAs (21–24 nucleotides) was approximately 0.66 μg/g (Ivashuta et al. 2009). Mature rice and corn grains were reported to have similar levels of small RNAs (Ivashuta et al. 2009). Strong expression of artificial small RNAs may be required to achieve virus resistance and other traits, affecting the abundance of RNA molecules released into the environment.

5.2.4 Effects of Mutations and Polymorphisms

Heritable genetic mutations (e.g., base changes, deletions, insertions) occur in all organisms including crop plants and their pests. In addition, polymorphisms (small variations in DNA sequences) occur in individuals within a population (Whangbo and Hunter 2008; Gordon and Waterhouse 2007). Given this natural background of mutations and polymorphisms, research is needed to characterize unintended effects from RNA-mediated processes in crop plants and the organisms that are exposed to them. Scenarios worth considering include the following: (1) mutations in the GE crop alter the nucleotide sequence of the artificial small RNAs, change patterns of gene silencing, and create off-target effects, (2) mutations and polymorphisms in plant pest populations (e.g., viruses, insects, nematodes) producing resistance to gene silencing and the pesticide properties of the HD-RNAi crop, and (3) mutations in nontarget organisms (e.g., beneficial insects) increase their susceptibility to the pesticidal properties of HD-RNAi crop plants. Research is needed to characterize potential hazards and exposures with regard to the following: (1) the probability of occurrence, (2) the effect of different time scales (e.g., one growing season, multiple years), (3) the scale and distribution pattern of the GE crop (e.g., small isolated fields, thousands of contiguous hectares), and (4) the severity of the potential negative impact to the environment.

5.2.5 Tracking Crops with RNA-Mediated Traits

Identity preservation, monitoring and segregation of GE crops are important to many stakeholders in the food chain including biotechnology companies, seed producers, farmers, food manufacturers, and exporters (Auer 2003). Regulators might also need the ability to monitor and track a GE crop if necessary. GE crops modified through artificial small RNAs will require a significant change in current crop testing and monitoring practices. At present, GE crops expressing novel proteins (e.g., herbicide-resistant soybeans, Bt maize) are often detected using an easy and inexpensive enzyme-linked immunosorbent assay (ELISA) procedure that detects the novel protein (Auer 2003). If crops are modified using artificial small RNAs, GE plants and plant parts (e.g., harvested seeds, tubers) will not be detectable using an ELISA because no protein is produced from the inserted coding sequence. Therefore, detection and monitoring of GE crops and food products would probably have to be done in a laboratory using PCR technology and prior knowledge of appropriate sequence-specific PCR primers. Artificial small RNAs might be detected using a RT-PCR strategy. Alternatively, marker genes and their proteins (e.g., antibiotic resistance genes, sugar isomerases) could be the basis for ELISA tests, but there may be problems with specificity because many GE crops contain similar marker genes. Clearly, new diagnostic tools would be highly desirable for tracking, identity preservation, monitoring, and segregation of GE crops with RNA-mediated traits. Ideally, these new analytical tools should be easy to use, have a low detection limit (<1%), have high specificity for each GE crop, should be low cost, and operate under field conditions.

5.2.6 Uncertainty

Epistemic uncertainty (what we do not know) is inherent in ERA and assessors are called upon to clearly indicate the extent of uncertainty in their analysis (Suter 2007). To the extent that there is greater uncertainty associated with new technologies compared to those with an established track record, this will likely have an impact on risk assessments for RNA-mediated traits in GE crops. Protein-based GE crops have been grown for nearly 20 years and risk assessment research preceded commercialization. Research has not only improved understanding of the mode of action of the novel proteins (e.g., *Bacillus thuringiensis* endotoxin) but also answered many ERA questions (Conner et al. 2003; Sanvido et al. 2007). Because RNA-mediated traits are relatively new and expanding, von Krauss et al. (2008) directly studied uncertainty using expert elicitation. The authors interviewed scientific experts and concluded there were competing interpretations of cause–effect relationships leading to gene silencing, and several sources of uncertainty. When making environmental decisions based on risk analysis, decision makers balance the level of risk against the uncertainty of risk assessments. It may be that RNAi technology will in most cases present low environmental risks, but a high level of uncertainty about risks could generate requirements for substantial testing and management controls before commercialization.

In the coming years, the process of predictive ERA will have to be flexible and adaptable to analyze the next generation of GE crops. Analysis will most likely continue to require detailed information on a case-by-case basis on three universal components: (1) the host crop species, (2) the inserted genetic sequences and their mechanism of action, and (3) the environment in which the crop will be cultivated.

Although an elusive concept, familiarity has been a cornerstone of risk assessment for GE crops for more than a decade (Hokanson et al. 1999). In future, regulators will likely be familiar with the biology and ecology of the host crop plants, but the action of small RNAs will be different compared with previous protein-based crops. National and international groups will need to ask the right questions about potential hazards and exposure pathways, collect relevant data, and assess potential environmental risks. Regulators will have to evaluate the design and implementation of research protocols to ensure safe conduct of confined experimental field trials. Analysts will also need to understand the optimal methods to identify and measure biologically active small RNAs.

6 Future Outlook: Applications and Challenges

Crops improvement through artificial small RNAs is already a familiar approach in the USA, and the scientific literature points to many potential applications for countries that are willing to adopt crop biotechnology. RNAi could be applied to algae so that solar energy is more efficiently transformed into chemical fuels (Mussgnug et al. 2007). Artificial small RNAs could generate male sterility in crops, thus assisting breeding programs or the containment of GE crops (Sandhu et al. 2007). Systemic gene silencing (transitivity) could allow grafted woody plants (e.g., grape vines, fruit trees) to transport small RNAs from their rootstocks to scion tissues (stems, leaves, fruits) to produce useful traits (Eamens et al. 2008; Dunoyer and Voinnet 2008). Target mimicry could be used to increase gene expression by antagonizing the effects of naturally occurring microRNAs (Franco-Zorrilla et al. 2007). Pharmaceutical-producing plants could express artificial small RNAs to suppress infectious viral diseases in humans or livestock (Zhou et al. 2004). Important targets for RNA therapies include HIV and hepatitis C viruses (Mahmood-ur-Rahman et al. 2008). RNAi could modify glycosylation pathways in plant cells, improving their manufacture of human monoclonal antibodies (Cox et al. 2006). The rate at which these ideas are converted into products depends on many variables including research funding. Consumer acceptance of foods created through biotechnology is another variable in the equation.

Technical barriers exist for optimal RNA-mediated crop improvement. Challenges exist in predicting, stabilizing, and controlling artificial small RNAs and phenotypes. The parallel RNAi pathways found in *Arabidopsis* have not been investigated in most other plants. Genomic and transcriptomic databases have yet to be developed for minor crops, horticultural crops, pest organisms (e.g., insects, nematodes), and nontarget organisms (e.g., beneficial insects, endangered species).

While some research has been done to assess human safety, more work needs to be done to predict environmental safety. Important questions exist about environmental persistence and abundance of artificial small RNAs in plant tissues, water, and soil systems. Unintended gene silencing in nontarget species needs to be studied using well-designed feeding studies and other complementary approaches. While this chapter emphasizes topics unique to RNA-mediated traits, risk analysts must still construct credible ERAs to evaluate the consequences of gene flow or the potential for new weeds. In future, risk will also have to be assessed for GE crops with stacked (polygenic) traits produced by a combination of mechanisms such as RNAi, expression of novel proteins, and modified transcription factors.

References

Allen R, Millgate A, Chitty J et al (2004) RNAi-mediated replacement of morphine with the nonnarcotic alkaloid reticuline in opium poppy. Nat Biotechnol 22:1559–1566

Andow DA, Zwalen C (2006) Assessing environmental risks of transgenic plants. Ecol Lett 9:196–214

Auer C (2003) Tracking genes from seed to supermarket. Trends Plant Sci 8:591–597

Auer C (2008) Ecological risk assessment and regulation for genetically-modified ornamental plants. Crit Rev Plant Sci 27:255–271

Auer C, Frederick R (2009) Crop improvement using small RNAs: applications and predictive ecological risk assessments. Trends Biotechnol 27:644–651

Baulcombe D (2004) RNA silencing in plants. Nature 431:356–363

Baum JA, Bogaert T, Clinton W et al (2007) Control of coleopteran insect pests through RNA interference. Nat Biotechnol 25:1322–1326

Chiang C, Wang J, Jan F et al (2001) Comparative reactions of recombinant papaya ringspot viruses with chimeric coat protein (CP) genes and wild-type viruses on CP-transgenic papaya. J Gen Virol 82:2827–2836

Conner AJ, Glare TR, Nap J-P (2003) The release of genetically modified crops into the environment. Part II. Overview of ecological risk assessment. Plant J 33:19–46

Cox KM, Sterling JD, Regan JT et al (2006) Glycan optimization of a human monoclonal antibody in the aquatic plant *Lemna minor*. Nat Biotechnol 24:1591–1597

Craig W, Tepfer M, Degrassi G et al (2008) An overview of general features of risk assessments of genetically modified crops. Euphytica 164:853–880

Davuluri GR, Tuinen A, Fraser PD et al (2005) Fruit-specific RNAi-mediated suppression of DET1 enhances carotenoid and flavonoid content in tomato. Nat Biotechnol 23:890–895

Dixon RA (2005) A two-for-one in tomato nutritional enhancement. Nat Biotechnol 23:825–826

Dunoyer P, Voinnet O (2008) Mixing and matching: the essence of plant systemic silencing? Trends Genet 24:151–154

Eamens A, Wang M-B, Smith NA et al (2008) RNA silencing in plants: yesterday, today, and tomorrow. Plant Physiol 147:456–468

Escobar MA, Civerolo EL, Summerfelt KR et al (2001) RNAi-mediated oncogene silencing confers resistance to crown gall tumorigenesis. Proc Natl Acad Sci USA 98:13437–13442

Fairbairn DJ, Cavallaro AS, Bernard M et al (2007) Host-delivered RNAi: an effective strategy to silence genes in plant parasitic nematodes. Planta 226:1525–1533

Franco-Zorrilla JM, Valli A, Todesco M et al (2007) Target mimicry provides a new mechanism for regulation of microRNA activity. Nat Genet 39:1033–1037

Fuentes A, Ramos PL, Fiallo E et al (2006) Intron-hairpin RNA derived from replication associated protein C1 gene confers immunity to Tomato Yellow Leaf Curl Virus infection in transgenic tomato plants. Transgenic Res 15:291–304

Fuller VL, Lilley CJ, Urwin PE (2008) Nematode resistance. New Phytol 180:27–44

Gatehouse JA (2008) Biotechnological prospects for engineering insect-resistant plants. Plant Physiol 146:881–887

Gheysen G, Vanholme B (2007) RNAi from plants to nematodes. Trends Biotechnol 25:89–92

Ghildiyal M, Zamore PD (2009) Small silencing RNAs: an expanding universe. Genetics 10:94–108

Giddings LV (2006) "Cisgenic" as a product designation. Nat Biotechnol 24:1329

Godfray HCJ, Beddington JR, Crute IR et al (2010) Food security: the challenge of feeding 9 billion people. Science 327:812–818

Gordon KHJ, Waterhouse PM (2007) RNAi for insect-proof plants. Nat Biotechnol 25:1231–1232

Hayes KR (2004) Best practice and current practice in ecological risk assessment for genetically modified organisms. CSIRO Division of Marine Research. http://www.environment.gov.au/settlements/biotechnology/publications/bestpractice.html (Last viewed 2 May 2011)

Hebert CG, Valdes JJ, Bentley WE (2008) Beyond silencing – engineering applications of RNA interference and antisense technology for altering cellular phenotypes. Curr Opin Biotechnol 19:500–505

Herman EM, Helm RM, Jung R et al (2003) Genetic modification removes an immunodominant allergen from soybean. Plant Physiol 132:36–43

Hokanson K, Heron D, Gupta S et al (1999) The concept of familiarity and pest resistant plants. In: Proceedings of a workshop on the ecological effects of pest resistance genes in managed ecosystems. http://digitalcommons.unl.edu/usdaarsfacpub/479/ (Last viewed 2 May 2011)

Houmard NM, Mainville JL, Bonin CP et al (2007) High-lysine corn generated by endosperm-specific suppression of lysine catabolism using RNAi. Plant Biotechnol J 5:605–614

Huang G, Allen R, Davis EL et al (2006) Engineering broad root-knot resistance in transgenic plants by RNAi silencing of a conserved and essential root-knot nematode parasitism gene. Proc Natl Acad Sci USA 103:14302–14306

Huvenne H, Smagghe G (2010) Mechanisms of dsRNA uptake in insects and potential for RNAi for pest control: a review. J Insect Physiol 56:227–235

Ivashuta SI, Petrick JS, Heisel SE et al (2009) Endogenous small RNAs in grain: semi-quantification and sequence homology to humans and animal genes. Food Chem Toxicol 47:353–360

Jin H (2008) Endogenous small RNAs and antibacterial immunity in plants. FEBS Lett 582: 2679–2684

Kang BC, Yeam I, Jahn MM (2005) Genetics of plant virus resistance. Ann Rev Phytopathol 43:581–621

Katsumoto Y, Fukuchi-Mizutani M, Fukui Y et al (2007) Engineering of the rose flavonoid biosynthetic pathway successfully generated blue-hued flowers accumulating delphinidin. Plant Cell Physiol 48:1589–1600

Kusaba M (2004) RNA interference in plants. Curr Opin Biotechnol 15:139–143

Kusaba M, Miyahara K, Iida S et al (2003) Low glutelin content 1: a dominant mutation that suppresses the glutelin multigene family via RNA silencing in rice. Plant Cell 15:1455–1467

Leonard E, Nielsen D, Solomon K et al (2008) Engineering microbes with synthetic biology frameworks. Trends Biotechnol 26:674–681

Levy A, Dafny-Yelin M, Tzfira T (2008) Attacking the defenders: plant viruses fight back. Trends Microbiol 16:194–197

Lindbo JA, Dougherty WG (2005) Plant pathology and RNAi: a brief history. Ann Rev Phytopathol 43:191–204

Lucioli A, Sallustio DE, Barboni D et al (2008) A cautionary note on pathogen-derived sequences. Nat Biotechnol 26:617–619

Mahmood-ur-Rahman IA, Husnain R, Riazuddin S (2008) RNA interference: the story of gene silencing in plants and humans. Biotechnol Adv 26:202–209

Mao Y-B, Cai W-J, Wang J-W et al (2007) Silencing a cotton bollworm P450 monoxygenase gene by plant-mediated RNAi impairs larval tolerance of gossypol. Nat Biotechnol 25:1307–1313

Marmiroli N, Agrimonti C, Visioli G et al (2000) Silencing of G1-1 and A2-1 genes. Effects on general plant phenotype and on tuber dormancy in Solanum tuberosum L. Potato Res 43:313–323

Martino-Catt S, Sachs ES (2008) Editors choice series: the next generation of biotech crops. Plant Physiol 147:3–5

Michaud V, Gil P, Kwiatek O et al (2007) Long-term storage at tropical temperature of dried-blood filter papers for detection and genotyping of RNA and DNA viruses by direct PCR. J Virol Meth 146:257–265

Mussgnug JH, Thomas-Hall S, Rupprecht J et al (2007) Engineering photosynthetic light capture: impacts on improved solar energy to biomass conversion. Plant Biotechnol J 5:802–814

Nakatsuka T, Mishiba K, Abe Y et al (2008) Flower color modification of gentian plants by RNAi-mediated gene silencing. Plant Biotechnol 25:61–68

Napoli C, Lemieux C, Jorgensen R (1990) Introduction of a chimeric chalcone synthase gene into petunia results in reversible co-suppression of homologous genes in trans. Plant Cell 2:279–289

Navarro L, Jay F, Nomura K et al (2008) Suppression of the microRNA pathway by bacterial effector proteins. Science 321:964–967

Newell-McGloughlin M (2008) Nutritionally improved agricultural crops. Plant Physiol 147:939–953

Nickson T (2008) Planning environmental risk assessment for genetically modified crops: problem formulation for stress-tolerant crops. Plant Physiol 147:494–502

Nielsen K (2003) Transgenic organisms – time for conceptual diversification? Nat Biotechnol 21:227–228

Ogita S, Uefuji H, Yamaguchi Y et al (2003) Producing decaffeinated coffee plants. Nature 423:823

Ogita S, Uefuji H, Morimoto M et al (2004) Application of RNAi to confirm theobromine as the major intermediate for caffeine biosynthesis in coffee plants with potential for construction of decaffeinated varieties. Plant Mol Biol 54:931–941

Oppermana CH, Birda DM, Williamsond VM et al (2008) Sequence and genetic map of *Meloidogyne hapla*: a compact nematode genome for plant parasitism. Proc Natl Acad Sci USA 105:14802–14807

Parrott W, Chassy B, Ligon J et al (2010) Application of food and feed safety assessment principles to evaluate transgenic approaches to gene modulation in crops. Food Chem Toxicol 48:1773–1790

Price DR, Gatehouse JA (2008) RNAi-mediated crop protection against insects. Trends Biotechnol 26:393–400

Raybould A (2007) Ecological versus ecotoxicological methods for assessing the environmental risks of transgenic crops. Plant Sci 173:589–602

Regina A, Bird A, Topping D et al (2006) High amylose wheat generated by RNAi interference improves indices of large bowel health in rats. Proc Natl Acad Sci USA 103:3546–3551

Rommens CM (2007) Intragenic crop improvement: combining the benefits of traditional breeding and genetic engineering. J Agric Food Chem 55:4281–4288

Sanders R, Hiatt W (2005) Tomato transgene structure and silencing. Nat Biotechnol 23:287–289

Sandhu APS, Abdelnoor RV, Mackenzie SA (2007) Transgenic induction of mitochondrial rearrangements for cytoplasmic male sterility in crop plants. Proc Natl Acad Sci USA 104:176–1770

Sanvido O, Romeis J, Bigler F (2007) Ecological impacts of genetically modified crops: ten years of field research and commercial cultivation. Adv Biochem Eng Biotechnol 107:235–278

Schouten HJ, Jacobsen E (2008) Cisgenesis and intragenesis, sisters in innovative plant breeding. Trends Plant Sci 13:260–261

Schouten HJ, Krens FA, Jacobsen E (2006) Cisgenic plants are similar to traditionally bred plants: international regulations for genetically modified organisms should be altered to exempt cisgenesis. EMBO Rep 7:750–753

Schwall GP, Safford R, Westcott RJ et al (2000) Production of very high amylose potato starch by inhibition of SBE A and B. Nat Biotechnol 18:551–554

Sheehy RE, Kramer M, Hiatt HR (1988) Reduction of polygalacturonase activity in tomato fruit by antisense RNA. Proc Natl Acad Sci USA 85:8805–8809

Shepherd DN, Martin DP, Thomson JA (2009) Transgenic strategies for developing crops resistant to geminiviruses. Plant Sci 176:1–11

Siritunga D, Sayre R (2003) Generation of cyanogen-free transgenic cassava. Planta 217:367–373

Small I (2007) RNAi for revealing and engineering plant gene functions. Curr Opin Biotechnol 18:148–153

Soosaar JLM, Burch-Smith TM, Dinesh-Kumar SP (2005) Mechanisms of resistance to viruses. Nat Rev Microbiol 3:789–798

Steeves RM, Todd TC, Essig JS et al (2006) Transgenic soybeans expressing siRNAs specific to a major sperm protein gene suppress *Heterodera glycines* reproduction. Funct Plant Biol 33:991–999

Stone R (2008) China plans $3.5 billion GE crops initiative. Science 321:1279

Stotzky G (2000) Persistence and biological activity in soil of insecticidal proteins from *Bacillus thuringiensis* and of bacterial DNA bound on clays and humic soils. J Environ Qual 26:691–705
Strauss SH (2003) Genomics, genetic engineering and domestication of crops. Science 300:61–62
Sunilkumar G, Campbell LM, Puckhaber L et al (2006) Engineering cottonseed for use in human nutrition by tissue-specific reduction of toxic gossypol. Proc Natl Acad Sci USA 103:18054–18059
Suter GW (2007) Ecological risk assessment, 2nd edn. CRC Press, Boca Raton
Tepfer M (2002) Risk assessment of virus-resistant transgenic plants. Ann Rev Phytopathol 40:467–491
Tuteja JH, Clough SJ, Chan W-C et al (2004) Tissue-specific gene silencing mediated by a naturally occurring chalcone synthase gene cluster in *Glycine max*. Plant Cell 16:819–835
Vance V, Vaucheret H (2001) RNA silencing in plants – defense and counterdefense. Science 292:2277–2280
Vanderschuren H, Alder A, Zhang P et al (2009) Dose-dependent RNAi-mediated geminivirus resistance in the tropical root crop cassava. Plant Mol Biol 70:265–272
Vaucheret H (2006) Post-transcriptional small RNA pathways in plants: mechanisms and regulations. Genes Dev 20:759–771
Verma AK (2008) RNAi technology for crop improvement. In: Rao GP, Zhao Y, Radchuk VV, Bhatnagar SK (eds) Advances in plant biotechnology. Studium Press, Houston
Vlassov VV, Laktionov PP, Rykova EY (2007) Extracellular nucleic acids. Bioessays 29:654–667
Von Krauss MP, Kaiser M, Almaas V et al (2008) Diagnosing and prioritizing uncertainties according to their relevance for policy: the case of transgene silencing. Sci Total Environ 390:23–34
Waterhouse PM, Graham MW, Wang M-B (1998) Virus resistance and gene silencing in plants can be induced by simultaneous expression of sense and antisense RNA. Proc Natl Acad Sci USA 95:13959–13964
Whangbo JS, Hunter CP (2008) Environmental RNA interference. Trends Genet 24:297–305
Xing S, Zachgo S (2007) Pollen lethality: a phenomenon in Arabidopsis RNA interference plants. Plant Physiol 145:330–333
Yadav BC, Veluthambi K, Subramaniam K (2006) Host-generated double-stranded RNA induces RNAi in plant-parasitic nematodes and protects the host from infection. Mol Biochem Parasitol 148:219–222
Zhou Y, Chan JH, Chan AY et al (2004) Transgenic plant-derived sRNAs can suppress propagation of influenza virus in mammalian cells. FEBS Lett 577:345–350

Biogenesis and Function of Virus-Derived Small Interfering RNAs in Plants

Zhixin Xie

Contents

Abstract RNA silencing is a deeply conserved mechanism that operates in most eukaryotes. A hallmark of RNA silencing is the processing of double-stranded RNA (dsRNA) precursors into 21–24-nucleotide (nt) small RNAs that function as sequence-specific cellular regulators. One of the well-established roles of RNA silencing is antiviral defense in plants. The virus-derived small RNAs found in the infected host cells serve as a manifestation that viral RNAs are targeted by the host RNA silencing machinery. On the other hand, many viruses encode proteins that suppress the activities of host silencing machinery, reflecting a viral counterdefense strategy evolved during the long-standing virus–host arms race. In many cases, viral disease symptoms are attributable, at least in part, to the interference of the host endogenous small RNA pathways by the virus-encoded silencing suppressors. During the last few years, significant advances have been made in our understanding of the host RNA silencing machinery involved in antiviral defense in plants,

Z. Xie (✉)
Department of Biological Sciences, Texas Tech University,
P.O. Box 43131, Lubbock, TX 79409, USA
e-mail: zhixin.xie@ttu.edu

V.A. Erdmann and J. Barciszewski (eds.), *Non Coding RNAs in Plants*,
RNA Technologies, DOI 10.1007/978-3-642-19454-2_29,

mainly through genetic analysis in *Arabidopsis*. Application of the next-generation DNA sequencing technology has enabled high-resolution profiling of the virus-derived small RNAs in the infected hosts. This review covers the most recent advances in our understanding of the biogenesis and cellular activity of virus-derived small interfering RNAs (siRNAs) in plants.

Keywords *Arabidopsis thaliana* • Argonautes (AGOs) • Dicer-like proteins (DCLs) • Host RNA-dependent RNA Polymerases (RDRs) • Silencing suppressors • Viral siRNAs

1 Introduction

RNA silencing is a recently discovered regulatory mechanism that operates in most eukaryotes (Chapman and Carrington 2007; Ghildiyal and Zamore 2009; Chen 2010). Central to RNA silencing is the generation of small RNA molecules that are typically 21–24-nucleotide (nt) in size and function as the guide in sequence-specific interactions with their regulatory targets (Chapman and Carrington 2007; Ghildiyal and Zamore 2009; Chen 2010). In general, RNA silencing is initiated by a double-stranded RNA (dsRNA) trigger, which could be either a bimolecular perfect dsRNA or a hairpin-like fold-back structure formed through extensive intramolecular base pairing. Recognition and subsequent processing of the dsRNA precursor into small RNAs rely on the activity of evolutionarily conserved Dicer (DCR) or Dicer-like (DCL) proteins that belong to the multidomain RNase III-like ribonuclease family (Chapman and Carrington 2007; Ghildiyal and Zamore 2009; Chen 2010). Functional small RNAs are recruited into members of another evolutionarily conserved family of proteins called Argonaute (AGO), forming multisubunit ribonucleoprotein effector complexes known as RNA-induced silencing complexes (RISCs). The small RNA molecule in a RISC serves as the specificity determinant for target recognition. Depending on the activity of a functional RISC, the outcome of the interaction between a small RNA and its target varies from target cleavage, translational repression, to RNA-directed DNA methylation (Chapman and Carrington 2007; Ghildiyal and Zamore 2009; Chen 2010). The RNA silencing-based regulatory mechanisms have been shown to collectively function in numerous essential biological processes including developmental timing, patterning, maintenance of genome integrity and stability, responses to environmental stress, and defense against invading viruses.

The role of RNA silencing as an antiviral defense mechanism in plants and certain invertebrates has been well established (Baulcombe 2004; Ding and Voinnet 2007), mainly based on two lines of compelling evidence. First, virus-derived small RNAs are found in the infected host cells, indicating that viral RNAs are targeted by the host RNA silencing machinery for destruction (Hamilton and Baulcombe 1999). Second, while viruses often have a small genome and code for only a handful of proteins, many viruses encode proteins that function as RNA silencing suppressors in their hosts, strongly suggesting that the capability to counteract the activity of host RNA silencing machinery is often required for a virus to be successful in terms of infection (Baulcombe 2004; Dunoyer and Voinnet 2005; Diaz-Pendon and Ding 2008). The discovery of viral RNA silencing suppressor proteins also helps to unveil the poorly understood molecular basis underlying the viral disease "symptoms". In many cases, such disease "symptoms" are attributable, at least in part, to perturbation of the host endogenous small RNA pathways caused by viral silencing suppressors (Dunoyer and Voinnet 2005; Diaz-Pendon and Ding 2008). Recent works in model systems involving genetically altered viruses and plant hosts have significantly advanced our current understanding of RNA silencing-based host defense and viral counterdefense strategies that have evolved during the everlasting molecular arms race between viruses and their plant hosts. In this chapter, I provide a brief review on the recent advances and development in this area, with a focus on the host RNA silencing machinery and viral factors that govern the biogenesis and cellular activity of the virus-derived small RNAs.

2 Biogenesis of Virus-Derived Small RNAs in Plants

Small RNAs derived from a replicating RNA virus were first observed in virus-infected plants (Hamilton and Baulcombe 1999), prior to the discovery of RNA silencing-associated endogenous small RNAs (i.e., microRNAs [miRNAs] and small interfering RNAs [siRNAs]). Since small RNAs were also associated with transgene-induced posttranscriptional gene silencing (PTGS), it was proposed that PTGS may in fact represent a natural antiviral defense mechanism (Hamilton and Baulcombe 1999), although the underlying molecular basis for small RNA generation was not clear at that time. With the discovery of endogenous small RNAs in diverse eukaryotes, extensive genetic and biochemical studies soon established the frameworks of small RNA biogenesis pathways in multiple model systems (Zamore and Haley 2005; Vaucheret 2006). Our current knowledge on virus-derived small RNA biogenesis in plants largely came from genetic analysis in *Arabidopsis*, which is discussed below.

2.1 Key RNA Silencing Factors and Endogenous Small RNA Pathways

The genome of *Arabidopsis thaliana* contains genes for four DCLs and ten AGOs. Genetic studies have revealed the association of all four DCLs and a subset of the ten AGOs with distinct endogenous small RNA pathways (Vaucheret 2006; Chen 2010). DCL1 functions in miRNA biogenesis, a process involving excision of an imperfect small RNA duplex from a hairpin-like fold-back precursor (Jones-Rhoades et al. 2006). The majority of mature miRNAs, which are typically 21-nt long and bear a uridine at their 5′ end, appear to be associated with AGO1 (Jones-Rhoades et al. 2006; Vaucheret 2008). DCL3 functions in the biogenesis of heterochromatic siRNAs which are typically 24-nt in size. This nuclear RNA silencing pathway also involves plant-specific RNA polymerases IV (pol IV) and pol V (also known as NRPD and NRPE for nuclear RNA polymerase D and E, respectively) (Lahmy et al. 2010; Pikaard et al. 2008), RNA-dependent RNA polymerase 2 (RDR2) (Vaucheret 2006; Chen 2010), and AGO4, AGO6, and AGO9 (Vaucheret 2008; Havecker et al. 2010; Olmedo-Monfil et al. 2010). DCL4 generates 21-nt phased array of *trans*-acting siRNAs (ta-siRNAs) from dsRNA precursors that arise from cleaved *TAS* transcripts through an RDR6- and suppressor of gene silencing 3 (SGS3)-dependent mechanism (Vaucheret 2006; Chen 2010). The initial miRNA-directed cleavage of a *TAS* transcript, which sets the register for subsequent DCL4 processing, involves either AGO1 or AGO7 (Allen et al. 2005; Axtell et al. 2006; Montgomery et al. 2008a, b). The role of DCL2 in the biogenesis of endogenous small RNAs is less well understood. DCL2 is required for the biogenesis of a salt-inducible, natural *cis*-antisense transcripts-derived siRNA (nat-siRNA) that is detectable as a 24-nt species in blot assays (Borsani et al. 2005). However, DCL1, instead of DCL2, is required for the biogenesis of another nat-siRNA that is inducible by a bacterial pathogen and accumulates as a 21-nt species (Katiyar-Agarwal et al. 2006). In addition, genetic dissection of ta-siRNA biogenesis suggests that in the absence of a functional DCL4, DCL2 can act to process the DCL4 substrates into 22-nt small RNAs (Gasciolli et al. 2005; Xie et al. 2005). Together, current data have collectively suggested several important features of the endogenous RNA silencing pathways in plants. First, unlike in some of the animals such as mammals, which have only one DCR, the DCL family has expanded and functionally diversified in multiple endogenous small RNA pathways in plants. DCL1 has apparently specialized in processing hairpin-like imperfect fold-back RNA substrates, whereas the other three DCLs show preference for perfect dsRNA substrates, although functional redundancy among the DCLs does exist (Xie et al. 2004, 2005; Gasciolli et al. 2005; Bouche et al. 2006). Second, in each of the endogenous small RNA pathways that involves a perfect dsRNA trigger, an RDR is also involved as a key factor and serves as an amplification step, presumably by converting a single-stranded RNA (ssRNA) template into dsRNA for dicing (Vaucheret 2006; Chen 2010). The *A. thaliana* genome contains six putative RDR genes, of which at least three (*RDR1*, *RDR2*, and *RDR6*) appear to encode functional RDR proteins. The remaining three (*RDR3*, *RDR4*, and *RDR5*) exist as tandem loci on chromosome II, and evidence for their activity has been lacking (Baulcombe 2004; Wassenegger and Krczal 2006). Implication of RDR proteins in RNA silencing is common in plants, fungi, and worms, but no RDR protein has been found in mammals or insects (Wassenegger and Krczal 2006), although it remains possible that other proteins may provide similar activities (Maida et al. 2009). Finally, the operation of multiple small RNA pathways in plants is facilitated not only by distinct substrate specificity of key enzymes such as the DCL family members but also by their unique subcellular localization. For example, DCL1 and DCL3, which function in the miRNA and heterochromatic siRNA pathways, respectively, are known to localize in distinct subnuclear bodies in vivo (Li et al. 2006; Pontes et al. 2006; Fang and Spector 2007; Song et al. 2007).

2.2 Host Factors Involved in the Biogenesis of Virus-Derived Small RNAs

2.2.1 The DCLs

Among the four *Arabidopsis* DCLs, DCL2 was the first family member shown to play a role in defense against Turnip Crinkle Virus (TCV). Accumulation of virus-derived small RNAs from TCV, but not Cucumber Mosaic Virus (CMV) or Turnip Mosaic Virus (TuMV), is impaired in *Arabidopsis dcl2* mutant, suggesting a role of DCL2 in viral small RNA biogenesis (Xie et al. 2004). Consistent with the idea of RNA silencing as a natural antiviral defense mechanism, the *dcl2* mutant is also more susceptible to TCV when compared with the wild-type control (Xie et al. 2004). The role of DCL2 in antiviral defense with an apparent specificity to TCV was rather puzzling, which prompted further genetic analysis involving the loss-of-function *dcl4* mutants.

Analysis of viral small RNA accumulation in a full spectrum of *Arabidopsis dcl* single, double and triple mutants revealed hierarchical and partially redundant roles for DCL4 and DCL2 in antiviral defense, with DCL4 being the major dicing activity in antiviral defense (Bouche et al. 2006; Deleris et al. 2006). In wild-type *Arabidopsis* plants infected with CMV or a modified Tobacco Rattle Virus (TRV-PDS), the DCL4-dependent 21-nt species accumulates as the major form of virus-derived small RNAs, as do the virus-infected *dcl2* and *dcl3* mutants, or the *dcl* mutant in case of the modified TRV. In the virus-infected *dcl4* mutant, however, DCL2-dependent 22-nt species becomes the major form of virus-derived small RNAs (Bouche et al. 2006; Deleris et al. 2006). The profile of TCV-derived small RNAs differs from those of CMV or TRV-PDS, in that the DCL2-dependent 22-nt species accumulates as the major form even in the infected wild-type plants. This profile remains unchanged in the TCV-infected *dcl1*, *dcl3*, and *dcl4* single mutants or the *dcl1dcl3* and *dcl3dcl4* double mutants. In the TCV-infected *dcl2* single mutant or any of the *dcl* double or triple mutants that involve a loss-of-function *dcl2* allele, however, the accumulation of viral small RNAs (mostly 22-nt) is substantially reduced (Bouche et al. 2006; Deleris et al. 2006). These observations suggest that the DCL4-dependent generation of 21-nt viral small RNA species is suppressed in TCV-infected plants, likely due to the silencing suppressor activity of the 38 kDa capsid protein (p38) (Qu et al. 2003; Bouche et al. 2006; Deleris et al. 2006). This hypothesis is supported by several lines of evidence. First, the DCL4-dependent ta-siRNA biogenesis, but not the DCL1-dependent miRNA biogenesis, is also suppressed in the TCV-infected plants (Bouche et al. 2006). Second, the DCL4-dependent siRNA generation from a transgenic inverted-repeat (IR) locus is suppressed in the TCV-infected plants or in transgenic plants expressing p38, coincident with the release of IR-induced silencing (Deleris et al. 2006). Finally, infection with a p38-deficient mutant TCV strain restores the accumulation of 21-nt viral small RNA in wild-type plants (Deleris et al. 2006). Suppression of DCL4 activity by the TCV-encoded silencing suppressor p38 would explain the increased susceptibility to TCV but not the other two viruses tested in the *dcl2* mutant, although it remains unclear if p38 interacts directly with DCL4. A recent study has showed that p38 interacts with AGO1 directly through its glycine/tryptophan (GW)-motif and inactivates AGO1 (Azevedo et al. 2010). The p38-mediated inactivation of AGO1 was shown to enhance the cellular level of DCL1, likely through suppression of miR162-directed cleavage of DCL1 mRNA by AGO1 (Azevedo et al. 2010), which in turn downregulates the cellular levels of DCL4 and DCL3 through an unknown mechanism (Qu et al. 2008; Azevedo et al. 2010).

In certain genetic backgrounds, minor DCL1- or DCL3-dependent viral small RNAs may be detected in infected plants, but those do not appear to play active roles in antiviral defense. By contrast, both the DCL4-dependent and the DCL2-dependent viral small RNAs are effective in antiviral defense, consistent with the highest susceptibility to viruses observed in the *dcl2dcl4* double mutant (Bouche et al. 2006; Deleris et al. 2006). Thus, current data support the idea that both DCL4 and DCL2 play important roles in antiviral defense, with DCL4 functioning at the forefront upon virus infection. In the absence of a functional DCL4, either in a loss-of-function *dcl4* mutant or when DCL4 activity is blocked by a silencing suppressor, either directly or indirectly, the role of DCL2 in antiviral defense is unmasked as the DCL2-dependent 22-nt viral small RNAs become the major species.

A subject that is highly relevant to the major DCL activities involved in the viral small RNA biogenesis is the nature of viral RNA substrates. In principle, there are at least three types of viral RNAs that can serve as DCL substrates: (1) viral dsRNAs that exist as replication intermediates; (2) viral dsRNAs generated by host RDR activity using single-stranded viral RNA as template; and (3) hairpin-like fold-back structures formed through extensive intramolecular base-pairing in a single-stranded viral RNA. Based on the observation that small RNAs derived from several positive-strand RNA viruses mostly map to the genomic sense strand, it has been proposed that viral small RNA are produced by DCL processing of highly structured viral ssRNA (Molnar et al. 2005). However, a direct correlation between small RNA-generating hotspots and highly structured regions in a viral genomic RNA has not been demonstrated (Donaire et al. 2008, 2009; Qi et al. 2009). Furthermore, the involvement of host RDRs in antiviral defense also supports bimolecular dsRNA as the precursor of viral small RNAs. In light of DCL4 and DCL2 being the major dicing activities in antiviral defense, and that DCL4 and DCL2 are known to produce endogenous siRNAs from dsRNA precursors (Borsani et al. 2005; Gasciolli et al. 2005; Xie et al. 2005; Yoshikawa et al. 2005; Rajagopalan et al. 2006), it is reasonable to assume that the majority of viral small RNAs are processed from dsRNA precursors. The virus-derived small RNAs should, therefore, be termed as viral siRNAs.

2.2.2 The Host RDRs

Among the three functional RDRs in *Arabidopsis*, both RDR1 and RDR6 have been implicated in antiviral defense (Baulcombe 2004). The plant *RDR1* gene and its viroid-inducible expression were first characterized in tomato (*Solanum lycopersicum*) (Schiebel et al. 1998). The orthologs of tomato *RDR1* in *Arabidopsis* and *Nicotiana tabacum* are inducible upon Tobacco Mosaic Virus (TMV) infection or in response to exogenous salicylic acid (SA) and play a role in antiviral defense (Xie et al. 2001; Yu et al. 2003). The tomato RDR1 ortholog in *Nicotiana benthamiana* is an inactive natural variant and may be the basis of *N. benthamiana*'s hypersusceptibility to many viral pathogens (Yang et al. 2004). The *Arabidopsis* RDR6 was recovered in genetic screens for mutants defective in sense transgene-induced PTGS. Interestingly, loss-of-function *Arabidopsis rdr6* mutants were shown to be more susceptible to CMV, but not to several other viruses tested (Dalmay et al. 2000; Mourrain et al. 2000). In *N. benthamiana*, RDR6 plays a role in limiting virus from entering shoot apical meristems, likely through promoting viral siRNA production in systemic tissues (Qu et al. 2005; Schwach et al. 2005). Based on the known roles of RDR6 in the DCL4-dependent ta-siRNA biogenesis, it is reasonable to assume that RDR6, and perhaps RDR1 as well, may function in viral siRNA formation by converting cleaved viral RNAs into dsRNA for DCL4- and DCL2-dependent processing.

Genetic analysis on the role of host RDR proteins in viral siRNA biogenesis has been facilitated by high-throughput sequencing of small RNAs in *rdr* mutants. Parallel analysis of TMV-Cg (a crucifer-infecting strain of TMV)-derived siRNAs from infected wild-type and *rdr1* or *rdr6* mutant plants revealed an important role for these host RDRs in viral siRNA biogenesis. Specifically, a loss-of-function mutation in RDR1 resulted in globally reduced viral siRNA generation at an early stage of viral infection (3 days post infection; 3 dpi). The reduced viral siRNA generation was also observed in the *rdr6* mutant, although to a much lesser extent (Qi et al. 2009). Interestingly, the *rdr1* and *rdr6* mutants also exhibited substantially reduced strand bias in viral siRNAs. For example, the ratio of 21-nt (78.4%) sense/antisense viral siRNA reads dropped from 3.77 in wild-type sample to 1.52 and 1.78 in *rdr1* and *rdr6* mutants, respectively. Analysis of viral siRNA populations derived from a silencing suppressor-deficient CMV (CMV-Δ2b) (Diaz-Pendon et al. 2007) or from TuMV (Garcia-Ruiz et al. 2010) also support a role of host RDR1 and RDR6 in viral siRNA biogenesis.

3 Cellular Activity of Viral siRNAs in Plants

The cellular fate and activities of viral siRNAs have not been extensively examined in the infected plant hosts. While recognition and subsequent processing of viral RNA into siRNAs by host DCL activities serve as an obvious frontline defense against an invading virus, effective RNA silencing-based antiviral defense appears to involve additional components of the host RNA silencing machinery.

3.1 Formation of Viral siRNA-Containing RISCs

Both genetic and biochemical data indicate that formation of viral siRNA-containing functional RISCs is important for effective antiviral defense. Hypomorphic *ago1* mutants and a loss-of-function *ago7* mutant are more susceptible to virus infection (Morel et al. 2002; Qu et al. 2008), suggesting that the slicer activities of AGO1 (Baumberger and Baulcombe 2005; Qi et al. 2005) and AGO7 (Montgomery et al. 2008a) are important for effective destruction of viral RNAs, presumably through formation of functional RISCs loaded with viral siRNAs. Indeed, viral siRNAs have been shown to coimmunoprecipitate with AGO1 in *Arabidopsis* (Zhang et al. 2006; Azevedo et al. 2010). Slicer activities from viral siRNA-containing putative RISCs have also been shown in cellular extracts prepared from virus-infected *N. benthamiana* (Omarov et al. 2007; Pantaleo et al. 2007). HUA ENANCER1 (HEN1) is a small RNA methyltransferase that methylates the 2′-hydroxyl group on the ribose of a small RNA 3′ terminal nucleotide (Yu et al. 2005). The HEN1-mediated methylation protects plant small RNAs from cellular nucleolytic activity or other type of modifications including oligouridylation (Li et al. 2005). Consistent with the idea of viral siRNAs functioning in RISCs, hypomorphic *hen1* mutants exhibit increased susceptibility to virus infection (Boutet et al. 2003). CMV-derived siRNAs are also found in the immunoprecipitates of *Arabidopsis* AGO2 and AGO5 (Takeda et al. 2008), although a slicer activity for these AGO family members has yet to be demonstrated.

It is currently unclear if all the viral siRNAs generated in the host cell incorporate into AGO-containing RISCs. It is also unknown if viral siRNAs are recruited into RISCs containing an AGO family member other than those mentioned above. Works from several groups have showed that the 5′-terminal base identity plays a role in sorting the endogenous small RNAs into distinct AGO complexes in plants (Mi et al. 2008; Montgomery et al. 2008a; Takeda et al. 2008). Among the ten *Arabidopsis* AGO family proteins, AGO1 is known to preferentially recruit small RNAs with a 5′-terminal uridine (U), whereas AGO2 and AGO5 preferentially recruit small RNAs with a 5′-terminal adenosine (A) and cytidine (C), respectively (Mi et al. 2008; Montgomery et al. 2008a; Takeda et al. 2008). Analysis of TMV-Cg-derived 21-nt siRNA populations in infected *Arabidopsis* revealed all four types of 5′-terminal nucleotides, with a moderate overall A/U to G/C preference (Qi et al. 2009). It remains to be examined if viral siRNAs with a 5′ terminal U, A, or C are preferentially recruited into AGO1-, AGO2-, and AGO5-containing RISCs, respectively.

3.2 Potential Roles of Viral siRNAs in Regulating Host Gene Expression

Virus infection can cause a wide range of disease symptoms in plant hosts, which is often associated with altered host gene expression (Golem and Culver 2003; Whitham et al. 2006; Agudelo-Romero et al. 2008; Havelda et al. 2008). As mentioned earlier, the viral disease symptoms are attributable, at least in part, to the activity of virus-encoded silencing suppressors that often perturb the developmentally important host endogenous small RNA pathways (Dunoyer and Voinnet 2005; Diaz-Pendon and Ding 2008; Alvarado and Scholthof 2009). Given the sequence complexity of viral siRNAs, it is also possible that some of the viral siRNAs may target host transcripts for posttranscriptional regulation. Indeed, bioinformatics analysis has identified host transcripts that can be potentially targeted by viral siRNAs. Experimental evidence for targeting of host transcripts by viral siRNAs has been reported in a few cases. Several siRNAs derived from the polycistronic 35S RNA leader sequence of Cauliflower Mosaic Virus (CaMV), a dsDNA virus belonging to the genus of Caulimovirus, were predicted to target an *Arabidopsis* transcript. One of these siRNAs was shown to effectively downregulate the predicted target in infected *Arabidopsis*, as well as to target an engineered reporter in a transient assay in turnip (Moissiard and Voinnet 2006). Using a target prediction algorithm that has a scoring stringency similar to those applied for plant miRNAs (Fahlgren et al. 2007), a large set of *Arabidopsis* transcripts were predicted as potential targets of TMV-Cg-derived siRNAs (Qi et al. 2009). The predicted potential targets cover a wide range of functional categories, including transcription factors, RNA processing factors, and defense-related proteins. A small subset of the predicted targets were subjected to experimental validation by modified RNA ligase-mediated rapid amplification of cDNA ends (RLM-5′RACE), a method that has been widely used for mapping the 5′ end of the 3′ cleavage product in vivo (Llave et al. 2002). Of the 16 predicted targets that were selected for validation, only two yielded a positive 5′RACE product. These are the transcripts for a cleavage and polyadenylation specificity factor (CPSF30, At1g30460) and an unknown protein similar to translocon-associated protein alpha (TRAP α, At2g16595), respectively (Qi et al. 2009). The fact that most of the tested host targets for TMV-Cg-derived siRNAs failed experimental validation suggests that there may be an unknown mechanism preventing the viral siRNAs from efficiently targeting a host transcript for cleavage. Nonetheless, the existence of viral siRNA-mediated virus–host interactions raises the interesting possibility for their contributions to viral pathogenicity and host specificity.

3.3 Virus-Encoded Silencing Suppressors Interfere with Viral siRNA Activity

Numerous virus-encoded, structurally diverse proteins have been identified as suppressors of RNA silencing (Dunoyer and Voinnet 2005; Diaz-Pendon and Ding 2008; Alvarado and Scholthof 2009). It is interesting to note that these viral proteins target the host RNA silencing pathways at specific steps, ranging from biogenesis to function of small RNAs, reflecting the diverse counterdefense strategies that have evolved among different viruses. Molecular analyses have revealed two common mechanisms by which viral silencing suppressors exert their cellular activity. In the first mechanism, viral silencing suppressors directly bind to the small RNA duplexes that are characteristic of DCL products (i.e., distinct size and the 2-nt 3′overhang). The p19 of Tomato Bushy Stunt Virus (TBSV) represents one of the best-characterized silencing suppressor proteins that act through this mechanism. The consequence of direct interaction between p19 and cellular small RNA duplexes is sequestration of cellular small RNAs, which prevents the small RNAs from being loaded onto AGO-containing complexes (Vargason et al. 2003; Ye et al. 2003; Chapman et al. 2004; Dunoyer et al. 2004; Lakatos et al. 2006). Such a sequestration can also prevent the nascent small RNA duplexes from being efficiently 3′methylated by HEN1, as has been shown for miRNAs in transgenic *Arabidopsis* expressing p19 (Yu et al. 2006). The potyvirus P1/HC-Pro (Kasschau et al. 2003; Chapman et al. 2004; Dunoyer et al. 2004; Lakatos et al. 2006) and the tobamovirus replication protein (Kubota et al. 2003; Ding et al. 2004; Csorba et al. 2007; Kurihara et al. 2007) are also known to act through a similar mechanism. The viral silencing suppressor-mediated sequestration also applies to viral siRNAs, as has been shown for P1/HC-Pro and p19 (Ebhardt et al. 2005; Omarov et al. 2007; Lozsa et al. 2008).

In the second mechanism, viral silencing suppressors inactivate a protein component of the host small RNA pathway through direct protein-protein interactions. The CMV 2b protein from a strain that causes severe symptoms was the first viral silencing suppressor shown to inhibit the slicer activity of AGO1 through direct interaction (Zhang et al. 2006). The Polerovirus-encoded F box protein P0, which is required for viral pathogenesis, has been shown to target *Arabidopsis* AGO1 for degradation (Baumberger et al. 2007; Bortolamiol et al. 2007; Csorba et al. 2010). Recent works have also indicated that the p25 of Potato virus X (PVX) physically interacts with AGO1 in *N. benthamiana* and promotes its degradation through the proteasome pathway (Chiu et al. 2010). The P1 protein of Sweet Potato Mild Mottle Virus (SPMMV) interacts with AGO1 through its WG/GW motifs and inhibits siRNA- and miRNA-programmed RISC activity (Giner et al. 2010). The fact that AGO1 is often the target of viral silencing suppressors is also consistent with the idea of AGO1 being the major RISC component in antiviral defense.

4 Prospects

The availability of genetically tractable virus–host model systems, along with the high-throughput small RNA profiling technology, has provided unprecedented opportunity for molecular dissection of RNA silencing-based virus–host interactions. A replicating virus is an inducer of the host RNA silencing machinery, which mounts an antiviral defense response. The virus is the target of the response, as evidenced by the generation of viral siRNAs in the infected host cells. The viral siRNAs, when fully functional, can become a “two-edged” sword. That is, once mature and assembled to form functional RISCs, they may target additional virus for effective destruction, but they can also become harmful to some of the host transcripts due to extensive sequence homology. However, a successful virus is unlikely to harm the host cells to an extent that would impede its own replication. A balance is most likely achieved through the activity of virus-encoded silencing suppressors so that the activity of viral siRNAs is kept in check.

Several outstanding questions remain. How do host RDRs affect the strand bias of viral siRNA populations? What are the host factors that govern the cellular fate of the viral siRNAs? Do viral siRNAs form functional RISCs with multiple AGO family members? Do all AGO proteins possess a slicer activity? To what extent does viral siRNA-directed targeting of host transcripts occur in the infected cell? Answers to some of these questions shall shed light on the molecular basis of viral pathogenicity and host specificity.

Acknowledgments I thank my colleague Chris Rock for critically reading the manuscript. I apologize to colleagues whose relevant work could not be covered here due to space limitations. The research on virus-induced RNA silencing in the author's laboratory was supported by a grant from the National Research Initiative (NRI) of the USDA Cooperative State Research, Education, and Extension Service (CSREES).

References

Agudelo-Romero P, Carbonell P, de la Iglesia F et al (2008) Changes in the gene expression profile of *Arabidopsis thaliana* after infection with Tobacco Etch Virus. Virol J 5:92

Allen E, Xie Z, Gustafson AM et al (2005) microRNA-directed phasing during *trans*-acting siRNA biogenesis in plants. Cell 121:207–221

Alvarado V, Scholthof HB (2009) Plant responses against invasive nucleic acids: RNA silencing and its suppression by plant viral pathogens. Semin Cell Dev Biol 20:1032–1040

Axtell MJ, Jan C, Rajagopalan R et al (2006) A two-hit trigger for siRNA biogenesis in plants. Cell 127:565–577

Azevedo J, Garcia D, Pontier D et al (2010) Argonaute quenching and global changes in Dicer homeostasis caused by a pathogen-encoded GW repeat protein. Genes Dev 24:904–915

Baulcombe D (2004) RNA silencing in plants. Nature 431:356–363

Baumberger N, Baulcombe DC (2005) Arabidopsis ARGONAUTE1 is an RNA slicer that selectively recruits microRNAs and short interfering RNAs. Proc Natl Acad Sci USA 102:11928–11933

Baumberger N, Tsai CH, Lie M et al (2007) The Polerovirus silencing suppressor P0 targets ARGONAUTE proteins for degradation. Curr Biol 17:1609–1614

Borsani O, Zhu J, Verslues PE et al (2005) Endogenous siRNAs derived from a pair of natural *cis*-antisense transcripts regulate salt tolerance in Arabidopsis. Cell 123:1279–1291

Bortolamiol D, Pazhouhandeh M, Marrocco K et al (2007) The Polerovirus F box protein P0 targets ARGONAUTE1 to suppress RNA silencing. Curr Biol 17:1615–1621

Bouche N, Lauressergues D, Gasciolli V et al (2006) An antagonistic function for Arabidopsis DCL2 in development and a new function for DCL4 in generating viral siRNAs. EMBO J 25:3347–3356

Boutet S, Vazquez F, Liu J et al (2003) *Arabidopsis* HEN1: a genetic link between endogenous miRNA controlling development and siRNA controlling transgene silencing and virus resistance. Curr Biol 13:843–848

Chapman EJ, Carrington JC (2007) Specialization and evolution of endogenous small RNA pathways. Nat Rev Genet 8:884–896

Chapman EJ, Prokhnevsky AI, Gopinath K et al (2004) Viral RNA silencing suppressors inhibit the microRNA pathway at an intermediate step. Genes Dev 18:1179–1186

Chen X (2010) Small RNAs – secrets and surprises of the genome. Plant J 61:941–958

Chiu MH, Chen IH, Baulcombe DC et al (2010) The silencing suppressor P25 of Potato virus X interacts with Argonaute1 and mediates its degradation through the proteasome pathway. Mol Plant Pathol 11:641–649

Csorba T, Bovi A, Dalmay T et al (2007) The p122 subunit of Tobacco Mosaic Virus replicase is a potent silencing suppressor and compromises both small interfering RNA- and microRNA-mediated pathways. J Virol 81:11768–11780

Csorba T, Lozsa R, Hutvagner G et al (2010) Polerovirus protein P0 prevents the assembly of small RNA-containing RISC complexes and leads to degradation of ARGONAUTE1. Plant J 62:463–472

Dalmay T, Hamilton A, Rudd S et al (2000) An RNA-dependent RNA polymerase gene in Arabidopsis is required for posttranscriptional gene silencing mediated by a transgene but not by a virus. Cell 101:543–553

Deleris A, Gallego-Bartolome J, Bao J et al (2006) Hierarchical action and inhibition of plant Dicer-like proteins in antiviral defense. Science 313:68–71

Diaz-Pendon JA, Ding SW (2008) Direct and indirect roles of viral suppressors of RNA silencing in pathogenesis. Annu Rev Phytopathol 46:303–326

Diaz-Pendon JA, Li F, Li WX et al (2007) Suppression of antiviral silencing by cucumber mosaic virus 2b protein in Arabidopsis is associated with drastically reduced accumulation of three classes of viral small interfering RNAs. Plant Cell 19:2053–2063

Ding SW, Voinnet O (2007) Antiviral immunity directed by small RNAs. Cell 130:413–426

Ding XS, Liu J, Cheng NH et al (2004) The Tobacco mosaic virus 126-kDa protein associated with virus replication and movement suppresses RNA silencing. Mol Plant Microbe Interact 17:583–592

Donaire L, Barajas D, Martinez-Garcia B et al (2008) Structural and genetic requirements for the biogenesis of tobacco rattle virus-derived small interfering RNAs. J Virol 82:5167–5177

Donaire L, Wang Y, Gonzalez-Ibeas D et al (2009) Deep-sequencing of plant viral small RNAs reveals effective and widespread targeting of viral genomes. Virology 392:203–214

Dunoyer P, Voinnet O (2005) The complex interplay between plant viruses and host RNA-silencing pathways. Curr Opin Plant Biol 8:415–423

Dunoyer P, Lecellier CH, Parizotto EA et al (2004) Probing the microRNA and small interfering RNA pathways with virus-encoded suppressors of RNA silencing. Plant Cell 16:1235–1250

Ebhardt HA, Thi EP, Wang MB et al (2005) Extensive 3′ modification of plant small RNAs is modulated by helper component-proteinase expression. Proc Natl Acad Sci USA 102: 13398–13403

Fahlgren N, Howell MD, Kasschau KD et al (2007) High-throughput sequencing of Arabidopsis microRNAs: evidence for frequent birth and death of *MIRNA* genes. PLoS ONE 2:e219

Fang Y, Spector DL (2007) Identification of nuclear dicing bodies containing proteins for microRNA biogenesis in living Arabidopsis plants. Curr Biol 17:818–823

Garcia-Ruiz H, Takeda A, Chapman EJ et al (2010) Arabidopsis RNA-dependent RNA polymerases and dicer-like proteins in antiviral defense and small interfering RNA biogenesis during Turnip Mosaic Virus infection. Plant Cell 22:481–496

Gasciolli V, Mallory AC, Bartel DP et al (2005) Partially redundant functions of Arabidopsis DICER-like enzymes and a role for DCL4 in producing *trans*-acting siRNAs. Curr Biol 15:1494–1500

Ghildiyal M, Zamore PD (2009) Small silencing RNAs: an expanding universe. Nat Rev Genet 10:94–108

Giner A, Lakatos L, Garcia-Chapa M et al (2010) Viral protein inhibits RISC activity by argonaute binding through conserved WG/GW motifs. PLoS Pathog 6:e1000996

Golem S, Culver JN (2003) Tobacco Mosaic Virus induced alterations in the gene expression profile of *Arabidopsis thaliana*. Mol Plant Microbe Interact 16:681–688

Hamilton AJ, Baulcombe DC (1999) A species of small antisense RNA in posttranscriptional gene silencing in plants. Science 286:950–952

Havecker ER, Wallbridge LM, Hardcastle TJ et al (2010) The Arabidopsis RNA-directed DNA methylation argonautes functionally diverge based on their expression and interaction with target loci. Plant Cell 22:321–334

Havelda Z, Varallyay E, Valoczi A et al (2008) Plant virus infection-induced persistent host gene downregulation in systemically infected leaves. Plant J 55:278–288

Jones-Rhoades MW, Bartel DP, Bartel B (2006) MicroRNAs and their regulatory roles in plants. Annu Rev Plant Biol 57:19–53

Kasschau KD, Xie Z, Allen E et al (2003) P1/HC-Pro, a viral suppressor of RNA silencing, interferes with *Arabidopsis* development and miRNA function. Dev Cell 4:205–217

Katiyar-Agarwal S, Morgan R, Dahlbeck D et al (2006) A pathogen-inducible endogenous siRNA in plant immunity. Proc Natl Acad Sci USA 103:18002–18007

Kubota K, Tsuda S, Tamai A et al (2003) Tomato mosaic virus replication protein suppressed virus-targeted posttranscriptional gene silencing. J Virol 77:11016–11026

Kurihara Y, Inaba N, Kutsuna N et al (2007) Binding of tobamovirus replication protein with small RNA duplexes. J Gen Virol 88:2347–2352

Lahmy S, Bies-Etheve N, Lagrange T (2010) Plant-specific multisubunit RNA polymerase in gene silencing. Epigenetics 5:4–8

Lakatos L, Csorba T, Pantaleo V et al (2006) Small RNA binding is a common strategy to suppress RNA silencing by several viral suppressors. EMBO J 25:2768–2780

Li J, Yang Z, Yu B et al (2005) Methylation protects miRNAs and siRNAs from a 3′-end uridylation activity in Arabidopsis. Curr Biol 15:1501–1507

Li CF, Pontes O, El-Shami M et al (2006) An ARGONAUTE4-containing nuclear processing center colocalized with Cajal bodies in *Arabidopsis thaliana*. Cell 126:93–106

Llave C, Xie Z, Kasschau KD et al (2002) Cleavage of Scarecrow-like mRNA targets directed by a class of Arabidopsis miRNA. Science 297:2053–2056

Lozsa R, Csorba T, Lakatos L et al (2008) Inhibition of 3′ modification of small RNAs in virus-infected plants require spatial and temporal co-expression of small RNAs and viral silencing-suppressor proteins. Nucleic Acids Res 36:4099–4107

Maida Y, Yasukawa M, Furuuchi M et al (2009) An RNA-dependent RNA polymerase formed by TERT and the RMRP RNA. Nature 461:230–235

Mi S, Cai T, Hu Y et al (2008) Sorting of small RNAs into Arabidopsis Argonaute complexes is directed by the 5′ terminal nucleotide. Cell 133:116–127

Moissiard G, Voinnet O (2006) RNA silencing of host transcripts by Cauliflower Mosaic Virus requires coordinated action of the four Arabidopsis Dicer-like proteins. Proc Natl Acad Sci USA 103:19593–19598

Molnar A, Csorba T, Lakatos L et al (2005) Plant virus-derived small interfering RNAs originate predominantly from highly structured single-stranded viral RNAs. J Virol 79:7812–7818

Montgomery TA, Howell MD, Cuperus JT et al (2008a) Specificity of ARGONAUTE7-miR390 interaction and dual functionality in *TAS3 trans*-acting siRNA formation. Cell 133:128–141

Montgomery TA, Yoo SJ, Fahlgren N et al (2008b) AGO1-miR173 complex initiates phased siRNA formation in plants. Proc Natl Acad Sci USA 105:20055–20062

Morel JB, Godon C, Mourrain P et al (2002) Fertile hypomorphic ARGONAUTE (*ago1*) mutants impaired in post-transcriptional gene silencing and virus resistance. Plant Cell 14:629–639

Mourrain P, Beclin C, Elmayan T et al (2000) Arabidopsis *SGS2* and *SGS3* genes are required for posttranscriptional gene silencing and natural virus resistance. Cell 101:533–542

Olmedo-Monfil V, Duran-Figueroa N, Arteaga-Vazquez M et al (2010) Control of female gamete formation by a small RNA pathway in Arabidopsis. Nature 464:628–632

Omarov RT, Ciomperlik JJ, Scholthof HB (2007) RNAi-associated ssRNA-specific ribonucleases in Tombusvirus P19 mutant-infected plants and evidence for a discrete siRNA-containing effector complex. Proc Natl Acad Sci USA 104:1714–1719

Pantaleo V, Szittya G, Burgyan J (2007) Molecular bases of viral RNA targeting by viral small interfering RNA-programmed RISC. J Virol 81:3797–3806

Pikaard CS, Haag JR, Ream T et al (2008) Roles of RNA polymerase IV in gene silencing. Trends Plant Sci 13:390–397

Pontes O, Li CF, Nunes PC et al (2006) The Arabidopsis chromatin-modifying nuclear siRNA pathway involves a nucleolar RNA processing center. Cell 126:79–92

Qi Y, Denli AM, Hannon GJ (2005) Biochemical specialization within Arabidopsis RNA silencing pathways. Mol Cell 19:421–428

Qi X, Bao FS, Xie Z (2009) Small RNA deep sequencing reveals role for *Arabidopsis thaliana* RNA-dependent RNA polymerases in viral siRNA biogenesis. PLoS ONE 4:e4971

Qu F, Ren T, Morris TJ (2003) The coat protein of *Turnip crinkle virus* suppresses posttranscriptional gene silencing at an early step. J Virol 77:511–522

Qu F, Ye X, Hou G et al (2005) RDR6 has a broad-spectrum but temperature-dependent antiviral defense role in *Nicotiana benthamiana.* J Virol 79:15209–15217

Qu F, Ye X, Morris TJ (2008) Arabidopsis DRB4, AGO1, AGO7, and RDR6 participate in a DCL4-initiated antiviral RNA silencing pathway negatively regulated by DCL1. Proc Natl Acad Sci USA 105:14732–14737

Rajagopalan R, Vaucheret H, Trejo J et al (2006) A diverse and evolutionarily fluid set of microRNAs in *Arabidopsis thaliana.* Genes Dev 20:3407–3425

Schiebel W, Pelissier T, Riedel L et al (1998) Isolation of an RNA-directed RNA polymerase-specific cDNA clone from tomato. Plant Cell 10:2087–2101

Schwach F, Vaistij FE, Jones L et al (2005) An RNA-dependent RNA polymerase prevents meristem invasion by potato virus X and is required for the activity but not the production of a systemic silencing signal. Plant Physiol 138:1842–1852

Song L, Han MH, Lesicka J et al (2007) Arabidopsis primary microRNA processing proteins HYL1 and DCL1 define a nuclear body distinct from the Cajal body. Proc Natl Acad Sci USA 104:5437–5442

Takeda A, Iwasaki S, Watanabe T et al (2008) The mechanism selecting the guide strand from small RNA duplexes is different among Argonaute proteins. Plant Cell Physiol 49:493–500

Vargason JM, Szittya G, Burgyan J et al (2003) Size selective recognition of siRNA by an RNA silencing suppressor. Cell 115:799–811

Vaucheret H (2006) Post-transcriptional small RNA pathways in plants: mechanisms and regulations. Genes Dev 20:759–771

Vaucheret H (2008) Plant ARGONAUTES. Trends Plant Sci 13:350–358

Wassenegger M, Krczal G (2006) Nomenclature and functions of RNA-directed RNA polymerases. Trends Plant Sci 11:142–151

Whitham SA, Yang C, Goodin MM (2006) Global impact: elucidating plant responses to viral infection. Mol Plant Microbe Interact 19:1207–1215

Xie Z, Fan B, Chen C et al (2001) An important role of an inducible RNA-dependent RNA polymerase in plant antiviral defense. Proc Natl Acad Sci USA 98:6516–6521

Xie Z, Johansen LK, Gustafson AM et al (2004) Genetic and functional diversification of small RNA pathways in plants. PLoS Biol 2:E104

Xie Z, Allen E, Wilken A et al (2005) DICER-LIKE 4 functions in *trans*-acting small interfering RNA biogenesis and vegetative phase change in *Arabidopsis thaliana.* Proc Natl Acad Sci USA 102:12984–12989

Yang SJ, Carter SA, Cole AB et al (2004) A natural variant of a host RNA-dependent RNA polymerase is associated with increased susceptibility to viruses by *Nicotiana benthamiana.* Proc Natl Acad Sci USA 101:6297–6302

Ye K, Malinina L, Patel DJ (2003) Recognition of small interfering RNA by a viral suppressor of RNA silencing. Nature 426:874–878

Yoshikawa M, Peragine A, Park MY et al (2005) A pathway for the biogenesis of *trans*-acting siRNAs in Arabidopsis. Genes Dev 19:2164–2175

Yu D, Fan B, MacFarlane SA et al (2003) Analysis of the involvement of an inducible Arabidopsis RNA-dependent RNA polymerase in antiviral defense. Mol Plant Microbe Interact 16:206–216

Yu B, Yang Z, Li J et al (2005) Methylation as a crucial step in plant microRNA biogenesis. Science 307:932–935

Yu B, Chapman EJ, Yang Z et al (2006) Transgenically expressed viral RNA silencing suppressors interfere with microRNA methylation in Arabidopsis. FEBS Lett 580:3117–3120

Zamore PD, Haley B (2005) Ribo-genome: the big world of small RNAs. Science 309:1519–1524

Zhang X, Yuan YR, Pei Y et al (2006) Cucumber mosaic virus-encoded 2b suppressor inhibits Arabidopsis Argonaute1 cleavage activity to counter plant defense. Genes Dev 20:3255–3268

Index

V.A. Erdmann and J. Barciszewski (eds.), *Non Coding RNAs in Plants*, RNA Technologies, DOI 10.1007/978-3-642-19454-2,

N

O

P